FUNCTIONS OF SEVERAL REAL VARIABLES

MATHEMATICS AND ITS APPLICATIONS
Series Editor: G. M. BELL,
Professor of Mathematics, King's College London, University of London

STATISTICS, OPERATIONAL RESEARCH AND COMPUTATIONAL MATHEMATICS Section
Editor: B. W. CONOLLY,
Emeritus Professor of Mathematics (Operational Research), Queen Mary College, University of London

Mathematics and its applications are now awe-inspiring in their scope, variety and depth. Not only is there rapid growth in pure mathematics and its applications to the traditional fields of the physical sciences, engineering and statistics, but new fields of application are emerging in biology, ecology and social organization. The user of mathematics must assimilate subtle new techniques and also learn to handle the great power of the computer efficiently and economically.

The need for clear, concise and authoritative texts is thus greater than ever and our series endeavours to supply this need. It aims to be comprehensive and yet flexible. Works surveying recent research will introduce new areas and up-to-date mathematical methods. Undergraduate texts on established topics will stimulate student interest by including applications relevant at the present day. The series will also include selected volumes of lecture notes which will enable certain important topics to be presented earlier than would otherwise be possible.

In all these ways it is hoped to render a valuable service to those who learn, teach, develop and use mathematics.

Mathematics and its Applications
Series Editor: G. M. BELL,
Professor of Mathematics, King's College London, University of London

Anderson, I. — **Combinatorial Designs: Construction Methods**
Artmann, B. — **Concept of Number: From Quaternions to Monads and Topological Fields**
Arczewski, K. & Pietrucha, J. — **Mathematical Modelling in Discrete Mechanical Systems**
Arczewski, K. and Pietrucha, J. — **Mathematical Modelling in Continuous Mechanical Systems**
Bainov, D.D. & Konstantinov, M. — **The Averaging Method and its Applications**
Baker, A.C. & Porteous, H.L. — **Linear Algebra and Differential Equations**
Balcerzyk, S. & Jösefiak, T. — **Commutative Rings**
Balcerzyk, S. & Jösefiak, T. — **Commutative Noetherian and Krull Rings**
Baldock, G.R. & Bridgeman, T. — **Mathematical Theory of Wave Motion**
Ball, M.A. — **Mathematics in the Social and Life Sciences: Theories, Models and Methods**
de Barra, G. — **Measure Theory and Integration**
Bartak, J., Herrmann, L., Lovicar, V. & Vejvoda, D. — **Partial Differential Equations of Evolution**
Bejancu, A. — **Finsler Geometry and Applications**
Bell, G.M. and Lavis, D.A. — **Statistical Mechanics of Lattice Models, Vols. 1 & 2**
Berry, J.S., Burghes, D.N., Huntley, I.D., James, D.J.G. & Moscardini, A.O. — **Mathematical Modelling Courses**
Berry, J.S., Burghes, D.N., Huntley, I.D., James, D.J.G. & Moscardini, A.O. — **Mathematical Modelling Methodology, Models and Micros**
Berry, J.S., Burghes, D.N., Huntley, I.D., James, D.J.G. & Moscardini, A.O. — **Teaching and Applying Mathematical Modelling**

Series continued at back of book

FUNCTIONS OF SEVERAL REAL VARIABLES

J. R. L. WEBB
Department of Mathematics, University of Glasgow

ELLIS HORWOOD
NEW YORK LONDON TORONTO SYDNEY TOKYO SINGAPORE

First published in 1991 by
ELLIS HORWOOD LIMITED
Market Cross House, Cooper Street,
Chichester, West Sussex, PO19 1EB, England

A division of
Simon & Schuster International Group
A Paramount Communications Company

Printed and bound in Great Britain
by Hartnolls Limited, Bodmin, Cornwall

British Library Cataloguing in Publication Data

Webb, J. R. L.
Functions of several real variables
CIP catalogue record for this book is available from the British Library
ISBN 0–13–763426–9 (Library Edn.)
ISBN 0–13–763434–X (Student Pbk. Edn.)

Library of Congress Cataloging-in-Publication Data available

To my family

Contents

Preface

This text treats the topics that usually fall under the general heading of "Advanced Calculus". This covers differentiation and integration of functions of several variables, including integrals along curves and over surfaces and connections between such integrals. In courses of Advanced Calculus, the aim is often to teach methods of computation. The theory justifying many rules is either omitted or treated rather sketchily. In this text we present a rigourous account of the underlying theory sufficient to bridge the gap between a first elementary course of Analysis and more advanced courses in Analysis.

Some mathematical maturity is expected of the readers of this book even though the prerequisites for reading it have been kept to a minimum; a first course in Analysis (a theoretical course of calculus) plus a little knowledge of Linear Algebra will suffice. The book is therefore aimed at senior undergraduate students (and beginning graduate students in the United States). Because it has been my experience that a first course in Analysis may not cover some important topics, possible gaps in the reader's knowledge are filled by some of the material presented in Chapter 1. The reader who is well versed in these topics will be able to quickly move on to other material. Some facts from Linear Algebra concerning linear operators and matrices are presented in Chapter 2. Again these may be familiar to many readers.

The study of functions of several variables requires a study of n-dimensional Euclidean space. Many of the important properties of Euclidean spaces can be studied in a more abstract setting. It would have been possible to study abstract spaces (particularly metric spaces) with very little extra effort. I have chosen not to pursue that course, even though I value the unifying effects of abstraction, because, once Euclidean space has been well understood, it is but a small step to grasp the ideas in the more abstract setting. The opposite approach runs the very real risk that the simple basic ideas will be lost because of the abstract setting when students encounter these ideas for the first time. However, even though the concrete approach has been adopted, the methods used are the same ones that apply in the more abstract setting so new methodology will not have to be learned later.

Let us say a few words concerning the contents. Most of the material presented in the text is standard and we make no claim for originality, but some topics are novel or are given new proofs. I have included rather more worked examples than is often found in texts at this level which hopefully will ease the reader's task. Proofs are given in detail and, when a choice of method of proof is available, I have chosen the method that I consider simpler or more intuitive.

Difficult topics have not been overlooked. The important Inverse and Implicit function theorems are given careful attention. A full treatment of the Riemann integral is given including a proof of the change of variables formula for multiple integrals. Topics from Vector Calculus, including integrals over surfaces and the theorem of Stokes are also studied. A full list of material covered is given in the Contents. Our choice of material has been guided by the desire to include material that is important but sometimes is omitted in the rush to greater abstraction. For example we include uniform convergence and the important Weierstrass approximation theorem.

What is new in this text? As already mentioned, most of the material is standard. However, Chapter 4 contains some novel features. The first is in the proof of the Inverse Function theorem. This is divided (in a natural way) into 3 parts each of which is easily manageable. I have simplified a previous proof of one of these parts using a well-known idea from more advanced work. I have not seen this proof previously used in this context. The Implicit Function theorem is illustrated by simple examples and by examples from the topic of bifurcation, which topic occurs in the study of differential equations. The Inverse Function theorem gives local information concerning invertibility of a function. We present a (possibly new) proof of a result concerning global invertibility, which requires only the elementary concepts of the present text. Finally in the same chapter, we include an elementary proof of the Brouwer fixed point theorem, based on the work of Milnor and Rogers (see the References). This may be appearing in a textbook for the first time.

Numbers of theorems, lemmas, propositions, definitions and examples run consecutively to ease the reader's search for a topic referred to later. Generally, theorems are more important than propositions, propositions more important than lemmas, but this rule of thumb is often broken.

Let us say a few words about the problems. Problems form an important part of any textbook and all students should attempt as many problems as possible. We make few claims to novelty here as some problems are almost part of the folklore and we have made use of questions that have been used at Glasgow University. Problems are set at the end of each chapter and the problem list is followed by a list of hints. The absence of a hint suggests that either the problem is so easy that no hint is needed, or a suitable hint was not apparent. The problems vary considerably in difficulty. Some are extremely easy, others routine while some are very demanding. However, all the problems can be solved by the methods presented in the text; it is not necessary to consult more advanced works. Complete solutions to all the problems are given at the end of the text. It is very important that the reader should make a determined effort to solve the problems before consulting their solution. Only then can a solution be properly appreciated. Alternative solutions are possible to many problems and the reader's own solution (if correct) is worth ten of mine.

The list of references and further reading is by no means complete but contains some of the texts that I have consulted and have found useful and partially reflects my own preferences.

The book undoubtedly contains errors. It is my hope that these are minor and that misprints will not confuse the reader.

It is a pleasure to thank my colleague Dr Michael Crabb for his valuable help. His careful reading of the manuscript enabled lack of precision to be corrected on a number of occasions. My colleagues, Dr Colin McGregor and Ms Weiyu Zhao, with whom I discussed some of the topics in the text, made valuable suggestions and I thank them for their help. I also thank the students at Glasgow University who have provided valuable feedback on the lectures on which part of the book is based.

I sincerely thank Jenny Beattie for her patience and skill in turning handwritten pages into the present manuscript using the Scientific word processor Chiwriter (Horstmann Software Design Corporation).

Finally, I appreciate the assistance (and persistence) of the staff at Ellis Horwood in bringing this project to fruition.

J. R. L. Webb
January 1991.

Chapter 1

Euclidean n-space

1.1. THE SPACE $\mathbb{R}^n$

We shall suppose that the reader has completed a first course of Mathematical Analysis as, for example, in the books of Spivak [12], or Stirling [13], listed on p. 239. The set $\mathbb{R}$ of all real numbers should therefore be familiar. Real numbers may be referred to as *scalars*.

For $n > 1$, $\mathbb{R}^n$ will denote the set of all ordered n-tuples (pairs, triples, quadruples, quintuples, $\cdots$) of real numbers, that is,

$$\mathbb{R}^n = \{(x_1, x_2, \cdots, x_n) \text{ where each } x_i \text{ is a real number, } i=1,2,\cdots,n\}.$$

It is important to regard an n-tuple as a single object. Accordingly we denote n-tuples by a single (**bold**) letter such as $\mathbf{x}$, $\mathbf{y}$, etc. We write $\mathbf{x} \in \mathbb{R}^n$ to mean $\mathbf{x}$ *belongs to* $\mathbb{R}^n$ and for $\mathbf{x} = (x_1, x_2, \cdots, x_n)$ we call the numbers $x_1, x_2, \cdots, x_n$ the first, second, $\cdots$, n-th *components* of $\mathbf{x}$. The n-tuple with every component zero is written $\mathbf{0}$, the *zero* element of $\mathbb{R}^n$. Following popular usage, when $n=2$ or $n=3$ we often write (x, y) or (x, y, z) rather than (x_1, x_2) or (x_1, x_2, x_3).

For $n=2$, (x_1, x_2) can be thought of as the point in the plane whose Cartesian co-ordinates are (x_1, x_2). Similarly we may identify $\mathbf{x}$ in $\mathbb{R}^3$ with a point in three-dimensional space. It is somewhat harder to visualise a point in n-dimensional space for $n > 3$. One can also think of elements of $\mathbb{R}^3$ as vectors (directed line segment from $\mathbf{0}$ to $\mathbf{x}$) and we shall refer to elements of $\mathbb{R}^n$ either as points or as vectors whichever interpretation seems appropriate.

Since components are ordered we have

Equality : $\mathbf{x} = \mathbf{y}$ if and only if $x_i = y_i$, $i = 1,2,\cdots,n$.

$\mathbb{R}^n$ is a *linear space* (or *vector space*) over $\mathbb{R}$ when the operations of addition and scalar multiplication are defined componentwise, that is,

Addition : $\mathbf{x} + \mathbf{y} := (x_1+y_1, x_2+y_2, \cdots, x_n+y_n)$

Scalar Multiplication : $\alpha\mathbf{x} := (\alpha x_1, \alpha x_2, \cdots, \alpha x_n)$, $\alpha \in \mathbb{R}$.

The *difference* is then given by $\mathbf{x} - \mathbf{y} := \mathbf{x} + (-1)\mathbf{y}$.

Linear spaces are studied in courses on linear algebra and there one proves that $\mathbb{R}^n$ has *dimension* n. This means that there is a set $\{\mathbf{e}_1,\mathbf{e}_2,\cdots,\mathbf{e}_n\}$ of minimal size n (a *basis*) such that every element of $\mathbb{R}^n$ can be written in precisely one way as a linear combination (a finite sum of scalar multiples) of the elements of the basis : the elements of a basis are *linearly independent*. The standard basis is obtained by taking $\mathbf{e}_i$ to be that n-tuple whose i-th component is 1 and every other component is 0. The unique linear combination is then $\mathbf{x} = x_1\mathbf{e}_1+\cdots+x_n\mathbf{e}_n$.

By way of reminder we say a little about the idea of a *Cartesian product* of sets. If A and B are two sets their Cartesian product $A\times B$ is the set of all ordered pairs (a,b) where $a\in A$ and $b\in B$. Similarly, the Cartesian product of m sets $A_1, A_2,\cdots, A_m$ is the set of all m-tuples $(a_1, a_2,\cdots, a_m)$ where $a_i\in A_i$. Thus $\mathbb{R}^2=\mathbb{R}\times\mathbb{R}$. Then $\mathbb{R}^n\times\mathbb{R}^m$ can then be thought of as the single space $\mathbb{R}^{n+m}$ or as the Cartesian product of $\mathbb{R}^n$ and $\mathbb{R}^m$ and we shall usually make this identification.

Since much of Analysis concerns the notions of continuity and differentiability, the concept of *limit* is crucial. In $\mathbb{R}^n$ this is made easier by the use of the concept of distance between points. We shall define this via the notion of inner product.

Definition 1.1. The *inner product* (or *scalar product*) of $\mathbf{x},\mathbf{y}$ is the real number

$$<\mathbf{x},\mathbf{y}> := x_1y_1+x_2y_2+\cdots+x_ny_n . \tag{1.1}$$

The inner product is a function from $\mathbb{R}^n\times\mathbb{R}^n$ to $\mathbb{R}$ with the properties (which the reader should verify) :

(*i*) $<\mathbf{x},\mathbf{x}> \geq 0$ with equality only if $\mathbf{x}=\mathbf{0}$,
(*ii*) $<\mathbf{x}+\mathbf{y},\mathbf{z}> = <\mathbf{x},\mathbf{z}> + <\mathbf{y},\mathbf{z}>$,
(*iii*) $<\alpha\mathbf{x},\mathbf{y}> = \alpha<\mathbf{x},\mathbf{y}>$, $\alpha\in\mathbb{R}$,
(*iv*) $<\mathbf{x},\mathbf{y}> = <\mathbf{y},\mathbf{x}>$.

It is possible to define other inner products on $\mathbb{R}^n$ with the above properties but we shall use only the standard one.

Definition 1.2. The *norm* of $\mathbf{x}\in\mathbb{R}^n$ is the nonnegative real number

$$\|\mathbf{x}\| := \sqrt{<\mathbf{x},\mathbf{x}>} = \left(\sum_{i=1}^{n} x_i^2\right)^{1/2} \quad \text{(positive square root).} \tag{1.2}$$

When $n=1$, the norm of $x\in\mathbb{R}$ is $|x|$, the absolute value of x, which measures the size irrespective of sign. It is the distance from the point x on the real line to the origin. Similarly in $\mathbb{R}^2$ or $\mathbb{R}^3$, $\|\mathbf{x}\|$ is the distance from the point with co-ordinates (x_1, x_2) or (x_1, x_2, x_3) to the origin (Pythagoras's theorem). We therefore make this interpretation in $\mathbb{R}^n$ too. We call this norm the *Euclidean* norm because it utilizes Euclidean geometry. The space $\mathbb{R}^n$ when endowed with this norm is called *n-dimensional Euclidean space*. The inner product $<\mathbf{x},\mathbf{y}>$ in $\mathbb{R}^3$ (or $\mathbb{R}^2$) is a measure of the angle θ between the vectors $\mathbf{x}$ and $\mathbf{y}$. Indeed in $\mathbb{R}^3$, $<\mathbf{x},\mathbf{y}> = \|\mathbf{x}\|\,\|\mathbf{y}\|\cos\theta$. For, in the triangle formed by the vectors $\mathbf{x},\mathbf{y}$ (see Figure 1.1), the cosine law for triangles gives

$$\|\mathbf{x}-\mathbf{y}\|^2 = \|\mathbf{x}\|^2 + \|\mathbf{y}\|^2 - 2\|\mathbf{x}\|\,\|\mathbf{y}\|\cos\theta.$$

But also $\|\mathbf{x}-\mathbf{y}\|^2 = <\mathbf{x}-\mathbf{y},\mathbf{x}-\mathbf{y}> = \|\mathbf{x}\|^2 + \|\mathbf{y}\|^2 - 2<\mathbf{x},\mathbf{y}>$, by expanding the inner product (properties (*ii*)-(*iv*)).

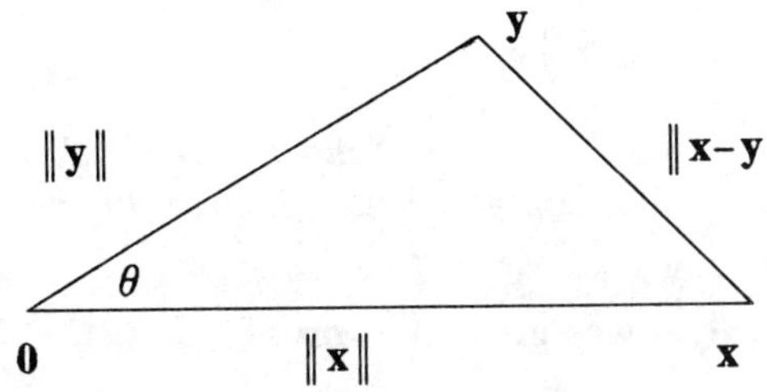

Figure 1.1.

Schwarz's inequality (Lemma 1.4 *infra*) shows that $<\mathbf{x},\mathbf{y}> \le \|\mathbf{x}\|\,\|\mathbf{y}\|$ so $<\mathbf{x},\mathbf{y}> = \|\mathbf{x}\|\,\|\mathbf{y}\|\cos\theta$ can be used as a definition of angle between elements of $\mathbb{R}^n$. In particular, it is natural to say that two elements of $\mathbb{R}^n$ are *orthogonal* (or perpendicular) if $<\mathbf{x},\mathbf{y}> = 0$, corresponding to $\theta = \pi/2$. Examples of orthogonal vectors in $\mathbb{R}^n$ are furnished by the standard basis elements. For, $<\mathbf{e}_i, \mathbf{e}_j> = 0$ unless $i=j$. It may be shown that

$$\|\mathbf{x}+\mathbf{y}\|^2 = \|\mathbf{x}\|^2 + \|\mathbf{y}\|^2 \Leftrightarrow <\mathbf{x},\mathbf{y}> = 0, \text{ (Pythagoras's theorem).}$$

The Euclidean norm also satisfies the so-called parallelogram law because it relates the lengths of the sides of a parallelogram in $\mathbb{R}^2$ to the lengths of the diagonals (see Figure 1.2) :

$$\|\mathbf{x}+\mathbf{y}\|^2 + \|\mathbf{x}-\mathbf{y}\|^2 = 2\|\mathbf{x}\|^2 + 2\|\mathbf{y}\|^2. \tag{1.3}$$

This result is easily proved using the properties of inner product. In fact, this result holds only if there is an inner product, see problem 1.3.

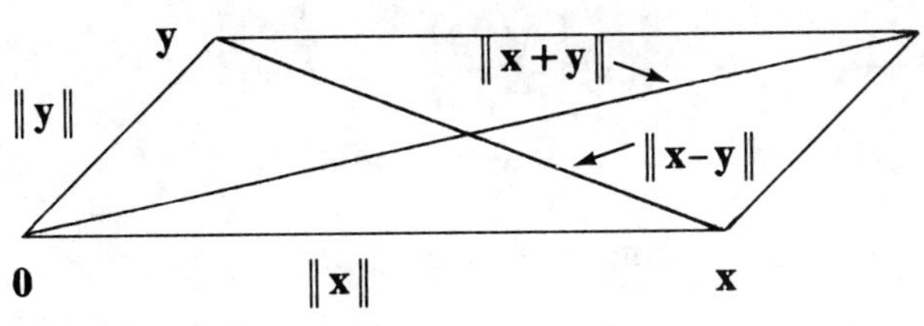

Figure 1.2.

The norm has the following properties

Theorem 1.3. *For all* $\mathbf{x},\mathbf{y}$ *in* $\mathbb{R}^n$ *and* $\alpha\in\mathbb{R}$,
(*a*) $\|\mathbf{x}\| \geq 0$ *and equality holds if and only if* $\mathbf{x}=\mathbf{0}$,
(*b*) $\|\alpha\mathbf{x}\| = |\alpha|\,\|\mathbf{x}\|$,
(*c*) $\|\mathbf{x}-\mathbf{y}\| = \|\mathbf{y}-\mathbf{x}\|$,
(*d*) $\|\mathbf{x}+\mathbf{y}\| \leq \|\mathbf{x}\| + \|\mathbf{y}\|$, *the triangle inequality.*

Proof (*a*) is immediate from (1.2) since a sum of squares is zero only if every term is zero. Also (*b*) follows at once because $\sqrt{\alpha^2}=|\alpha|$. Since $\|\mathbf{y}-\mathbf{x}\|=\|-(\mathbf{x}-\mathbf{y})\|$, (*c*) is a consequence of (*b*). The triangle inequality is nontrivial and we prove it via a useful inequality.

Lemma 1.4. *For* $\mathbf{x},\mathbf{y}\in\mathbb{R}^n$, $|<\mathbf{x},\mathbf{y}>| \leq \|\mathbf{x}\|\,\|\mathbf{y}\|$.

We shall refer to this Lemma as the *Schwarz inequality*, though it is also called the Cauchy or the Bunyakovskii inequality or the names may be listed together.

Assuming this temporarily, we give the proof of (*d*). We have

$$\begin{aligned}\|\mathbf{x}+\mathbf{y}\|^2 &= <\mathbf{x}+\mathbf{y},\mathbf{x}+\mathbf{y}>\\ &= <\mathbf{x},\mathbf{x}> +2<\mathbf{x},\mathbf{y}> + <\mathbf{y},\mathbf{y}>\\ &\leq \|\mathbf{x}\|^2+2\|\mathbf{x}\|\,\|\mathbf{y}\|+\|\mathbf{y}\|^2, \quad \text{using Lemma 1.4,}\\ &=(\|\mathbf{x}\|+\|\mathbf{y}\|)^2.\end{aligned}$$

Taking square roots gives the result.

Proof of Lemma 1.4. If $\mathbf{x}=\mathbf{0}$, both sides are zero so the inequality is certainly valid. If $\mathbf{x}\neq\mathbf{0}$, for any real number t,

$$0\leq \|t\mathbf{x}+\mathbf{y}\|^2 = <t\mathbf{x}+\mathbf{y},\, t\mathbf{x}+\mathbf{y}> = <\mathbf{x},\mathbf{x}>t^2+2<\mathbf{x},\mathbf{y}>t+<\mathbf{y},\mathbf{y}>.$$

This expression is a quadratic in t of the form

$$at^2+bt+c = a(t+b/2a)^2+c-b^2/4a \quad (\text{since } a\neq 0).$$

Its minimum value is when $t=-b/2a$ and so is nonnegative for all t if and only if $b^2\leq 4ac$. This gives $4<\mathbf{x},\mathbf{y}>^2\leq 4<\mathbf{x},\mathbf{x}>\,<\mathbf{y},\mathbf{y}>$. Cancelling 4 and taking the square root proves the result.

Lemma 1.4 says that $\displaystyle\left|\sum_{i=1}^{n} x_i y_i\right| \leq \left(\sum_{i=1}^{n} x_i^2\right)^{1/2}\left(\sum_{i=1}^{n} y_i^2\right)^{1/2}$,
a nontrivial fact usually referred to as Cauchy's inequality.

A useful consequence of the triangle inequality is :

Corollary 1.5. *For all* $\mathbf{x},\mathbf{y}\in\mathbb{R}^n$, $\|\mathbf{x}-\mathbf{y}\| \geq |\ \|\mathbf{x}\|-\|\mathbf{y}\|\ |$.

Proof. $\|\mathbf{x}\|=\|(\mathbf{x}-\mathbf{y})+\mathbf{y}\|\leq\|\mathbf{x}-\mathbf{y}\|+\|\mathbf{y}\|$, by the triangle inequality. Therefore, $\|\mathbf{x}-\mathbf{y}\|\geq\|\mathbf{x}\|-\|\mathbf{y}\|$. Exactly similarly (merely a change of symbols) $\|\mathbf{y}-\mathbf{x}\|\geq\|\mathbf{y}\|-\|\mathbf{x}\|=-(\|\mathbf{x}\|-\|\mathbf{y}\|)$. Since $\|\mathbf{y}-\mathbf{x}\|=\|\mathbf{x}-\mathbf{y}\|$ this proves the result.

Note that the best one can say in the other direction is that $\|\mathbf{x}-\mathbf{y}\|\leq\|\mathbf{x}\|+\|-\mathbf{y}\|=\|\mathbf{x}\|+\|\mathbf{y}\|$.

For $\mathbf{x},\mathbf{y}\in\mathbb{R}^n$, let $d(\mathbf{x},\mathbf{y}):=\|\mathbf{x}-\mathbf{y}\|$. We call d the *Euclidean metric.* It is the usual distance between points when $n=2$, or $n=3$ and has the expected properties of a distance function for any n.

Theorem 1.6. *For all* $\mathbf{x}, \mathbf{y}, \mathbf{z}\in\mathbb{R}^n$, *we have* :
(*a*) $d(\mathbf{x},\mathbf{y})\geq 0$ *with equality if and only if* $\mathbf{x}=\mathbf{y}$,
(*b*) $d(\mathbf{x},\mathbf{y}) = d(\mathbf{y},\mathbf{x})$,
(*c*) $d(\mathbf{x},\mathbf{y})\leq d(\mathbf{x},\mathbf{z})+d(\mathbf{z},\mathbf{y})$, *the triangle inequality.*

These properties follow readily from Theorem 1.3. Part (*c*), interpreted in two dimensions, explains the terminology. It says that the length of one side of a triangle cannot exceed the sum of the lengths of the other two sides, or, it is shorter to go directly from $\mathbf{x}$ to $\mathbf{y}$ than to go via any third point $\mathbf{z}$. This interpretation makes plausible the fact that equality holds in the triangle inequality in Theorem 1.3 exactly when $\mathbf{x}$ is a scalar multiple of $\mathbf{y}$, corresponding to equality holding in Theorem 1.6 if and only if $\mathbf{x}$, $\mathbf{y}$, and $\mathbf{z}$ are collinear ($\mathbf{x}-\mathbf{z}$ is a scalar multiple of $\mathbf{z}-\mathbf{y}$).

Any function which satisfies the properties listed in Theorem 1.3 is called a norm (similarly a metric is any function as in Theorem 1.6) but is not usually given by a inner product. The proof of the triangle inequality for other norms must be shown by some other method. A linear space endowed with a norm is called a *normed linear space.* Of the infinitely many possible norms on $\mathbb{R}^n$ we mention the following which are of interest in different situations and may be termed p-norms. For $\mathbf{x}\in\mathbb{R}^n$ and $1\leq p<\infty$ let

$$\|\mathbf{x}\|_p := \left(\sum_{i=1}^{n} |x_i|^p\right)^{1/p} \quad\text{and let } \|\mathbf{x}\|_\infty := \max\{\ |x_i|\ : 1\leq i\leq n\}\ . \qquad (1.4)$$

These expressions do define norms with the properties given in Theorem 1.3. The Euclidean norm fits into this family as the 2-norm. The 1-norm may be described as the New York taxicab norm since the distance between two points in $\mathbb{R}^2$ is the sum of the lengths of the sides of a rectangle with the points at opposite corners ; the taxicab must drive round the block. Many of the properties of the space $\mathbb{R}^n$, and of functions acting between two such spaces, do not depend on the choice of norm. We shall, therefore, unless specified to the contrary, use the Euclidean norm. The advantage is that we can utilize the inner product. Moreover, our geometric intuition, based on the geometry of space in which we live, and particularly the two dimensional pictures we draw, cannot lead us astray. The use of more exotic norms does not always fit our intuitive ideas so closely.

1.2. TOPOLOGY OF $\mathbb{R}^n$

Elementary topological ideas are concerned with the idea of neighbourhoods. These express the notion of one point being close to another which is essential in the discussion of concepts such as limits and continuity of functions. It is natural to use distance as a measure of nearness. Therefore we use the metric to define the basic concepts of *open* and *closed* sets.

Definition 1.7. For $\mathbf{x}\in\mathbb{R}^n$ and a scalar $\delta>0$, the *open ball with centre* $\mathbf{x}$ *and radius* δ is the set $B(\mathbf{x},\delta) := \{\mathbf{y}\in\mathbb{R}^n : \|\mathbf{y}-\mathbf{x}\| < \delta\}$.

We call $B(\mathbf{x},\delta)$ the *δ-neighbourhood* of $\mathbf{x}$. We shall abbreviate neighbourhood to *nbd* and by neighbourhood of a point we shall mean a suitable δ-nbd with δ not specified.

We use the name *sphere* for the set $S(\mathbf{x},\delta) := \{\mathbf{y}\in\mathbb{R}^n : \|\mathbf{y}-\mathbf{x}\| = \delta\}$, which can be thought of as the surface of $B(\mathbf{x},\delta)$. The union of these two sets is $\overline{B}(\mathbf{x},\delta) := \{\mathbf{y} : \|\mathbf{y}-\mathbf{x}\| \le \delta\}$, and is called the *closed ball*. The terminology open and closed will be justified shortly.

When $n=1$, the δ-nbd of x is the open interval $(x-\delta, x+\delta)$ and the sphere is the two point set $\{x-\delta, x+\delta\}$, the endpoints of the interval. For $n=2$, the δ-nbd of $(0,0)$ is the disk $\{(x,y) : x^2+y^2<\delta^2\}$ and the sphere is the circle of radius δ. The terminology ball and sphere is that appropriate to $\mathbb{R}^3$. If we use norms other than the Euclidean one, balls may not be of the familiar shape, for example in $\mathbb{R}^2$, the balls with respect to the norm $\|\cdot\|_\infty$ are squares.

Definition 1.8. A point $\mathbf{x}$ is said to be an *interior point* of a set A if there is $\delta>0$ such that $B(\mathbf{x},\delta)$ lies entirely in A. The set of all interior points of A is called the *interior* of A and is written $\mathrm{int}(A)$. A set A is *open* if every point in A is an interior point, that is, $A=\mathrm{int}(A)$.

Thus, if a set A is open and if $\mathbf{x}\in A$ then all points "sufficiently close" to $\mathbf{x}$ also lie in A.

The whole space $\mathbb{R}^n$ is clearly an open set. Also the empty set $\emptyset$ is an open set for every point in $\emptyset$ satisfies the requirements of an interior point for there are no points at all.

Example 1.9. For $r>0$, the open ball $B(\mathbf{0},r)$ is an open set.

Solution. Let $\mathbf{x}$ be any point of $B(\mathbf{0},r)$. Then $\|\mathbf{x}\|<r$. The point $\mathbf{x}$ is now fixed. We have to find $\delta>0$ so that *every* $\mathbf{y}$ with $\|\mathbf{y}-\mathbf{x}\|<\delta$ also lies in $B(\mathbf{0},r)$. Since, by the triangle inequality, $\|\mathbf{y}\|\leq\|\mathbf{y}-\mathbf{x}\|+\|\mathbf{x}\|$, we see that it suffices to take $\delta=r-\|\mathbf{x}\|$. Then $\delta>0$ and $\|\mathbf{y}\|<\delta+\|\mathbf{x}\|=r$ for $\|\mathbf{y}-\mathbf{x}\|<\delta$.

Figure 1.3

A simple modification of Example 1.9 shows that the open ball $B(\mathbf{z},r)$ is an open set for any $\mathbf{z}$ and $r>0$.

For a set A in $\mathbb{R}^n$ its *complement* is the set $A^c := \{\mathbf{x}\in\mathbb{R}^n : \mathbf{x}\notin A\}$. Thus $\mathbb{R}^n$ is the disjoint union of A with A^c : $\mathbb{R}^n = A\cup A^c$ with $A\cap A^c=\emptyset$, the empty set. Note that $(A^c)^c=A$ and $A=B \Leftrightarrow A^c=B^c$. The interior of A^c may be called the *exterior* of A.

Definition 1.10. A set A is called *closed* if its complement A^c is open.

Warning. This is NOT the same as *not open.*

Since $(\mathbb{R}^n)^c=\emptyset$, $\mathbb{R}^n$ is closed. Similarly the empty set is closed. Thus, the sets $\mathbb{R}^n$ and $\emptyset$ are both open and closed. These are the only subsets of $\mathbb{R}^n$ that are both open and closed. This is a statement concerning the connectedness of $\mathbb{R}^n$, a topic treated in Chapter 2. However, sets need be neither open nor closed : this will be seen more easily when we have discussed the concepts further.

Example 1.11. The closed ball $\overline{B}(\mathbf{0},r)$ is a closed set.

Solution. We have to show that the set $A=\overline{B}(\mathbf{0},r)^c=\{\mathbf{x}\in\mathbb{R}^n\colon \|\mathbf{x}\|>r\}$ is open. Fix any $\mathbf{x}$ with $\|\mathbf{x}\|>r$ and let $\delta=\|\mathbf{x}\|-r$. Then for every $\mathbf{y}$ with

$\|\mathbf{y}-\mathbf{x}\| < \delta$ we have $\|\mathbf{y}\| = \|\mathbf{x}-(\mathbf{x}-\mathbf{y})\| \geq \|\mathbf{x}\| - \|\mathbf{x}-\mathbf{y}\| > \|\mathbf{x}\| - \delta = r$, using Corollary 1.5. Thus, $B(\mathbf{x},\delta)$ lies in A and A is indeed open so $\overline{B}(\mathbf{0},r)$ is closed. See the right half of Figure 1.3.

Definition 1.12. A point $\mathbf{x}$ is said to be a *boundary point* of a set A if every neighbourhood of $\mathbf{x}$ contains points of both A and A^c. The set of all boundary points is called the *boundary* of A and will be denoted bdy(A).

Note that the point $\mathbf{x}$ need not belong to A in this definition. Roughly speaking, the boundary of A separates A from A^c. Another common notation for bdy(A) is ∂A.

Example 1.13. For $r > 0$, let $A_1 = B(\mathbf{0},r)$, and let $A_2 = \overline{B}(\mathbf{0},r)$. Then

$$\text{bdy}(A_1) = \text{bdy}(A_2) = S(\mathbf{0},r), \qquad \text{int}(A_2) = A_1 .$$

Solution. We give the details for A_1. If $\mathbf{x} \in \mathbb{R}^n$ has $\|\mathbf{x}\| < r$ then, by Example 1.9, there is a nbd of $\mathbf{x}$ lying entirely in A_1. Therefore $\mathbf{x}$ cannot be a boundary point of A_1. If $\|\mathbf{x}\| > r$, by Example 1.11, there is a nbd of $\mathbf{x}$ lying entirely outside A_2, so certainly outside A_1. Thus, the only possible boundary points have $\|\mathbf{x}\| = r$. This does not yet prove the result ; such points could conceivably fail to satisfy the definition of boundary point. So let $\mathbf{x}$ have $\|\mathbf{x}\| = r$. For any $\delta > 0$ we have to show that there is at least one point of $B(\mathbf{x},\delta)$ which lies in A_1 and at least one other in A_1^c. We take such points on the line from $\mathbf{0}$ to $\mathbf{x}$, that is, of the form $\alpha\mathbf{x}$ with $\alpha > 0$. Then, for $0 < 1-\alpha < \delta/r$, $\|\alpha\mathbf{x}\| = \alpha\|\mathbf{x}\| < r$ and $\|\mathbf{x}-\alpha\mathbf{x}\| = (1-\alpha)\|\mathbf{x}\| < \delta$, so that $\alpha\mathbf{x} \in B(\mathbf{x},\delta) \cap A_1$. Similarly, $\beta\mathbf{x}$ lies in $B(\mathbf{x},\delta) \cap A_1^c$ for $0 < \beta - 1 < \delta/r$.

Definition 1.14. The *closure* of A, written cl(A), is the set $A \cup \text{bdy}(A)$.

Another common notation for cl(A) is $\overline{A}$. This is consistent with the notation for closed balls. It is readily seen that we may also write $\text{cl}(A) = \text{int}(A) \cup \text{bdy}(A)$.

For $\mathbf{x} \in \text{cl}(A)$, every neighbourhood of $\mathbf{x}$ contains at least one point of A, ($\mathbf{x}$ itself if $\mathbf{x} \in A$). Thus, the closure of A consists of all points that are not exterior to A : $[\text{cl}(A)]^c = \text{int}(A^c)$. We use this to prove :

Theorem 1.15. *A set A is closed if and only if $A = \text{cl}(A)$.*

Proof. A is closed $\Leftrightarrow$ A^c is open $\Leftrightarrow$ $A^c = \text{int}(A^c)$ $\Leftrightarrow$ $A^c = [\text{cl}(A)]^c$ $\Leftrightarrow$ $A = \text{cl}(A)$.

Remarks. A set need be neither open nor closed. For example, the set $\{x\in\mathbb{R} : 0<x\leq 1\}$ is not open because 1 is not an interior point, and is not closed as 0 is a boundary point. A similar example in $\mathbb{R}^2$ is $A=\{(x,y) : 0\leq x\leq 1, 0<y<1\}$, a square. The two vertical edges are in the set but the horizontal ones are not. For this example we have

$$\mathrm{int}(A)=\{(x,y) : 0<x<1, 0<y<1\}, \text{ and } \mathrm{cl}(A)=\{(x,y) : 0\leq x\leq 1, 0\leq y\leq 1\}.$$

The family of all open subsets constitutes the topology of $\mathbb{R}^n$. We have defined the open subsets via neighbourhoods using the Euclidean norm. However, if we had used any other norm, the same subsets would be called open. In other words, all norms on $\mathbb{R}^n$ are equivalent in that they define the same topology. This is not obvious at this stage in our development but, we will be able to prove this later (Example 2.24). It is intuitively clear that in $\mathbb{R}^2$ the Euclidean norm and the taxicab norm are equivalent. For we can always fit a square inside a disk and fit a disk inside a square, so that a point is an interior point of a set whether this is decided using circular or rectangular neighbourhoods.

The property of openness is preserved under unions, while that of closedness is preserved under intersections. Let $\mathscr{F}$ be a collection of subsets of $\mathbb{R}^n$. We write $\bigcup\{A : A\in\mathscr{F}\}$, or $\bigcup_{A\in\mathscr{F}} A$, or simply $\bigcup A$, for the *union* of all the sets in $\mathscr{F}$. This consists of all points which belong to *at least one* of the sets in $\mathscr{F}$. When $\mathscr{F}$ is a finite or a countable collection we adopt the notations $\bigcup_{i=1}^{m} A_i = A_1\cup A_2\cup\cdots\cup A_m$, or $\bigcup_{i=1}^{\infty} A_i$. The *intersection* of all the sets in $\mathscr{F}$ consists of all points that belong to *every* set in $\mathscr{F}$ and is written $\bigcap\{A : A\in\mathscr{F}\}$, or $\bigcap A$, with a corresponding notation for a finite or countable family. Of course the intersection of sets may well be empty. For readers who have not encountered the notion of countability, details may be found in section 1.5 *infra*.

The following useful facts concerning the interaction between unions, intersections and complements are called the De Morgan laws.

Lemma 1.16. *For any collection $\mathscr{F}$ of sets ,*

$$\left[\bigcup\{A : A\in\mathscr{F}\}\right]^c = \bigcap\{A^c : A\in\mathscr{F}\}, \text{ and } \left[\bigcap\{A : A\in\mathscr{F}\}\right]^c = \bigcup\{A^c : A\in\mathscr{F}\}. \quad (1.5)$$

That is, the complement of a union of sets is the intersection of their complements and the complement of an intersection of sets is the union of their complements.

Proof. Let $\mathbf{x}\in\left[\bigcup\{A : A\in\mathscr{F}\}\right]^c$, equivalently $\mathbf{x}\notin\bigcup A$. This means that $\mathbf{x}$ cannot belong to any A in $\mathscr{F}$ and therefore $\mathbf{x}\in A^c$ for all $A\in\mathscr{F}$, that is,

$\mathbf{x} \in \bigcap\{A^c : A \in \mathscr{F}\}$. As the steps in this argument can be reversed, this shows that the two sets coincide. The proof for the other case is exactly similar but may also be proved by applying the case just shown to the sets A^c and taking complements.

Theorem 1.17. (*a*) *The union of any collection of open sets is open.*
(*b*) *The intersection of a finite collection of open sets is open.*
(*c*) *The intersection of any collection of closed sets is closed.*
(*d*) *The union of a finite collection of closed sets is closed.*

Proof. (*a*). Let $S=\bigcup\{G : G\in\mathscr{F}\}$ where each G is an open set. For any $\mathbf{x}\in S$, there is some G in the collection for which $\mathbf{x}\in G$. Since G is open, some δ-nbd of $\mathbf{x}$ lies entirely within G and hence also in the larger set S. Therefore S is open.

(*b*). Let $S=\bigcap_{i=1}^{m} G_i$, and let $\mathbf{x}\in S$. Then $\mathbf{x}\in G_i$ for $i=1,2,\cdots,m$. As each G_i is open there are $\delta_i>0$ so that $\mathrm{B}(\mathbf{x},\delta_i)\subseteq G_i$ Take $\delta=\min\{\delta_1, \delta_2, \cdots, \delta_m\}$. Then $\delta>0$ and $\mathrm{B}(\mathbf{x},\delta)\subseteq G_i$ for all i, that is $\mathrm{B}(\mathbf{x},\delta)\subseteq S$ and S is open.

(*c*) and (*d*) follow by taking complements, using the De Morgan laws. We leave the details to the reader.

Theorem 1.17 says that the family of open sets is closed under arbitrary unions and finite intersections. This is described by saying that they define a topology on $\mathbb{R}^n$. Generally, a topology is a family $\{U : U\in\mathscr{F}\}$ of sets that contains $\emptyset$ and the whole space and is closed under arbitrary unions and finite intersections, and these sets are then called the open sets. Different families of sets may define different topologies.

1.3. SEQUENCES IN $\mathbb{R}^n$.

If for every positive integer k there is an associated point $\mathbf{x}^{(k)}$ in $\mathbb{R}^n$, the ordered list $\{\mathbf{x}^{(1)}, \mathbf{x}^{(2)}, \cdots\}$ is a *sequence*. We use superscripts to avoid confusion with components. It is to be noted that a sequence is not simply a set. In a sequence the order of the terms is important ; in a set it is irrelevant. Terms in a sequence can be repeated ; in a set only distinct elements are listed. For example, $\{\mathbf{x}, \mathbf{x}, \mathbf{x}, \cdots\}$, is a constant sequence while the set of its elements is the singleton set $\{\mathbf{x}\}$. A more precise definition of sequence is as a function $\mathbf{f} : \mathbb{N} \longrightarrow \mathbb{R}^n$, where we write $\mathbf{x}^{(k)}$ rather than $\mathbf{f}(k)$. A precise definition of the concept of function is given in section 2.1 *infra*.

The most important concept involving sequences is that of

convergence.

Definition 1.18. A sequence $\{\mathbf{x}^{(k)}\}$ is said to *converge* to a point $\mathbf{x}$ as $k \longrightarrow \infty$ if, for every $\varepsilon > 0$ there exists a positive integer N such that $\|\mathbf{x}^{(k)} - \mathbf{x}\| < \varepsilon$ for all $k \geq N$.

The value of N may depend on $\mathbf{x}$ as well as on ε. We call $\mathbf{x}$ the *limit* of the sequence and write $\mathbf{x}^{(k)} \longrightarrow \mathbf{x}$ as $k \longrightarrow \infty$, or $\lim_{k \to \infty} \mathbf{x}^{(k)} = \mathbf{x}$. We may omit "$k \longrightarrow \infty$" when this is clear from the context. The definition of convergence may be phrased as follows. Every ε-nbd of $\mathbf{x}$ contains $\mathbf{x}^{(k)}$ for all sufficiently large k. Here "all sufficiently large" means for all but finitely many k, all $k \geq N$ for some N. Some abbreviations are often employed, namely $\forall$ for "for all" or "for every" , $\exists$ for "there exists", and s.t. for "such that". Using these, the definition of convergence takes the succinct form :

$$\forall\ \varepsilon > 0\,,\ \exists\ N \in \mathbb{N} \text{ s.t. } \forall\ k \geq N,\ \|\mathbf{x}^{(k)} - \mathbf{x}\| < \varepsilon\,. \tag{1.6}$$

The notion of convergence for sequences of real numbers is given in a first course in analysis and the above definition should have a familiar appearance with $\|\ \|$ replacing $|\ |$.

It is worth observing that to prove (1.6) it is sufficient to show that, for an arbitrary $\varepsilon > 0$, there is N such that $\|\mathbf{x}^{(k)} - \mathbf{x}\| < C\varepsilon$ for all $k \geq N$ where C is a constant (C must not depend on k, ε, or N). For if $\varepsilon' > 0$ we may take $\varepsilon = \varepsilon'/C$ and obtain (1.6) with ε replaced by ε'.

In giving a proof we have to show that certain hypotheses (some statement P) imply ($\Rightarrow$) certain conclusions (Q). This can be written as $P \Rightarrow Q$, and means *if P is true then Q is also true*. The direct method of proof involves assuming the truth of P and by a direct, logical argument deducing that Q is also true. Another equivalent method is to prove the so-called *contrapositive* statement, that is, show that (not Q) $\Rightarrow$ (not P), where (not P) is the *negation* of P and (not P) is true $\Leftrightarrow$ P is false. Yet another method of proof is the important method of *proof by contradiction*. Here we assume that P is true but that Q is false and then deduce some impossibility. This contradiction shows that some hypothesis is false. This must be (not Q) and therefore Q is indeed true. Using $\forall$ and $\exists$, so-called *quantifiers*, makes the process of negating more complicated statements routine. If a statement has the form

$$P = (\forall\ s \in S \text{ some property holds}),$$

then this statement is false if the property does not hold for every $s \in S$, that is, if there is at least one $s \in S$ for which the property fails. In symbols :

not($\forall$ $s \in S$ some property holds) = ($\exists$ $s \in S$ for which the property fails).

If a statement has the form

$$Q = (\exists\ s \in S \text{ some property holds})$$

then this is false if no $s \in S$ has the property, that is, the property fails for all $s \in S$:

$$\text{not}(\exists\ s \in S \text{ some property holds}) = (\forall\ s \in S \text{ the property fails}).$$

A more complicated statement involving several quantifiers can be tackled piece by piece and we can conclude that to negate such a statement we should begin at the left, replace every $\forall$ by $\exists$, replace every $\exists$ by $\forall$, and change "property holds" to "property fails". The order of the quantifiers is very important; changing the order changes the meaning.

We illustrate these comments by writing what it means for a sequence $\{\mathbf{x}^{(k)}\}$ not to converge to $\mathbf{x}$, $\mathbf{x}^{(k)} \not\to \mathbf{x}$. The sequence might not converge, but, even if it does, the limit cannot equal $\mathbf{x}$. Negating the above definition gives

$$\exists\ \varepsilon > 0\ ,\ \forall\ N \in \mathbb{N},\ \exists\ k \geq N \text{ with } \|\mathbf{x}^{(k)} - \mathbf{x}\| \geq \varepsilon. \tag{1.7}$$

Discussion of convergence of sequences in $\mathbb{R}^n$ can be done either directly or by using the following result.

Theorem 1.19. $\mathbf{x}^{(k)} \longrightarrow \mathbf{x}$ *if and only if* $x_i^{(k)} \longrightarrow x_i$ *for each* i, $1 \leq i \leq n$.

Proof. Suppose that $\mathbf{x}^{(k)} \longrightarrow \mathbf{x}$. Then for any $\varepsilon > 0$ there is N such that $\|\mathbf{x}^{(k)} - \mathbf{x}\| < \varepsilon$ for all $k \geq N$. For any fixed j, $1 \leq j \leq n$, and for $k \geq N$, we have

$$|x_j^{(k)} - x_j| = \left[|x_j^{(k)} - x_j|^2\right]^{1/2} \leq \left[\sum_{i=1}^{n} |x_i^{(k)} - x_i|^2\right]^{1/2} = \|\mathbf{x}^{(k)} - \mathbf{x}\| < \varepsilon.$$

Conversely, suppose that $x_i^{(k)} \longrightarrow x_i$ for every i, $1 \leq i \leq n$. Then for any $\varepsilon > 0$ and for each i there exists N_i such that $|x_i^{(k)} - x_i| < \varepsilon$ for all $k \geq N_i$. [For the slickest proof we should use $\varepsilon/\sqrt{n}$ here but we illustrate the comment following Definition 1.18.] Let $N = \max\{N_i : 1 \leq i \leq n\}$. Then these inequalities hold for all $k \geq N$ and

$$\|\mathbf{x}^{(k)} - \mathbf{x}\| = \left[\sum_{i=1}^{n} |x_i^{(k)} - x_i|^2\right]^{1/2} < \left[\sum_{i=1}^{n} \varepsilon^2\right]^{1/2} = \sqrt{n}\,\varepsilon.$$

This proves that $\mathbf{x}^{(k)} \longrightarrow \mathbf{x}$, since $\sqrt{n}$ is a fixed constant.

This result says that a sequence converges to $\mathbf{x}$ in $\mathbb{R}^n$ if and only if each sequence of components converges to the corresponding component of $\mathbf{x}$. Therefore the well known general results for sequences of real numbers can be brought to bear, such as the theorems on limits of sums, products and quotients, which we take as known. However, one cannot have products and quotients of elements of $\mathbb{R}^n$, for example, $\mathbf{x}/\mathbf{y}$ is meaningless for $n>1$; (we do not consider complex numbers here).

Example 1.20. (*a*) The sequence $\{(1-1/k,\ 1/(1+k))\}$ converges to $(1,0)$ in $\mathbb{R}^2$ since $1-1/k$ converges to 1 and $1/(1+k)$ converges to 0 in $\mathbb{R}$.

(*b*) The sequence $\{(1-1/k,\ (-1)^k)\}$ has no limit in $\mathbb{R}^2$ as the second component has no limit in $\mathbb{R}$ as $k \longrightarrow \infty$.

Definition 1.21. We say that a set A is *bounded* if there is a real number $M>0$ such that $\|\mathbf{a}\| \leq M$ for all $\mathbf{a} \in A$. A sequence is bounded if the set of its elements is bounded.

It is readily shown that a convergent sequence is bounded but bounded sequences need not be convergent (problem 1.20). The behaviour of limits with regard to the arithmetical operations is given by the following result.

Theorem 1.22. *Suppose that* $\mathbf{x}^{(k)} \longrightarrow \mathbf{x}$ *and* $\mathbf{y}^{(k)} \longrightarrow \mathbf{y}$ *as* $k \longrightarrow \infty$. *Then*
(*a*) $\mathbf{x}^{(k)} + \mathbf{y}^{(k)} \longrightarrow \mathbf{x} + \mathbf{y}$; (*b*) *for* $\alpha \in R$, $\alpha\mathbf{x}^{(k)} \longrightarrow \alpha\mathbf{x}$;
(*c*) $<\mathbf{x}^{(k)}, \mathbf{y}^{(k)}> \longrightarrow <\mathbf{x}, \mathbf{y}>$; (*d*) *limits are unique.*

Proof. (*a*) For any $\varepsilon>0$ there exists N such that $\|\mathbf{x}^{(k)} - \mathbf{x}\| < \varepsilon/2$ and $\|\mathbf{y}^{(k)} - \mathbf{y}\| < \varepsilon/2$ for all $k \geq N$. Therefore, for $k \geq N$, by the triangle inequality, $\|\mathbf{x}^{(k)} + \mathbf{y}^{(k)} - (\mathbf{x} + \mathbf{y})\| \leq \|\mathbf{x}^{(k)} - \mathbf{x}\| + \|\mathbf{y}^{(k)} - \mathbf{y}\| < \varepsilon$. The simple proof of (*b*) is left to the reader. To prove (*c*) let N be as in (*a*) and suppose that $\|\mathbf{x}^{(k)}\| \leq K$, ($\|\mathbf{x}^{(k)}\|$ is bounded). Then we have

$$
\begin{aligned}
|<\mathbf{x}^{(k)}, \mathbf{y}^{(k)}> - <\mathbf{x}, \mathbf{y}>| &= |<\mathbf{x}^{(k)}, \mathbf{y}^{(k)}> - <\mathbf{x}^{(k)}, \mathbf{y}> + <\mathbf{x}^{(k)}, \mathbf{y}> - <\mathbf{x}, \mathbf{y}>| \\
&\leq |<\mathbf{x}^{(k)}, \mathbf{y}^{(k)} - \mathbf{y}>| + |<\mathbf{x}^{(k)} - \mathbf{x}, \mathbf{y}>| \\
&\leq \|\mathbf{x}^{(k)}\| \|\mathbf{y}^{(k)} - \mathbf{y}\| + \|\mathbf{y}\| \|\mathbf{x}^{(k)} - \mathbf{x}\| \\
&\leq (K + \|\mathbf{y}\|)\, \varepsilon/2 , \text{ for } k \geq N.
\end{aligned}
$$

(*d*). Let $\mathbf{x}^{(k)} \longrightarrow \mathbf{x}$. For $\mathbf{z} \neq \mathbf{x}$, taking $\varepsilon = \|\mathbf{x}-\mathbf{z}\|/2$, the ε-nbds of $\mathbf{x}$ and $\mathbf{z}$ have no points in common so the ε-nbd of $\mathbf{z}$ can contain at most finitely many $\mathbf{x}^{(k)}$. Therefore, $\mathbf{z}$ cannot be the limit and the limit is unique.

We shall use the following fact many times in the sequel. It corresponds to the result in $\mathbb{R}$ that says : if $x^{(k)} \longrightarrow x$ and $\alpha < x^{(k)} < \beta$ for every k, then $\alpha \leq x \leq \beta$; the inequality is weakened in general. The same conclusion may be drawn if $\alpha \leq x^{(k)} \leq \beta$.

Proposition 1.23. *Let* $\mathbf{x}^{(k)} \longrightarrow \mathbf{x}$ *as* $k \longrightarrow \infty$ *and suppose that* $\mathbf{x}^{(k)} \in A$ *for every k. Then* $\mathbf{x} \in \mathrm{cl}(A)$.

Proof. We show that $\mathbf{x}$ does not lie in $\mathrm{int}(A^c)$. The proof is by contradiction. Thus, suppose that $\mathbf{x} \in \mathrm{int}(A^c)$. Then there is an ε-nbd of $\mathbf{x}$ lying entirely within A^c. In particular this nbd contains none of the points $\mathbf{x}^{(k)}$. This contradicts the fact that $\mathbf{x}^{(k)} \longrightarrow \mathbf{x}$.

To prove that a sequence converges using the definition requires us to know, or at least guess, the limit. It is useful to have a criterion that determines convergence even if the limit cannot be guessed. This criterion is the Cauchy criterion for sequences.

Definition 1.24. We say that $\{\mathbf{x}^{(k)}\}$ is a *Cauchy sequence* if for any $\varepsilon > 0$ there exists N such that $\|\mathbf{x}^{(j)} - \mathbf{x}^{(k)}\| < \varepsilon$ for all $j,k \geq N$.

Note that j and k are unrelated but it can always be supposed that $j \geq k$ without changing anything.

Our intention is to prove that the Cauchy condition is equivalent to convergence. Since the Cauchy condition says that all but finitely many terms of the sequence are very close together, it is reasonable that they should be close to a single point. This will be proved via the corresponding result for sequences in $\mathbb{R}$. We therefore temporarily study real sequences, where we can utilize the ordering of real numbers, and obtain some useful results which are not always covered in a first course of Analysis.

Let A be a set of real numbers. We say that A is bounded above if there is $M \in \mathbb{R}$ such that $a \leq M$ for all $a \in A$, M is called an *upper bound* of A. Upper bounds are not unique but there is a unique least upper bound or *supremum*. For a set A that is bounded above its supremum (denoted $\sup A$) is the unique number M such that :

(S_1) $a \leq M$ for all $a \in A$; (M is an upper bound),

(S_2) for every $\varepsilon > 0$ there is $a_\varepsilon \in A$ such that $a_\varepsilon > M - \varepsilon$; (no smaller number is an upper bound).

Note that $\sup A - \varepsilon$ need not be an element of A.

The fact that if $A \subseteq \mathbb{R}$ is bounded above then $\sup A$ exists is a basic property of $\mathbb{R}$ and may be taken as an *axiom*. The supremum of a set is a replacement for a maximal element ($M \in A$ such that $a \leq M$ for all $a \in A$). A set need have no maximum, for example, the open interval (0,1) has no maximal element but it does have a supremum, namely 1. If a set does have a maximal element then it coincides with the supremum. From the two properties above one may show that $\sup A \in \text{cl}(A)$ so $\sup A \in A$ if A is closed. There is the dual notion of the *infimum*, $\inf A$, of a set that is bounded below obtained by reversing the inequalities above. Observe that for a set $A \subseteq \mathbb{R}$, A is bounded ($|a| \leq M$ for all $a \in A$) if and only if A is bounded above and bounded below.

For sequences of real numbers the idea of a monotone sequence is important. We say that $\{x^{(k)}\}$ is *monotone increasing* if $x^{(k)} \leq x^{(k+1)}$ for every k, and *decreasing* if the inequality is reversed. The word monotone alone means that the sequence is either monotone increasing or monotone decreasing. A sequence is ***bounded above*** if the set of its elements is bounded above . The basic result for such sequences is :

Theorem 1.25. *A monotone increasing sequence (of real numbers) that is bounded above converges.*

There is an analogous result for decreasing sequences that are bounded below which is equivalent to this, as is seen by considering the sequence $\{-x^{(k)}\}$. This result is a basic property of $\mathbb{R}$ and could be taken as an axiom. It is equivalent to sets that are bounded above having suprema (see problem 1.23). The limit of the monotone increasing sequence $\{x^{(k)}\}$ is $\sup \{x^{(k)}\}$. Discussion of these matters may be found in many books on Analysis, for example, Spivak [12] or Stirling [13].

After these preliminaries, which hopefully are already familiar to the reader, we prove a useful consequence.

Theorem 1.26. (*The nested intervals theorem.*)
For each $m \in \mathbb{N}$ let I_m be a closed interval, $I_m = [a_m, b_m]$, such that
(i) $I_{m+1} \subseteq I_m$, for each m, (ii) $b_m - a_m \longrightarrow 0$ as $m \longrightarrow \infty$.
Then there is precisely one point x_0 (say) which belongs to every I_m, that is $\bigcap_{m=1}^{\infty} I_m = \{x_0\}$. Moreover, if $x^{(m)} \in I_m$, then $x^{(m)} \longrightarrow x_0$ as $m \longrightarrow \infty$.

Proof. By (*i*), $a_m \leq a_{m+1} \leq b_{m+1} \leq b_m$ for all m, so that $\{a_m\}$ is monotone increasing and bounded above (by b_1 for example) and $\{b_m\}$ is decreasing and bounded below (by a_1). Therefore $a_m \to a$ and $b_m \to b$. Hence, $b_m - a_m \longrightarrow b - a$ as $m \to \infty$. By (*ii*) and uniqueness of limits,

$b-a=0$. Writing x_0 as the common value we see that $a_m \le x_0 \le b_m$ for every m so x_0 belongs to every I_m. If $x^{(m)} \in I_m$ then $a_m \le x^{(m)} \le b_m$ and $x^{(m)} \to x_0$ by a sandwiching (or sqeezing) argument. Also x_0 is unique for if y were some other point in every I_m, $|y-x_0| \le b_m - a_m \longrightarrow 0$, therefore $y=x_0$.

A *subsequence* of a sequence $\{\mathbf{x}^{(k)}\}$ consists of a sequence of some of the terms taken in the same order. We write a subsequence as $\{\mathbf{x}^{k(j)}\}$ where $k(1) < k(2) < \cdots$; of course $k(j) \ge j$. More precisely, k is an increasing function from $\mathbb{N}$ to $\mathbb{N}$. Many properties of sequences are inherited by subsequences, for example, convergence and boundedness. Generally, properties of a subsequence say little about the full sequence. Monotone sequences provide an exception. If a monotone sequence has a convergent subsequence then the whole sequence converges to the same limit. We leave this as an exercise; see problem 1.22. The same holds for Cauchy sequences.

Lemma 1.27. *Let $\{\mathbf{x}^{(k)}\}$ be a Cauchy sequence and suppose there is a convergent subsequence $\{\mathbf{x}^{k(j)}\}$ with limit $\mathbf{x}$. Then $\mathbf{x}^{(k)} \longrightarrow \mathbf{x}$.*

Proof. For any $\varepsilon > 0$, since the sequence is Cauchy, there is N such that $\|\mathbf{x}^{(m)} - \mathbf{x}^{(p)}\| < \varepsilon/2$ for all $m,p \ge N$. Since the subsequence converges to $\mathbf{x}$ there is M such that $\|\mathbf{x}^{k(j)} - \mathbf{x}\| < \varepsilon/2$ for all $k(j) \ge M$. For a fixed $k(j) \ge \max\{N,M\}$ we may use this as p to conclude that $\|\mathbf{x}^{(m)} - \mathbf{x}^{k(j)}\| < \varepsilon/2$ for $m \ge N$. Therefore, $\|\mathbf{x}^{(m)} - \mathbf{x}\| \le \|\mathbf{x}^{(m)} - \mathbf{x}^{k(j)}\| + \|\mathbf{x}^{k(j)} - \mathbf{x}\| < \varepsilon$ for all $m \ge N$.

Another property of Cauchy sequences is :

Lemma 1.28. *A Cauchy sequence is bounded.*

Proof. Take $\varepsilon = 1$ in the definition of Cauchy sequence. Then there is N such that $\|\mathbf{x}^{(m)} - \mathbf{x}^{(p)}\| < 1$ for all $m, p \ge N$. In particular, $\|\mathbf{x}^{(m)} - \mathbf{x}^{(N)}\| < 1$ for all $m > N$. This implies that $\|\mathbf{x}^{(m)}\| \le \|\mathbf{x}^{(N)}\| + 1$, for all $m \ge N$. Take $M := \max\{\|\mathbf{x}^{(1)}\|, \cdots, \|\mathbf{x}^{(N-1)}\|, \|\mathbf{x}^{(N)}\| + 1\}$. Then $\|\mathbf{x}^{(m)}\| \le M$ for all $m \ge 1$.

We shall use this result, together with the previous one, to prove that a Cauchy sequence converges. To do this we shall show that a bounded sequence contains a convergent subsequence. We do this for sequences of real numbers using the nested intervals theorem.

Theorem 1.29. *Any bounded sequence of real numbers contains convergent subsequences.*

Proof. Let $\{x^{(k)}\}$ be the bounded sequence. Then all $x^{(k)}$ lie in some closed interval I_1, say. Choose $x^{k(1)}$ from I_1. Bisect I_1. One of the two closed halves must contain $x^{(k)}$ for infinitely many values of k, for otherwise there would be only finitely many terms of the sequence in all of I_1. Let I_2 be that half or, if both halves will do, the right-hand half for definiteness. Choose $x^{k(2)}$ from I_2 with $k(2) > k(1)$. This is possible as I_2 contains $x^{(k)}$ for infinitely many k. Bisect I_2 and choose $x^{k(3)}$ from I_3 with $k(3) > k(2)$ in the same fashion. This process continues and by the nested intervals theorem the subsequence so formed converges to the unique point common to all the intervals.

By consistently choosing the right-hand half when possible, the subsequence obtained converges to the largest possible limit, for the left half is chosen only when there are only a finite number of points of the sequence in the right half. If the left-hand half were chosen, the subsequence found would have a limit as small as possible. We shall have more to say on the matter of such *limit points* shortly.

Theorem 1.30. *(The General Principle of Convergence (GPC).)*
A sequence in $\mathbb{R}^n$ converges if and only if it is a Cauchy sequence.

Proof. Suppose that $\mathbf{x}^{(k)} \longrightarrow \mathbf{x}$. Then, for any $\varepsilon > 0$, there is N such that $\|\mathbf{x}^{(k)} - \mathbf{x}\| < \varepsilon/2$ for all $k \geq N$. In particular, for $j \geq N$ and $k \geq N$ we have $\|\mathbf{x}^{(j)} - \mathbf{x}\| < \varepsilon/2$ and $\|\mathbf{x}^{(k)} - \mathbf{x}\| < \varepsilon/2$. Therefore, by the triangle inequality, $\|\mathbf{x}^{(j)} - \mathbf{x}^{(k)}\| \leq \|\mathbf{x}^{(j)} - \mathbf{x}\| + \|\mathbf{x}^{(k)} - \mathbf{x}\| < \varepsilon$, and $\{\mathbf{x}^{(k)}\}$ is Cauchy. Conversely, let $\{\mathbf{x}^{(k)}\}$ be Cauchy. Then each sequence of components is a Cauchy sequence of real numbers. By Lemma 1.28 and Theorem 1.29, each component sequence contains a convergent subsequence and hence, by Lemma 1.27, each component sequence is convergent. Theorem 1.19 completes the proof.

The importance of this result is that it enables us to prove that a sequence converges without *a priori* knowledge of the limit. It can therefore be used to prove existence of the limit which enables us to prove important existence theorems in the theory of differential equations and elsewhere. We shall see such ideas in later chapters.

Definition 1.31. A set E is said to be *complete* if every Cauchy sequence of elements of E is convergent to a point of E.

The GPC says that $\mathbb{R}^n$ is complete. When $n=1$ completeness is usually phrased loosely as "the real line has no gaps" ; this is often made more precise via Dedekind's axiom. Again this is equivalent to the existence of least upper bounds and to the convergence of bounded monotone sequences. Since these properties use the idea of an ordering of the real numbers they cannot be used in $\mathbb{R}^n$.

Lemma 1.32. *A subset of* $\mathbb{R}^n$ *is complete* $\Leftrightarrow$ *it is closed.*

Proof. Let E be a closed subset of $\mathbb{R}^n$ and let $\{\mathbf{x}^{(m)}\}$ be a Cauchy sequence of points of E. Then $\{\mathbf{x}^{(m)}\}$ is certainly a Cauchy sequence of points of $\mathbb{R}^n$ and so converges to $\mathbf{x}\in\mathbb{R}^n$. By Proposition 1.23 we have $\mathbf{x}\in \mathrm{cl}(E)=E$ so E is complete. The converse is a simple exercise.

We now give a generalization of the nested intervals theorem. In it we use the notion of the *diameter* of a set A,

$$\mathrm{diam}(A) := \sup\{\|\mathbf{x}-\mathbf{y}\| : \mathbf{x},\mathbf{y}\in A\}$$

It may be checked that $\mathrm{diam}(A)<\infty \Leftrightarrow A$ is bounded.

Theorem 1.33. (*Cantor's intersection theorem.*)
Let $\{A_m : m=1,2,\cdots\}$ *be closed sets in* $\mathbb{R}^n$ *such that*

(*a*) $A_{m+1} \subseteq A_m$ *for every* m, (*b*) $\mathrm{diam}(A_m) \longrightarrow 0$ *as* $m \longrightarrow \infty$.

Then $\bigcap_{m=1}^{\infty} A_m$ *consists of a single point.*

Proof. For each $m\in\mathbb{N}$ let $\mathbf{x}^{(m)}$ be a point of A_m. We will show that $\{\mathbf{x}^{(m)}\}$ is a Cauchy sequence. For any $\varepsilon>0$, by (*b*), there is N such that $\mathrm{diam}(A_N)<\varepsilon$. Since by (*a*), $\mathbf{x}^{(j)}$ and $\mathbf{x}^{(k)}$ both lie in A_N for $j,k\geq N$, $\|\mathbf{x}^{(j)}-\mathbf{x}^{(k)}\| \leq \mathrm{diam}(A_N)<\varepsilon$, that is, $\{\mathbf{x}^{(m)}\}$ is Cauchy, therefore convergent to $\mathbf{x}$ (say). For every k, $\mathbf{x}^{(m)}\in A_k$ for $m\geq k$, and so $\mathbf{x}\in A_k$, since A_k is closed, (Proposition 1.23). This proves that $\mathbf{x}$ lies in the given intersection. Finally, if $\mathbf{y}$ were any other point in every A_k then $\|\mathbf{x}-\mathbf{y}\| \leq \mathrm{diam}(A_k)$ for all k. By (*b*) we must have $\|\mathbf{x}-\mathbf{y}\|=0$.

We shall see later that if (*b*) is replaced by the assumption that the sets are bounded, then the intersection is nonempty (but need no longer be a singleton).

1.4. LIMIT POINTS OF SETS

If infinitely many points lie in a bounded region it is intuitively clear that they cannot be well spaced out; some ε-nbd must contain infinitely many of the points. Such bunching together gives us the concept of limit point.

Definition 1.34. An *isolated point* of a set A is a point $\mathbf{x}\in A$ such that there is a neighbourhood U of $\mathbf{x}$ with $A\cap U=\{\mathbf{x}\}$. A *limit point* of A is a point $\mathbf{x}$ (not necessarily in A) such that every neighbourhood of $\mathbf{x}$ contains infinitely many points of A.

As will be apparent from the proof of Proposition 1.36 *infra*, in the definition of limit point we could equally well say that every neighbourhood of $\mathbf{x}$ contains at least one point of A which is unequal to $\mathbf{x}$. Hence points of A are either isolated points or limit points. We feel that the version given conveys the idea of limit point rather better. Limit points are also termed points of accumulation or cluster points.

Example 1.35. Every interior point of A is a limit point of A.

Solution. For $\mathbf{x}\in\mathrm{int}(A)$, there is some δ-nbd lying entirely in A. If $\varepsilon\leq\delta$, the ε-nbd then lies in A and every ε-nbd certainly contains infinitely many points. If $\varepsilon>\delta$, the ε-nbd contains the δ-nbd which contains infinitely many points of A.

The next few results establish which points are limit points.

Proposition 1.36. *A point* $\mathbf{x}$ *is a limit point of* A *if and only if* $\mathbf{x}\in\mathrm{cl}(A)$ *and* $\mathbf{x}$ *is not an isolated point of* A.

Proof. Let $\mathbf{x}$ be a limit point of A. Then $\mathbf{x}$ cannot belong to $\mathrm{int}(A^c)$ because no nbd of $\mathbf{x}$ can lie entirely within A^c; thus $\mathbf{x}\in\mathrm{cl}(A)$. Also, $\mathbf{x}$ is not an isolated point of A by definition. Conversely, suppose that $\mathbf{x}\in\mathrm{cl}(A)$ and is not an isolated point of A. Then, by definition of isolated point and closure, every nbd of $\mathbf{x}$ contains at least one point of A other than $\mathbf{x}$. For any $\varepsilon>0$ let U be the ε-nbd of $\mathbf{x}$. We shall show that U contains infinitely many points of A. Indeed, otherwise $U\cap A$ consists of only finitely many points distinct from $\mathbf{x}$, say $\mathbf{x}^{(1)},\mathbf{x}^{(2)},\cdots,\mathbf{x}^{(m)}$. Now take $\delta=\min\{\varepsilon,\ \|\mathbf{x}^{(j)}-\mathbf{x}\| : j=1,2,\cdots,m\}$. Then $\mathrm{B}(\mathbf{x},\delta)$ contains no point of A distinct from $\mathbf{x}$, a contradiction.

Proposition 1.37. *A point* $\mathbf{x}$ *is a limit point of* A *if and only if there is a sequence* $\{\mathbf{x}^{(m)}\}$, *of distinct points of* A, *with* $\mathbf{x}^{(m)} \neq \mathbf{x}$ *for all* m *and* $\mathbf{x}^{(m)} \longrightarrow \mathbf{x}$.

Proof. Let $\mathbf{x}$ be a limit point of A and, for each m, let $U_m = B(\mathbf{x}, 2^{-m})$. Each of these neighbourhoods contains infinitely many points of A. Choose $\mathbf{x}^{(1)} \in U_1 \cap A$ with $\mathbf{x}^{(1)} \neq \mathbf{x}$. If $\mathbf{x}^{(1)}, \mathbf{x}^{(2)}, \cdots, \mathbf{x}^{(p)}$ $(p \geq 1)$ have been selected, choose $\mathbf{x}^{(p+1)}$ from $U_{p+1} \cap A$ which differs from each of $\mathbf{x}, \mathbf{x}^{(1)}, \mathbf{x}^{(2)}, \cdots, \mathbf{x}^{(p)}$. This is possible since U_{p+1} contains infinitely many points of A. Now $\|\mathbf{x}^{(m)} - \mathbf{x}\| \leq 2^{-m}$ so the sequence converges to $\mathbf{x}$. The converse is simpler. For, if $\mathbf{x}^{(m)} \longrightarrow \mathbf{x}$, every nbd of $\mathbf{x}$ contains $\mathbf{x}^{(m)}$ for all sufficiently large m.

The next result provides a useful characterization of closedness.

Theorem 1.38. *A set is closed if and only if it contains all its limit points.*

Proof. If $\mathbf{x}$ is a limit point of A then $\mathbf{x} \in \mathrm{cl}(A)$ by Proposition 1.36, therefore $\mathbf{x} \in A$ when A is closed. Conversely, if all limit points of A belong to A, then for any $\mathbf{y} \in A^c$, there is a nbd U of $\mathbf{y}$ containing at most finitely many points of A. As in the proof of Proposition 1.36, there is a (smaller) nbd V containing no points of A. Thus, V is a nbd of $\mathbf{y}$ lying entirely within A^c, so that A^c is open, that is, A is closed.

The question as to which sets have limit points is answered by the following important result.

Theorem 1.39. (*The Bolzano-Weierstrass theorem.*)
Every infinite, bounded set in $\mathbb{R}^n$ *has at least one limit point.*

Proof. Let A be the bounded set with infinitely many elements. Since A is bounded there is a closed *n-cube* $K_1 = \{\mathbf{x} \in \mathbb{R}^n : -a \leq x_i \leq a,\ i = 1, 2, \cdots, n\}$ containing A. K_1 is the Cartesian product $I_1 \times I_2 \times \cdots \times I_n$ of intervals, that is the set of points $\{x_1, x_2, \cdots, x_n\}$ where $x_i \in I_i$ and each $I_i = [-a, a]$. Every interval can be bisected to form two closed intervals $I_{i,1}$ and $I_{i,2}$. A Cartesian product of the form

$I_{1,\alpha(1)} \times I_{2,\alpha(2)} \times \cdots \times I_{n,\alpha(n)}$, with each $\alpha(i)$ equal to either 1 or 2, is a *subcube* of K_1. There are 2^n such products so that K_1 can be divided into 2^n subcubes. At least one of this finite number must contain infinitely many points of A, K_2 (say). Continue this subdivision process and obtain a decreasing family of closed cubes such that each one contains infinitely many points of A. Since the diameter is halved at each stage, Cantor's theorem shows that there is exactly one point $\mathbf{x}$, say, common to every cube. For any $\varepsilon > 0$ let m be so large that $\text{diam}(K_m) < \varepsilon$. Then $K_m \subseteq B(\mathbf{x},\varepsilon)$ and so the ε-nbd of $\mathbf{x}$ contains infinitely many points of A; this proves that $\mathbf{x}$ is a limit point of A.

Corollary 1.40. *Every bounded sequence in* $\mathbb{R}^n$ *contains convergent subsequences.*

Proof. Let $\{\mathbf{x}^{(m)}\}$ be any bounded sequence and let A be the set of its elements. If A is a finite set, say $A = \{\mathbf{a}^{(1)}, \mathbf{a}^{(2)}, \cdots, \mathbf{a}^{(q)}\}$, some term in this set, $\mathbf{a}^{(p)}$ (say), must be repeated infinitely often in the sequence $\{\mathbf{x}^{(m)}\}$ and then $\{\mathbf{a}^{(p)}, \mathbf{a}^{(p)}, \cdots\}$ is a subsequence which is plainly convergent. If A is an infinite set, then it has a limit point $\mathbf{x}$ by the Bolzano-Weierstrass theorem. As in the proof of Proposition 1.36, there is a sequence of distinct points of A converging to $\mathbf{x}$. To obtain a *subsequence* of $\{\mathbf{x}^{(m)}\}$ the selection process must be modified slightly. If $\mathbf{x}^{k(1)}, \mathbf{x}^{k(2)}, \cdots, \mathbf{x}^{k(j)}$ have been selected let $\mathbf{x}^{k(j+1)}$ be chosen so that $k(j+1) > k(j)$. This gives a desired subsequence.

Corollary 1.41. *Let* $\{A_m : m = 1,2, \cdots\}$ *be nonempty, closed, bounded sets such that* $A_{m+1} \subseteq A_m$ *for every* m. *Then* $\bigcap_{m=1}^{\infty} A_m$ *is nonempty.*

Proof. For $k \in \mathbb{N}$ let $\mathbf{x}^{(k)}$ be a point of A_k. By Corollary 1.40, there is a subsequence $\{\mathbf{x}^{k(j)}\}$ converging to a point $\mathbf{x}$. Since $\mathbf{x}^{k(j)} \in A_m$ for all $k(j) \geq m$, $\mathbf{x} \in A_m$ as A_m is closed. This is true for every m so $\mathbf{x}$ lies in the intersection.

Remark. Of course, the intersection need no longer be a singleton. For example, consider $A_m = \{\mathbf{x} : \|\mathbf{x}\| \leq 1 + 1/m\}$; the intersection is the closed unit ball.

1.5. COUNTABILITY

If A and B are two sets, we say that there is a *one-to-one correspondence* between them if the elements of A can be paired off with the elements of B. More precisely a one-to-one correspondence is a *bijection* $f: A \longrightarrow B$ (recalled at the beginning of Chapter 2). Then elements can be paired either as $(a, f(a))$ or as $(f^{-1}(b), b)$. A set A is said to be *finite* if it is in one-to-one correspondence with the set $\{1,2,\cdots,m\}$ for some $m \in \mathbb{N}$, and is then said to have m members. The empty set $\emptyset$ is regarded as a finite set with 0 members. A set that is not finite is called *infinite*. The set $\mathbb{N}$ is a rather special infinite set, its terms can be written in a single list $\{1,2,3,\cdots\}$ or more simply as the sequence $\{n\}$. Any set A whose elements can be listed in a single sequence is said to be *enumerable*. This means precisely that A is in one-to-one correspondence with $\mathbb{N}$. For if $A=\{a_1,a_2,a_3,\cdots\}$ then $n \leftrightarrow a_n$ provides the pairing. A set that is either finite or enumerable is said to be *countable*.

The notion of countability is a measure of how large the set may be. We shall see that sets such as $\mathbb{Q}$, the rational numbers, are countable, while others, such as $\mathbb{R}$, are not. This fact shows that, in some sense, there are "many more" irrational numbers than there are rational ones.

Example 1.42. The even numbers $\{2,4,6,\cdots\}$ can be put in one-to-one correspondence with $\{1,2,3,\cdots\}$.

Example 1.43. $\mathbb{Z}=\{0,1,-1,2,-2,3,-3,\cdots\}$ can be put in one-to-one correspondence with $\{1,2,3,\cdots\}$. This listing of $\mathbb{Z}$ is already a proof corresponding to the function from $\mathbb{N}$ to $\mathbb{Z}$ given by $f(2k-1)=1-k$, $f(2k)=k$, $k=1,2,\cdots$.

We first prove a result that is intuitively rather obvious but whose proof rests on basic properties of $\mathbb{N}$.

Proposition 1.44. *Any subset of* $\mathbb{N}$ *is countable.*

Proof. Let A be any subset of $\mathbb{N}$. Obviously we may suppose A is non-empty. By the well-ordering principle of $\mathbb{N}$, A contains a smallest integer, say n_0. Let $A_1=A\backslash\{n_0\}$. If $A_1=\emptyset$ then S consists of the single point n_0 and is finite. If $A_1 \neq \emptyset$ it likewise contains a smallest integer n_1 and $n_1>n_0$. Continue this process taking $A_2=A_1\backslash\{n_1\}$. Either the process stops at some k where $A_{k+1}=\emptyset$ and then $A=\{n_0, n_1, \cdots, n_k\}$ or it continues indefinitely. We claim that, in the latter case, the sequence $\{n_0, n_1, \cdots\}$ exhausts A. If not, there is some $p\in A$ with $p\neq n_k$ for all $k\in\mathbb{N}$. Now, only finitely many n_i's can be less than p (indeed, at most $p-1$) so there exists k such that $n_k<p<n_{k+1}$. But, n_{k+1} was chosen to be the smallest element of S that was larger than $n_0, n_1, \cdots, n_k$, contradicting the above inequality.

Corollary 1.45. *If A is an infinite subset of $\mathbb{N}$ then there is a bijection $f: A \longrightarrow \mathbb{N}$.*

Corollary 1.46. *If S is a countable set and $T\subseteq S$ then T is countable.*

Proof. This is evident if T is finite. So suppose T and S are infinite. Then, as S is enumerable, there is a bijection $g: S\rightarrow\mathbb{N}$. Now $g(T)$ is an infinite subset of $\mathbb{N}$ so there exists a bijection $f: g(T)\longrightarrow\mathbb{N}$. Then the composite map $f\circ g: T\longrightarrow\mathbb{N}$ is a bijection so T is countable.

The idea behind this result is that S is enumerated in a single list and one goes through the list in order deleting terms that are not in T. The resulting list is an enumeration of T.

Example 1.47. The set $\mathbb{Q}$ of all rational numbers is enumerable.

Solution. We arrange the positive rational numbers in an array as follows.

$$\begin{array}{cccc} 1/1 & 1/2 & 1/3 & 1/4 \cdots \\ 2/1 & 2/2 & 2/3 & 2/4 \cdots \\ 3/1 & 3/2 & 3/3 & 3/4 \cdots \\ 4/1 & 4/2 & 4/3 & 4/4 \cdots \\ \vdots & \vdots & \vdots & \vdots \end{array}$$

We then enumerate these diagonally beginning at the top left and proceeding along from top right to bottom left, that is, we write as the single list, where the diagonals have been separated by semicolons :

$$1/1;\ 1/2,\ 2/1;\ 1/3,\ 2/2,\ 3/1;\ 1/4,\ 2/3,\ 3/2,\ 4/1;\ \cdots$$

Then a/b arises from the $(a+b-1)$-th diagonal and occupies the a-th place on that diagonal so occupies the

$[1+2+3+\cdots+(a+b-2)]+a=\left[\frac{1}{2}(a+b-2)(a+b-1)+a\right]$-th place in the list.

(Recall that $1+2+3+\cdots+n = \frac{1}{2}n(n+1)$). This sets up a bijection from the set $\{\frac{a}{b} : a,b \in \mathbb{N}\}$ to $\mathbb{N}$, so this set is enumerable. Since the set of all positive rationals is an infinite subset of this (there are infinitely many repetitions, for example $2/3=4/6=6/9=\cdots$), the positive rationals are enumerable. Thus, we can write these in a single list $\{r_1, r_2, r_3, \cdots\}$ and then enumerate $\mathbb{Q}$ as $\{0, r_1, -r_1, r_2, -r_2, \cdots\}$.

Theorem 1.48. *A countable union of countable sets is countable, that is, if* $\{A_1, A_2, \cdots\}$ *is a finite or infinite sequence of sets each of which is countable then* $\bigcup_{j=1}^{\infty} A_j$ *is also countable.*

Proof It suffices to prove the result when there is an infinite sequence of enumerable sets. Since each A_i is countable, the elements of A_i can be written in a single sequence $A_i = \{a_1^i, a_2^i, a_3^i, \cdots\}$. We write out the elements of the sets in an array :

$$\begin{array}{llllll} A_1: & a_1^1 & a_2^1 & a_3^1 & a_4^1 & \cdots \\ A_2: & a_1^2 & a_2^2 & a_3^2 & a_4^2 & \cdots \\ A_3: & a_1^3 & a_2^3 & a_3^3 & a_4^3 & \cdots \quad \text{etc.} \end{array}$$

and then we may enumerate these in a single list as follows: $a_1^1, a_2^1, a_1^2, a_3^1, a_2^2, a_1^3, \cdots$. (Compare the rationals in Example 1.47.)
This list certainly includes $\bigcup_{j=1}^{\infty} A_j$ so $\bigcup_{j=1}^{\infty} A_j$ is countable.

Proposition 1.49. *The Cartesian product of a finite number of countable sets is countable.*

Proof. Let A and B be countable sets. Then $A \times B = \{(a,b) : a \in A, b \in B\}$. Now, for each fixed $a \in A$, the set $B_a := \{(a,b) : b \in B\}$ is in one-to-one correspondence with B so is countable. Then $A \times B = \bigcup_{a \in A} B_a$ is a countable union of countable sets. This shows that the result is true for two sets. A simple induction argument proves the result for an arbitrary finite number of sets.

It might be supposed that with sufficient ingenuity it would be possible to prove that all infinite sets are enumerable. However, this is not so as we now prove by showing that $\mathbb{R}$ is uncountable.

Example 1.50. The set S of all numbers of the form $\sum_{n=1}^{\infty} \frac{a_n}{10^n}$, where each $a_n \in \{0,1,2,\cdots,9\}$, is not enumerable.

Some remarks are in order. First, the series is convergent since $\frac{a_n}{10^n} \leq \left(\frac{9}{10}\right)^n$ and $\sum_{n=1}^{\infty} \left(\frac{9}{10}\right)^n$ converges to 1, being a geometric series with ratio 9/10, so the comparison test applies. Second, it is usual to write $x = \sum_{n=1}^{\infty} \frac{a_n}{10^n}$ in the form $x = 0.a_1a_2a_3\cdots$, the *decimal expansion* of the number x. Therefore S is really the interval $[0,1]$. It is known that, given a number x, either its decimal expansion is unique or it has two expansions, a typical example being $0.25 = 0.24999\cdots$. This problem of recurring nines is the only one where the expansion is not unique.

Solution. Suppose S is enumerable say $S = \{s_1, s_2, s_3, \cdots\}$. Then we have

$$s_1 = 0.a_1^1 a_2^1 a_3^1 \cdots$$

$$s_2 = 0.a_1^2 a_2^2 a_3^2 \cdots$$

$$s_3 = 0.a_1^3 a_2^3 a_3^3 \cdots \qquad \text{etc.}$$

We let $t \in S$ be defined by choosing t_n to be some number between 0 and 8 that is not equal to a_n^n (along the main diagonal). For definiteness we may choose $t_n = \begin{cases} 1 & \text{if } a_n^n \neq 1 \\ 2 & \text{if } a_n^n = 1 \end{cases}$. Then t is not equal to any of the s_n, a contradiction. Since $\mathbb{R}$ is a larger set, $\mathbb{R}$ is certainly uncountable.

Note that this proves that the irrational numbers must be uncountable. As the rational numbers are countable while the real numbers are not, we may consider the real numbers to be more numerous. Nevertheless, real numbers can be approximated arbitrarily closely by rationals as we now show.

Lemma 1.51. *For any $x \in \mathbb{R}$ and any $n \in \mathbb{N}$ there exists $m \in \mathbb{Z}$ such that $0 \leq x - m/n < 1/n$.*

Proof. Let m be the largest integer less than or equal to nx. Then $m \leq nx < m+1$ or $0 \leq nx - m < 1$. Dividing by n gives the result.

This result proves that $\mathrm{cl}(\mathbb{Q})=\mathbb{R}$. We say $\mathbb{Q}$ is *dense* in $\mathbb{R}$. A simple consequence is

Theorem 1.52. *$\mathbb{Q}^n$ is dense in $\mathbb{R}^n$ for any $n \geq 1$.*

Proof. For $\mathbf{x} \in \mathbb{R}^n$ and $\varepsilon > 0$ let $r_i \in \mathbb{Q}$ be chosen so that $|x_i - r_i| < \varepsilon/\sqrt{n}$. Set $\mathbf{r} = (r_1, r_2, \cdots, r_n) \in \mathbb{Q}^n$. Then $\|\mathbf{x}-\mathbf{r}\| = (\sum_{i=1}^{n} |x_i - r_i|^2)^{1/2} < \varepsilon$.

Since $\mathbb{Q}^n$ is countable, $\mathbb{R}^n$ contains a countable dense subset. Spaces with this property are called *separable.*

1.6. COMPACTNESS

The notion of compactness plays a prominent part in analysis. There are several possible definitions but these coincide in $\mathbb{R}^n$. Also there is a simple characterisation of the compact sets in $\mathbb{R}^n$: they are the closed, bounded sets.

Definition 1.53. A set C is said to be *sequentially compact* if every sequence of points of C contains a subsequence that converges to a point of C.

Proposition 1.54. *A sequentially compact set is bounded.*

Proof. Suppose the set C is not bounded. Then, for every $k \in \mathbb{N}$, there is at least one point $\mathbf{x} \in C$ with $\|\mathbf{x}\| > k$. Select one and label it $\mathbf{x}^{(k)}$. By the sequential compactness, $\{\mathbf{x}^{(k)}\}$ has a convergent subsequence $\{\mathbf{x}^{k(j)}\}$, say. Since a convergent sequence is bounded, there is M such that $\|\mathbf{x}^{k(j)}\| \leq M$ for all $k(j)$. This contradicts $\|\mathbf{x}^{k(j)}\| > k(j)$.

Proposition 1.55. *A sequentially compact set C is closed.*

Proof. Let $\mathbf{x}$ be any limit point of C. Then there is a sequence $\{\mathbf{x}^{(k)}\}$ of points of C with $\mathbf{x}^{(k)} \longrightarrow \mathbf{x}$. By compactness, $\{\mathbf{x}^{(k)}\}$ has a subsequence which converges to a point of C. However, every subsequence of a convergent sequence converges to the same limit as the whole sequence. Therefore the subsequence has limit $\mathbf{x}$ and $\mathbf{x} \in C$.

In $\mathbb{R}^n$ the above results characterize sequentially compact sets.

Theorem 1.56. *A set C in $\mathbb{R}^n$ is sequentially compact if and only if it is closed and bounded.*

Proof. The implication $\Rightarrow$ has been shown above. For the reverse, suppose C is closed and bounded. By Corollary 1.40, any sequence of points of C has a convergent subsequence whose limit belongs to C, since C is closed. Thus, C is sequentially compact.

Definition 1.57. We say that a set C has the *Bolzano-Weierstrass property* if every infinite subset of C has a limit point in C.

Proposition 1.58. *A set C is sequentially compact if and only if it has the Bolzano-Weierstrass property.*

Proof. Since sequential compactness is the same as closed and bounded, C has the B-W property by Theorems 1.38 and 1.39. The converse is proved in exactly the same way as Corollary 1.40.

There is yet another definition of compactness which has proved to be most appropriate in spaces more general than $\mathbb{R}^n$. It is more topological in nature and relies on the notion of an *open cover* of a set. A collection of sets $\{V : V \in \mathcal{F}\}$ is said to *cover* a set A if $A \subseteq \cup\{V : V \in \mathcal{F}\}$. The collection is called an *open cover* if each $V \in \mathcal{F}$ is an open set. A *subcover* is a subfamily which also covers A.

Definition 1.59. A set C is said to be *compact* if every open cover of C contains a finite subcover.

We shall shortly prove that the two concepts of compactness and sequential compactness coincide in $\mathbb{R}^n$ (this is actually true in any metric space, but may fail in other settings).

There is a useful equivalent formulation of compactness in terms of (relatively) closed sets. A set E is said to be *closed relative to a set C* if there is a closed set F such that $E = F \cap C$. A family $\{F : F \in \mathcal{F}\}$ of sets is said to have the *finite intersection property* (f.i.p.) if every finite subfamily has a nonempty intersection, that is, if $F_1, F_2, \cdots, F_m$ are members of $\mathcal{F}$ then $\bigcap_{i=1}^{m} F_i \neq \emptyset$.

Proposition 1.60. *A set C is compact $\Leftrightarrow$ every family of relatively closed subsets of C with the f.i.p. has nonempty intersection.*

Proof. This is an exercise in the use of the De Morgan laws (Lemma 1.16). Indeed, suppose C is compact and let $\mathcal{F}$ be a family of relatively closed subsets of C with the f.i.p. and let $\mathscr{F}=\{F\cap C\}$ where each F is closed. Then $V=F^c$ is open and $\cup V=\cup F^c=(\cap F)^c$. If we had $\cap(F\cap C)=\emptyset$ then $\cap F\subseteq C^c$ and $\{V : V\in\mathscr{F}\}$ would be an open cover of C so would contain a finite subcover say $C\subseteq\bigcup_{i=1}^{m}V_i$, or $C^c\supseteq\bigcap_{i=1}^{m}V_i^c$. This shows that $\bigcap_{i=1}^{m}(F_i\cap C)=\emptyset$ which contradicts the f.i.p. Conversely, suppose every family of relatively closed sets with the f.i.p. has nonempty intersection. Then, for any open cover $\{V : V\in\mathscr{F}\}$ of C, the family $\{F=C\cap V^c\}$ consists of relatively closed subsets of C and $\cap F=C\cap(\cup V)^c=\emptyset$. Therefore this family cannot have the f.i.p. so some finite subfamily must have empty intersection $F_1\cap F_2\cap\cdots\cap F_m=\emptyset$. This shows that $V_1\cup V_2\cup\cdots\cup V_m$ is a finite subcover of C so C is compact.

Theorem 1.61. *A compact set C is closed and bounded.*

Proof. $\{B(\mathbf{0},m) : m=1,2,3,\cdots\}$ is an open cover of C so there exists a finite subcover, $\{B(\mathbf{0},m_i)\}$. Let $M=\max\{m_i\}$. Then $C\subseteq B(\mathbf{0},M)$ so C is bounded. Now let $\mathbf{y}\in C^c$: we want to show that there is a neighbourhood of $\mathbf{y}$ lying entirely within C^c. For $\mathbf{x}\in C$ let $r_{\mathbf{x}}=\|\mathbf{x}-\mathbf{y}\|/3$; $r_{\mathbf{x}}>0$ since $\mathbf{y}\notin C$. Then $\{B(\mathbf{x}, r_{\mathbf{x}}) : \mathbf{x}\in C\}$ is an open cover of C so has a finite subcover say $\{B(\mathbf{x}^{(i)}, r_i) : i=1,2,\cdots,m\}$. Let $r=\min\{r_i\}$. We show that $B(\mathbf{y},r)\cap C=\emptyset$. Indeed, if $\mathbf{x}\in B(\mathbf{y},r)\cap C$ then $\|\mathbf{x}-\mathbf{y}\|<r$. Also $\|\mathbf{x}-\mathbf{x}^{(i)}\|<r_i$ for some i and thus $\|\mathbf{x}^{(i)}-\mathbf{y}\|\le\|\mathbf{x}^{(i)}-\mathbf{x}\|+\|\mathbf{x}-\mathbf{y}\|<r_i+r\le 2r_i$. This contradicts $\|\mathbf{x}^{(i)}-\mathbf{y}\|=3r_i$ and shows that $B(\mathbf{y},r)\subseteq C^c$ so C^c is open and C is closed.

The converse result is true in $\mathbb{R}^n$, a result known as the Heine-Borel theorem. This is less easy to prove. It is convenient to prove a preliminary lemma.

Lemma 1.62. *Let $\{U : U\in\mathscr{F}\}$ be an open covering of a set $A\subseteq\mathbb{R}^n$. Then there is a countable subcovering of A.*

Proof. The set of all "rational neighbourhoods", neighbourhoods with centres in $\mathbb{Q}^n$ and radii in $\mathbb{Q}$, is countable (problem 1.34), say $B_1,B_2,\cdots$. We show that for any $\mathbf{x}\in A$, there is k and a set U from the family $\mathscr{F}$ such that $\mathbf{x}\in B_k\subseteq U$. Indeed, $\mathbf{x}$ belongs to some open set $U\in\mathscr{F}$, so there is $\varepsilon>0$ such that $B(\mathbf{x},\varepsilon)\subseteq U$. As $\mathbb{Q}^n$ is dense in $\mathbb{R}^n$ (Thm.1.52), there exists $\mathbf{r}\in\mathbb{Q}^n$ such that $\|\mathbf{x}-\mathbf{r}\|<\varepsilon/3$. Let $s\in\mathbb{Q}$ be taken so that

$\varepsilon/3 < s < 2\varepsilon/3$. Then $\mathbf{x} \in B(\mathbf{r},s)$ and $B(\mathbf{r},s) \subseteq B(\mathbf{x},\varepsilon) \subseteq U$, that is, $\mathbf{x} \in B_k \subseteq U$ for some $k = k(\mathbf{x})$. As there are infinitely many k for which $\mathbf{x} \in B_k$ we choose $k(\mathbf{x})$ to be the smallest possible one for definiteness. Let $\{U_1, U_2, \cdots\}$ be sets from the family $\mathscr{F}$ such that $B_i \subseteq U_i$. Then $\{U_{k(\mathbf{x})} : \mathbf{x} \in A\}$ is a countable subcover of A.

Theorem 1.63. (*The Heine-Borel theorem.*)
Let C be a closed, bounded subset of $\mathbb{R}^n$, then C is compact.

Proof. Let $\{U : U \in \mathscr{F}\}$ be an open covering of C. Then there is a countable subcovering. We prove that there is a finite subcover by showing, equivalently, that every countable collection of closed sets with the finite intersection property (f.i.p.) has nonempty intersection (as in Proposition 1.60). Given $\{F_i\}$ with the f.i.p. define sets K_m by $K_m = \bigcap_{i=1}^{m} F_i$. Then each K_m is closed and bounded and K_m is nonempty by the f.i.p.; also $K_{m+1} \subseteq K_m$. By Corollary 1.41, there exists $\mathbf{x} \in C$ with $\mathbf{x} \in K_m$ for every m. This proves that $\mathbf{x} \in F_i$ for every i so $\bigcap F_i \neq \emptyset$.

This result shows that the compact sets in $\mathbb{R}^n$ are precisely the same whether they are defined using the notions of open covers, or by the existence of convergent sequences, or as simply closed and bounded sets.

1.7. METRIC SPACES

Many of the results we have given for the space $\mathbb{R}^n$ are valid for other spaces. The important feature needed is that there should be a metric defined on the space. If X is a nonempty set, a *metric* is a real-valued function d defined on $X \times X$ with the properties listed in Theorem 1.6. When X is endowed with a metric we say that X, more precisely the pair (X,d), is a *metric space*. The concept of metric space is therefore very general but, nevertheless, occurs sufficiently often to have important applications. Such spaces are studied in courses on Topology and Functional Analysis.

Various concepts can be defined exactly as in $\mathbb{R}^n$. The δ-nbd of a point $x \in X$ is the set $B(x, \delta) = \{y \in X : d(y,x) < \delta\}$. Open and closed sets are then defined exactly as in $\mathbb{R}^n$. Convergence of sequences is defined either using neighbourhoods or, equivalently, by $x^{(k)} \longrightarrow x \Leftrightarrow d(x^{(k)}, x) \longrightarrow 0$. A set $Y \subseteq X$ is bounded if there is M such that

$d(x, x_0) \le M$ for all $x \in Y$, where x_0 is some fixed element of X. Compactness and sequential compactness are defined as above, they are equivalent concepts, Theorem 1.61 remains valid but, its converse may well be false.

A sequence in a metric space is called Cauchy if for any $\varepsilon > 0$ there is N such that $d(x^{(k)}, x^{(p)}) < \varepsilon$ for all $k,p \ge N$. Analogously to Lemma 1.28 it is proved that Cauchy sequences are bounded. Also convergent sequences are Cauchy. The space X is called complete if every Cauchy sequence converges to a point of X. Proving that a given metric space is complete may be a nontrivial task.

Proposition 1.64. *A compact metric space is complete.*

Proof. The proof is rather similar to that of Theorem 1.30 ; we give it to illustrate the arguments. Let $\{x^{(k)}\}$ be a Cauchy sequence in X. Then, by compactness, there is a subsequence $\{x^{k(j)}\}$ converging to a point x of X. For any $\varepsilon > 0$, there is N such that $d(x^{k(j)}, x) < \varepsilon/2$ for all $k(j) \ge N$. As the sequence is Cauchy, there is M so that $d(x^{(k)}, x^{(p)}) < \varepsilon/2$ for all $k,p \ge M$. Take a fixed $k(j) \ge \max\{N,M\}$. Then, for any $k > M$, by the triangle inequality, $d(x^{(k)}, x) \le d(x^{(k)}, x^{k(j)}) + d(x^{k(j)}, x) < \varepsilon$. This proves that $x^{(k)} \longrightarrow x \in X$.

Problems.

1. Prove that equality holds in the Schwarz inequality for $\mathbf{x}$ and $\mathbf{y}$ both nonzero if and only if $\mathbf{y} = \alpha\mathbf{x}$ for some nonzero $\alpha \in \mathbb{R}$. Hence show that equality holds in the triangle inequality for the Euclidean norm for $\mathbf{x},\mathbf{y}$ nonzero if and only if $\mathbf{y} = \alpha\mathbf{x}$ for $\alpha > 0$.

2. For $\mathbf{x},\mathbf{y} \in \mathbb{R}^n$ let $<\mathbf{x},\mathbf{y}>_{\mathbf{w}} := \sum_{i=1}^{n} x_i y_i w_i$ where $w_i > 0$ for every i. Show that this defines an inner product, that is, satisfies the properties (*i*)-(*iv*) listed after Definition 1.2.

3. Suppose that $|\cdot|$ is any norm that satisfies the parallelogram law (1.3). Show that $<\mathbf{x},\mathbf{y}> := \{|\mathbf{x}+\mathbf{y}|^2 - |\mathbf{x}-\mathbf{y}|^2\}/4$ defines an inner product with the properties (*i*), (*ii*), (*iv*) and (*iii*) for $\alpha \in \mathbb{Q}$. [Continuity of the norm (Chapter 2) then proves (*iii*) for $\alpha \in \mathbb{R}$.]

4. (*i*) For $\mathbf{x}=(x_1,x_2)\in\mathbb{R}^2$, let $\|\mathbf{x}\|_1:=|x_1|+|x_2|$. Show that this defines a norm satisfying the properties listed in Theorem 1.3. Extend to $\mathbb{R}^n$. Sketch the set $\{\mathbf{x}\in\mathbb{R}^2 : \|\mathbf{x}\|_1=1\}$. (*ii*) Repeat when $\|\mathbf{x}\|_\infty:=\max\{|x_1|,|x_2|\}$ in $\mathbb{R}^2$ and extend to $\mathbb{R}^n$.

5. With the norms as defined in problem 4 prove that $\|\mathbf{x}\|_\infty\le\|\mathbf{x}\|\le\sqrt{n}\|\mathbf{x}\|_\infty$, and $\|\mathbf{x}\|_1\le\sqrt{n}\|\mathbf{x}\|\le\sqrt{n}\|\mathbf{x}\|_1$, for all $\mathbf{x}\in\mathbb{R}^n$. Hence show that a set is open relative to the Euclidean norm $\Leftrightarrow$ it is open relative to either of the other norms.

6. For $1<p<\infty$ and $a>0$, $b>0$, show by elementary calculus that

$$(a+b)^p = \min\{\, t^{1-p} a^p + (1-t)^{1-p} b^p : 0<t<1 \,\}.$$

Deduce that for $\mathbf{x}=(x_i)$, $\mathbf{y}=(y_i)$, we have *Minkowski's inequality* :

$$\left(\sum_{i=1}^n |x_i+y_i|^p\right)^{1/p} \le \left(\sum_{i=1}^n |x_i|^p\right)^{1/p} + \left(\sum_{i=1}^n |y_i|^p\right)^{1/p}.$$

Hence show that $\|\mathbf{x}\|_p := \left(\sum_{i=1}^n |x_i|^p\right)^{1/p}$ defines a norm on $\mathbb{R}^n$.

(I learned this elementary proof from Prof. Hans König.)

7. Let $A:=\{(x,y)\in\mathbb{R}^2 :\ 0\le x\le 1,\ 0<y<1\}$. Determine int($A$), cl($A$), and bdy($A$). Is A open? Is A closed?

8. Let $A:=\{(x,y,0)\in\mathbb{R}^3 : 0<x<1,\ 0<y<1\}$. Prove that A is not an open subset of $\mathbb{R}^3$. Is A closed ?

9. Show that $\mathrm{cl}(B(\mathbf{0},r))=\bar{B}(\mathbf{0},r)$ and that $\mathrm{int}(\bar{B}(\mathbf{0},r))=B(\mathbf{0},r)$.

10. Show that $\mathrm{int}(A\cap B)=\mathrm{int}(A)\cap\mathrm{int}(B)$ and that $\mathrm{int}(A)\cup\mathrm{int}(B)\subseteq\mathrm{int}(A\cup B)$. Give an example to show that strict inclusion may hold here. Are there corresponding results for closures?

11 Show that for any set A, int(A) is open and cl(A) is closed.

12. Show that int(A) is the largest open set contained in A, that is, if $\mathrm{int}(A)\subseteq V\subseteq A$ and V is open then $V=\mathrm{int}(A)$. Hence, or otherwise, prove that cl(A) is the smallest closed set containing A. Deduce that cl(A) is the intersection of all closed sets that contain A.

13. Prove that $\mathrm{bdy}(A)=\mathrm{cl}(A)\cap\mathrm{cl}(A^c)$. Deduce that $\mathrm{bdy}(A)=\mathrm{bdy}(A^c)$ and that bdy(A) is a closed set.

14. Show that any finite set is closed.

15. Prove that $\text{cl}(A)=\text{int}(A)\cup\text{bdy}(A)$. Deduce that A is closed $\Leftrightarrow$ $\text{bdy}(A)\subseteq A$.

16. Let V be an open set and let $\mathbf{a}\in V$. Define the translate V^{t} of V by $V^{t} := \{\mathbf{x} : \mathbf{x}+\mathbf{a}\in V\}$. Show that V^{t} is an open set containing $\mathbf{0}$.

17. Are the following statements true or false? If true, provide a proof, if false, give a counterexample.
(*a*) If A is closed then $\text{cl}(\text{int}(A))=A$.
(*b*) If A is open, $\text{int}(\text{cl}(A))=A$.

18. Let G be an open subset of $\mathbb{R}^n$ and suppose $\mathbf{a}\in G$ and $\mathbf{h}\neq\mathbf{0}$. Let $E=\{\, t\in\mathbb{R} : \mathbf{a}+t\mathbf{h}\in G\}$. Show that E is open in $\mathbb{R}$.

19. Give an example of a countable family of closed sets whose union is not closed. Give an example of a countable family of open sets whose intersection is not open.

20. Prove that a convergent sequence in $\mathbb{R}^n$ is bounded. Give an example of a bounded sequence that is not convergent.

21. Let $\{x^{(k)}\}$ be a monotone increasing sequence of real numbers which has a convergent subsequence with limit x. Show that the whole sequence converges to x.

22. *Property 1* Every monotone increasing sequence $\{x^{(k)}\}$ of real numbers which is bounded above converges.

Property 2 Every nonempty set of real numbers that is bounded above has a least upper bound.
Prove that these two properties of real numbers are equivalent.

23. Prove that if $A\subseteq\mathbb{R}$ is bounded above then $\sup A\in\text{cl}(A)$.

24. Let $\{\mathbf{x}^{(m)}\}$ be a Cauchy sequence. Show directly from the definition of Cauchy sequence that the set of its elements has at most one limit point.

25. Let $\{\mathbf{x}^{(m)}\}$ be a convergent sequence with infinitely many distinct points. Prove that its limit is a limit point of its set of elements.

26. Let $\{\mathbf{x}^{(m)}\}$ be a convergent sequence. Show that every subsequence converges to the same limit. Further prove that if $\{\mathbf{x}^{(m)}\}$ is a sequence with the property that there exists a point $\mathbf{x}$ and every

subsequence contains a further subsequence thet converges to $\mathbf{x}$, then $\mathbf{x}^{(m)} \longrightarrow \mathbf{x}$.

27. Suppose that $\mathbf{x}^{(m)} \longrightarrow \mathbf{x}$ and there exist $\mathbf{a} \in \mathbb{R}^n$ and $\delta > 0$ such that $\|\mathbf{x}^{(m)} - \mathbf{a}\| \leq \delta$ for all m. Prove that $\|\mathbf{x} - \mathbf{a}\| \leq \delta$.

28. Prove that a set has a finite diameter $\Leftrightarrow$ it is bounded.

29. Give an example of sets A_m with $A_{m+1} \subseteq A_m$ and $\mathrm{diam}(A_m) \longrightarrow 0$ but $\cap A_m = \emptyset$. Give an example of closed sets A_m with $A_{m+1} \subseteq A_m$ and $\cap A_m = \emptyset$.

30. Find the limit points of the following sets : $A = \{1/2^m : m = 1, 2, \cdots\}$; $B = \{(1 + 1/m, \sin(m\pi/2))\}$; $C = \{$The rational numbers in $(0,1)\}$; $D = \{((1 + 1/m)\cos(m\pi), 1/2^m)\}$.

31.(*a*) Let S be a dense subset of $\mathbb{R}^n$ [$\mathrm{cl}(S) = \mathbb{R}^n$] and let U be open. Prove that $S \cap U \neq \emptyset$.
(*b*) Let $\{V_i\}$ be a sequence of open dense sets. Prove that $\cap V_i$ is dense.
(*c*) (Baires theorem.) A set X is said to be *nowhere dense* if $\mathrm{cl}(X)$ has empty interior. Prove that $\mathbb{R}^n$ cannot be written as a countable union of nowhere dense sets.
(*d*) Prove that the irrational numbers, $\mathbb{R} \setminus \mathbb{Q}$, cannot be written as a countable union of nowhere dense sets. Deduce that $\mathbb{Q}$ cannot be written as a countable intersection of open dense sets.

32. For any sequence $\{x^{(m)}\}$ in $\mathbb{R}$ and any interval I use the nested intervals theorem to show that there exists $y \in I$ with $y \neq x^{(m)}$ for every m. Deduce that intervals are uncountable. [This gives an alternative method of proving that $\mathbb{R}$ is uncountable.]

33. Prove that the irrational numbers ($\mathbb{R} \setminus \mathbb{Q}$) are uncountable.

34. Show that the set of all "rational neighbourhoods" in $\mathbb{R}^n$ (Lemma 1.62) is a countable set. Deduce that the set of isolated points of a set $A \subseteq \mathbb{R}^n$ is countable.

35. If F is a finite subset of $\mathbb{R}$ write $\sum_F x$ for the sum of the elements of F. For an infinite subset E of nonnegative real numbers define $\sum_E x := \sup\{\sum_F x : F$ is a nonempty finite subset of $E\}$. Prove that if $\sum_E x$ is finite then $\{x \in E : x > 0\}$ is countable.

36. Let S denote the set of all functions $f: \mathbb{N} \longrightarrow \{0,1\}$, that is, the set of all sequences consisting of only zeros and ones. Show that S is uncountable.

37. Starting from $[0,1]$ let P_1 be obtained by deleting the open middle third, thus $P_1=[0,1/3]\cup[2/3,1]$. P_2 is obtained from P_1 by deleting the open middle third of each closed interval in P_1, thus $P_2=[0,1/9]\cup[2/9,1/3]\cup[2/3,7/9]\cup[8/9,1]$. Repeat this process of deleting middle thirds of intervals and let $P=\bigcap_{n=1}^{\infty} P_n$. P is called the *Cantor set*. Show that P is closed. Let S be as in problem 36 and let $\phi : S \longrightarrow [0,1]$ be given by $\phi(f)= \sum_{i=1}^{\infty} \frac{2f(i)}{3^i}$. Show that $\phi(f)\in P$ for each $f\in S$ and deduce that P is uncountable.

38. Are the following statements true or false? If true, provide a proof, if false, give a counterexample.
(*a*) If C and K are compact so is $C\cap K$.
(*b*) If C and K are compact so is $C\cup K$.
(*c*) An infinite union of compact sets is compact.
(*d*) An infinite intersection of compact sets is compact.

39. Let A be a subset of $\mathbb{R}^n$ and define the distance from a point $\mathbf{x}\in\mathbb{R}^n$ to the set A by $\text{dist}(\mathbf{x},A) := \inf\{\|\mathbf{x}-\mathbf{a}\| : \mathbf{a}\in A\}$. Prove that $\text{dist}(\mathbf{x},A)=0$ if and only if $\mathbf{x}\in\text{cl}(A)$. Show also, that if A is compact, there exists $\mathbf{a}\in A$ such that $\text{dist}(\mathbf{x},A)=\|\mathbf{x}-\mathbf{a}\|$.

40. Find an open cover of $(0,1)$ that has no finite subcover.

41. Let K be a compact set and let $\{U : U\in\mathcal{F}\}$ be an open cover of K. Show that there exist a finite subcover $\{U_i : 1\leq i\leq m\}$ of K and $\delta>0$ such that for every set $S\subseteq K$ with diameter less than δ there is j with $S\subseteq U_j$. [δ is called the Lebesgue number of the covering.]

Hints on the problems.

1. If equality holds then $b^2=4ac$ in the proof of Lemma 1.4.
3. Show that $<\mathbf{x}+\mathbf{z},\mathbf{y}> + <\mathbf{x}-\mathbf{z},\mathbf{y}> = 2<\mathbf{x},\mathbf{y}>$ for all $\mathbf{x},\mathbf{y},\mathbf{z}$. Take $\mathbf{z}=\mathbf{x}$, then deduce $<\mathbf{u},\mathbf{y}> + <\mathbf{v},\mathbf{y}> = <\mathbf{u}+\mathbf{v},\mathbf{y}>$. To get $<\alpha\mathbf{z},\mathbf{y}> = \alpha<\mathbf{z},\mathbf{y}>$ show this for $\alpha=m\in\mathbb{N}$ by taking $\mathbf{x}=m\mathbf{z}$ and using induction. Then do $\alpha=1/m$.
5. Cauchy's inequality with $y_i=1$ and then $\sum|x_i|^2\leq\left[\sum|x_i|\right]^2$.
7. A sketch will show what the answers must be.
10. Consider rationals and irrationals. Take complements.

11. Second part : what is $[\mathrm{cl}(A)]^c$?
12. Show that if $\mathbf{x}\in V$ then a nbd of $\mathbf{x}$ lies in A.
13. For $\mathbf{x}\in\mathrm{cl}(A)\cap\mathrm{cl}(A^c)$ every nbd of $\mathbf{x}$ contains at least one point of A and one of A^c.
14. If $A=\{\mathbf{a}^{(1)},\cdots,\mathbf{a}^{(m)}\}$ and $\mathbf{x}\in A^c$ let $\delta=\min\{\|\mathbf{x}-\mathbf{a}^{(i)}\| : 1\le i\le m\}$.
17. (*a*) False, int(A) might be empty. (*b*) False, delete some points from an open set.
18. Compare the proof of boundedness of Cauchy sequences.
21. For any m, consider $n(k)<m<n(p)$.
22. Let $M=\sup\{x^{(m)}\}$. For any $\varepsilon>0$ $\exists$ m such that $M-\varepsilon<x^{(m)}\le M$, by definition of supremum. With monotonicity this shows that the sequence converges to M. If a set S is bounded above by K take $s\in S$. Let $I_1=[s,K]$ and bisect I_1. Let I_2 be the the right-hand half if this contains points of S, otherwise the left half. Repeat this process and consider the decreasing sequence of right-hand endpoints.
24. If there were two, take disjoint nbds containing them.
26. Second part, if $\{\mathbf{x}^{(m)}\}$ does not converge to $\mathbf{x}$, use (1.7) to get a subsequence outside an ε-nbd of $\mathbf{x}$.
27. $\overline{\mathrm{B}}(\mathbf{a},\delta)$ is a closed set.
29. Should not be closed. Should not be bounded.
31. (*a*) For $\mathbf{u}\in U$ there are points of S arbitrarily close to $\mathbf{u}$.
(*b*) For a fixed $\mathbf{x}$ and fixed $\varepsilon>0$ find a ball B_1 with radius less than ε with closure A_1 in $V_1\cap\mathrm{B}(\mathbf{x},\varepsilon)$, then a closed ball $A_2\subseteq V_2\cap B_1$ with radius less than $\varepsilon/2$, etc. Then apply Cantor's theorem. (*c*) Show that if X is nowhere dense then $[\mathrm{cl}(X)]^c$ is open and dense; apply (*b*).
(*d*) Let $\{r_i : i=1,2,\cdots\}$ be an enumeration of $\mathbb{Q}$. If $\mathbb{R}\setminus\mathbb{Q}=\cup X_i$ then $\mathbb{R}=\cup(X_i\cup\{r_i\})$.
32. Take I_0 to be a closed subinterval of I and trisect. One of the 3 subintervals does not contain $x^{(1)}$. Continue this process.
34. In one-to-one correspondence with $\mathbb{Q}^{n+1}$. Each point can be put in a rational neighbourhood.
35. If $\sum_E x=K$ there are at most mK elements of E larger than $1/m$.
37. For $x\in[0,1]$, write $x=\sum_{i=1}^{\infty} a^{(i)}/3^i$, $a^{(i)}\in\{0,1,2\}$, the ternary expansion of x. Show that $x\in P \Leftrightarrow x$ has a ternary expansion containing no 1's. Show that ϕ is one-to-one.
39. There exists $\mathbf{a}^{(m)}\in A$ so that $\mathrm{dist}(\mathbf{x},A)\le\|\mathbf{x}-\mathbf{a}^{(m)}\|\le\mathrm{dist}(\mathbf{x},A)+1/m$.
41. For $\mathbf{x}\in K$, there is $\delta(\mathbf{x})>0$ and $U(\mathbf{x})\in\mathscr{F}$ such that $\mathrm{B}(\mathbf{x},2\delta(\mathbf{x}))\subseteq U$. The open cover of K consisting of all such nbds has a finite subcover.
Or set up a contradiction argument.

Chapter 2

Continuous Functions

2.1. FUNCTIONS AND CONTINUITY

We want to discuss functions acting between Euclidean spaces. A function $\mathbf{f}$ from a subset $\mathrm{dom}(\mathbf{f}) \subseteq \mathbb{R}^n$, the *domain* of $\mathbf{f}$, into $\mathbb{R}^m$, written $\mathbf{f} : \mathrm{dom}(\mathbf{f}) \longrightarrow \mathbb{R}^m$, is a rule that assigns to every $\mathbf{x} \in \mathrm{dom}(\mathbf{f})$ a unique point $\mathbf{y} \in \mathbb{R}^m$. We write $\mathbf{y}=\mathbf{f}(\mathbf{x})$ and call $\mathbf{f}(\mathbf{x})$ the *image of* $\mathbf{x}$ *under* $\mathbf{f}$. We may say that $\mathbf{f}$ takes its values in $\mathbb{R}^m$. We shall write $\mathbf{f} : A \longrightarrow \mathbb{R}^m$ to mean that $\mathbf{f}$ is considered with domain A, that is, $\mathbf{f}(\mathbf{x})$ is well defined for all $\mathbf{x}$ in A, possibly also for points outside A too; we do not insist that A is the maximal domain of $\mathbf{f}$ in any sense. The set of all images is called the *image set* of $\mathbf{f}$, or *range* of $\mathbf{f}$, and is written $\mathbf{f}(A) := \{\mathbf{f}(\mathbf{x}) : \mathbf{x} \in A\}$. For $\mathbf{f} : A \longrightarrow \mathbb{R}^m$ we have $\mathbf{f}(\mathbf{x})=\mathbf{y}=(y_1, y_2, \cdots, y_m)$. The components y_i are uniquely determined by $\mathbf{x}$ and $\mathbf{f}$ so define functions $f_i : A \longrightarrow \mathbb{R}$, $1 \le i \le m$, by the rules $f_i(\mathbf{x})=y_i$. These functions are called *component functions* or simply *components* of $\mathbf{f}$. We may then write $\mathbf{f}(\mathbf{x})=(f_1(\mathbf{x}), f_2(\mathbf{x}), \cdots, f_m(\mathbf{x}))$, or briefly, $\mathbf{f}=(f_1, f_2, \cdots, f_m)$. We use bold face for functions when $m > 1$.

Examples 2.1. (*a*) $\mathbf{f}(x,y) = (y, x, x+y)$, defines $\mathbf{f} : \mathbb{R}^2 \longrightarrow \mathbb{R}^3$.

(*b*) $f(x, y, z) = xyz$, defines $f : \mathbb{R}^3 \longrightarrow \mathbb{R}$.

(*c*) $\mathbf{f}(x) = (x, x, \sqrt{x})$, defines $\mathbf{f} : [0, \infty) \subset \mathbb{R} \longrightarrow \mathbb{R}^3$.

(*d*) $\mathbf{f}(x_1, x_2, x_3, x_4) = (x_1+x_2, x_3+x_4, x_1x_2, x_3x_4)$, defines $\mathbf{f} : \mathbb{R}^4 \longrightarrow \mathbb{R}^4$. Here, $f_1(\mathbf{x})=x_1+x_2$, $f_3(\mathbf{x})=x_1x_2$, etc.

In these examples, the functions are defined via formulae. However, insisting that a function should be so defined is too restrictive. For example, we can define a function $f : [1, \infty) \longrightarrow \mathbb{R}$ by the "rule" : $f(x)$ is the solution y of the equation $(y^2+1)e^y=x$. This is perfectly valid since $(y^2+1)e^y$ is strictly increasing in y so the equation does have a unique solution, but, there is no formula for $f(x)$. This is an example of the inverse of a function, a topic to be discussed further in Chapter 4. We may therefore ask the pertinent question as to what exactly constitutes a valid rule in the above naive definition of function. You are now at a stage of your mathematical development where a more precise definition can be appreciated. The precise definition is given using the concept of the graph of a function.

If $\mathbf{f} : A \subseteq \mathbb{R}^n \longrightarrow \mathbb{R}^m$ is a function, its *graph* G is the set of all ordered pairs, $G := \{(\mathbf{x},\mathbf{f}(\mathbf{x})) : \mathbf{x} \in A\}$, a subset of $\mathbb{R}^n \times \mathbb{R}^m$. An ordered pair $(\mathbf{x},\mathbf{y}) \in \mathbb{R}^n \times \mathbb{R}^m$ belongs to $G \Leftrightarrow \mathbf{x} \in A$ and $\mathbf{y} = \mathbf{f}(\mathbf{x})$. Knowledge of the graph of f tells us everything about f so the function f can be identified with the graph G. Therefore, to define a function precisely is to specify the set of ordered pairs which is to be its graph. Which sets of ordered pairs can be graphs? Let H be any set of ordered pairs of $\mathbb{R}^n \times \mathbb{R}^m$. To be the graph of some function, H must have the property that $(\mathbf{x},\mathbf{y}) \in H$ and $(\mathbf{x},\mathbf{z}) \in H$ imply that $\mathbf{y} = \mathbf{z}$, for otherwise $\mathbf{f}(\mathbf{x})$ would not be uniquely determined.

Definition 2.2. A function f is a set of ordered pairs $G \subseteq \mathbb{R}^n \times \mathbb{R}^m$ with the property that $(\mathbf{x},\mathbf{y}) \in G$ and $(\mathbf{x},\mathbf{z}) \in G \Rightarrow \mathbf{y} = \mathbf{z}$. The *domain* of f is $\text{dom}(\mathbf{f}) := \{\mathbf{x} \in \mathbb{R}^n : (\mathbf{x},\mathbf{y}) \in G \text{ for some } \mathbf{y} \in \mathbb{R}^m\}$ and we write

$$\mathbf{y} = \mathbf{f}(\mathbf{x}) \Leftrightarrow (\mathbf{x},\mathbf{y}) \in G.$$

We shall, however, use the intuitive definition of function given at the beginning of this chapter with the understanding that this can be made precise if necessary. We shall also use, by way of variety, the terms *map* or *mapping* synonymously with function.

If $\mathbf{f} : A \longrightarrow \mathbb{R}^m$ is a function and if $A_1 \subseteq A$ the function $\mathbf{f}_1 : A_1 \longrightarrow \mathbb{R}^n$ defined by $\mathbf{f}_1(\mathbf{x}) = \mathbf{f}(\mathbf{x})$ for $\mathbf{x} \in A_1$ is called the *restriction of* f *to* A_1 which may be written $\mathbf{f}|_{A_1}$. Similarly, if $A_2 \supseteq A$ and $\mathbf{g} : A_2 \longrightarrow \mathbb{R}^n$ is such that $\mathbf{g}(\mathbf{x}) = \mathbf{f}(\mathbf{x})$ for all $\mathbf{x} \in A$, then g is called an *extension* of f. Note that extensions are not unique.

Given a set of ordered pairs $G = \{(\mathbf{x},\mathbf{y})\}$, the set of ordered pairs $G^{\sim} = \{(\mathbf{y},\mathbf{x})\}$ might also serve as the graph of a function. For this we need $(\mathbf{y},\mathbf{x}_1) \in G^{\sim}$ and $(\mathbf{y},\mathbf{x}_2) \in G^{\sim}$ to imply $\mathbf{x}_1 = \mathbf{x}_2$. If G is the graph of a function f then f is said to be *injective* (or *one-to-one*) if $\mathbf{x}_1 \neq \mathbf{x}_2 \Rightarrow \mathbf{f}(\mathbf{x}_1) \neq \mathbf{f}(\mathbf{x}_2)$. $G^{\sim}$ is the graph of a function precisely when f is injective. We call this function the *inverse of* f, denoted $\mathbf{f}^{-1}$. In short

$$\mathbf{x} = \mathbf{f}^{-1}(\mathbf{y}) \Leftrightarrow \mathbf{y} = \mathbf{f}(\mathbf{x})$$

If $\mathbf{f} : A \longrightarrow S$ we say f is *surjective* (or maps A *onto* S) if

$S=\mathbf{f}(A)$. When $\mathbf{f}: A \longrightarrow S$ is one-to-one and onto (injective and surjective) it is said to be *bijective* (a *bijection*). Then $\mathbf{f}^{-1}: S \longrightarrow A$ is also a bijection. For a function in $\mathbb{R}$ it is a relatively simple matter to determine whether or not f is injective (f^{-1} exists); for example, if f either strictly increasing or strictly decreasing on an interval then f is one-to-one on that interval. This is often determined from the sign of the derivative. In Chapter 4 we shall see that it is possible to use the derivative (to be studied in Chapter 3) to get information concerning the existence of an inverse function.

We now turn to the important concept of continuity.

Definition 2.3. Let $\mathbf{f}: A \subseteq \mathbb{R}^n \longrightarrow \mathbb{R}^m$ be a function. We say that $\mathbf{f}$ is *continuous* at $\mathbf{a} \in A$ if, for any $\varepsilon > 0$, there exists $\delta > 0$ such that $\mathbf{x} \in A$ and $\|\mathbf{x}-\mathbf{a}\| < \delta$ imply that $\|\mathbf{f}(\mathbf{x})-\mathbf{f}(\mathbf{a})\| < \varepsilon$. More briefly :

$\forall \varepsilon > 0,\ \exists\ \delta > 0$ so that $\mathbf{x} \in A$ and $\|\mathbf{x}-\mathbf{a}\| < \delta \Rightarrow \|\mathbf{f}(\mathbf{x})-\mathbf{f}(\mathbf{a})\| < \varepsilon$

We say that $\mathbf{f}$ is *continuous on A* if $\mathbf{f}$ is continuous at $\mathbf{a}$ for every $\mathbf{a} \in A$.

We refer to this as the *ε-δ condition.* Note that δ may depend on $\mathbf{a}$ as well as ε.

In terms of neighbourhoods, continuity of $\mathbf{f}$ at $\mathbf{a}$ says that, for any ε-neighbourhood of $\mathbf{f}(\mathbf{a})$ there is a δ-neighbourhood of $\mathbf{a}$ such that $\mathbf{f}(A \cap B(\mathbf{a},\delta)) \subseteq B(\mathbf{f}(\mathbf{a}),\varepsilon)$. This makes precise the intuitive idea that $\mathbf{f}$ is continuous at $\mathbf{a}$ when points $\mathbf{x}$ that are close to $\mathbf{a}$ have images $\mathbf{f}(\mathbf{x})$ close to $\mathbf{f}(\mathbf{a})$. If we imagine the point $\mathbf{x}$ approaching $\mathbf{a}$ then the corresponding point $\mathbf{f}(\mathbf{x})$ must approach $\mathbf{f}(\mathbf{a})$. It is important to realize that $\mathbf{x}$ may approach $\mathbf{a}$ in any manner whatever, provided $\mathbf{x}$ always stays in A. Continuity at $\mathbf{a}$ is a *local* property : it depends only on the behaviour of $\mathbf{f}$ in a (small) neighbourhood of $\mathbf{a}$.

If $\mathbf{a}$ is an isolated point of A, δ can be taken so small that the only point $\mathbf{x} \in A$ with $\|\mathbf{x}-\mathbf{a}\| < \delta$ is $\mathbf{a}$ itself. Then the definition is satisfied so $\mathbf{f}$ is automatically continuous at isolated points. However, such points are usually of little interest as far as continuity is concerned.

Example 2.4. Let $f: \mathbb{R}^n \longrightarrow \mathbb{R}$ be defined by $f(\mathbf{x})=\|\mathbf{x}\|$. Then f is continuous on $\mathbb{R}^n$, that is, the norm is a continuous function.

Solution. By Corollary 1.5, $|f(\mathbf{x})-f(\mathbf{y})| \leq \|\mathbf{x}-\mathbf{y}\|$, so that for any $\varepsilon > 0$, taking $\delta=\varepsilon$ yields $|f(\mathbf{x})-f(\mathbf{y})| < \varepsilon$ for $\|\mathbf{x}-\mathbf{y}\| < \delta$, and f is continuous at $\mathbf{y}$, for any $\mathbf{y} \in \mathbb{R}^n$.

Example 2.5. Let $p : \mathbb{R}^n \longrightarrow \mathbb{R}$ be the i-th co-ordinate function (or projection), that is, $p(\mathbf{x})=x_i$. Then p is continuous on $\mathbb{R}^n$.

Solution. $|p(\mathbf{x})-p(\mathbf{y})| = |x_i-y_i| \leq \|\mathbf{x}-\mathbf{y}\|$, (as in Theorem 1.19). Again, for any $\varepsilon>0$, taking $\delta=\varepsilon$ establishes the ε-δ property and proves continuity at all points $\mathbf{y}$.

The function $f : \mathbb{R}^3 \longrightarrow \mathbb{R}$ given by $f(x,y,z)=x$, the first co-ordinate function, is sometimes called "the function x". Similarly, if we have a formula for $\mathbf{f}(\mathbf{x})$, "the function $\mathbf{f}(\mathbf{x})$" often suffices to describe the function $\mathbf{f}$. This is a misuse of language but should not lead to error.

The notion of continuity can be defined equivalently in terms of sequences. This equivalent definition is often easier to check.

Theorem 2.6. *Let* $\mathbf{f} : A \subseteq \mathbb{R}^n \longrightarrow \mathbb{R}^m$. *The following statements are equivalent.*

(*i*) $\mathbf{f}$ *is continuous at a point* $\mathbf{a} \in A$.

(*ii*) *For any sequence* $\{\mathbf{x}^{(k)}\}$ *of points of* A, $\mathbf{x}^{(k)} \longrightarrow \mathbf{a} \Rightarrow \mathbf{f}(\mathbf{x}^{(k)}) \longrightarrow \mathbf{f}(\mathbf{a})$.

Proof. We first show that (*i*) ⇒ (*ii*). Thus, let $\mathbf{f}$ be continuous at $\mathbf{a}$, and let $\{\mathbf{x}^{(k)}\}$ be any sequence of points of A with $\mathbf{x}^{(k)} \longrightarrow \mathbf{a}$. Then,

$$\forall \varepsilon>0,\ \exists\ \delta>0 \text{ so that } \mathbf{x}\in A \text{ and } \|\mathbf{x}-\mathbf{a}\| < \delta \Rightarrow \|\mathbf{f}(\mathbf{x})-\mathbf{f}(\mathbf{a})\| < \varepsilon. \qquad (2.1)$$

Since $\mathbf{x}^{(k)} \longrightarrow \mathbf{a}$, there exists N such that $\|\mathbf{x}^{(k)}-\mathbf{a}\| < \delta$ for all $k \geq N$. Therefore, for $k \geq N$, $\mathbf{x}^{(k)}$ is a suitable $\mathbf{x}$ to be used in (2.1) and hence $\|\mathbf{f}(\mathbf{x}^{(k)})-\mathbf{f}(\mathbf{a})\| < \varepsilon$, for all $k \geq N$, that is, $\mathbf{f}(\mathbf{x}^{(k)}) \longrightarrow \mathbf{f}(\mathbf{a})$ and (*ii*) holds. We now show that (*ii*) ⇒ (*i*). The proof is by showing that, equivalently (the contrapositive), if (*i*) is false then (*ii*) is also false. So suppose that (*i*) is false. Then there must exist $\varepsilon>0$ so that (2.1) fails for every possible $\delta>0$:

$$\text{not}(\forall\varepsilon>0, \exists\delta : \text{property holds}) \Leftrightarrow (\exists\varepsilon, \forall\delta : \text{property fails}).$$

Fix this ε and take $\delta=1/k$, $k=1,2,\cdots$. Then for every k there exists $\mathbf{x}^{(k)} \in A$ with $\|\mathbf{x}^{(k)}-\mathbf{a}\| < \delta$ but $\|\mathbf{f}(\mathbf{x}^{(k)})-\mathbf{f}(\mathbf{a})\| \geq \varepsilon$. We then have a sequence $\{\mathbf{x}^{(k)}\}$, with $\mathbf{x}^{(k)} \longrightarrow \mathbf{a}$ but $\mathbf{f}(\mathbf{x}^{(k)}) \not\longrightarrow \mathbf{f}(\mathbf{a})$, showing that (*ii*) also fails.

Continuity considerations can be reduced to the case when $m=1$, as we now show.

Theorem 2.7. *Let* $\mathbf{f} : A \longrightarrow \mathbb{R}^m$ *have components* $(f_1, f_2, \cdots, f_m)$. *Then* $\mathbf{f}$ *is continuous at* $\mathbf{a} \in A \Leftrightarrow$ *each* f_i *is continuous at* $\mathbf{a}$.

Proof. By Theorem 2.6, $\mathbf{f}$ is continuous at $\mathbf{a}$ if, and only if, for any sequence $\{\mathbf{x}^{(k)}\} \subseteq A$ with $\mathbf{x}^{(k)} \longrightarrow \mathbf{a}$, we have $\mathbf{f}(\mathbf{x}^{(k)}) \longrightarrow \mathbf{f}(\mathbf{a})$. By Theorem 1.19, this is equivalent to every sequence of components converging, that is, to each f_i being continuous.

This result means that we can often call on the standard results concerning continuity of functions of one variable. When combined with the following result, continuity is frequently simple to verify.

Theorem 2.8. *Let* $\mathbf{f}, \mathbf{g} : A \subseteq \mathbb{R}^n \longrightarrow \mathbb{R}^m$ *be continuous at* $\mathbf{a} \in A$ *and let* $\alpha \in \mathbb{R}$. *Then* $\alpha\mathbf{f}$, $\mathbf{f}+\mathbf{g}$, *and, if* $m=1$, *the pointwise product* fg *are each continuous at* $\mathbf{a}$.

Proof. These may all be proved using *(ii)* of Theorem 2.6. We leave these arguments to the reader and prove the trickiest case using the ε-δ condition. Recall that the pointwise product fg is given by $(f\,g)(\mathbf{x}) = f(\mathbf{x})\,g(\mathbf{x})$, defined only for $m=1$. We have to ensure that $|f(\mathbf{x})\,g(\mathbf{x}) - f(\mathbf{a})\,g(\mathbf{a})| < \varepsilon$, for $\|\mathbf{x}-\mathbf{a}\| < \delta$. Using the triangle inequality we have

$$|f(\mathbf{x})g(\mathbf{x}) - f(\mathbf{a})g(\mathbf{a})| = |f(\mathbf{x})g(\mathbf{x}) - f(\mathbf{a})g(\mathbf{x}) + f(\mathbf{a})g(\mathbf{x}) - f(\mathbf{a})g(\mathbf{a})|$$
$$\leq |f(\mathbf{x}) - f(\mathbf{a})|\,|g(\mathbf{x})| + |f(\mathbf{a})|\,|g(\mathbf{x}) - g(\mathbf{a})|$$

Write $M = 1 + |f(\mathbf{a})| + |g(\mathbf{a})|$. For any $\varepsilon > 0$ with $\varepsilon/M < 1$, there is $\delta > 0$ so that $|f(\mathbf{x}) - f(\mathbf{a})| < \varepsilon/M$ and $|g(\mathbf{x}) - g(\mathbf{a})| < \varepsilon/M$ for $\|\mathbf{x}-\mathbf{a}\| < \delta$. Then, again by the triangle inequality, $|g(\mathbf{x})| \leq |g(\mathbf{a})| + \varepsilon/M < |g(\mathbf{a})| + 1$, for $\|\mathbf{x}-\mathbf{a}\| < \delta$. This yields, $|f(\mathbf{x})g(\mathbf{x}) - f(\mathbf{a})g(\mathbf{a})| < (|g(\mathbf{a})| + 1)\varepsilon/M + |f(\mathbf{a})|\,\varepsilon/M = \varepsilon$, for $\|\mathbf{x}-\mathbf{a}\| < \delta$, and completes the proof.

Remark. As we see in this proof, if we wish to end up with an expression such as $\|\mathbf{f}(\mathbf{x}) - \mathbf{f}(\mathbf{a})\| < \varepsilon$, we have to choose δ rather carefully. By observing that it suffices to prove that $\|\mathbf{f}(\mathbf{x}) - \mathbf{f}(\mathbf{a})\| < C\varepsilon$, where C is a constant independent of ε, the process can be simplified.

Example 2.9. Let $f : \mathbb{R}^2 \longrightarrow \mathbb{R}$ be defined by $f(x, y) = x\,y$, then f is continuous on $\mathbb{R}^2$.

Solution. f is the product of the two functions $g(x,y) = x$, $h(x,y) = y$, each of which is continuous by Example 2.5.

There is one other important way of combining functions, by composition. This means that we apply one function $\mathbf{f}$ to a point $\mathbf{x}$ to get its image $\mathbf{f}(\mathbf{x})$ and then apply another function $\mathbf{g}$ to that image to get $\mathbf{g}(\mathbf{f}(\mathbf{x}))$. However, this is only defined if every $\mathbf{f}(\mathbf{x})$ belongs to the domain of g. Thus, if $\mathbf{f} : A \subseteq \mathbb{R}^n \longrightarrow \mathbb{R}^p$ and $\mathbf{g} : B \subseteq \mathbb{R}^p \longrightarrow \mathbb{R}^m$, and if $\mathbf{f}(A) \subseteq B$, the composite map $\mathbf{g} \circ \mathbf{f} : A \longrightarrow \mathbb{R}^m$ is defined by $(\mathbf{g} \circ \mathbf{f})(\mathbf{x}) = \mathbf{g}(\mathbf{f}(\mathbf{x}))$, $\mathbf{x} \in A$. Note that functions are applied starting on the right : $\mathbf{g} \circ \mathbf{f}$ means first apply $\mathbf{f}$ and then $\mathbf{g}$, but $\mathbf{f} \circ \mathbf{g}$ (if it is defined at all) means first apply $\mathbf{g}$ and then $\mathbf{f}$ and may be quite different : composition of functions is not commutative.

Theorem 2.10. *Let* $\mathbf{f} : A \subseteq \mathbb{R}^n \longrightarrow \mathbb{R}^p$, $\mathbf{g} : B \subseteq \mathbb{R}^p \longrightarrow \mathbb{R}^m$ *where* $\mathbf{f}(A) \subseteq B$. *Suppose that* $\mathbf{f}$ *is continuous at* $\mathbf{a} \in A$ *and* $\mathbf{g}$ *is continuous at* $\mathbf{b} = \mathbf{f}(\mathbf{a}) \in B$. *Then* $\mathbf{g} \circ \mathbf{f}$ *is continuous at* $\mathbf{a}$.

Proof. For any $\varepsilon > 0$, as $\mathbf{g}$ is continuous at $\mathbf{f}(\mathbf{a})$, there exists $\eta > 0$ such that $\|\mathbf{g}(\mathbf{y}) - \mathbf{g}(\mathbf{f}(\mathbf{a}))\| < \varepsilon$ whenever $\mathbf{y} \in B$ and $\|\mathbf{y} - \mathbf{f}(\mathbf{a})\| < \eta$. As $\mathbf{f}$ is continuous at $\mathbf{a}$, there exists $\delta > 0$ such that $\|\mathbf{f}(\mathbf{x}) - \mathbf{f}(\mathbf{a})\| < \eta$ whenever $\mathbf{x} \in A$ and $\|\mathbf{x} - \mathbf{a}\| < \delta$. Therefore, for $\mathbf{x} \in A$ and $\|\mathbf{x} - \mathbf{a}\| < \delta$, $\mathbf{f}(\mathbf{x})$ is a suitable $\mathbf{y}$ and so $\|\mathbf{g}(\mathbf{f}(\mathbf{x})) - \mathbf{g}(\mathbf{f}(\mathbf{a}))\| < \varepsilon$.

Corollary 2.11. *Let* $g : A \subseteq \mathbb{R}^n \longrightarrow \mathbb{R}$ $(m=1)$ *be continuous at* $\mathbf{a} \in A$, *and suppose* $g(\mathbf{a}) \neq 0$. *Then its reciprocal* $\left(\frac{1}{g}\right)(\mathbf{x}) := \frac{1}{g(\mathbf{x})}$ *is continuous at* $\mathbf{a}$.

Proof. Let $h : \mathbb{R} \backslash \{0\} \longrightarrow \mathbb{R}$ be the function $h(u) = 1/u$; h is continuous on its domain. As g is continuous at $\mathbf{a}$ and $g(\mathbf{a}) \neq 0$, there is a neighbourhood $\mathrm{B}(\mathbf{a},\delta)$ with $g(\mathbf{x}) \neq 0$ on $\mathrm{B}(\mathbf{a},\delta)$; see problem 2.5. Then $\frac{1}{g}(\mathbf{x}) = h(g(\mathbf{x}))$ for $\mathbf{x} \in \mathrm{B}(\mathbf{a},\delta)$, a composition of continuous functions.

By applying the Corollary together with the result on products, gives a continuity result for quotients, of course only for $m=1$.

Example 2.12. Let $\mathbf{f} : \mathbb{R}^2 \longrightarrow \mathbb{R}^2$ be defined by $\mathbf{f}(x,y) = (x^2 y,\ y/(1+x^2))$. Then $\mathbf{f}$ is continuous everywhere in $\mathbb{R}^2$.

Solution. $f_1(x,y) = x^2 y$, and is continuous being a product of co-ordinate functions (Example 2.5). $f_2(x,y) = y/(1+x^2)$ is a product of "the function y" and $g(x,y) = 1/(1+x^2)$. As "the function $1+x^2$" is continuous, g is continuous by Corollary 2.11. In all, $\mathbf{f}$ is continuous.

Example 2.13. Discuss the continuity of the function $f : \mathbb{R}^2 \longrightarrow \mathbb{R}$ defined by

$$f(x,y) = \begin{cases} xy/(x^2+y^2), & \text{for } (x,y) \neq (0,0) \\ 0 \quad , & \text{when } (x,y) = (0,0) \end{cases}.$$

Solution. f is certainly continuous at all points other than (0,0) by the results on products and reciprocals given above. At (0,0) we have to investigate whether $|f(x,y)-f(0,0)| = \frac{|xy|}{x^2+y^2} < \varepsilon$ for (x,y) in a δ-nbd of (0,0). For any $x \neq 0$, and $y=0$, we have $f(x,0)=0$, so the above ε-δ condition holds. Similarly it holds for points of the form $(0,y)$, $y \neq 0$. However, along the line $y=x$, $x \neq 0$, $f(x,x)=1/2$ and so $|f(x,x)-f(0,0)| = 1/2$ for all $x \neq 0$. Therefore, for $\varepsilon < 1/2$, the ε-δ condition can never be satisfied and f is not continuous at (0,0).

It is often convenient to think of points $\mathbf{x}$ approaching $\mathbf{a}$ and considering whether $\mathbf{f}(\mathbf{x})$ approaches $\mathbf{f}(\mathbf{a})$. We now make this idea precise.

Definition 2.14. A *punctured ε-neighbourhood* of $\mathbf{a} \in \mathbb{R}^n$ is the set

$$B'(\mathbf{a},\varepsilon) = \{\mathbf{x} : 0 < \|\mathbf{x}-\mathbf{a}\| < \varepsilon\}.$$

Thus, $B'(\mathbf{a},\varepsilon) = B(\mathbf{a},\varepsilon) \setminus \{\mathbf{a}\}$, the open ball with $\mathbf{a}$ deleted.

Definition 2.15. Let $\mathbf{f} : A \subseteq \mathbb{R}^n \longrightarrow \mathbb{R}^m$. If $\mathbf{a}$ is a limit point of A we say that the limit of $\mathbf{f}(\mathbf{x})$ as $\mathbf{x}$ tends to $\mathbf{a}$ equals $\mathbf{b} \in \mathbb{R}^m$ and we write $\lim_{\mathbf{x} \to \mathbf{a}} \mathbf{f}(\mathbf{x}) = \mathbf{b}$, if, and only if, for any $\varepsilon > 0$ there is $\delta > 0$ such that

$$\mathbf{x} \in B'(\mathbf{a},\delta) \cap A \Rightarrow \mathbf{f}(\mathbf{x}) \in B(\mathbf{b},\varepsilon), \quad \text{that is,}$$

$$\mathbf{x} \in A,\ 0 < \|\mathbf{x}-\mathbf{a}\| < \delta \Rightarrow \|\mathbf{f}(\mathbf{x})-\mathbf{b}\| < \varepsilon. \tag{2.2}$$

We need $\mathbf{a}$ to be a limit point of A in this definition to ensure that there is a point in $B'(\mathbf{a},\delta) \cap A$ for every positive δ. However $\mathbf{a}$ need not be in A. There is an equivalent definition involving sequences which we give as a proposition.

Proposition 2.16. *Let $\mathbf{a}$ be a limit point of $A \subseteq \mathbb{R}^n$ and let $\mathbf{f} : A \longrightarrow \mathbb{R}^m$. Then $\lim_{\mathbf{x} \to \mathbf{a}} \mathbf{f}(\mathbf{x}) = \mathbf{b} \Leftrightarrow$ for every sequence $\{\mathbf{x}^{(k)}\} \subseteq A$ with $\mathbf{x}^{(k)} \neq \mathbf{a}$ for every k and such that $\mathbf{x}^{(k)} \longrightarrow \mathbf{a}$, it follows that $\mathbf{f}(\mathbf{x}^{(k)}) \longrightarrow \mathbf{b}$.*

Proof. The proof is exactly similar to that of Theorem 2.6 and we omit it.

Of course, $\mathbf{f}$ has limit $\mathbf{b}$ if and only if each component function f_i has limit b_i .

Useful properties of such limits are given in the following result.

Proposition 2.17. *Let* $\mathbf{f},\mathbf{g} : A \subseteq \mathbb{R}^n \longrightarrow \mathbb{R}^m$ *and suppose* $\lim_{\mathbf{x}\to\mathbf{a}} \mathbf{f}(\mathbf{x}) = \mathbf{b}$, *and* $\lim_{\mathbf{x}\to\mathbf{a}} \mathbf{g}(\mathbf{x}) = \mathbf{c}$. *Then,*

(*a*) $\lim_{\mathbf{x}\to\mathbf{a}} [\mathbf{f}(\mathbf{x}) + \mathbf{g}(\mathbf{x})] = \mathbf{b}+\mathbf{c}$, (*b*) $\lim_{\mathbf{x}\to\mathbf{a}} (\alpha\mathbf{f})(\mathbf{x}) = \alpha \lim_{\mathbf{x}\to\mathbf{a}} \mathbf{f}(\mathbf{x}) = \alpha\mathbf{b}$, $\alpha \in \mathbb{R}$,

(*c*) *for* $m=1$, $\lim_{\mathbf{x}\to\mathbf{a}} f(\mathbf{x})\, g(\mathbf{x}) = \lim_{\mathbf{x}\to\mathbf{a}} f(\mathbf{x}) \lim_{\mathbf{x}\to\mathbf{a}} g(\mathbf{x}) = bc$,

(*d*) *for* $m=1$, *if also* $c \neq 0$, *then* $\lim_{\mathbf{x}\to\mathbf{a}} f(\mathbf{x})/g(\mathbf{x}) = b/c$.

Proof. The proofs of (*a*)-(*c*) are exactly similar to those of Theorem 2.8. We prove (*d*) which provides an alternative proof for Corollary 2.11. It suffices to prove only the reciprocal rule ($f(\mathbf{x})=1$) and then apply the result for products. Thus, suppose $\lim_{\mathbf{x}\to\mathbf{a}} g(\mathbf{x}) = c \neq 0$. We first show that there is a punctured δ-neighbourhood of $\mathbf{a}$ on which $|g(\mathbf{x})| \geq |c|/2$. Indeed, taking $\varepsilon = |c|/2$ in the definition of limit, there is $\delta_1 > 0$ such that $\mathbf{x} \in B'(\mathbf{a},\delta_1) \cap A \Rightarrow |g(\mathbf{x}) - c| < |c|/2$. Therefore, by Corollary 1.5, $|g(\mathbf{x})| \geq |c| - |g(\mathbf{x}) - c| > |c|/2$ for $\mathbf{x} \in B'(\mathbf{a},\delta_1) \cap A$. Now

$$\left| \frac{1}{g(\mathbf{x})} - \frac{1}{c} \right| = \frac{|c - g(\mathbf{x})|}{|g(\mathbf{x})|\;|c|} < \frac{2|c - g(\mathbf{x})|}{|c|^2}, \quad \text{for } \mathbf{x} \in B'(\mathbf{a},\delta_1) \cap A .$$

For any $\varepsilon > 0$, take $\delta < \delta_1$ so that $\mathbf{x} \in B'(\mathbf{a},\delta) \cap A$ implies $|g(\mathbf{x}) - c| < \varepsilon |c|^2/2$.

This gives $\left| \frac{1}{g(\mathbf{x})} - \frac{1}{c} \right| < \varepsilon$, for $\mathbf{x} \in B'(\mathbf{a},\delta) \cap A$ and completes the proof.

Using the notion of limit we can say $\mathbf{f} : A \subseteq \mathbb{R}^n \longrightarrow \mathbb{R}^m$ is continuous at a limit point $\mathbf{a}$ of $A \Leftrightarrow \mathbf{f}$ is defined at $\mathbf{a}$ and $\lim_{\mathbf{x}\to\mathbf{a}} \mathbf{f}(\mathbf{x}) = \mathbf{f}(\mathbf{a})$.

In the definition of continuity at a point $\mathbf{a} \in A$, we included the expression "$\mathbf{x} \in A$". If $\mathbf{a}$ is an interior point of A this is automatically satisfied for $\delta > 0$ sufficiently small. When applied to functions on closed intervals in $\mathbb{R}$, continuity at the endpoints is then defined as the one-sided limit, as is usual. We do not need to specify this explicitly. Similarly, when discussing continuity of a function defined on the closed disk $\bar{B}(0,r)$ in $\mathbb{R}^2$, continuity at points on the boundary is defined by taking the limit from inside the disk only. When A has no interior points the situation can be confusing. For example, let $A = \{x \in [0,1] : x \text{ is rational}\}$ and let $f : A \longrightarrow \mathbb{R}$ be given by $f(x) = 1$. Then f is certainly continuous on A. However we may consider f as the restriction to A of the function $g : [0,1] \longrightarrow [0,1]$ defined by $g(x) = \begin{cases} 1, & \text{if } x \in A \\ 0, & \text{if } x \notin A \end{cases}$. The function g is not continuous at any point of $[0,1]$. *The restriction* $g|_A$ *is continuous on* A *even though* g

itself is discontinuous at every point of A. Of course if a function is continuous then so is any restriction. What this example shows is that extensions need not be continuous.

We mentioned earlier that, for $\mathbf{f}(\mathbf{x})$ to have a limit as $\mathbf{x}$ approaches $\mathbf{a}$, the point $\mathbf{a}$ may be approached in any manner. In particular, approaching $\mathbf{a}$ along all straight lines must give the same answer. A straight line in $\mathbb{R}^n$ containing $\mathbf{a}$ and with direction $\mathbf{u}$ is the set $\{\mathbf{x}=\mathbf{a}+t\mathbf{u},\ t\in\mathbb{R}\}$. When we use an element $\mathbf{u}$ as a direction we also add the requirement $\|\mathbf{u}\|=1$, which, of course, changes nothing.

Proposition 2.18. *Let* $\mathbf{f}: A\subseteq\mathbb{R}^n \longrightarrow \mathbb{R}^m$ *and* $\mathbf{a}\in\mathrm{int}(A)$. *Then* $\lim_{\mathbf{x}\to\mathbf{a}} f(\mathbf{x})=\mathbf{b}$ *implies that* $\lim_{t\to 0} \mathbf{f}(\mathbf{a}+t\mathbf{u})=\mathbf{b}$ *for every direction* $\mathbf{u}$.

Proof. From the ε-δ condition, since $\mathbf{a}+t\mathbf{u}\in B'(\mathbf{a},\delta)$ for $0<|t|<\delta$, we have $\mathbf{f}(\mathbf{a}+t\mathbf{u})\in B(\mathbf{b},\varepsilon)$, and this is the result.

The converse of this result is *false* : $\mathbf{f}$ can have a limit along lines but need not have a limit ; see problem 2.4.

2.2 CONTINUITY AND COMPACTNESS

An important property of continuous functions is that they map compact sets into compact sets. Recall that, for subsets of $\mathbb{R}^n$, compact means closed and bounded.

Theorem 2.19. *Let K be a compact subset of* $\mathbb{R}^n$ *and let* $\mathbf{f}: K\longrightarrow\mathbb{R}^m$ *be continuous on K. Then* $\mathbf{f}(K)$ *is a compact subset of* $\mathbb{R}^m$.

Proof. We first show that $\mathbf{f}(K)$ is bounded. If not, for every $m\in\mathbb{N}$, there exists $\mathbf{x}^{(m)}\in K$ such that $\|\mathbf{f}(\mathbf{x}^{(m)})\|>m$. As $\{\mathbf{x}^{(m)}\}$ lies in the compact set K it has a convergent subsequence $\mathbf{x}^{m(j)}\longrightarrow\mathbf{x}\in K$. By continuity of $\mathbf{f}$, this gives $\mathbf{f}(\mathbf{x}^{m(j)})\longrightarrow\mathbf{f}(\mathbf{x})$. In particular, $\{\mathbf{f}(\mathbf{x}^{m(j)})\}$ is bounded (problem 1.20). This contradicts the fact that $\|\mathbf{f}(\mathbf{x}^{m(j)})\|>m(j)\longrightarrow\infty$ as $m(j)\to\infty$. Next we show that $\mathbf{f}(K)$ is closed. Therefore let $\mathbf{b}$ be any limit point of $\mathbf{f}(K)$; we will show that $\mathbf{b}\in\mathbf{f}(K)$. There is a sequence $\{\mathbf{x}^{(m)}\}\subseteq K$ such that $\mathbf{f}(\mathbf{x}^{(m)})\longrightarrow\mathbf{b}$. By compactness of K, $\{\mathbf{x}^{(m)}\}$ has a convergent subsequence, $\mathbf{x}^{m(j)}\longrightarrow\mathbf{x}\in K$. Continuity of $\mathbf{f}$ then yields $\mathbf{f}(\mathbf{x}^{m(j)})\longrightarrow\mathbf{f}(\mathbf{x})$. Since a subsequence of a convergent sequence converges to the same limit as the full sequence, (problem 1.26) $\mathbf{f}(\mathbf{x}^{m(j)}) \longrightarrow \mathbf{b}$. Uniqueness of limits gives $\mathbf{b}=\mathbf{f}(\mathbf{x})\in\mathbf{f}(K)$.

Note. The image of a closed set under a continuous mapping need not be closed nor need the image of a bounded set be bounded.

Example 2.20. (*a*) Let $f:\mathbb{R}\longrightarrow\mathbb{R}$ be defined by $f(x)=x^2/(1+x^2)$. Then $f(\mathbb{R})=[0,1)$, neither open nor closed. This also shows that the image of an open set need not be open

(*b*) Let $f:(0,1)\longrightarrow\mathbb{R}$ be defined by $f(x)=1/x$. Then the image of $(0,1)$ is the set $(1,\infty)$.

Using ideas similar to that of Theorem 2.19 we can prove the following result concerning the continuity of inverses of continuous functions.

Theorem 2.21. *Let K be a compact subset of $\mathbb{R}^n$ and let $\mathbf{f}:K\longrightarrow\mathbb{R}^m$ be continuous and one-to-one on K. Then $\mathbf{f}^{-1}:\mathbf{f}(K)\longrightarrow K$ is continuous.*

Proof. Let $\mathbf{y}^{(k)}\longrightarrow\mathbf{y}$ in $\mathbf{f}(K)$. We want to prove that $\mathbf{x}^{(k)}\longrightarrow\mathbf{x}$ in K where $\mathbf{f}(\mathbf{x}^{(k)})=\mathbf{y}^{(k)}$ and $\mathbf{f}(\mathbf{x})=\mathbf{y}$. Suppose not. Then there is $\varepsilon>0$ and a subsequence $\{\mathbf{x}^{(p)}\}$ (say) such that $\|\mathbf{x}^{(p)}-\mathbf{x}\|\geq\varepsilon$ for all p. As K is compact $\{\mathbf{x}^{(p)}\}$ has a convergent subsequence, say $\mathbf{x}^{(q)}\longrightarrow\mathbf{z}\in K$. By continuity of $\mathbf{f}$, $\mathbf{f}(\mathbf{x}^{(q)})\longrightarrow\mathbf{f}(\mathbf{z})$. As $\{\mathbf{f}(\mathbf{x}^{(q)})\}$ is a subsequence of $\{\mathbf{y}^{(k)}\}$ the limit must be $\mathbf{y}$, and therefore $\mathbf{z}=\mathbf{x}$ since $\mathbf{f}$ is one-to-one. So $\mathbf{x}^{(q)}\longrightarrow\mathbf{x}$ and there exists N such that $\|\mathbf{x}^{(q)}-\mathbf{x}\|<\varepsilon$ for all $q\geq N$. This contradicts $\|\mathbf{x}^{(p)}-\mathbf{x}\|\geq\varepsilon$ for all p, and establishes the required contradiction.

This result may be *false* if K is not compact. You are invited to construct an example in problem 2.13.

When the image set lies in $\mathbb{R}$ one can discuss maxima and minima of functions.

Definition 2.22. A function $f:A\subseteq\mathbb{R}^n\longrightarrow\mathbb{R}$. is said to have a *maximum over A* (or a *global maximum*) at $\mathbf{a}\in A$ if $f(\mathbf{x})\leq f(\mathbf{a})$ for all $\mathbf{x}\in A$. f is said to have a *local maximum* at $\mathbf{a}$ if there is a δ-neighbourhood of $\mathbf{a}$ such that $f(\mathbf{x})\leq f(\mathbf{a})$ for all $\mathbf{x}\in B(\mathbf{a},\delta)$. If strict inequality holds in these definitions we use the term *strict* global (or local) maximum. Minima are defined by replacing $\leq$ by $\geq$; similarly for strict minima.

The study of maxima and minima is a familiar topic from the calculus and we shall also use the techniques of advanced calculus later in the book. For the present we have an existence result.

Theorem 2.23. *Let $f: K \subseteq \mathbb{R}^n \longrightarrow \mathbb{R}$ be continuous where K is a nonempty compact set. Then f has a global maximum and a global minimum over K.*

Proof. Let $M = \sup\{f(\mathbf{x}) : \mathbf{x} \in K\}$, the supremum or least upper bound of $f(K)$; M exists (and is finite) since every nonempty bounded set of real numbers has a supremum. Obviously $f(\mathbf{x}) \le M$ for all $\mathbf{x} \in K$ and M is the smallest number with this property. We claim that M belongs to $f(K)$, a *compact* set (Theorem 2.19). Indeed, if not, M belongs to the open set $[f(K)]^c$ and so there is an ε-neighbourhood, $(M-\varepsilon, M+\varepsilon)$, which contains no point of $f(K)$. Therefore $f(\mathbf{x}) \le M-\varepsilon$, for all $\mathbf{x} \in K$, that is, $M-\varepsilon$ is an upper bound. This contradicts the fact that M is the least upper bound. Thus, $M \in f(K)$, that is, there is $\mathbf{x} \in K$ with $f(\mathbf{x}) = M$. The proof for the minimum is exactly similar.

The above result can be summed up as : a continuous function on a compact set attains its bounds.

Example 2.24. Every norm on $\mathbb{R}^n$ is equivalent to the Euclidean norm in the sense that, for any norm $|\cdot|$, there are positive constants α, β such that $\alpha|\mathbf{x}| \le \|\mathbf{x}\| \le \beta|\mathbf{x}|$ for all $\mathbf{x} \in \mathbb{R}^n$.

Solution. To say that $|\cdot|$ is a norm is to say that it satisfies the properties listed in Theorem 1.3. Let $\{\mathbf{e}_i\}$ be the standard basis of $\mathbb{R}^n$ and let $M = \max\{|\mathbf{e}_i| : 1 \le i \le n\}$. Since $\mathbf{x} = x_1\mathbf{e}_1 + x_2\mathbf{e}_2 + \cdots + x_n\mathbf{e}_n$, by the triangle inequality for $|\cdot|$ we have,

$$\begin{aligned} |\mathbf{x}| &\le |x_1|\,|\mathbf{e}_1| + |x_2|\,|\mathbf{e}_2| + \cdots + |x_n|\,|\mathbf{e}_n| \\ &\le M(|x_1| + \cdots + |x_n|) \le M\sqrt{n}\|\mathbf{x}\|, \text{ (Problem 1.5).} \end{aligned}$$

This proves the left-hand inequality with $\alpha = 1/M\sqrt{n}$. It also shows that $|\mathbf{x}-\mathbf{y}| \le M\sqrt{n}\|\mathbf{x}-\mathbf{y}\|$ for all $\mathbf{x},\mathbf{y}$. Hence, $|\cdot|$ is continuous. Therefore the norm $|\cdot|$ attains its infimum m on the compact set $S = \{\mathbf{x} : \|\mathbf{x}\| = 1\}$, at $\mathbf{z}$ (say) and thus $m = |\mathbf{z}| > 0$ since $\mathbf{z} \ne \mathbf{0}$. For any $\mathbf{x} \ne \mathbf{0}$, we can write $\mathbf{x} = \|\mathbf{x}\|\dfrac{\mathbf{x}}{\|\mathbf{x}\|}$ and obtain (by property (*b*) of Theorem 1.3 for $|\cdot|$) $|\mathbf{x}| = \|\mathbf{x}\|\left|\dfrac{\mathbf{x}}{\|\mathbf{x}\|}\right| \ge m\|\mathbf{x}\|$. Taking $\beta = 1/m$ proves the right-hand inequality since it obviously holds when $\mathbf{x} = \mathbf{0}$.

The distance of a point $\mathbf{x}$ to a set A is defined to be $\operatorname{dist}(\mathbf{x}, A) := \inf\{\|\mathbf{x}-\mathbf{a}\| : \mathbf{a} \in A\}$. There may be no point $\mathbf{a} \in A$ at which the infimum is attained, in general. However, if A is compact the infimum is always attained. This is a consequence of continuity of the norm and hence holds whatever norm is employed, by the proof in Example 2.24. When a point $\mathbf{a} \in A$ nearest to $\mathbf{x}$ exists it need not be unique, for example consider $\mathbf{x}=\mathbf{0}$ and $A=\{\mathbf{a} : \|\mathbf{a}\|=1\}$. We might expect the shape of A to play a part. An important geometrical idea is that of convexity.

Definition 2.25. A set C in $\mathbb{R}^n$ is said to be *convex* if, whenever $\mathbf{x}$ and $\mathbf{y}$ are two points of C, the line segment $\{(1-t)\mathbf{x}+t\mathbf{y}\ ,\ 0 \le t \le 1\}$ also lies in C.

This geometrical interpretation is that the boundary of C is rounded away from the interior of C and is not indented. Straight line segments are allowed.

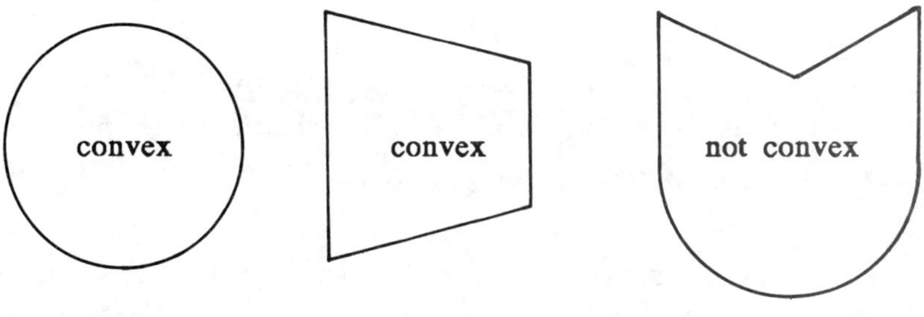

Figure 2.1.

Example 2.26. Open and closed balls are convex.

Solution. Consider $C=\mathbf{B}(\mathbf{z},r)$ and let $\mathbf{x},\mathbf{y} \in C$. Then $\|\mathbf{x}-\mathbf{z}\| < r$ and $\|\mathbf{y}-\mathbf{z}\| < r$, so that, by the triangle inequality, for $0 \le t \le 1$,

$$\|(1-t)\mathbf{x}+t\mathbf{y}-\mathbf{z}\| = \|(1-t)\mathbf{x}-(1-t)\mathbf{z}+t\mathbf{y}-t\mathbf{z}\| \le (1-t)\|\mathbf{x}-\mathbf{z}\| + t\|\mathbf{y}-\mathbf{z}\| < r.$$

The proof for the closed ball is exactly similar.

Proposition 2.27. *Let C be a closed convex set in $\mathbb{R}^n$. Then there exists a unique point $\mathbf{c} \in C$ of minimum norm, that is, nearest to $\mathbf{0}$.*

Proof. Let $d=\inf\{\|\mathbf{y}\| : \mathbf{y} \in C\}$. If $d=0$, $\mathbf{0} \in C$ and there is nothing to prove so we suppose that $\mathbf{0} \notin C$. We wish to show that there is exactly one $\mathbf{c} \in C$ such that $\|\mathbf{c}\|=d$. If C were bounded, existence would be immediate, since a continuous function attains its infimum on a compact set (Problem 2.7). However, even though C need not be bounded here, we can reduce the situation to this case by considering $C' := C \cap \overline{\mathbf{B}}(\mathbf{0}, d+1)$. Then C' is compact so there is $\mathbf{c}' \in C'$ of minimal norm and this clearly is a point of C of minimal norm. Existence is thus shown for any norm. Uniqueness depends not only on convexity of C but also on the fact that neighbourhoods in the Euclidean norm have a stronger convexity property : if $\|\mathbf{x}\|=\|\mathbf{y}\|=r$ and $\mathbf{x}\neq\mathbf{y}$ then $\|(\mathbf{x}+\mathbf{y})/2\|<r$, so-called *strict convexity*. This is an immediate consequence of the parallelogram law $\|\mathbf{x}+\mathbf{y}\|^2+\|\mathbf{x}-\mathbf{y}\|^2=2\|\mathbf{x}\|^2+2\|\mathbf{y}\|^2$, (Chapter 1). Now to prove uniqueness suppose $\|\mathbf{x}\|=\|\mathbf{y}\|=d$ and $\mathbf{x}\neq\mathbf{y}$. Then $\|(\mathbf{x}+\mathbf{y})/2\|<d$ by strict convexity of the norm. As C is convex $(\mathbf{x}+\mathbf{y})/2\in C$ so its norm cannot be less than d. This shows that $\mathbf{x}=\mathbf{y}$ and establishes uniqueness.

Remark. When the Euclidean norm is used, compactness is not necessary for existence. The following argument shows how the parallelogram law alone suffices. Since $d=\inf\{\|\mathbf{y}\| : \mathbf{y}\in C\}$, for every $i\in\mathbb{N}$ there is $\mathbf{c}^{(i)}\in C$ such that $d\leq\|\mathbf{c}^{(i)}\|<d+1/i$. We show that this is a Cauchy sequence. Indeed for $i\neq j$, noting that $(\mathbf{c}^{(i)}+\mathbf{c}^{(j)})/2\in C$, we have

$$\begin{aligned}\|\mathbf{c}^{(i)}-\mathbf{c}^{(j)}\|^2&=2\|\mathbf{c}^{(i)}\|^2+2\|\mathbf{c}^{(j)}\|^2-\|\mathbf{c}^{(i)}+\mathbf{c}^{(j)}\|^2\\&\leq 2(d+1/i)^2+2(d+1/j)^2-4d^2\\&=2(1/i^2+1/j^2)+4d(1/i+1/j)\,.\end{aligned}$$

Since the space is complete and C is closed this proves that $\mathbf{c}^{(i)}\longrightarrow\mathbf{c}\in C$. Therefore $\|\mathbf{c}^{(i)}\|\longrightarrow\|\mathbf{c}\|$ and so $\|\mathbf{c}\|=d$.

The ε-δ condition at $\mathbf{a}$ says that for any $\varepsilon>0$ there is a $\delta>0$ (depending on $\mathbf{a}$ and ε) such that $\mathbf{x}\in A$ and $\|\mathbf{x}-\mathbf{a}\|<\delta \Rightarrow \|\mathbf{f}(\mathbf{x})-\mathbf{f}(\mathbf{a})\|<\varepsilon$. It is conceivable that for a particular ε, the same δ might work for every $\mathbf{a}\in A$. This stronger property is called *uniform continuity*, and is a global property of $\mathbf{f}$.

Definition 2.28. A function $\mathbf{f}: A\subseteq\mathbb{R}^n\longrightarrow\mathbb{R}^m$ is said to be *uniformly continuous on A* if for any $\varepsilon>0$, there exists $\delta>0$ (depending only on ε) such that

$$\mathbf{x},\mathbf{y}\in A \text{ and } \|\mathbf{x}-\mathbf{y}\|<\delta \Rightarrow \|\mathbf{f}(\mathbf{x})-\mathbf{f}(\mathbf{y})\|<\varepsilon. \qquad (2.3)$$

Example 2.24 above shows that any norm is, in fact, uniformly continuous. It is evident that a function that is uniformly continuous on A is continuous at each point of A. The converse is generally false. The simplest example is $f:(0,2)\longrightarrow\mathbb{R}$ where $f(x)=1/x$. For any δ with $0<\delta<1$, take $x=\delta$, $y=2\delta$. Then $|f(x)-f(y)|=\left|\frac{1}{\delta}-\frac{1}{2\delta}\right|=\frac{1}{2\delta}>\frac{1}{2}$, so it is impossible to satisfy the uniform continuity requirement. This function is continuous at points $a>0$ by the quotient rule (see Proposition 2.17(d)), but the δ depends on a (see problem 2.10).

However, when the domain is a *compact* set, continuity is equivalent to uniform continuity.

Theorem 2.29. *Let* $\mathbf{f} : K \subseteq \mathbb{R}^n \longrightarrow \mathbb{R}^m$ *be continuous on the compact set K. Then* **f** *is uniformly continuous on K.*

Proof. We shall first prove this by contradiction using a sequential compactness argument. The negation of uniform continuity is

$$\exists\ \varepsilon > 0,\ \forall\ \delta > 0, \text{ there exist } \mathbf{x},\mathbf{y} \in K \text{ with } \|\mathbf{x}-\mathbf{y}\| < \delta \text{ but } \|\mathbf{f}(\mathbf{x})-\mathbf{f}(\mathbf{y})\| \geq \varepsilon.$$

Fix this ε. For every $m \in \mathbb{N}$, there are points $\mathbf{x}^{(m)}$, $\mathbf{y}^{(m)}$ in K with $\|\mathbf{x}^{(m)}-\mathbf{y}^{(m)}\| < 1/m$ but $\|\mathbf{f}(\mathbf{x}^{(m)})-\mathbf{f}(\mathbf{y}^{(m)})\| \geq \varepsilon$. As K is compact, there are subsequences $\mathbf{x}^{m(j)} \longrightarrow \mathbf{x} \in K$, $\mathbf{y}^{m(j)} \longrightarrow \mathbf{y} \in K$. Since $\|\mathbf{x}^{m(j)}-\mathbf{y}^{m(j)}\| < 1/m(j)$, we have $\mathbf{x}=\mathbf{y}$. As $\mathbf{f}$ is continuous at $\mathbf{x}$, there exists an integer N such that $\|\mathbf{f}(\mathbf{x}^{m(j)})-\mathbf{f}(\mathbf{x})\| < \varepsilon/2$ and $\|\mathbf{f}(\mathbf{y}^{m(j)})-\mathbf{f}(\mathbf{x})\| < \varepsilon/2$ for all $m(j) \geq N$. Thus, $\|\mathbf{f}(\mathbf{x}^{m(j)})-\mathbf{f}(\mathbf{y}^{m(j)})\| \leq \|\mathbf{f}(\mathbf{x}^{m(j)})-\mathbf{f}(\mathbf{x})\| + \|\mathbf{f}(\mathbf{x})-\mathbf{f}(\mathbf{y}^{m(j)})\| < \varepsilon$, for $m(j) \geq N$, a contradiction.

We now give a proof using the ideas of open covers. We do so because the type of argument is an important one. Let $\varepsilon > 0$ be given. Then for every $\mathbf{x} \in K$ there exists $\delta(\mathbf{x}) > 0$ such that $\mathbf{y} \in K$ and $\|\mathbf{x}-\mathbf{y}\| < \delta(\mathbf{x})$ implies that $\|\mathbf{f}(\mathbf{x})-\mathbf{f}(\mathbf{y})\| < \varepsilon/2$. The collection $\{\, B(\mathbf{x}, \delta(\mathbf{x})/2) : \mathbf{x} \in K\}$ is an open cover of the compact set K so there is a finite collection $\{B(\mathbf{x}^{(i)},\delta^{(i)}/2) : 1 \leq i \leq p\}$ which covers K. Let $\delta = \min\{\delta^{(i)}/2 : 1 \leq i \leq p\}$. Then, for $\mathbf{x},\mathbf{y} \in K$ and $\|\mathbf{x}-\mathbf{y}\| < \delta$, there is some i with $\|\mathbf{x}-\mathbf{x}^{(i)}\| < \delta^{(i)}/2$ and therefore $\|\mathbf{y}-\| < \delta + \delta^{(i)}/2 < \delta^{(i)}$. This gives $\|\mathbf{f}(\mathbf{x})-\mathbf{f}(\mathbf{x}^{(i)})\| < \varepsilon/2$ and $\|\mathbf{f}(\mathbf{y})-\mathbf{f}(\mathbf{x}^{(i)})\| < \varepsilon/2$ and the triangle inequality yields $\|\mathbf{f}(\mathbf{x})-\mathbf{f}(\mathbf{y})\| < \varepsilon$.

2.3. SEQUENCES OF FUNCTIONS

Let A be a subset of $\mathbb{R}^n$ and suppose that for each $k \in \mathbb{N}$ there is a function $\mathbf{f}^{(k)} : A \longrightarrow \mathbb{R}^m$. We say that $\{\mathbf{f}^{(k)}\}$ is a sequence of functions. For every $\mathbf{x} \in A$, $\{\mathbf{f}^{(k)}(\mathbf{x})\}$ is a sequence of points in $\mathbb{R}^m$. If this sequence converges, the limit depends, in general, on $\mathbf{x}$ and we may write the limit as $\mathbf{f}(\mathbf{x})$. This process defines a function $\mathbf{f}$ called the pointwise limit of the sequence $\{\mathbf{f}^{(k)}\}$ whose domain is the set of all $\mathbf{x}$ for which $\mathbf{f}^{(k)}$ has a limit.

Definition 2.30. Let $\{\mathbf{f}^{(k)}\}$ be a sequence of functions defined on a subset A of $\mathbb{R}^n$ with values in $\mathbb{R}^m$. We say that $\mathbf{f}^{(k)}$ *converges pointwise* to a function $\mathbf{f}$ on $A_1 \subseteq A$ and write $\mathbf{f}^{(k)} \longrightarrow \mathbf{f}$, if, for every $\mathbf{x} \in A_1$, $\mathbf{f}^{(k)}(\mathbf{x}) \longrightarrow \mathbf{f}(\mathbf{x})$ as $k \longrightarrow \infty$.

This means precisely that, for every $\varepsilon>0$ and each $\mathbf{x}\in A_1$ there exists $N(\varepsilon,\mathbf{x})$ such that

$$\|\mathbf{f}^{(k)}(\mathbf{x})-\mathbf{f}(\mathbf{x})\|<\varepsilon \text{ for } k\geq N(\varepsilon,\mathbf{x}).$$

Example 2.31. Let $f^{(k)}\colon [0,2]\longrightarrow\mathbb{R}$ be defined by $f^{(k)}(x)=x^k$. Determine the pointwise limit function.

Solution. For $0\leq x<1$, $x^k\longrightarrow 0$ as $k\longrightarrow\infty$. For $x=1$ the limit is obviously 1 while for $x>1$, there is no limit. So the domain of f is $[0,1]$ and $f(x)=\begin{cases}0, & 0\leq x<1\\ 1, & x=1\end{cases}$. Note that, even though each function $f^{(k)}$ is uniformly continuous, f itself is not continuous.

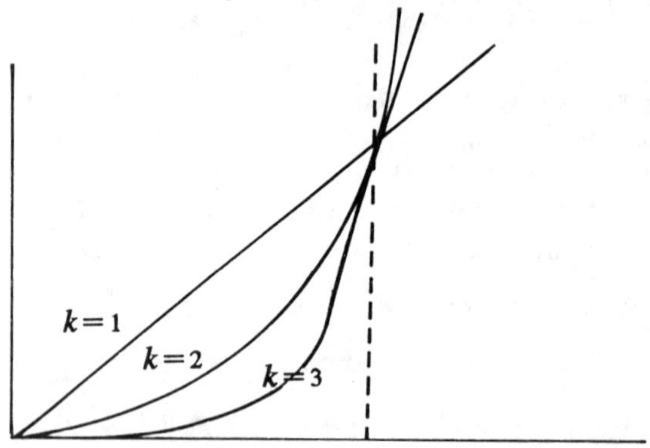

Figure 2.2

Example 2.32. Let $f^{(k)}\colon [0,2]\longrightarrow\mathbb{R}$ be given by $f^{(k)}(x)=1/(1+x^k)$. Then the pointwise limit function is

$$f(x)=\begin{cases}1, & 0\leq x<1\\ 1/2, & x=1\\ 0, & 1<x\leq 2\end{cases}.$$

These examples show that the pointwise limit of continuous functions need not be continuous. Suppose we wished to prove that $\mathbf{f}$ is continuous at some point $\mathbf{a}$ of the domain. Then certainly $\mathbf{f}^{(k)}(\mathbf{a})$ can be made arbitrarily close to $\mathbf{f}(\mathbf{a})$ for k large enough, but to prove continuity we also need $\mathbf{f}^{(k)}(\mathbf{x})$ to be close to $\mathbf{f}(\mathbf{x})$ for $\mathbf{x}$ in a neighbourhood of $\mathbf{a}$. The problem is that k depends on $\mathbf{x}$, and to make progress we need there to be some k that works for all $\mathbf{x}$. This gives rise to the concept of uniform convergence.

Definition 2.33. Let $\{\mathbf{f}^{(k)}\}$ be a sequence of functions defined on a set $A \subseteq \mathbb{R}^n$ and with values in $\mathbb{R}^m$. We say that $\{\mathbf{f}^{(k)}\}$ *converges uniformly* on A to a function $\mathbf{f}$ if, for every $\varepsilon > 0$ there exists $N(\varepsilon) \in \mathbb{N}$ such that

$$\|\mathbf{f}^{(k)}(\mathbf{x}) - \mathbf{f}(\mathbf{x})\| < \varepsilon \text{ for all } k \geq N(\varepsilon) \text{ and for all } \mathbf{x} \in A.$$

This may conveniently be written:

$$\forall \varepsilon > 0 \quad \exists\, N(\varepsilon) \text{ s.t. } \sup_{\mathbf{x} \in A} \|\mathbf{f}^{(k)}(\mathbf{x}) - \mathbf{f}(\mathbf{x})\| < \varepsilon \quad \forall k \geq N(\varepsilon).$$

and we write $\mathbf{f}^{(k)} \longrightarrow \mathbf{f}$ *uniformly on* A. The function $\mathbf{f}$ is called the *uniform limit* of $\{\mathbf{f}^{(k)}\}$ on A. It is evident that if $\mathbf{f}^{(k)} \longrightarrow \mathbf{f}$ uniformly on A then $\mathbf{f}$ is the pointwise limit of $\{\mathbf{f}^{(k)}\}$.

Uniform convergence of $\mathbf{f}^{(k)}$ to $\mathbf{f}$ means that the whole graph of $\mathbf{f}^{(k)}$ lies within distance ε of the graph of $\mathbf{f}$. For $n = m = 1$ this is illustrated in figure 2.3.

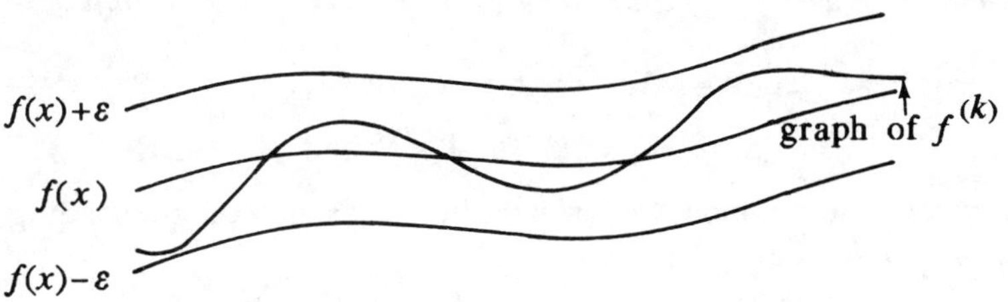

Figure 2.3

Example 2.34. Let $f^{(k)}(x) = x^k(1-x)$, $0 \leq x < 1$. Show that $\{f^{(k)}\}$ converges uniformly to the zero function on $[0,1]$.

Solution. Since $x^k \longrightarrow 0$ for $0 \leq x < 1$ and $f^{(k)}(1) = 0$, the pointwise limit is 0. For an arbitrary $\varepsilon > 0$ we want to make $f^{(k)}(x) < \varepsilon$ for all $x \in [0,1]$ and all k sufficiently large. We find the maximum of $f^{(k)}$ on $[0,1]$. Since $\frac{d}{dx} f^{(k)}(x) = x^{k-1}[k-(k+1)x]$, we see that $f^{(k)}$ has a maximum at $x = k/(k+1)$. Therefore, it suffices to make this maximum value less that ε. Thus, for all $x \in [0,1]$,

$$f^{(k)}(x) \leq f^{(k)}(k/(k+1)) = \left(\frac{k}{(k+1)}\right)^k \frac{1}{k+1} < \frac{1}{k+1}.$$

Taking $N \geq 1/\varepsilon$ shows that $\sup f^{(k)}(x) < \varepsilon$ for all $k \geq N$, that is,

$$f^{(k)} \longrightarrow 0 \text{ uniformly on } [0,1].$$

Example 2.35. Let $f^{(k)}(x)=x^k(1-x^k)$, $0\leq x\leq 1$. Show that $\{f^{(k)}\}$ is not uniformly convergent.

Solution. The pointwise limit is the zero function. For any k, $f^{(k)}$ has a maximum when $x^k = 1/2$ and the maximum value is 1/4. Thus $\sup|f^{(k)}(x)-f(x)| = \max|f^{(k)}(x)| = 1/4$ and $f^{(k)}$ does not converge uniformly to 0. For each x we can find $k(x)$ so that $f^{(k(x))}(x)$ is arbitrarily small, but whatever k we choose there exists x where $f^{(k)}(x)$ is not close to 0.

An important consequence of uniform convergence is that continuity is inherited by the limit.

Theorem 2.36. *Let* $\mathbf{f}^{(k)}: A \longrightarrow \mathbb{R}^m$, $k = 1,2,\cdots$, *be continuous at* $\mathbf{a}\in A$ *and suppose that* $\mathbf{f}^{(k)} \longrightarrow \mathbf{f}$ *uniformly on* A. *Then* $\mathbf{f}$ *is continuous at* $\mathbf{a}$.

Proof. For $\varepsilon>0$ let $N = N(\varepsilon)$ be such that

$$\|\mathbf{f}^{(k)}(\mathbf{x})-\mathbf{f}(\mathbf{x})\| < \varepsilon/3 \text{ for } k\geq N \text{ and all } \mathbf{x}\in A. \tag{2.4}$$

Since $\mathbf{f}^{(N)}$ is continuous at $\mathbf{a}$, there exists $\delta>0$, (depending on ε and the fixed N) so that

$$\mathbf{x}\in A \text{ and } \|\mathbf{x}-\mathbf{a}\| < \delta \Rightarrow \|\mathbf{f}^{(N)}(\mathbf{x})-\mathbf{f}^{(N)}(\mathbf{a})\| < \varepsilon/3. \tag{2.5}$$

Then for $\mathbf{x}\in A\cap B(\mathbf{a},\delta)$,

$$\begin{aligned}\|\mathbf{f}(\mathbf{x})-\mathbf{f}(\mathbf{a})\| &\leq \|\mathbf{f}(\mathbf{x})-\mathbf{f}^{(N)}(\mathbf{x})\| + \|\mathbf{f}^{(N)}(\mathbf{x})-\mathbf{f}^{(N)}(\mathbf{a})\| + \|\mathbf{f}^{(N)}(\mathbf{a})-\mathbf{f}(\mathbf{a})\| \\ &< \varepsilon/3 + \varepsilon/3 + \varepsilon/3 = \varepsilon, \quad \text{by (2.4), (2.5), and (2.4).}\end{aligned}$$

When $\mathbf{f}^{(k)} \longrightarrow \mathbf{f}$ uniformly on a set A we say that $\mathbf{f}$ is *uniformly approximated* by $\{\mathbf{f}^{(k)}\}$. Given a function $\mathbf{f}$ it is often useful to know whether $\mathbf{f}$ can be uniformly approximated by functions that have certain extra properties. The Weierstrass approximation theorem provides an affirmative answer.

Theorem 2.37. (*Weierstrass approximation theorem.*)
Let $f: [a,b] \longrightarrow \mathbb{R}$ *be continuous on* $[a,b]$. *Then there exists a sequence of polynomials* $\{P_k\}$ *which converges uniformly to* f *on* $[a,b]$.

We shall prove this using polynomials of S. Bernstein.

Definition 2.38. For $f: [0,1] \longrightarrow \mathbb{R}$, the m-th *Bernstein polynomial* of

f is $B_m(x) := \sum_{r=0}^{m} f(r/m) \binom{m}{r} x^r (1-x)^{m-r}$.

Here $\binom{m}{r}$ are the Binomial coefficients which occur in probability theory and in the Binomial expansion of $(x+y)^m$. Indeed

$$(x+y)^m = \sum_{r=0}^{m} \binom{m}{r} x^r y^{m-r}. \tag{2.6}$$

They may be defined by $\binom{m}{r} = \frac{m!}{r!(m-r)!}$, where ! denotes factorial, $m! = m(m-1)\cdots(2)(1)$, for $m \geq 1$ and $0! := 1$. The Bernstein polynomial $B_m(x)$ is calculated from the values of f at the points $0, 1/m, 2/m, \cdots, 1$, multiplied by certain weights $w_r(x) := \binom{m}{r} x^r (1-x)^{m-r}$. We require some identities involving these weights.

Setting $y = 1-x$ in (2.6) gives

$$\sum_{r=0}^{m} w_r(x) = 1. \tag{2.7}$$

Differentiate (2.6) with respect to x and then multiply by x to get

$$mx(x+y)^{m-1} = \sum_{r=0}^{m} r \binom{m}{r} x^r y^{m-r}.$$

Hence, setting $y=1-x$,

$$\sum_{r=0}^{m} r\, w_r(x) = mx. \tag{2.8}$$

Similarly, differentiating (2.6) twice with respect to x and multiplying by x^2 yields

$$\sum_{r=0}^{m} r(r-1) w_r(x) = m(m-1)x^2. \tag{2.9}$$

Hence by (2.7), (2.8) and (2.9)

$$\sum_{r=0}^{m} (r-mx)^2 w_r(x) = m^2x^2 \sum_{r=0}^{m} w_r(x) - 2mx \sum_{r=0}^{m} r\, w_r(x) + \sum_{r=0}^{m} r^2 w_r(x)$$

$$= m^2x^2 - 2mx\, mx + [mx + m(m-1)x^2], \text{ that is,}$$

$$\sum_{r=0}^{m} (r-mx)^2 w_r(x) = mx(1-x). \tag{2.10}$$

With these preliminaries we can now prove the approximation theorem.

Theorem 2.39. (*Bernstein's approximation theorem.*)
Let $f:[0,1]\longrightarrow\mathbb{R}$ be continuous. Then the sequence $\{B_m\}$ of Bernstein polynomials converges uniformly to f on $[0,1]$.

Proof. Since f is continuous on $[0,1]$ it is bounded, say $|f(x)|\leq M$ for all x, and uniformly continuous (Theorem 2.29). Therefore, for any $\varepsilon>0$ there exists $\delta>0$ such that $|f(x)-f(y)|<\varepsilon$ whenever $x,y\in[0,1]$ and $|x-y|<\delta$. Now, using (2.7), we have

$$\Big|f(x)-\sum_{r=0}^{m} f(r/m)w_r(x)\Big| = \Big|\sum_{r=0}^{m}[f(x)-f(r/m)]w_r(x)\Big|$$

$$\leq \sum_{r=0}^{m}|f(x)-f(r/m)|\,w_r(x).$$

For those r with $|x-r/m|<\delta$ we may use the uniform continuity of f to estimate the sum. Hence we split the summation and write

$$\sum_{r=0}^{m} = \sum_{|r-mx|\leq m\delta} + \sum_{|r-mx|>m\delta} := S_1 + S_2 .$$

Then $S_1\leq \sum_{|r-mx|\leq m\delta} |f(x)-f(r/m)|\,w_r(x)<\varepsilon\sum_{|r-mx|\leq m\delta}|w_r(x)|\leq\varepsilon$, using (2.7).

For S_2 we can only use the estimate $|f(x)-f(r/m)|\leq|f(x)|+|f(r/m)|\leq 2M$.

Then we have $S_2\leq 2M\sum_{|r-mx|>m\delta}|w_r(x)|\leq 2M\sum_{|r-mx|>m\delta}\dfrac{(r-mx)^2}{m^2\delta^2}w_r(x)$

$$\leq \frac{2M}{m^2\delta^2}\sum_{r=0}^{m}(r-mx)^2w_r(x)$$

$$= 2Mx(1-x)/m\delta^2 \text{, by (2.10)}$$

$$\leq M/(2m\delta^2), \text{ since } x(1-x)\leq 1/4.$$

Thus $|f(x)-B_m(x)|<\varepsilon+M/(2m\delta^2)$. Hence there exists N independent of x such that $|f(x)-B_m(x)|<2\varepsilon$ for $m\geq N$ and for all $x\in[0,1]$.

The Weierstrass approximation theorem is a simple consequence of this. For $x\in[0,1]$ let $\phi(x)=a+(b-a)x$. Then ϕ is a bijection of $[0,1]$ onto $[a,b]$, a change of variable. Given $f:[a,b]\longrightarrow\mathbb{R}$ consider $g=f\circ\phi:[0,1]\longrightarrow\mathbb{R}$. Since g can be uniformly approximated by Bernstein polynomials, reversing the change of variable gives polynomials approximating f.

There is an important generalization of the Weierstrass theorem, the Stone-Weierstrass theorem, which allows, among other things, for $[a,b]$ to be replaced by a compact subset of $\mathbb{R}^n$. We shall not prove this extended result. We refer to references [2], [5], or [11].

2.4. SPACES OF FUNCTIONS

Let A be a subset of $\mathbb{R}^n$ and let $\mathbf{f} : A \longrightarrow \mathbb{R}^m$ be a function. We say that $\mathbf{f}$ is *bounded* on A if there exists $M > 0$ such that $\|\mathbf{f}(\mathbf{x})\| \leq M$ for all $\mathbf{x} \in A$. The set of all such bounded functions is denoted $B(A\,;\mathbb{R}^m)$, or simply $B(A)$ when $m = 1$. We define addition and scalar multiplication pointwise, that is,

$$(\mathbf{f}+\mathbf{g})(\mathbf{x}) := \mathbf{f}(\mathbf{x})+\mathbf{g}(\mathbf{x}), \ (\alpha\mathbf{f})(\mathbf{x}) := \alpha\mathbf{f}(\mathbf{x}), \ \alpha \in \mathbb{R}.$$

With these definitions $B(A\,;\mathbb{R}^m)$ is a linear space.

Definition 2.40. For $\mathbf{f} \in B(A\,;\mathbb{R}^m)$ we define the *uniform norm* (or supremum norm) of $\mathbf{f}$ by $\|\mathbf{f}\|_u := \sup\{\|\mathbf{f}(\mathbf{x})\| : \mathbf{x} \in A\}$.

One reason for the name uniform norm is that

$$\|\mathbf{f}^{(k)} - \mathbf{f}\|_u \longrightarrow 0 \Leftrightarrow \mathbf{f}^{(k)} \longrightarrow \mathbf{f} \text{ uniformly on } A.$$

Proposition 2.41. *$B(A\,;\mathbb{R}^m)$ endowed with the uniform norm is a normed linear space.*

Proof. We have to show that the uniform norm satisfies the properties of a norm listed in Theorem 1.3. We prove only the triangle inequality.

$$\begin{aligned}\|\mathbf{f}+\mathbf{g}\|_u &= \sup\{\|\mathbf{f}(\mathbf{x})+\mathbf{g}(\mathbf{x})\| : \mathbf{x} \in A\} \\ &\leq \sup\{\|\mathbf{f}(\mathbf{x})\| + \|\mathbf{g}(\mathbf{x})\| : \mathbf{x} \in A\} \\ &\leq \sup\{\|\mathbf{f}(\mathbf{x})\| : \mathbf{x} \in A\} + \sup\{\|\mathbf{g}(\mathbf{x})\| : \mathbf{x} \in A\} \\ &= \|\mathbf{f}\|_u + \|\mathbf{g}\|_u .\end{aligned}$$

Now let $BC(A\,;\mathbb{R}^m)$ denote the set of bounded, continuous functions from A to $\mathbb{R}^m$. If A is compact, continuity implies boundedness but if A is not compact there exist unbounded continuous functions. The space $BC(A\,;\mathbb{R}^m)$ endowed with the uniform norm is certainly a normed linear space and therefore is a metric space when endowed with the metric inherited from the norm, namely $d(\mathbf{f},\mathbf{g}) := \|\mathbf{f}-\mathbf{g}\|_u$. Convergence in $BC(A\,;\mathbb{R}^m)$ is defined via the metric so that $\mathbf{f}^{(k)} \longrightarrow \mathbf{f}$ in $BC(A\,;\mathbb{R}^m) \Leftrightarrow \|\mathbf{f}^{(k)} - \mathbf{f}\|_u \longrightarrow 0$.

Thus,

$$\mathbf{f}^{(k)} \longrightarrow \mathbf{f} \text{ in } BC(A\,;\mathbb{R}^m) \Leftrightarrow \mathbf{f}^{(k)} \longrightarrow \mathbf{f} \text{ uniformly on } A.$$

A *Cauchy sequence* in $BC(A\,;\mathbb{R}^m)$ is defined by demanding that for every

$\varepsilon > 0$ there is $N(\varepsilon) > 0$ so that $\|\mathbf{f}^{(k)} - \mathbf{f}^{(j)}\|_u < \varepsilon$ for $j,k \geq N$.

Theorem 2.42. *$BC(A\,;\mathbb{R}^m)$ endowed with the uniform norm is a complete normed linear space.*

Proof. We have to prove that if $\{\mathbf{f}^{(k)}\}$ is a Cauchy sequence in BC then there exists $\mathbf{f} \in BC$ such that $\|\mathbf{f}^{(k)} - \mathbf{f}\|_u \longrightarrow 0$. Let $\{\mathbf{f}^{(k)}\}$ be a Cauchy sequence in $BC(A\,;\mathbb{R}^m)$. Then, for each $\mathbf{x} \in A$, $\{\mathbf{f}^{(k)}(\mathbf{x})\}$ is a Cauchy sequence in $\mathbb{R}^m$, hence convergent by completeness of $\mathbb{R}^m$. This shows that the pointwise limit $\mathbf{f}$ exists. We have to show that $\|\mathbf{f}^{(k)} - \mathbf{f}\|_u \longrightarrow 0$ and that $\mathbf{f}$ is continuous on A. To prove the former, take $\varepsilon > 0$. Since $\{\mathbf{f}^{(k)}\}$ is Cauchy, there exists $N(\varepsilon)$ such that $\|\mathbf{f}^{(k)}(\mathbf{x}) - \mathbf{f}^{(j)}(\mathbf{x})\| < \varepsilon/2$ for $k,j > N$ and for all $\mathbf{x} \in A$. Since $\mathbf{f}$ is the pointwise limit of $\mathbf{f}^{(k)}$, for an arbitrary $\mathbf{x} \in A$ there exists $k(\mathbf{x})$ (which we can suppose is greater than $N(\varepsilon)$) such that $\|\mathbf{f}^{k(\mathbf{x})}(\mathbf{x}) - \mathbf{f}(\mathbf{x})\| < \varepsilon/2$. This yields, by the triangle inequality, $\|\mathbf{f}^{(j)}(\mathbf{x}) - \mathbf{f}(\mathbf{x})\| < \varepsilon$ for all $j \geq N(\varepsilon)$. As this holds for every $\mathbf{x}$ and N is independent of $\mathbf{x}$, we have shown that $\|\mathbf{f}^{(j)} - \mathbf{f}\|_u \longrightarrow 0$. Thus $\mathbf{f}$ is the uniform limit of continuous functions and hence is indeed continuous by Theorem 2.36 . Finally, $\mathbf{f}$ is bounded since there exists j so that $\|\mathbf{f}^{(j)} - \mathbf{f}\|_u < 1$ and then

$$\|\mathbf{f}(\mathbf{x})\| \leq \|\mathbf{f}^{(j)}(\mathbf{x}) - \mathbf{f}(\mathbf{x})\| + \|\mathbf{f}^{(j)}(\mathbf{x})\| < 1 + \|\mathbf{f}^{(j)}\|_u.$$

Proposition 2.43. (*The Cauchy criterion for uniform convergence.*)
$\mathbf{f}^{(k)} \longrightarrow \mathbf{f}$ *uniformly on* A $\Leftrightarrow$

$$\forall\ \varepsilon > 0\ \exists\ N(\varepsilon) \text{ such that } \|\mathbf{f}^{(k)} - \mathbf{f}^{(j)}\|_u < \varepsilon \text{ for all } j,k \geq N.$$

Proof. This is contained in the proof of Theorem 2.42.

2.5. SERIES IN $\mathbb{R}^n$ AND SERIES OF FUNCTIONS

Given a sequence $\{\mathbf{x}^{(k)}\}$ in $\mathbb{R}^n$ there is an associated series denoted by the symbolism $\sum_{k=1}^{\infty} \mathbf{x}^{(k)}$. Since we cannot carry out infinitely many additions, this requires definition. We can sum over any finite number of terms so $\mathbf{S}^{(p)} = \sum_{k=1}^{p} \mathbf{x}^{(k)}$ is well defined.

Definition 2.44. If $\{\mathbf{x}^{(k)}\}$ is a sequence in $\mathbb{R}^n$, then the associated infinite series is defined to be the sequence $\{\mathbf{S}^{(p)}\}$, the so-called sequence of *partial sums*. If $\mathbf{S}^{(p)} \longrightarrow \mathbf{S}$ as $p \longrightarrow \infty$ we say that the series *converges* and has *sum* $\mathbf{S}$. The series *diverges* if $\{\mathbf{S}^{(p)}\}$ has no limit.

We shall write $\sum_{k=1}^{\infty} \mathbf{x}^{(k)}$, or more simply $\sum \mathbf{x}^{(k)}$, to denote both the series (whether it is convergent or divergent) and its sum when it converges. It should be noted that in forming an infinite series, the terms have to be taken in the given ordering of the sequence. Changing the order of the terms changes the partial sum sequence and can change the sum or, even worse, change convergence into divergence, or vice versa.

Proposition 2.45. (*The Cauchy criterion (GPC) for series.*)
A series $\sum_{k=1}^{\infty} \mathbf{x}^{(k)}$ *in* $\mathbb{R}^n$ *converges if and only if, for every* $\varepsilon > 0$ *there is* $N(\varepsilon)$ *such that*

$$\|\mathbf{S}^{(k)} - \mathbf{S}^{(j)}\| = \|\mathbf{x}^{(j+1)} + \mathbf{x}^{(j+2)} + \cdots + \mathbf{x}^{(k)}\| < \varepsilon \textit{ for all } k \geq j \geq N(\varepsilon).$$

Proof. $\{\mathbf{S}^{(p)}\}$ converges $\Leftrightarrow$ it is a Cauchy sequence.

Definition 2.46. A series $\sum_{k=1}^{\infty} \mathbf{x}^{(k)}$ is said to be *absolutely convergent* if the series $\sum_{k=1}^{\infty} \|\mathbf{x}^{(k)}\|$ is convergent.

Note that $\sum_{k=1}^{\infty} \|\mathbf{x}^{(k)}\|$ is just a series of positive real numbers and various well-known tests may be applied such as the ratio test, comparison test, or integral test. We assume the reader is familiar with these tests from a first course in Analysis.

Proposition 2.47. *An absolutely convergent series is convergent.*

Proof. Since $\sum_{k=1}^{\infty} \|\mathbf{x}^{(k)}\|$ converges, by the Cauchy criterion (GPC) there exists $N(\varepsilon)$ such that $\|\mathbf{x}^{(j+1)}\| + \cdots + \|\mathbf{x}^{(k)}\| < \varepsilon$ for $k \geq j \geq N(\varepsilon)$.

But $\|\mathbf{x}^{(j+1)}+\cdots+\mathbf{x}^{(k)}\| \le \|\mathbf{x}^{(j+1)}\|+\cdots+\|\mathbf{x}^{(k)}\|$ by the triangle inequality, which proves convergence of $\sum_{k=1}^{\infty} \mathbf{x}^{(k)}$ by the sufficiency of the GPC.

Example 2.48. The exponential of an $n\times n$ matrix. If A is an $n\times n$ matrix we can regard A and its powers A^k as elements of $\mathbb{R}^{n^2}$. Then the series $\sum_{k=0}^{\infty} \frac{A^k}{k!}$ converges absolutely. Its limit is denoted e^A.

In fact $\frac{\|A^k\|}{k!} \le \frac{\|A\|^k}{k!}$ and $\sum_{k=0}^{\infty} \frac{\|A\|^k}{k!} = e^{\|A\|}$ so the series is absolutely convergent. This series is useful in the study of systems of linear ordinary differential equations.

Now let $\{\mathbf{f}^{(k)}\}$ be a sequence of functions defined on some set A in $\mathbb{R}^n$ and with values in $\mathbb{R}^m$. For each $\mathbf{x}\in A$, $\{\mathbf{f}^{(k)}(\mathbf{x})\}$ is a sequence in $\mathbb{R}^m$ and we may form the corresponding series $\sum_{k=1}^{\infty} \mathbf{f}^{(k)}(\mathbf{x})$. If this series converges, that is, the sequence $\{\mathbf{F}^{(p)}(\mathbf{x})\}$ of partial sums converges, we write $\mathbf{F}(\mathbf{x})$ for its sum and say that the series $\sum_{k=1}^{\infty} \mathbf{f}^{(k)}(\mathbf{x})$ *converges* to $\mathbf{F}(\mathbf{x})$. If the sequence of partial sums $\{\mathbf{F}^{(p)}\}$ converges uniformly to $\mathbf{F}$ on A, we say that $\sum_{k=1}^{\infty} \mathbf{f}^{(k)}$ *converges uniformly* to $\mathbf{F}$ on A.

Theorem 2.49. *If $A\subseteq\mathbb{R}^n$ and $\mathbf{f}^{(k)}: A \longrightarrow \mathbb{R}^m$ is continuous on A and if $\sum_{k=1}^{\infty} \mathbf{f}^{(k)}$ converges uniformly to $\mathbf{F}$ on A, then $\mathbf{F}$ is continuous on A.*

Proof. Each partial sum is continuous so the result follows immediately from Theorem 2.36.

Example 2.50. The series $\sum_{k=1}^{\infty} x^{2k} \sin(1-x)$, is not uniformly convergent on $[0,1]$.

Solution. Let $F^{(m)}(x) := \sum_{k=1}^{m} x^{2k} \sin(1-x) = \sin(1-x) \sum_{k=1}^{m} x^{2k}$.

Since $\sum_{k=1}^{m} x^{2k}$ is a finite geometric series its sum is $\frac{x^2(1-x^{2m})}{1-x^2}$. For $0 \le x < 1$, this converges to $x^2/(1-x^2)$. Therefore, the sum of the series is $F(x)=\begin{cases}\sin(1-x)\, x^2/(1-x^2), & \text{if } 0 \le x < 1 \\ 0 & , \text{ for } x=1\end{cases}$. This function is not continuous at $x=1$, since $\lim_{x\to 1-} F(x) = \lim_{x\to 1-} \frac{\sin(1-x)}{(1-x)} \frac{x^2}{(1+x)} = \frac{1}{2} \ne F(1)$. Therefore the convergence is not uniform else Theorem 2.49 would be contradicted.

It is useful to have tests which can be used to establish the uniform convergence of a series of functions. A necessary and sufficient condition arises from the Cauchy condition for uniform convergence of the partial sum sequence. It is useful as knowledge of the limit is not required.

Proposition 2.51. *Let* $\mathbf{f}^{(k)}: A \subseteq \mathbb{R}^n \longrightarrow \mathbb{R}^m$. *The series* $\sum_{k=1}^{m} \mathbf{f}^{(k)}$ *is uniformly convergent on* A *if and only if, for every* $\varepsilon > 0$ *there exists* $N(\varepsilon)$ *such that* $\|\mathbf{f}^{(j)} + \mathbf{f}^{(j+1)} + \cdots + \mathbf{f}^{(k)}\|_u < \varepsilon$ *for all* $k \ge j \ge N(\varepsilon)$.

Proof. This is an immediate consequence of the criterion for sequences, Proposition 2.43.

We can now give a test which is easy to apply and is often successful in practical examples.

Theorem 2.52. (*The Weierstrass M-test.*)
Let $\{\mathbf{f}^{(k)}\}$ *be a sequence of functions from* $A \subseteq \mathbb{R}^n$ *to* $\mathbb{R}^m$. *Let* $\{M^{(k)}\}$ *be a sequence of non-negative real numbers and suppose that* $\|\mathbf{f}^{(k)}\|_u \le M^{(k)}$ *for each* $k \in \mathbb{N}$. *If* $\sum_{k=1}^{\infty} M^{(k)}$ *converges then* $\sum_{k=1}^{\infty} \mathbf{f}^{(k)}$ *converges uniformly on* A.

Proof. For $k > j$, by the triangle inequality for the uniform norm, we have

$$\|\mathbf{f}^{(j)} + \mathbf{f}^{(j+1)} + \cdots + \mathbf{f}^{(k)}\|_u \le \|\mathbf{f}^{(j)}\|_u + \cdots + \|\mathbf{f}^{(k)}\|_u \le M^{(j)} + \cdots + M^{(k)}.$$

Since $\sum M^{(k)}$ converges, the GPC in $\mathbb{R}$ shows that the Cauchy criterion of Proposition 2.51 holds for $\sum_{k=1}^{\infty} \mathbf{f}^{(k)}$.

Example 2.53. The series $\sum_{k=1}^{\infty} \cos(kx)/k^{\alpha}$ is uniformly convergent on $[0,2\pi]$ for $\alpha > 1$.

Solution. $|\cos kx \,/\, k^{\alpha}| \leq 1/k^{\alpha} := M^{(k)}$ and $\sum M^{(k)} = \sum k^{-\alpha}$ is convergent.

When $\alpha=1$ the M-test fails on this example and indeed the series with $\alpha=1$ diverges when $x=2\pi$.

Some other important tests are based on the following identity.

Lemma 2.54. (*Abel's partial summation formula.*)
Let $\{\alpha^{(k)}\}$ *be a sequence of real numbers and* $\{\mathbf{u}^{(k)}\}$ *be a sequence in* $\mathbb{R}^n$. *Let* $\mathbf{U}^{(k)} = \sum_{j=1}^{k} \mathbf{u}^{(j)}$ *be the k-th partial sum of* $\{\mathbf{u}^{(j)}\}$. *Then*

$$\sum_{k=1}^{p} \alpha^{(k)}\mathbf{u}^{(k)} = \alpha^{(p)}\mathbf{U}^{(p)} + \sum_{k=1}^{p-1} \mathbf{U}^{(k)}(\alpha^{(k)} - \alpha^{(k+1)}).$$

Proof. Each side is $\alpha^{(1)}\mathbf{U}^{(1)} + \alpha^{(2)}(\mathbf{U}^{(2)} - \mathbf{U}^{(1)}) + \cdots + \alpha^{(p)}(\mathbf{U}^{(p)} - \mathbf{U}^{(p-1)})$.

Corollary 2.55. $\sum_{k=1}^{\infty} \alpha^{(k)}\mathbf{u}^{(k)}$ *is convergent if the sequence* $\{\alpha^{(k)}\mathbf{U}^{(k)}\}$ *and the series* $\sum_{k=1}^{\infty} \mathbf{U}^{(k)}(\alpha^{(k)} - \alpha^{(k+1)})$ *both converge.*

Corollary 2.56. *If* $\alpha^{(k)}$ *is a decreasing sequence of positive numbers and if* $\|\mathbf{U}^{(k)}\| \leq M$ *for all k, then* $\|\alpha^{(1)}\mathbf{u}^{(1)} + \cdots + \alpha^{(p)}\mathbf{u}^{(p)}\| \leq M\|\alpha^{(1)}\|$.

Proof. By Abel's formula and the triangle inequality,

$$\begin{aligned}\|\alpha^{(1)}\mathbf{u}^{(1)} + \cdots + \alpha^{(p)}\mathbf{u}^{(p)}\| &\leq \|\alpha^{(p)}\mathbf{U}^{(p)}\| + \sum_{k=1}^{p-1} \|\mathbf{U}^{(k)}(\alpha^{(k)} - \alpha^{(k+1)})\| \\ &\leq M\alpha^{(p)} + M\sum_{k=1}^{p-1} (\alpha^{(k)} - \alpha^{(k+1)}) = M\alpha^{(1)}.\end{aligned}$$

Using the partial summation formula allows another useful test for uniform convergence.

Theorem 2.57. (*Dirichlet's test for uniform convergence.*)

Let $\{\mathbf{f}^{(k)}\}$ be a sequence of functions from $A\subseteq\mathbb{R}^n$ to $\mathbb{R}^m$ and let $\{\mathbf{F}^{(k)}\}$ be the corresponding sequence of partial sums. Suppose that $\mathbf{F}^{(k)}$ are uniformly bounded, that is, $\|\mathbf{F}^{(k)}\|_u\le M$ for all k. Let $\{g^{(k)}\}$ be a sequence of non-negative functions from A to $\mathbb{R}$ such that $g^{(k+1)}(\mathbf{x})\le g^{(k)}(\mathbf{x})$ for every $\mathbf{x}$ and suppose that $g^{(k)}\longrightarrow 0$ uniformly on A. Then the series $\sum g^{(k)}\mathbf{f}^{(k)}$ is uniformly convergent on A.

Proof. Since $\|\mathbf{f}^{(i+1)}(\mathbf{x})+\cdots+\mathbf{f}^{(i+m)}(\mathbf{x})\| = \|\mathbf{F}^{(i+m)}(\mathbf{x}) - \mathbf{F}^{(i)}(\mathbf{x})\|$

$$\le \|\mathbf{F}^{(i+m)}(\mathbf{x})\| + \|\mathbf{F}^{(i)}(\mathbf{x})\| \le 2M,$$

using Corollary 2.56 (relabelled to start at $i+1$) gives

$$\|g^{(i+1)}(\mathbf{x})\mathbf{f}^{(i+1)}(\mathbf{x})+\cdots+g^{(i+m)}(\mathbf{x})\mathbf{f}^{(i+m)}(\mathbf{x})\| \le 2M\, g^{(i+1)}(\mathbf{x}).$$

As $g^{(i)}\longrightarrow 0$ uniformly on A, this shows that the series satisfies the Cauchy criterion for uniform convergence.

Every convergent series has uniformly bounded partial sums but there are also divergent series with the same property. The following example illustrates this.

Example 2.58. Show that the series $\sum_{k=1}^{\infty}\cos(kx)/k$ is uniformly convergent on $[\delta,2\pi-\delta]$ for any δ with $0<\delta<\pi$.

Solution. We cannot apply the M-test since $\sum_{k=1}^{\infty}\frac{1}{k}$ is divergent. Since the sequence $\{1/k\}$ decreases to 0 we can apply the Dirichlet test if we can prove that the partial sums $C^{(p)}(x)=\sum_{k=1}^{p}\cos(kx)$ are uniformly bounded (this series is not convergent since $\cos(kx)\not\longrightarrow 0$ as $k\longrightarrow\infty$).
We employ the identity

$$2\cos(kx)\sin(x/2) = \sin(k+1/2)x - \sin(k-1/2)x.$$

Summing these from $k=1$ to $k=p$ gives

$$2\sin(x/2)\sum_{k=1}^{p}\cos(kx)=\sin(p+1/2)x - \sin(x/2).$$

Hence, if x is not an integral multiple of 2π, we have

$$C^{(p)}(x)=[\sin(p+1/2)x - \sin(x/2)]/2\sin(x/2).$$

In particular, $|C^{(p)}(x)|\le 1/|\sin(x/2)|\le 1/\sin(\delta/2)$, for $\delta\le x\le 2\pi-\delta$. This shows that Dirichlet's test applies.

It is possible to prove a similar result, known as Abel's theorem where the hypotheses on the sequence $\{g^{(k)}\}$ are weakened while the hypothesis on the series $\sum \mathbf{f}^{(k)}$ is strengthened, see problem 2.22

2.6. TOPOLOGICAL ASPECTS OF CONTINUITY

We have seen that the images of open (closed) sets under continuous functions are not necessarily open (respectively closed). However, *inverse images* behave much better. Let $\mathbf{f} : A \subseteq \mathbb{R}^n \longrightarrow B \subseteq \mathbb{R}^m$ be a function and let T be a subset of B. The *inverse image of T under* $\mathbf{f}$, denoted $\mathbf{f}^{-1}(T)$, is the set $\mathbf{f}^{-1}(T) := \{\mathbf{x} \in A : \mathbf{f}(\mathbf{x}) \in T\}$, that is, the set of all elements of A whose images under $\mathbf{f}$ lie in T. When $\mathbf{f}$ is one-to-one ($\mathbf{f}^{-1}$ exists) and T is contained in $\mathbf{f}(A)$, $\mathbf{f}^{-1}(T)$ coincides with the image of T under $\mathbf{f}^{-1}$, so the notation should be unambiguous.

Lemma 2.59. *Let* $\mathbf{f} : A \subseteq \mathbb{R}^n \longrightarrow \mathbb{R}^m$ *be a function and let* $S \subseteq A$, $T \subseteq \mathbb{R}^m$. *Then we have* (*a*) $S = \mathbf{f}^{-1}(T) \Rightarrow \mathbf{f}(S) \subseteq T$, *that is,* $\mathbf{f}(\mathbf{f}^{-1}(T)) \subseteq T$;

(*b*) $T = \mathbf{f}(S) \Rightarrow S \subseteq \mathbf{f}^{-1}(T)$, *that is,* $S \subseteq \mathbf{f}^{-1}(\mathbf{f}(S))$.

Proof. (*a*) Let $\mathbf{y} \in \mathbf{f}(\mathbf{f}^{-1}(T))$ so that $\mathbf{y} = \mathbf{f}(\mathbf{x})$ for some $\mathbf{x} \in \mathbf{f}^{-1}(T)$. By the definition of $\mathbf{f}^{-1}(T)$ this says that $\mathbf{x} \in A$ and $\mathbf{f}(\mathbf{x}) \in T$, that is $\mathbf{y} \in T$.

(*b*) Note that $S \subseteq A$. Let $\mathbf{x} \in S$, then $\mathbf{f}(\mathbf{x}) \in \mathbf{f}(S)$ so, by the definition, $\mathbf{x} \in \mathbf{f}^{-1}(\mathbf{f}(S))$.

Note that equality does not hold in (*a*) or (*b*) in general. To see this for (*a*) consider a set T strictly containing $\mathbf{f}(A)$. Equality does hold in (*a*) if $T \subseteq \mathbf{f}(A)$, that is, when $\mathbf{f}$ is onto. For (*b*) consider the example $f : \mathbb{R} \longrightarrow \mathbb{R}$ defined by $f(x) = x^2$. Then, $[0,1] = T = f([0,1])$ but $f^{-1}([0,1]) = [-1,1]$. Equality holds in (*b*) if $\mathbf{f}$ is one-to-one.

In courses on topology the definition of continuity on a set is usually phrased as, "the inverse image of every open set is open". The topology on the domain of $\mathbf{f}$ is that inherited from the space $\mathbb{R}^n$, the so-called relative topology. We now explain these concepts.

Definition 2.60. A set $U \subseteq A$ is said to be *open relative to* $A \subseteq \mathbb{R}^n$ if there is an open set V in $\mathbb{R}^n$ such that $U = A \cap V$.

Lemma 2.61. $U \subseteq A$ *is open relative to* $A \Leftrightarrow$ *for every* $\mathbf{u} \in U$ *there exists* $\delta > 0$ *such that* $\mathrm{B}(\mathbf{u}, \delta) \cap A \subseteq U$.

Proof. If $U = A \cap V$, then for any $\mathbf{u} \in U$, $\mathbf{u}$ belongs to the open set V so there is $\delta>0$ such that $B(\mathbf{u},\delta)\subseteq V$; hence $B(\mathbf{u},\delta)\cap A\subseteq U$. For the converse, suppose that, for every $\mathbf{u}\in U$, there is $\delta(\mathbf{u})>0$ such that $B(\mathbf{u},\delta(\mathbf{u}))\cap A\subseteq U$. Let $V=\cup\{B(\mathbf{u},\delta(\mathbf{u})) : \mathbf{u}\in U\}$, a union of open sets, therefore open. Then $V\cap A=\cup\{B(\mathbf{u},\delta(\mathbf{u}))\cap A\}\subseteq U$ and clearly V contains every point of U so $U\subseteq V\cap A$ and the desired equality holds.

Theorem 2.62. *Let* $\mathbf{f} : A\subseteq\mathbb{R}^n \longrightarrow \mathbb{R}^m$. *Then* $\mathbf{f}$ *is continuous on* A *if and only if for every* $T\subseteq\mathbf{f}(A)$ *that is open relative to* $\mathbf{f}(A)$, $\overset{-1}{\mathbf{f}}(T)$ *is open relative to* A.

Proof. Suppose $\mathbf{f}$ is continuous on A and let T be open relative to $\mathbf{f}(A)$. Let $\mathbf{u}\in\overset{-1}{\mathbf{f}}(T)$ so that $\mathbf{f}(\mathbf{u})\in T$. Then there is an ε-neighbourhood of $\mathbf{f}(\mathbf{u})$ such that $B(\mathbf{f}(\mathbf{u}),\varepsilon)\cap\mathbf{f}(A)\subseteq T$. By the ε-δ condition, there is $\delta>0$ such that $\mathbf{x}\in B(\mathbf{u},\delta)\cap A\Rightarrow\mathbf{f}(\mathbf{x})\in B(\mathbf{f}(\mathbf{u}),\varepsilon)$. This shows that for $\mathbf{x}\in B(\mathbf{u},\delta)\cap A$, $\mathbf{f}(\mathbf{x})\in T$. Therefore $B(\mathbf{u},\delta)\cap A\subseteq\overset{-1}{\mathbf{f}}(T)$ and $\overset{-1}{\mathbf{f}}(T)$ is open relative to $\mathbf{f}(A)$.

Conversely, suppose that for every $T\subseteq\mathbf{f}(A)$ that is open relative to $\mathbf{f}(A)$, $\overset{-1}{\mathbf{f}}(T)$ is open relative to A. Take any fixed $\mathbf{a}\in A$ and any $\varepsilon>0$. Then $T=B(\mathbf{f}(\mathbf{a}),\varepsilon)\cap\mathbf{f}(A)$ is open relative to $\mathbf{f}(A)$, by Lemma 2.61. By the hypothesis, $\overset{-1}{\mathbf{f}}(T)$ is open relative to A so there is $\delta>0$ such that $B(\mathbf{a},\delta)\cap A\subseteq\overset{-1}{\mathbf{f}}(T)$. This means that, if $\mathbf{x}\in B(\mathbf{a},\delta)\cap A$ then $\mathbf{f}(\mathbf{x})\in T$. Since $\mathbf{f}(\mathbf{x})\in\mathbf{f}(A)$ is automatically satisfied for such $\mathbf{x}$, this has verified the ε-δ condition at $\mathbf{a}$.

One of the properties of intervals in $\mathbb{R}$ is that they consist of a single piece. An interval I may be any of the following types : (a,b), $[a,b)$, $(a,b]$, $[a,b]$, $(-\infty,b)$, $(-\infty,b]$, (a,∞) and $[a,\infty)$. A single point may be considered as the degenerate interval $[a,a]$. The important property of an interval is that, if $x,y\in I$ and $x<y$ then $[x,y]$ is contained in I.

The notion of a set consisting of one piece gives the important concept of connectedness.

Definition 2.63. A set A in $\mathbb{R}^n$ is said to be *disconnected* or *separated* if A can be written $A=U\cup V$ where U and V are nonempty disjoint sets that are both open relative to A. A set is *connected* if it is not disconnected.

Example 2.64. The set $A=[1,2)\cup(2,3]$ is disconnected in $\mathbb{R}$, for $[1,2)$ and $(2,3]$ are open relative to A since for example $[1,2)=A\cap(0,2)$, and $(2,3]=A\cap(2,4)$ and these sets are disjoint.

The reader should consider why the same argument fails when applied to $B=[1,2]\cup(2,3]$. In fact $B=[1,3]$ and the next example shows this must be connected.

Example 2.65. The set $A=[0,1]$ is connected.

Solution. Suppose A is disconnected and let $[0,1]=U\cup V$ where U, V are nonempty disjoint sets of the form $U=U_1\cap[0,1]$, $V=V_1\cap[0,1]$, where U_1, V_1 are open in $\mathbb{R}$. Obviously we may assume, by a change of notation if necessary, that $0\in U_1$. As U_1 is open, there is $\varepsilon>0$ such that $[0,\varepsilon)\in U$. Let $x=\sup\{u\in[0,1]:[0,u)\subseteq U\}$. Since $x\in[0,1]$, either $x\in U$ or $x\in V$. If $x\in U$, then $[0,x]\subseteq U$ and we must have $x<1$, else V is empty. As x belongs to the open set U_1 there exists δ such that $(x-\delta,x+\delta)\subseteq U_1$. In particular, $[0,x+\delta)$ lies in U for δ sufficiently small, contradicting the fact that x is an upper bound. Hence, $x\in V$. By the same arguments, there is $\delta_1>0$ so that $(x-\delta_1,x+\delta_1)$ lies in V. However $x-\delta_1/2$ lies in U for $\delta_1<\varepsilon/2$, from the definition of x. This contradicts $U\cap V=\emptyset$.

This example is typical for we have :

Theorem 2.66. *A subset I of $\mathbb{R}$ is connected $\Leftrightarrow$ I is an interval.*

Proof. We use the fact that a subset I of $\mathbb{R}$ is an interval if and only if I has the property that if $x\in I, y\in I$ and $x<z<y$ then $z\in I$. First suppose that I is not an interval. Then there exist points $x,y\in I$ with $x<y$ and a point $z\notin I$ with $x<z<y$. Then $I=[I\cap(-\infty,z)]\cup[I\cap(z,\infty)]$ proves that I is disconnected. Notice that x,y belong to the separate pieces. Thus I connected $\Rightarrow$ I is an interval.

Next we prove that if I is an interval then I is connected. The proof is by contradiction so suppose that I is an interval and that $I=U\cap V$ where U, V are nonempty disjoint sets that are open relative to I. Since they are nonempty there is $a\in U$ and $b\in V$ with $a\neq b$. We can always suppose that $a<b$, by a change of notation if necessary. Since I is an interval, $[a,b]\subseteq I$ and each point of $[a,b]$ either belongs to U or else to V. Let $x=\sup\{u\in[a,b]:[a,u)\subseteq U\}$; evidently $x\in[a,b]$ so $x\in I$. We repeat the arguments given in Example 2.65 almost verbatim and obtain a contradiction exactly as there.

Connectedness is preserved by continuous functions.

Theorem 2.67. *Let $\mathbf{f}:A\subseteq\mathbb{R}^n\longrightarrow\mathbb{R}^m$ be continuous on the connected set A. Then $\mathbf{f}(A)$ is connected.*

Proof. Suppose $\mathbf{f}(A)$ is disconnected. Then $\mathbf{f}(A)=U\cap V$ where U, V are nonempty, disjoint, and open relative to $\mathbf{f}(A)$. Then we have $A=\mathbf{f}^{-1}(\mathbf{f}(A))=\mathbf{f}^{-1}(U\cap V)=\mathbf{f}^{-1}(U)\cap\mathbf{f}^{-1}(V)$ (why do we have equalities here?).

By Theorem 2.62, $\overset{-1}{\mathbf{f}}(U)$ and $\overset{-1}{\mathbf{f}}(V)$ are open relative to A, they are nonempty since U, V are, and are disjoint for the same reason. This shows that A is disconnected, a contradiction establishing the result.

Corollary 2.68. *Let* $\mathbf{f} : A \subseteq \mathbb{R}^n \longrightarrow \mathbb{R}$ *be continuous on the connected set* A. *Then* $\mathbf{f}(A)$ *is an interval.*

Proof. An immediate consequence of Theorems 2.66 and 2.67.

A subset of $\mathbb{R}^n$ need not be connected but it can be written as a union of connected pieces.

Definition 2.69. A maximal connected subset of a set A is called a *component* of A. Thus C is a component of A if C is connected and C is not properly contained in any larger connected subset of A.

Obviously a connected set has only one component, namely the set itself. The following results are needed to establish the main facts about components.

Proposition 2.70. *If* $\{C : C \in \mathcal{F}\}$ *is a family of connected sets such that* $\bigcap\{C : C \in \mathcal{F}\} \neq \emptyset$, *then* $\bigcup\{C : C \in \mathcal{F}\}$ *is also connected.*

Proof. Write A for the union of the sets. Suppose A is disconnected. Then there exist open sets U,V such that $A \cap U$ and $A \cap V$ are nonempty, disjoint sets and $A \subseteq U \cup V$. Since every set C is connected and $C \subseteq A$, each C lies either entirely inside U or else inside V; no set C can intersect both U and V. Since $\bigcap C \neq \emptyset$ either all of the sets lie entirely within U or else entirely within V. This then shows that A lies either in U or in V, contradicting the assumption that A is disconnected.

Proposition 2.71. *If* A *is connected and* $A \subseteq B \subseteq \mathrm{cl}(A)$, *then* B *is also connected. In particular, the closure of a connected set is connected.*

Proof. Suppose that B is disconnected and let U,V be open sets with $B \subseteq U \cup V$ where U,V have nonempty disjoint intersections with B. Since $A \subseteq B$ and A is connected, either A lies entirely within U or else entirely within V. Suppose $A \subseteq U$ and $A \cap V = \emptyset$. We show that $\mathrm{cl}(A) \cap V = \emptyset$ also. For if $\mathbf{x} \in \mathrm{cl}(A) \cap V$, every neighbourhood of $\mathbf{x}$ contains points of A and points of V. As V is open, there is a neighbourhood of $\mathbf{x}$ entirely inside V which thus contains some point of A, impossible. Thus, we see that $B \cap V \subseteq \mathrm{cl}(A) \cap V = \emptyset$ contradicting the disconnectedness of B.

We can now prove the basic facts concerning components. We say that a set C is closed relative to A if $C = A \cap F$ where F is closed.

Theorem 2.72. *Let A be any set. Then*
(a) each component of A is a relatively closed set;
(b) each point $\mathbf{a} \in A$ lies in precisely one component of A, that is, A is the disjoint union of its components.

Proof. (*a*) Since a component C is a maximal connected set and $C \subseteq \mathrm{cl}(C) \cap A \subseteq \mathrm{cl}(C)$ we see from Proposition 2.71 that $C = \mathrm{cl}(C) \cap A$, so C is closed relative to A.
(*b*) Let $\mathbf{a} \in A$ and consider the family $\{C\}$ of all connected subsets of A that contain $\mathbf{a}$. This is a nonempty collection since $\{\mathbf{a}\}$ is a connected set. By Proposition 2.70, $E = \bigcup C$ is a connected set that contains $\mathbf{a}$ and is clearly maximal. Finally components are disjoint for if $\mathbf{a}$ belongs to two different components their union is a larger connected set containing $\mathbf{a}$, contradicting the maximality property of components.

Components need not be open sets in general but we have :

Theorem 2.73. *If A is an open subset of $\mathbb{R}^n$ then each component of A is also open.*

Proof. Let C be a component of A and let $\mathbf{c} \in C$. Since A is open there is a neighbourhood $\mathrm{B}(\mathbf{c},\varepsilon) \subseteq A$. But $\mathrm{B}(\mathbf{c},\varepsilon)$ is connected and since C is a maximal connected set containing $\mathbf{c}$ we must have $\mathrm{B}(\mathbf{c},\varepsilon) \subseteq C$. This proves that C is open.

In this proof we use the "obvious" fact that neighbourhoods are connected. The simplest way to prove this is via the notion of pathwise connected sets which has uses elsewhere too.

Definition 2.74. A *path* is a continuous function $\mathbf{f} : [0,1] \longrightarrow \mathbb{R}^n$, its image set is a *curve*. A set A is called *pathwise connected* if for every $\mathbf{x},\mathbf{y} \in A$ there is a path $\mathbf{f} : [0,1] \longrightarrow A$ such that $\mathbf{f}(0) = \mathbf{x}$ and $\mathbf{f}(1) = \mathbf{y}$, that is, $\mathbf{x}$ and $\mathbf{y}$ can be joined by a curve lying entirely within A.

Example 2.75. A convex set is pathwise connected, for any two points $\mathbf{x}$, $\mathbf{y}$, can be joined by the line segment $(1-t)\mathbf{x} + t\mathbf{y}$, $0 \leq t \leq 1$.

We first show that this concept is more restrictive than connectedness.

Proposition 2.76. *A pathwise connected set is connected.*

Proof. Suppose that A is a pathwise connected set that is not connected. Then there are relatively open sets U, V with $U \cap V = \emptyset$ and

$A=U\cup V$. Let $\mathbf{x}\in U$, $\mathbf{y}\in V$ and let $\mathbf{f}$ be a path in A with $\mathbf{f}(0)=\mathbf{x}$, $\mathbf{f}(1)=\mathbf{y}$. Then $I=\mathbf{f}^{-1}(U)$ and $J=\mathbf{f}^{-1}(V)$ are nonempty ($0\in I$, $1\in J$) relatively open subsets of $[0,1]$. Furthermore, $I\cap J=\mathbf{f}^{-1}(U\cap V)=\emptyset$, and $I\cup J=\mathbf{f}^{-1}(U\cup V)=[0,1]$. This contradicts the connectedness of $[0,1]$.

The set in $\mathbb{R}^2$, $A=\{(x,\sin(1/x) : x>0\}\cup\{(0,y) : -1\le y\le 1\}$ is an example of a connected set that is not pathwise connected. This shows that the two concepts do not coincide. However, we do have :

Proposition 2.77. *Let A be an open connected set in $\mathbb{R}^n$. Then A is pathwise connected.*

Proof. Let $\mathbf{a}\in A$ and let $U=\{\mathbf{z}\in A : \mathbf{z}$ can be joined to $\mathbf{a}$ by a path in $A\}$, and let $V=\{\mathbf{x}\in A : \mathbf{x}\notin U\}$. Then A is the union of the disjoint sets U and V. We first show that U is open. Let $\mathbf{z}\in U\subseteq A$. Since A is open there is $\delta>0$ such that $B(\mathbf{z},\delta)\subseteq A$. Any point $\mathbf{x}$ in $B(\mathbf{z},\delta)$ can be joined to $\mathbf{a}$ by taking the path from $\mathbf{a}$ to $\mathbf{z}$ and then the line segment from $\mathbf{z}$ to $\mathbf{x}$. Thus $B(\mathbf{z},\delta)\subseteq U$ and U is open. Next we show that V is also open. If $\mathbf{v}\in V$ there is a nbd $B(\mathbf{v},\varepsilon)\subseteq A$ and, by the agument just given, if any point of $B(\mathbf{v},\varepsilon)$ can be joined by a path to $\mathbf{a}$ so can $\mathbf{v}$ itself. Since $\mathbf{v}\notin U$ we have $B(\mathbf{v},\varepsilon)\subseteq V$ and V is also open. Since A is connected and $\mathbf{a}\in U$, V must be empty. However we then have $A=U$ and U is pathwise connected because any two points $\mathbf{x}$, $\mathbf{y}$ can be joined by taking the path from $\mathbf{x}$ to $\mathbf{a}$ and then from $\mathbf{a}$ to $\mathbf{y}$.

In fact it is possible to join any two points in an open connected set by a *polygonal path* : a finite collection of line segments. We say A is polygonally connected in this case.

Proposition 2.78. *If A is an open connected set, any two points of A can be joined by a polygonal path lying in A.*

Proof. Let $\mathbf{a}$, $\mathbf{b}$ be any two points of A. Then they can be joined by a path $\mathbf{f} : [0,1] \longrightarrow A$ with $\mathbf{f}(0)=\mathbf{a}$, $\mathbf{f}(1)=\mathbf{b}$. Write C for the image set, the curve joining $\mathbf{a}$ to $\mathbf{b}$. Since $\mathbf{f}$ is continuous, C is compact. For each $\mathbf{x}\in C\subseteq A$, as A is open, there is $\varepsilon(\mathbf{x})>0$ so that $B(\mathbf{x},\varepsilon(\mathbf{x}))\subseteq A$. By compactness of C there is a finite collection $\{B(\mathbf{x}^{(i)},\varepsilon^{(i)}/2) : 1\le i\le m\}$ that covers C. Let $\varepsilon=\min\{\varepsilon^{(i)}\}$.

Then for any $\mathbf{c}\in C$, there is i so that $B(\mathbf{c},\varepsilon/2)\subseteq B(\mathbf{x}^{(i)},\varepsilon^{(i)})\subseteq A$. (2.11)

Since $\mathbf{f}$ is uniformly continuous on $[0,1]$, there exists $\delta>0$ so that $\|\mathbf{f}(t)-\mathbf{f}(s)\|<\varepsilon/2$ for $|t-s|<\delta$. Take $0=t_0<t_1<\cdots<t_k=1$ so that $|t_{i+1}-t_i|<\delta$. We now assert that the polygonal curve joining the points $\mathbf{f}(t_i)$ and $\mathbf{f}(t_{i+1})$ lies in A. Indeed, if $\mathbf{x}$ lies on the line segment joining these, then $\mathbf{x}=(1-\alpha)\mathbf{f}(t_i)+\alpha\mathbf{f}(t_{i+1})$ for $0\leq\alpha\leq1$ and $\|\mathbf{x}-\mathbf{f}(t_i)\| = \alpha\|\mathbf{f}(t_i)-\mathbf{f}(t_{i+1})\| < \varepsilon/2$. Then (2.11) shows that $\mathbf{x}\in A$.

Some of these results are needed in the discussion of integrals along curves in Chapter 7.

2.7. LINEAR OPERATORS AND MATRICES

Functions that preserve the linear structure of the space are called linear operators and play an important role in differentiation (Chapter 3). They may well be familiar to the reader from a course on linear algebra. However we are particularly interested in their continuity properties.

Definition 2.79. A function $\mathbf{L}:\mathbb{R}^n \longrightarrow \mathbb{R}^m$ is said to be *linear* if, for all $\mathbf{x},\mathbf{y}\in\mathbb{R}^n$ and $\alpha\in\mathbb{R}$, $\mathbf{L}(\mathbf{x}+\mathbf{y})=\mathbf{L}(\mathbf{x})+\mathbf{L}(\mathbf{y})$ and $\mathbf{L}(\alpha\mathbf{x})=\alpha\mathbf{L}(\mathbf{x})$.

These two conditions are equivalent to the single one

$$\mathbf{L}(\alpha\mathbf{x}+\beta\mathbf{y})=\alpha\mathbf{L}(\mathbf{x})+\beta\mathbf{L}(\mathbf{y}),\ \mathbf{x},\mathbf{y}\in\mathbb{R}^n,\ \alpha,\beta\in\mathbb{R}.$$

Indeed this extends to any finite sum (proof by induction). This shows that the image set of $\mathbf{L}$ is a linear subspace of $\mathbb{R}^n$. Observe that for a linear function $\mathbf{L}$, it is always the case that $\mathbf{L}(\mathbf{0})=\mathbf{0}$. We usually say *linear operator* rather than linear function and we also use the terminology *linear functional* when $m=1$.

Linear operators are uniquely determined by their action on the basis elements. We shall generally use only the standard bases in $\mathbb{R}^n$ and $\mathbb{R}^m$ and we can then classify linear operators as follows.

Proposition 2.80. *$L:\mathbb{R}^n \longrightarrow \mathbb{R}$ is a linear functional if, and only if, there exists $\mathbf{a}\in\mathbb{R}^n$ such that $L(\mathbf{x})=\langle\mathbf{a},\mathbf{x}\rangle$ for all $\mathbf{x}\in\mathbb{R}^n$.*

Proof. It is easy to prove that, for a given $\mathbf{a}\in\mathbb{R}^n$, $L(\mathbf{x}):=\langle\mathbf{a},\mathbf{x}\rangle$ defines a linear functional. For the converse, let $\{\mathbf{e}_1,\mathbf{e}_2,\cdots,\mathbf{e}_n\}$ be the standard basis of $\mathbb{R}^n$: $\mathbf{e}_i$ has 1 in the i-th place and 0 elsewhere. Then $\mathbf{x}\in\mathbb{R}^n$ can be written (uniquely) as $\mathbf{x}=x_1\mathbf{e}_1+x_2\mathbf{e}_2+\cdots+x_n\mathbf{e}_n$. By linearity, $L(\mathbf{x})=x_1L(\mathbf{e}_1)+x_2L(\mathbf{e}_2)+\cdots+x_nL(\mathbf{e}_n)=\langle\mathbf{x},\mathbf{a}\rangle$ where $a_j:=L(\mathbf{e}_j)$.

Proposition 2.81. $\mathbf{L}:\mathbb{R}^n \longrightarrow \mathbb{R}^m$ *is a linear operator* $\Leftrightarrow$ *every component function $L_i:\mathbb{R}^n \longrightarrow \mathbb{R}$, $1\leq i\leq m$, is a linear functional.*

Proof. The routine proof is left to the reader.

Theorem 2.82. *A function* $\mathbf{L}:\mathbb{R}^n \longrightarrow \mathbb{R}^m$ *is linear if and only if there is an* $m\times n$ *matrix of real numbers* $A=(a_{ij})$, $1\le i\le m$, $1\le j\le n$, *such that*

$$\mathbf{L}(\mathbf{x})=\mathbf{y} \Leftrightarrow y_i=\sum_{k=1}^{n} a_{ik}x_k\ ,\ i=1,2,\cdots,m. \quad \textit{Moreover, } a_{ij}=L_i(\mathbf{e}_j).$$

Proof. By Propositions 2.80 and 2.81, $\mathbf{L}(\mathbf{x})=(L_1(\mathbf{x}), L_2(\mathbf{x}),\cdots,L_m(\mathbf{x}))$ where $L_i(\mathbf{x})=<\mathbf{a}^{(i)},\mathbf{x}>=\sum_{k=1}^{n} a_k^{(i)}x_k$. Let $a_{ij}:=a_j^{(i)}$. Then we have

$$L_i(\mathbf{e}_j)=<\mathbf{a}^{(i)},\mathbf{e}_j>=a_j^{(i)}=a_{ij}\ .$$

Note that a linear operator $\mathbf{L}:\mathbb{R}^n \longrightarrow \mathbb{R}^m$ (n first, m second) corresponds to an $m\times n$ matrix, not $n\times m$.

When A is an $m\times n$ matrix and B is an $n\times p$ matrix, the product matrix AB is defined as the $m\times p$ matrix C whose elements c_{ij}, $1\le i\le m$, $1\le j\le p$ are given by $c_{ij}=\sum_{k=1}^{n} a_{ik}b_{kj}$. Here, $(a_{i1},a_{i2},\cdots,a_{in})$ is the i-th row of A and $(b_{1j},b_{2j},\cdots,b_{nj})$ is the j-th column of B. Writing the i-th row of A as $\mathbf{a}^{(i)}$ and the j-th column of B as $\mathbf{b}^{(j)}$, regarded as elements of $\mathbb{R}^n$, $c_{ij}=<\mathbf{a}^{(i)},\mathbf{b}^{(j)}>$, the inner product.

The expression for $\mathbf{y}$ in Theorem 2.82 can be written $\mathbf{y}^T=A\mathbf{x}^T$, where $\mathbf{x}^T$ (T stands for transpose) denotes $\mathbf{x}$ regarded as a column vector, an $n\times 1$ matrix, and $A\mathbf{x}^T$ is the matrix product. Since A is $m\times n$ and $\mathbf{x}^T$ is $n\times 1$ this defines an $m\times 1$ matrix $\mathbf{y}^T$, equivalently, an element $\mathbf{y}$ of $\mathbb{R}^m$. We say that $\mathbf{L}$ is *represented by* A. A depends on the bases used in the relevant spaces, but we continue to use only the standard ones. Theorem 2.82 shows that $\mathbf{L}(\mathbf{e}_i)$ defines the i-th column of A. The action of $\mathbf{L}$ on $\mathbf{x}$ is thus equivalent to multiplication of $\mathbf{x}^T$ by the matrix A. It is convenient to *identify* $\mathbf{L}$ and its matrix representation with the understanding that, when using matrices, $\mathbf{x}\in\mathbb{R}^n$ must be regarded as the column vector $\mathbf{x}^T$. We often write $\mathbf{Lx}$ rather than $\mathbf{L}(\mathbf{x})$ when dealing with linear operators for this reason. A 1×1 matrix is identified with a scalar. If $\mathbf{L},\mathbf{M}$ are linear operators (matrices) $\mathbf{LM}$ can be thought of either as composition of functions or as multiplication of matrices; this is the major reason why matrix multiplication is defined in the above manner.

Example 2.83. (*a*) Let $\mathbf{L}(x,y)=(x, y, x+y)$. Then $\mathbf{L}$ is a linear operator from $\mathbb{R}^2 \longrightarrow \mathbb{R}^3$. It is represented by the matrix

$\mathbf{L}=\begin{bmatrix}1 & 0\\ 0 & 1\\ 1 & 1\end{bmatrix}$ since $\begin{bmatrix}1 & 0\\ 0 & 1\\ 1 & 1\end{bmatrix}\begin{bmatrix}x\\ y\end{bmatrix}=\begin{bmatrix}x\\ y\\ x+y\end{bmatrix}$. Note that $\mathbf{L}(\mathbf{e}_1)=(1,0,1)$ gives the first column of the matrix $\mathbf{L}$. Similarly for $\mathbf{L}(\mathbf{e}_2)$.

(*b*) $L(x,y,z)=x+y+2z$ is a linear functional on $\mathbb{R}^3$, or a linear operator from $\mathbb{R}^3$ to $\mathbb{R}$. The matrix representing L is therefore 1×3. In fact, $L=[1 \quad 1 \quad 2]$ since $L(\mathbf{e}_1)=1$, $L(\mathbf{e}_2)=1$ and $L(\mathbf{e}_3)=2$. Also we see that $<(1,1,2),(x,y,z)> = x+y+2z$.

Theorem 2.84. *Every linear operator* $\mathbf{L}:\mathbb{R}^n \longrightarrow \mathbb{R}^m$ *is uniformly continuous on* $\mathbb{R}^n$.

Proof. It suffices to prove that each component function is uniformly continuous so consider $L:\mathbb{R}^n \longrightarrow \mathbb{R}$. By Proposition 2.80, there exists $\mathbf{a}\in\mathbb{R}^n$ such that $L\mathbf{x}=<\mathbf{a},\mathbf{x}>$. By Schwarz's inequality, $|L\mathbf{x}| \le \|\mathbf{a}\|\,\|\mathbf{x}\|$. As this is true for any $\mathbf{x}$, $|L\mathbf{x}-L\mathbf{y}|=|L(\mathbf{x}-\mathbf{y})| \le \|\mathbf{a}\|\,\|\mathbf{x}-\mathbf{y}\|$. Now, for any $\varepsilon>0$, let $\delta=\varepsilon/\|\mathbf{a}\|$. Then $|L\mathbf{x}-L\mathbf{y}|<\varepsilon$ for $\|\mathbf{x}-\mathbf{y}\|<\delta$ and L is uniformly continuous on $\mathbb{R}^n$. If $\mathbf{a}=\mathbf{0}$ the choice of δ is impossible, but when $\mathbf{a}=\mathbf{0}$, $L\mathbf{x}=\mathbf{0}$ for all $\mathbf{x}$ and uniform continuity is trivial.

Definition 2.85. A linear operator $\mathbf{L}:\mathbb{R}^n \longrightarrow \mathbb{R}^m$ is said to be *bounded* if there is a constant $M\ge 0$ such that $\|\mathbf{L}\mathbf{x}\| \le M\|\mathbf{x}\|$, for all $\mathbf{x}\in\mathbb{R}^n$.

This is equivalent to $\left\|\mathbf{L}\left(\frac{\mathbf{x}}{\|\mathbf{x}\|}\right)\right\| \le M$ for all $\mathbf{x}\ne\mathbf{0}$, which says that the image of the unit sphere $S(\mathbf{0},1)$ is a bounded set. For a bounded linear operator $\mathbf{L}$ we define a real number $\|\mathbf{L}\|$, called the (*operator*) *norm* of $\mathbf{L}$, by $\|\mathbf{L}\| := \sup\{\|\mathbf{L}(\mathbf{x})\| : \|\mathbf{x}\|=1\}$. Then $\|\mathbf{L}(\mathbf{x})\| \le \|\mathbf{L}\|\,\|\mathbf{x}\|$ for all $\mathbf{x}\in\mathbb{R}^n$ and $\|\mathbf{L}\|$ is the smallest number with this property.

Theorem 2.86. *A linear operator is bounded if and only if it is continuous. Hence,* $\mathbf{L}:\mathbb{R}^n \longrightarrow \mathbb{R}^m$ *is automatically bounded.*

Proof. If $\mathbf{L}$ is bounded then $\|\mathbf{L}\mathbf{x}-\mathbf{L}\mathbf{y}\|=\|\mathbf{L}(\mathbf{x}-\mathbf{y})\| \le \|\mathbf{L}\|\,\|\mathbf{x}-\mathbf{y}\|$ and uniform continuity follows as in the proof of Theorem 2.84. Now suppose that $\mathbf{L}$ is continuous but not bounded. Then for $k=1,2,\cdots$, there exists $\mathbf{x}^{(k)}$ such that $\|\mathbf{L}\mathbf{x}^{(k)}\| > k\|\mathbf{x}^{(k)}\|$; obviously $\mathbf{x}^{(k)}\ne\mathbf{0}$. Let $\mathbf{y}^{(k)}=\mathbf{x}^{(k)}/k\|\mathbf{x}^{(k)}\|$. Then $\|\mathbf{y}^{(k)}\|=1/k$ and $\|\mathbf{L}\mathbf{y}^{(k)}\|=\dfrac{1}{k\|\mathbf{x}^{(k)}\|}\|\mathbf{L}\mathbf{x}^{(k)}\|>1$ for all k. This shows that $\mathbf{y}^{(k)} \longrightarrow \mathbf{0}$ but $\mathbf{Ly}^{(k)} \not\longrightarrow \mathbf{0}=\mathbf{L0}$, contradicting the fact that $\mathbf{L}$ is continuous at $\mathbf{0}$.

Remark. It is possible to give a direct proof of the boundedness of linear operators from $\mathbb{R}^n$ to $\mathbb{R}^m$ using Theorem 2.82. Indeed, let (a_{ij}) be the $m\times n$ matrix that represents $\mathbf{L}$. Then $\mathbf{Lx}=\mathbf{y}$ where

$$y_i = \sum_{j=1}^{n} a_{ij}x_j = <\mathbf{a}^{(i)}, \mathbf{x}>,\ i=1,2,\cdots,m, \text{ and } \|\mathbf{Lx}\| = \left(\sum_{i=1}^{m} |y_i|^2\right)^{1/2}.$$

By Schwarz's inequality we have $|y_i| \le \|\mathbf{a}^{(i)}\|\,\|\mathbf{x}\| = \left(\sum_{j=1}^{n} |a_{ij}|^2\right)^{1/2} \|\mathbf{x}\|.$

Therefore, $\sum_{i=1}^{m} |y_i|^2 \le \sum_{i=1}^{m} \left(\|\mathbf{x}\|^2 \sum_{j=1}^{n} |a_{ij}|^2\right) = \|\mathbf{x}\|^2 \sum_{i=1}^{m} \sum_{j=1}^{n} |a_{ij}|^2 .$

This shows that $\|\mathbf{Lx}\| \le \left(\sum_{i=1}^{m} \sum_{j=1}^{n} |a_{ij}|^2\right)^{1/2} \|\mathbf{x}\|$, that is, $\mathbf{L}$ is bounded and

$\|\mathbf{L}\| \le \left(\sum_{i=1}^{m} \sum_{j=1}^{n} |a_{ij}|^2\right)^{1/2}$. This expression is not generally equal to the norm of $\mathbf{L}$, for instance if $\mathbf{L}=\mathbf{I}=\begin{bmatrix}1 & 0\\ 0 & 1\end{bmatrix}$ the identity operator on $\mathbb{R}^2$,

obviously $\|\mathbf{Lx}\| = \|\mathbf{x}\|$ so that $\|\mathbf{L}\|=1$. However, $\left(\sum_{i,j=1}^{2} |a_{ij}|^2\right)^{1/2} = \sqrt{2}.$

A lower bound for the $\|\mathbf{L}\|$ can also be given in terms of the elements of the matrix representing it. Indeed $\mathbf{Le}_j=(a_{1j}, a_{2j}, \cdots, a_{mj})$ and therefore $\|\mathbf{Le}_j\| \ge |a_{ij}|$ for every i. Since $\|\mathbf{e}_j\|=1$, this shows that $\|\mathbf{L}\| \ge \max_{i,j} |a_{ij}|$.

When $\mathbf{L} : \mathbb{R}^n \longrightarrow \mathbb{R}^m$ is a linear operator the set $N(\mathbf{L}) := \{\mathbf{x}\in\mathbb{R}^n : \mathbf{Lx}=\mathbf{0}\}$ is a subspace of $\mathbb{R}^n$ and is called the *nullspace* (or *kernel*) of $\mathbf{L}$. The dimension of $N(\mathbf{L})$, $\dim(N(\mathbf{L}))$, is called the *nullity* of $\mathbf{L}$. The *image set* (or *range*) is $\text{im}(\mathbf{L}) := \{\mathbf{Lx} : \mathbf{x}\in\mathbb{R}^n\} = \mathbf{L}(\mathbb{R}^n)$. The image set is a subspace of $\mathbb{R}^m$ and its dimension is called the *rank* of $\mathbf{L}$. One of the more important facts about linear operators in finite dimensional spaces is the so-called *rank-nullity theorem* :

Theorem 2.87. *If* $\mathbf{L} : \mathbb{R}^n \longrightarrow \mathbb{R}^m$ *is a linear operator then*

$$\dim(N(\mathbf{L})) + \dim(\text{im}(\mathbf{L})) = n. \tag{2.12}$$

Proof. Let $\{\mathbf{u}_1,\mathbf{u}_2,\cdots,\mathbf{u}_p\}$ be a basis for $N(\mathbf{L})$ and extend this to a basis $\{\mathbf{u}_1,\mathbf{u}_2,\cdots,\mathbf{u}_n\}$ for $\mathbb{R}^n$. Then any $\mathbf{x}\in\mathbb{R}^n$ can be written in exactly one way as $\mathbf{x}=\alpha_1\mathbf{u}_1+\alpha_2\mathbf{u}_2+\cdots+\alpha_n\mathbf{u}_n$ and then

$$\mathbf{L}(\mathbf{x})=\alpha_{p+1}\mathbf{L}(\mathbf{u}_{p+1})+\cdots+\alpha_n\mathbf{L}(\mathbf{u}_n).$$

The vectors $\mathbf{Lu}_{p+1}, \mathbf{Lu}_{p+2},\cdots, \mathbf{Lu}_n$ are linearly independent since if $\lambda_{p+1}\mathbf{Lu}_{p+1}+\cdots+\lambda_n\mathbf{Lu}_n=\mathbf{0}$ we would have $\lambda_{p+1}\mathbf{u}_{p+1}+\cdots+\lambda_n\mathbf{u}_n\in N(\mathbf{L})$ which is impossible since $\{\mathbf{u}_1,\mathbf{u}_2,\cdots,\mathbf{u}_p\}$ is a basis for $N(\mathbf{L})$ and $\mathbf{u}_{p+1},\cdots,\mathbf{u}_n$ are linearly independent of these. This shows that $\{\mathbf{Lu}_{p+1}, \mathbf{Lu}_{p+2},\cdots, \mathbf{Lu}_n\}$ is a basis of im(**L**) and that dim(im(**L**))$=n-p$.

Corollary 2.88. *If $p\leq n$ and $\mathbf{L}:\mathbb{R}^n\longrightarrow\mathbb{R}^p$ is one-to-one (injective) then $p=n$ and* **L** *is invertible.*

Proof. Since **L** is one-to-one, we have $N(\mathbf{L})=\{\mathbf{0}\}$ so dim(N(**L**))$=0$ and therefore dim(im(**L**))$=n$. This shows that **L** is also onto (surjective).

There is the concept of rank of a matrix which is usually defined as follows . For an $m\times n$ matrix consider the columns as vectors in $\mathbb{R}^m$. The column rank is then the maximum number of these linearly independent columns. It is shown in courses on linear algebra that this is the same as the maximum number of linearly independent rows, the common value being the rank of the matrix. Moreover, the rank of a linear operator coincides with the rank of the matrix that represents it (irrespective of basis). A consequence of the rank-nullity theorem is that an $n\times n$ matrix is invertible if and only if its rank is n.

2.8. DETERMINANTS

We assume that the reader has some familiarity with matrices but we briefly mention the notion of determinant of a square matrix and the basic properties that we need. We only consider matrices with real entries.

For a 2×2 matrix $A=\begin{bmatrix} a & b \\ c & d\end{bmatrix}$, the *determinant* of A is the real number $\det A=ad-bc$. The notation $|A|$ is also employed for $\det A$. For a 3×3 matrix $A=\begin{bmatrix} a_{11} & a_{12} & a_{13} \\ a_{21} & a_{22} & a_{23} \\ a_{31} & a_{32} & a_{33}\end{bmatrix}$, $\det A$ may be defined via the process of "expansion by the first row" as $\det A = a_{11}|A_{11}| - a_{12}|A_{12}| + a_{13}|A_{13}|$ where A_{ij} denotes the 2×2 matrix obtained from A by deleting the i-th row and j-th column and $|A_{ij}| \equiv \det A_{ij}$. For larger matrices the same process can be employed to recursively define the determinant. Thus, if $A = [a_{ij}]$, $1 \le i \le n$, $1 \le j \le n$, denotes an $n\times n$ matrix,

$$\det A = a_{11}|A_{11}| - a_{12}|A_{12}| + a_{13}|A_{13}| - \cdots + (-1)^{n+1}|A_{1n}|.$$

This process gives an extremely inefficient method of calculation for large matrices but is perfectly satisfactory for 2×2 and 3×3 matrices. The following properties (which we shall not prove) are particularly useful.

(D1) If B is obtained from A by interchanging 2 rows then $\det B = -\det A$.

(D2) If A has rows $\{\mathbf{r}_1, \mathbf{r}_2, \cdots, \mathbf{r}_n\}$ and B has rows $\{\mathbf{r}_1 - \alpha\mathbf{r}_k, \mathbf{r}_2, \cdots, \mathbf{r}_n\}$ for some $\alpha \in \mathbb{R}$, then $\det B = \det A$.

The same holds for any row in place of $\mathbf{r}_1$: just use (D1) twice.

It can be shown that $\det A^{\mathrm{T}} = \det A$, where A^{T} denotes the transpose of A obtained by interchanging rows and columns, so properties (D1) and (D2) also hold if "row" is replaced by "column".

One other useful result is that, for $n\times n$ matrices A, B $\det(AB) = (\det A)(\det B)$.

Of particular interest is that a matrix has an inverse if and only if $\det A \neq 0$, and in this case, if A^{-1} denotes the inverse, $(\det A)(\det A^{-1}) = \det I = 1$ so $\det A^{-1} = 1/\det A$. In fact A^{-1} can be written $A^{-1} = (1/\det A)\,\mathrm{adj}(A)$ where $\mathrm{adj}(A)$, the *adjutant* (sometimes adjoint) of A, is a matrix whose entries are determinants of $(n-1)\times(n-1)$ submatrices of A. This has some theoretical interest but is not an efficient method of calculation.

Problems.

1. Show that, if $\mathbf{f},\mathbf{g} : A \subseteq \mathbb{R}^n \longrightarrow \mathbb{R}^m$ are continuous at $\mathbf{a} \in A$ and $\alpha \in \mathbb{R}$ then $\mathbf{f}+\mathbf{g}$ and $\alpha\mathbf{f}$, are continuous at $\mathbf{a}$.

2. Let $f : \mathbb{R}^3 \longrightarrow \mathbb{R}$ be given by $f(x,y,z) = (x^2+y^2+z^2) + \sin(xyz)$. Explain why f is continuous at all points.

3. Let $f:\mathbb{R}^2 \longrightarrow \mathbb{R}$ be defined by $f(x,y)=\begin{cases} \dfrac{xy^2}{x^2+y^2}, & \text{for } (x,y)\neq(0,0),\\ 0, & \text{when } (x,y)=(0,0).\end{cases}$
Show that f is continuous at all points, (treat $(0,0)$ separately).

4. Let $f:\mathbb{R}^2 \longrightarrow \mathbb{R}$ be defined by $f(x,y)=\begin{cases} \dfrac{xy^2}{x^2+y^4}, & \text{for } (x,y)\neq(0,0),\\ 0, & \text{when } (x,y)=(0,0).\end{cases}$
Show that $\lim f(x,y)=0$ as $(x,y)\longrightarrow 0$ along *any straight line* but that the limit of $f(x,y)$ as $(x,y)\longrightarrow 0$ does not exist.

5. Let $\mathbf{f}: A\subseteq\mathbb{R}^n \longrightarrow \mathbb{R}^m$ be continuous at $\mathbf{a}\in A$ and suppose that $\mathbf{f}(\mathbf{a})\neq\mathbf{0}$. Prove that there is a neighbourhood U of $\mathbf{a}$ with $\mathbf{f}(\mathbf{x})\neq\mathbf{0}$ for all $\mathbf{x}\in A\cap U$. Deduce that, if $\mathbf{f}$ and $\mathbf{g}$ are continuous at $\mathbf{a}$ and $\mathbf{f}(\mathbf{a})\neq\mathbf{g}(\mathbf{a})$, there is a neighbourhood V of $\mathbf{a}$ with $\mathbf{f}(\mathbf{x})\neq\mathbf{g}(\mathbf{x})$ for $\mathbf{x}\in A\cap V$.

6. Prove (*a*)–(*c*) of Proposition 2.17.

7. Prove that, if K is a compact subset of $\mathbb{R}^n$ and $f: K\longrightarrow\mathbb{R}$ is continuous on K, then f attains its minimum over K.

8. (*a*) Let A be a nonempty subset of $\mathbb{R}^n$ and let $\phi:\mathbb{R}^n\longrightarrow\mathbb{R}$ be defined by $\phi(\mathbf{x}):=\text{dist}(\mathbf{x},A)=\inf\{\|\mathbf{x}-\mathbf{a}\| : \mathbf{a}\in A\}$. Show that $|\phi(\mathbf{x})-\phi(\mathbf{y})|\leq\|\mathbf{x}-\mathbf{y}\|$.
(*b*) Now let A and B be two nonempty, closed, disjoint subsets of $\mathbb{R}^n$ and let $f(\mathbf{x})=\text{dist}(\mathbf{x},B)/\{\text{dist}(\mathbf{x},A)+\text{dist}(\mathbf{x},B)\}$, $\mathbf{x}\in\mathbb{R}^n$. Show that f is continuous, that $f(\mathbf{x})=1$ for all $\mathbf{x}\in A$, and that $f(\mathbf{x})=0$ for all $\mathbf{x}\in B$.

9. Let K be a compact subset of $\mathbb{R}^n$ and $\mathbf{f}: K\longrightarrow\mathbb{R}^m$ be continuous and *open*, that is, the image of each open set is an open set. Prove that
(*a*) if $\|\mathbf{f}\|$ attains its minimum at $\mathbf{a}\in\text{int}(K)$ then $\|\mathbf{f}(\mathbf{a})\|=0$;
(*b*) if $\|\mathbf{f}\|$ attains its maximum at $\mathbf{b}\in K$ then $\mathbf{b}\notin\text{int}(K)$.

10. Let $f:(0,1)\longrightarrow\mathbb{R}$ be defined by $f(x)=1/x$. For $a>0$ and $\varepsilon>0$ find δ, in terms of a and ε, so that $x>0$ and $|x-a|<\delta \Rightarrow |f(x)-f(a)|<\varepsilon$.

11. Let $\mathbf{f}: A\subseteq\mathbb{R}^n\longrightarrow\mathbb{R}^m$ be continuous on the closed set A and let $Z=\{\mathbf{x}\in A : \mathbf{f}(\mathbf{x})=\mathbf{0}\}$. Prove that Z is closed.

12. Let $\mathbf{f}: A\subseteq\mathbb{R}^n\longrightarrow\mathbb{R}^m$ be continuous at $\mathbf{a}=(a_1, a_2,\cdots,a_n)\in A$. Let $A_1\subseteq\mathbb{R}$ be the set of first components of A, that is, $A_1=\{x_1 : \exists\, x_2, x_3,\cdots,x_n$ such that $(x_1,x_2,\cdots,x_n)\in A\}$. Prove that $\mathbf{g}: A_1\subseteq\mathbb{R}\longrightarrow\mathbb{R}^m$ defined by $\mathbf{g}(x)=\mathbf{f}(x, a_2,\cdots,a_n)$ is continuous at a_1. (A continuous function is continuous in each variable separately).

Give an example to show that the converse is false.

13. Give an example of a set $A \subseteq \mathbb{R}^2$ and a continuous one-to-one function $\mathbf{f} : A \longrightarrow \mathbb{R}^2$ such that $\mathbf{f}^{-1}$ is not continuous on $\mathbf{f}(A)$.

14. Let S be a bounded (but not closed) subset of $\mathbb{R}^n$ and let $f : S \rightarrow \mathbb{R}$ be uniformly continuous on S. Prove that f is bounded on S, that is, there exists $M \in \mathbb{R}$ such that $|f(\mathbf{x})| \leq M$ for all $\mathbf{x} \in S$.

15. Let $f : \mathbb{R} \longrightarrow \mathbb{R}$ be a function. For each $x \in \mathbb{R}$ the *oscillation of f at x* is defined to be the number $\omega_f(x) := \inf\{\operatorname{diam}(f(x-\delta, x+\delta)) : \delta > 0\}$.

(*i*) Prove that a function $f : \mathbb{R} \longrightarrow \mathbb{R}$ is continuous at c if, and only if, $\omega_f(c) = 0$.

(*ii*) Show that for each $\alpha \in \mathbb{R}$, the set $\{x : \omega_f(x) < \alpha\}$ is open.

(*iii*) There is no function $f : \mathbb{R} \longrightarrow \mathbb{R}$ such that

$$\{x : f \text{ is continuous at } x\} = \mathbb{Q}.$$

16. Let K be a compact subset of $\mathbb{R}^n$ and let $f^{(m)} : K \longrightarrow \mathbb{R}$ be continuous on K. Suppose also that $f^{(m+1)}(\mathbf{x}) \leq f^{(m)}(\mathbf{x})$ for all $\mathbf{x}$ and that $\{f^{(m)}(\mathbf{x})\}$ converges pointwise to 0. Prove that $f^{(m)} \longrightarrow 0$ uniformly on K. Deduce that if $g^{(m+1)}(\mathbf{x}) \leq g^{(m)}(\mathbf{x})$ and $g^{(m)}(\mathbf{x}) \longrightarrow g(\mathbf{x})$ (pointwise) for each $\mathbf{x}$ where $g^{(m)}$ and g are continuous on K, then $g^{(m)} \longrightarrow g$ uniformly on K.

17. Let $f^{(m)}(x) = 1/(1+mx)$, for $0 < x < 1$. Show that $f^{(m)}(x) \longrightarrow 0$ for each x, that $f^{(m+1)}(x) \leq f^{(m)}(x)$ for all x but that the convergence is not uniform on $(0,1)$. Compare problem 16.

18. Show that $f^{(m)}(x) = x/(1+mx)$ converges uniformly on $[0,1]$.

19. Let $A = [1,2] \times [1,2]$ in $\mathbb{R}^2$ and let $f^{(m)} : A \longrightarrow \mathbb{R}$ be defined by $f^{(m)}(x,y) = \dfrac{1+mx}{1+my}$. Show that $f^{(m)}$ converges uniformly on A.

20. Suppose that $\mathbf{f}^{(m)} \longrightarrow \mathbf{f}$ uniformly on a set A and that $\{\mathbf{x}^{(m)}\}$ is a sequence in A. If $\mathbf{x}^{(m)} \longrightarrow \mathbf{x}$, does it follow that $\mathbf{f}^{(m)}(\mathbf{x}^{(m)}) \longrightarrow \mathbf{f}(\mathbf{x})$?

21. Let $f^{(m)}(x) = mx/(1+m^2x^2)$. Show that $f^{(m)}(x) \longrightarrow 0$ for each $x \in [0,1]$. Does $f^{(m)} \longrightarrow 0$ uniformly on $[0,1]$?

22. Suppose that $\sum_{m=1}^{\infty} \mathbf{f}^{(m)}$ is uniformly convergent on A and that $\{g^{(m)}\}$

is a uniformly bounded sequence on A such that $g^{(m+1)}(x) \leq g^{(m)}(x)$ for all $x \in A$, and $m \in \mathbb{N}$. Show that $\sum g^{(k)}(x) f^{(k)}(x)$ is uniformly convergent on A. (Abel's test).

23. If $\sum_{n=0}^{\infty} a_n x^n$ is absolutely convergent for $|x| < R$, prove that the series is uniformly convergent on $[-R', R']$, for any $R' < R$.

24. Suppose $\sum_{n=0}^{\infty} a_n x^n$ converges to $f(x)$ for $|x| < 1$. Show that if $\sum_{n=0}^{\infty} a_n$ converges to α then $\sum_{n=0}^{\infty} a_n x^n$ is uniformly convergent on $[0,1]$ and that $\lim_{x \to 1-} f(x) = \alpha$. (Abel's theorem). Deduce that if $\sum_{n=0}^{\infty} \frac{a_n}{n+1}$ is convergent (but $\sum_{n=0}^{\infty} a_n$ may be divergent) then $\int_0^1 f(x)\,dx = \sum_{n=0}^{\infty} \frac{a_n}{n+1}$.

25. Prove that, for $\alpha > 0$, $\sum_{n=1}^{\infty} \frac{\cos nx}{n^{\alpha}}$ is uniformly convergent on any closed subinterval of $(0, 2\pi)$. Deduce that $\sum_{n=1}^{\infty} \frac{\sin^2 nx}{n}$ is divergent for $x \in (0, \pi)$.

26. Show that $\sum_{n=1}^{\infty} \frac{(-1)^{n+1}}{n+x}$ is uniformly convergent on $[0, \infty)$ but is not absolutely convergent for any $x \geq 0$.

27. Show that $\sum_{n=1}^{\infty} \frac{x^n(1-x)}{\ln(n+1)}$ is uniformly convergent on $[0,1]$. Let $M_n := \sup\{ \frac{x^n(1-x)}{\ln(n+1)} : 0 \leq x \leq 1 \}$. Show that $\sum_{n=1}^{\infty} M_n$ is divergent.

28. Let f be a function from $A \subseteq \mathbb{R}^n$ into $B \subseteq \mathbb{R}^m$ and let $S \subseteq A$, $T, W \subseteq B$.
(*a*) Prove that $f^{-1}(T \cup W) = f^{-1}(T) \cup f^{-1}(W)$. Is there a corresponding result for intersections?
(*b*) Prove that $f(f^{-1}(T) = T \cap f(A)$. If f is also injective (one-to-one) show that, $f^{-1}(f(S)) = S$.

29. Let $f : A \longrightarrow B$ and let $S \subseteq A$, $T \in B$. Prove that $f(A \setminus S) \supseteq f(A) \setminus f(S)$. (Here $A \setminus S := \{a \in A : a \notin S\} = A \cap S^c$ is the complement of S relative to A.)

Give an example to show that strict inclusion is possible here. Show also that $\mathbf{f}^{-1}(B\setminus T)=A\setminus \mathbf{f}^{-1}(T)$.

30. Let A,E be subsets of $\mathbb{R}^n$. E is said to be *closed relative to A* if there is a closed set F with $E=A\cap F$. Prove that

$$E \text{ is closed relative to } A \Leftrightarrow A\setminus E \text{ is open relative to } A.$$

Show that $\mathbf{f}:A\longrightarrow B$ is continuous on A $\Leftrightarrow$ for any set T that is closed relative to $\mathbf{f}(A)$ the inverse image $\mathbf{f}^{-1}(T)$ is closed relative to A

31. Let C be a convex set with nonempty interior. Show that if $\mathbf{x}\in\text{int}(C)$ and $\mathbf{y}\in C$ then $(1-t)\mathbf{x}+t\mathbf{y}\in\text{int}(C)$ for $0\le t<1$. Deduce that $\text{int}(C)$ is convex.

32. Let $A\subseteq\mathbb{R}^n$ and define $\text{co}(A)$, the *convex hull* of A, to be the set of all convex combinations of elements of A, that is,

$$\text{co}(A):=\{\textstyle\sum\lambda^{(i)}\mathbf{a}^{(i)} : \mathbf{a}^{(i)}\in A,\ \lambda^{(i)}\ge 0,\ \sum\lambda^{(i)}=1\}$$

where the sums contain only finitely many terms. Show that $\text{co}(A)$ is convex and is the smallest convex set containing A.

33. Let $A\subseteq\mathbb{R}^n$ be disconnected with $A=U\cup V$, a disjoint union of nonempty, relatively open sets. Prove that U and V are closed relative to A. Deduce that a set $A\subseteq\mathbb{R}^n$ is connected if, and only if, the only subsets of A that are both open and closed relative to A are A and $\varnothing$.

34. Let $f:\overline{\mathbf{B}}(\mathbf{0},1)\subseteq\mathbb{R}^2\longrightarrow\mathbb{R}$ be continuous. Show that f cannot be one-to-one.

35. Let $A\subseteq\mathbb{R}^2$ be the union of the open disk centred at (0,0) of radius 1 and the closed disk centred at (2,0) of radius 1. Is A connected?

36. Let A,B be subsets of $\mathbb{R}^n$ and let $\rho(A,B):=\inf(\|\mathbf{x}-\mathbf{y}\| : \mathbf{x}\in A, \mathbf{y}\in B\}$. Show that, *(i)* if A is compact then there exists $\mathbf{a}\in A$ such that $\text{dist}(\mathbf{a},B)=\rho(A,B)$, *(ii)* if A and B are both compact there exist $\mathbf{a}\in A$ and $\mathbf{b}\in B$ such that $\|\mathbf{a}-\mathbf{b}\|=\rho(A,B)$, *(iii)* if A,B are compact and $\rho(A,B)>0$ then $A\cup B$ is disconnected.

37. Let A be a bounded set with $\text{int}(A)\ne\varnothing$. Let C be a connected set such that $C\cap\text{bdy}(A)=\varnothing$. Show that either $C\subseteq\text{int}(A)$ or else $C\subseteq\text{int}(A^c)$.

38. Prove that $L\mathbf{x}=\langle\mathbf{a},\mathbf{x}\rangle$, $\mathbf{a},\mathbf{x}\in\mathbb{R}^n$, defines a linear functional on $\mathbb{R}^n$.

39. *(a)* Let $\mathbf{f}(x,y,z)=(x+y+z,\ x,\ x-y)$. Is $\mathbf{f}$ linear?

(b) Let $\mathbf{f}(x,y,z)=(e^x-1,\ x,\ x-y)$. Is $\mathbf{f}$ linear?

40. Show that $\mathbf{L}:\mathbb{R}^n\longrightarrow\mathbb{R}^m$ is linear if and only if each component L_i

is linear, $i=1,2,\cdots,m$.

41. (*a*) Show that $\|\mathbf{L}\|=\sup\{\|\mathbf{Lx}\| : \|\mathbf{x}\|\leq 1\}$ and that $\|\mathbf{Lx}\|\leq\|\mathbf{L}\|\,\|\mathbf{x}\|$ for all $\mathbf{x}$.
(*b*) Let $\mathbf{L},\mathbf{M}$ be linear operators. Show that $\|\mathbf{L}+\mathbf{M}\|\leq\|\mathbf{L}\|+\|\mathbf{M}\|$ and $\|\mathbf{LM}\|\leq\|\mathbf{L}\|\,\|\mathbf{M}\|$.

42. For a linear operator $\mathbf{L}:\mathbb{R}^n\longrightarrow\mathbb{R}^m$ show that the following statements are equivalent.
(1) $\mathbf{L}$ is one-to-one.
(2) The nullspace of $\mathbf{L}$ is trivial, that is, $N(\mathbf{L})=\{\mathbf{0}\}$.
(3) There is $k>0$ such that $\|\mathbf{Lx}\|\geq k\|\mathbf{x}\|$ for all $\mathbf{x}\in\mathbb{R}^n$.

43. Let $\mathbf{L}:\mathbb{R}^n\longrightarrow\mathbb{R}^n$ be linear and suppose there are linear operators $\mathbf{P},\mathbf{Q}$ such that $\mathbf{LQ}=\mathbf{PL}=\mathbf{I}$ (the identity operator). Show that $\mathbf{Q}=\mathbf{P}$. Show also that $\mathbf{L}$ is invertible and $\mathbf{P}=\mathbf{Q}=\mathbf{L}^{-1}$.

44. Let $\mathbf{L},\mathbf{M}:\mathbb{R}^n\longrightarrow\mathbb{R}^n$ be invertible linear operators. Show that $\mathbf{LM}$ is also invertible with $(\mathbf{LM})^{-1}=\mathbf{M}^{-1}\mathbf{L}^{-1}$.

45. Let $\mathbf{L}:\mathbb{R}^n\longrightarrow\mathbb{R}^m$ be a linear operator and let C be a convex set in $\mathbb{R}^n$. Show that $\mathbf{L}(C)$ is a convex set.

Hints on the problems

3. Estimate $|f(x,y)-f(0,0)|$ by writing $x=r\cos\theta$, $y=r\sin\theta$ (polars).

4. $r^3|\cos\theta\sin^2\theta|/(r^2\cos^2\theta+r^4\sin^4\theta)\leq r^3|\cos\theta\sin^2\theta|/r^2\cos^2\theta:=C(\theta)r$ where $C(\theta)$ is a finite constant if $\cos(\theta)\neq 0$. Check that one can set $C(\theta)=0$ in the latter case. Limits along lines have θ constant.

5. Compare the proof of Prop.2.7.

7. Consider $-f$.

8. (*a*) $\|\mathbf{x}-\mathbf{a}\|\leq\|\mathbf{x}-\mathbf{y}\|+\|\mathbf{y}-\mathbf{a}\|$. (*b*) Show denominator is positive (recall problem 1.39). Then f is a quotient of continuous functions by (*a*).

9. Let $M=\|\mathbf{f}(\mathbf{a})\|$. $\mathbf{f}(\mathbf{B}(\mathbf{a},\delta))$ is open so contains an ε-nbd of $\mathbf{f}(\mathbf{a})$, in particular, $\mathbf{f}(\mathbf{a})\pm\varepsilon\mathbf{f}(\mathbf{a})/2M=\mathbf{f}(\mathbf{x})$ for some $\mathbf{x}\in\mathbf{B}(\mathbf{a},\delta)$, if $M\neq 0$.

11. Show that Z contains all its limit points.

12. The ε-δ condition for $\mathbf{f}$ says something about $\mathbf{g}$.

13. An example from $\mathbb{R}$ to $\mathbb{R}^2$ would be the mapping that takes $[0,2\pi)$ into the unit circle. Work on the same idea. For example, consider a domain such as $A=\{(x,y): x=0,\ 0\leq y\leq 1\}\cup\{0\leq x\leq 1,\ 0\leq y<x\}$, a line plus a nonclosed triangle and make $\mathbf{f}$ pull the bits together.

14. As f is uniformly continuous on S it takes Cauchy sequences into

Cauchy squences. Set up a contradiction argument.

15. The set $\{x : f$ is continuous at $x\}$ is a countable intersection of open sets. By problem 1.31, $\mathbb{Q}$ is not of this form.

16. For any $\varepsilon > 0$, there is $N(\varepsilon)$ so that $0 \leq f^{(N)}(\mathbf{x}) < \varepsilon$. Then $\exists\ \delta(x)$ such that $0 \leq f^{(N)}(\mathbf{y}) < 2\varepsilon$ for all $\mathbf{y} \in \mathrm{B}(\mathbf{x}, \delta(\mathbf{x}))$.

20. No, in general, give an example. Prove it is YES if also **f** is *continuous.*

22. Suppose $|g^{(k)}(\mathbf{x})| \leq M$ for all x, all k. Given $\varepsilon > 0$ there exists $N(\varepsilon)$ such that $\sup \|f^{(k+1)}(\mathbf{x}) + \cdots + f^{(k+p)}(\mathbf{x})\| < \varepsilon$ for all $k \geq N(\varepsilon)$, $p > 0$.
Apply Abel's partial summation formulae.

23. M–test.

24. Apply Abel's test taking $g^{(m)}(x) = x^m$, $f^{(m)}(x) = a_m$.

25. $\sin^2\theta = (1 - \cos 2\theta)/2$.

27. Example 2.34. and $[n/(n+1)]^n$ decreases to $1/e$.

29. Example : **f** constant.

31. For $\mathbf{x} \in \mathrm{int}(C)$ there is $\delta > 0$ so that $\mathbf{x}+\mathbf{z} \in C$ for $\mathbf{z} \in \mathbb{R}^n$ with $\|\mathbf{z}\| < \delta$. Then $(1-t)(\mathbf{x}+\mathbf{z}) + t\mathbf{y} \in C$, by convexity.

32. Prove $\mathrm{co}(A)$ is convex and if C is a convex set containing A then C also contains $\mathrm{co}(A)$.

33. $U = A \setminus V$. If U is open and closed relative to A write $A = U \cup (A \setminus U)$.

34. Let $f(\mathbf{a})$ be the maximum and $f(\mathbf{b})$ be the minimum of f on $\overline{\mathrm{B}}(\mathbf{0},1)$ and consider f on the line segment between **a** and **b**.

36. If $\|\mathbf{a}-\mathbf{b}\| = d > 0$ consider the open sets $U = \{\mathbf{x} \in \mathbb{R}^n : \mathrm{dist}(\mathbf{x}, A) < d/3\}$ and $V = \{\mathbf{y} \in \mathbb{R}^n : \mathrm{dist}(\mathbf{y}, A) < d/3\}$.

37. Note that $\mathbb{R}^n = \mathrm{int}(A) \cup \mathrm{bdy}(A) \cup \mathrm{int}(A^c)$.

42. $\|\mathbf{Lx}\|$ attains its minimum k on $\{\mathbf{x} : \|\mathbf{x}\| = 1\}$.

43. **L** is one-to-one since $\mathbf{Lx} = \mathbf{Ly} \Rightarrow \mathbf{PLx} = \mathbf{PLy}$.

44. Use 43.

45. By problem 32, C is convex $\Leftrightarrow$ $C = \mathrm{co}(C)$.

Chapter 3.

Differentiable functions

3.1. THE DERIVATIVE

For a function $f : \mathbb{R} \longrightarrow \mathbb{R}$, the derivative of f at a point a is defined to be $f'(a) = \lim_{h\to 0} \dfrac{f(a+h)-f(a)}{h}$, whenever this limit exists. This definition cannot be used for functions defined on $\mathbb{R}^n$ with $n>1$ since division by elements of $\mathbb{R}^n$ is impossible. However, $f'(a)$ gives information on the local behaviour of f near the point a, for example, it is a measure of the rate of increase of f. The derivative is also the gradient, or slope, of the curve $y=f(x)$ (the graph of f in the x-y plane) at the point $(a, f(a))$. This gradient defines the tangent to the curve as that line with equation $y=f(a)+f'(a)(x-a)$.

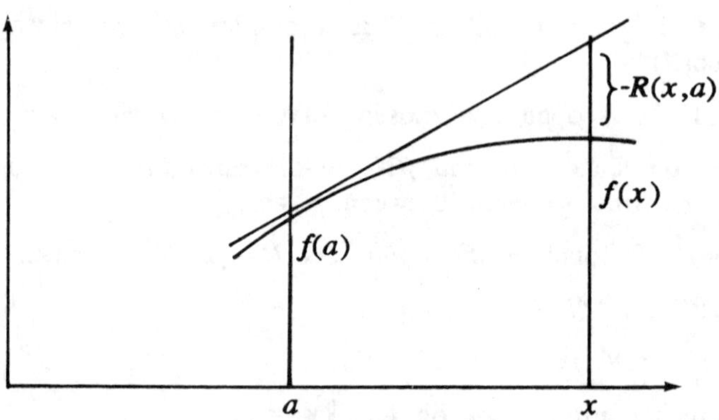

Figure 3.1

When f is differentiable at a, the error made by approximating a point on the graph by a point on the tangent line with the same x-coordinate is, for $x \neq a$,

$$R(x,a) = f(x) - [f(a)+f'(a)(x-a)] = (x-a)\left\{\frac{f(x)-f(a)}{x-a} - f'(a)\right\} .$$

By definition of $f'(a)$, the term inside { } has limit 0 as $x \longrightarrow a$. Thus, $\dfrac{|R(x,a)|}{|x-a|} \longrightarrow 0$ as $|x-a| \longrightarrow 0$. Roughly speaking this says that

$R(x,a)$ approaches 0 "faster than" $|x-a|$. In particular, $f(a)+f'(a)(x-a)$ is a good approximation to $f(x)$ for $|x-a|$ small. The local behaviour of f is captured by this approximation.

It is this idea that generalizes to $\mathbb{R}^n$. Of course, we cannot use tangent line in our description, but we can use the terminology of linear operators. Observe that $Lx := f'(a)x$ defines a linear operator and the good approximation to $f(x)$ is $f(a)+L(x-a)$.

Definition 3.1. Let G be an open subset of $\mathbb{R}^n$. We say that a function $\mathbf{f} : G \longrightarrow \mathbb{R}^m$ is *differentiable* at $\mathbf{a} \in G$ if there is a linear operator $\mathbf{L} : \mathbb{R}^n \longrightarrow \mathbb{R}^m$ such that $\mathbf{f}(\mathbf{x}) = \mathbf{f}(\mathbf{a}) + \mathbf{L}(\mathbf{x}-\mathbf{a}) + \mathbf{R}(\mathbf{x},\mathbf{a})$, $\mathbf{x} \in G$, where $\dfrac{\|\mathbf{R}(\mathbf{x},\mathbf{a})\|}{\|\mathbf{x}-\mathbf{a}\|} \longrightarrow 0$ as $\|\mathbf{x}-\mathbf{a}\| \longrightarrow 0$. We say $\mathbf{f}$ is *differentiable* on G if $\mathbf{f}$ is differentiable at each point of G.

$\mathbf{L}$ depends on $\mathbf{a}$ as well as $\mathbf{f}$. We are not saying that there is an $\mathbf{L}$ which works for all $\mathbf{a}$, but that for a fixed $\mathbf{a}$ an $\mathbf{L}$ exists.

$\mathbf{L}$ is called the *derivative of* $\mathbf{f}$ *at* $\mathbf{a}$ and is written $D\mathbf{f}(\mathbf{a})$. It is implicit that $D\mathbf{f}(\mathbf{a})$ is uniquely determined by the definition. This may be done directly (see problem 3.1) but we will shortly show this by determining $D\mathbf{f}(\mathbf{a})$ in terms of more familiar quantities. In fact, we shall determine the matrix that represents (is identified with) $D\mathbf{f}(\mathbf{a})$. One may ask why we do not simply define $D\mathbf{f}(\mathbf{a})$ to be the relevant matrix. The answer is that the matrix depends on the bases chosen in the relevant spaces, linear operators do not. Of course, we only use the standard bases; nevertheless it is sometimes useful to change bases (a change of variable), and it is useful to know that the derivative is invariant. Generally, we compute derivatives by forming the matrix, just as in one variable we do not differentiate from first principles to calculate derivatives.

There are two simple cases where the definition can be used directly.

Example 3.2. Let $\mathbf{f} : \mathbb{R}^n \longrightarrow \mathbb{R}^m$ be defined by $\mathbf{f}(\mathbf{x}) = \mathbf{c}$, for all $\mathbf{x}$ (a constant function). Then $D\mathbf{f}(\mathbf{a}) = \mathbf{0}$, the zero operator.

Solution. We have to verify that $\mathbf{f}(\mathbf{x}) = \mathbf{f}(\mathbf{a}) + \mathbf{0} + \mathbf{R}$ where $\dfrac{\|\mathbf{R}\|}{\|\mathbf{x}-\mathbf{a}\|} \longrightarrow 0$ as $\|\mathbf{x}-\mathbf{a}\| \to 0$; this is satisfied by $\mathbf{R} = \mathbf{0}$.

Example 3.3. If $\mathbf{L} : \mathbb{R}^n \longrightarrow \mathbb{R}^m$ is linear then $D\mathbf{L}(\mathbf{a}) = \mathbf{L}$ for every $\mathbf{a}$.

Solution. We have $\mathbf{L}(\mathbf{x}) = \mathbf{L}(\mathbf{a}) + \mathbf{L}(\mathbf{x}-\mathbf{a}) + \mathbf{R}$ with $\mathbf{R} = \mathbf{0}$, by linearity. In fact, Example 3.2 is a special case of this.

Remark. This is, at first sight, paradoxical. For, if $f(x)=mx$ is a real linear function then $f'(x)=m$ for all x. However, this is purely notational confusion; $Df(a)$ is the linear function L defined by $Lx=f'(a)x=mx$, so indeed the derivative of f at a (as defined above) is the function f itself.

For a function $f:\mathbb{R}^2\longrightarrow\mathbb{R}$, the graph of f is the subset of $\mathbb{R}^2\times\mathbb{R}$ given by $\{(\mathbf{x},f(\mathbf{x})):\mathbf{x}\in\mathbb{R}^2\}$. If we write $\mathbf{x}=(x,y)$ and $z=f(x,y)$, the graph of f is the set $\{(x,y,z):z=f(x,y)\}$. This is a *surface* in $\mathbb{R}^3$: for each point (x,y) the point (x,y,z) sits "immediately above" the point $(x,y,0)$ at a "height" z.

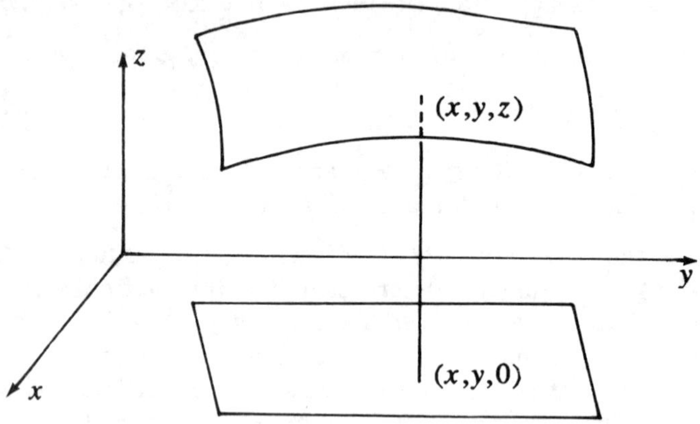

Figure 3.2

Suppose $f:\mathbb{R}^2\longrightarrow\mathbb{R}$ is differentiable at a point $\mathbf{a}=(a,b)$. Then $Df(\mathbf{a})$ is a linear operator from $\mathbb{R}^2$ to $\mathbb{R}$ so is (represented by) a 1×2 matrix, say $Df(\mathbf{a})=(d_1,d_2)$. Then $f(\mathbf{x})=f(\mathbf{a})+(d_1,d_2)\begin{bmatrix}x-a\\y-b\end{bmatrix}+R(\mathbf{x},\mathbf{a})$, so $f(\mathbf{x})$ is well approximated by $f(\mathbf{a})+d_1(x-a)+d_2(y-b)$. In other words, points on the graph of f are well approximated by the points on the *plane* $z=f(\mathbf{a})+d_1(x-a)+d_2(y-b)$. This plane is called the *tangent plane* at $(\mathbf{a},f(\mathbf{a}))$. This is a complete generalization of the familiar notion of tangent to the curve $y=f(x)$ intimated earlier. In general, no such geometric picture is available, but the picture for maps from $\mathbb{R}^2$ to $\mathbb{R}$ is a good guide to intuition.

Differentiability of $\mathbf{f}$ at $\mathbf{a}$ says that $\mathbf{f}(\mathbf{x})$ is well approximated by $\mathbf{g}(\mathbf{x})=\mathbf{f}(\mathbf{a})+\mathbf{L}(\mathbf{x}-\mathbf{a})$ near $\mathbf{a}$; here $\mathbf{g}(\mathbf{x})$ is of the form $\mathbf{Lx}+\mathbf{c}$ (linear operator+constant) and is termed *affine*. The derivative leads to the best affine approximation to $\mathbf{f}$ near $\mathbf{a}$. The use of "best" is justified in problem 3.12.

As we shall be defining derivatives only at points of open sets we shall assume from now on that G *always denotes an open subset of* $\mathbb{R}^n$.

One simple, but important, consequence of differentiability is continuity.

Theorem 3.4. *Let* $\mathbf{f} : G \longrightarrow \mathbb{R}^m$ *be differentiable at* $\mathbf{a} \in G$. *Then* $\mathbf{f}$ *is continuous at* $\mathbf{a}$.

Proof. Write $\mathbf{L} = D\mathbf{f}(\mathbf{a})$. Then we have $\mathbf{f}(\mathbf{x}) - \mathbf{f}(\mathbf{a}) = \mathbf{L}(\mathbf{x} - \mathbf{a}) + \mathbf{R}(\mathbf{x},\mathbf{a})$. For any $\varepsilon > 0$, let $\delta > 0$ be so small that the δ-neighbourhood of $\mathbf{a}$ lies in G and such that $\|\mathbf{R}(\mathbf{x},\mathbf{a})\| < \varepsilon\|\mathbf{x} - \mathbf{a}\|$ for $\|\mathbf{x} - \mathbf{a}\| < \delta$, and with $\delta < \varepsilon/(\varepsilon + \|\mathbf{L}\|)$. Then, for $\|\mathbf{x} - \mathbf{a}\| < \delta$, we have

$$\|\mathbf{f}(\mathbf{x}) - \mathbf{f}(\mathbf{a})\| \leq \|\mathbf{L}\|\,\|\mathbf{x} - \mathbf{a}\| + \|\mathbf{R}(\mathbf{x},\mathbf{a})\| < \|\mathbf{L}\|\delta + \varepsilon\delta < \varepsilon.$$

We observe the following fact which is contained in this proof.

Corollary 3.5. *Let* $\mathbf{f} : G \longrightarrow \mathbb{R}^m$ *be differentiable at* $\mathbf{a} \in G$. *Then, for any* $\varepsilon > 0$ *there is* $\delta > 0$ *(which may depend on* $\mathbf{a}$*) such that*

$$\|\mathbf{f}(\mathbf{x}) - \mathbf{f}(\mathbf{a})\| \leq (\|D\mathbf{f}(\mathbf{a})\| + \varepsilon)\|\mathbf{x} - \mathbf{a}\| \text{ for every } \mathbf{x} \text{ with } \|\mathbf{x} - \mathbf{a}\| < \delta. \tag{3.1}$$

Now let $\mathbf{e} \in \mathbb{R}^n$ have $\|\mathbf{e}\| = 1$. We consider $\mathbf{e}$ as a *direction*: it determines a *straight line* through $\mathbf{a}$ consisting of all points of the form $\mathbf{a} + t\mathbf{e}$, $t \in \mathbb{R}$. If $\mathbf{f} : G \longrightarrow \mathbb{R}^m$ is differentiable at $\mathbf{a}$, we have $\mathbf{f}(\mathbf{a} + t\mathbf{e}) = \mathbf{f}(\mathbf{a}) + t\mathbf{L}(\mathbf{e}) + \mathbf{R}(t)$ where $\|\mathbf{R}(t)\|/t \longrightarrow 0$ as $t \longrightarrow 0$. In particular, for each component function

$$\lim_{t \longrightarrow 0} \frac{f_i(\mathbf{a} + t\mathbf{e}) - f_i(\mathbf{a})}{t} = L_i(\mathbf{e}). \tag{3.2}$$

Definition 3.6. Let $g : G \subseteq \mathbb{R}^n \longrightarrow \mathbb{R}$. *The derivative of* g *in the direction* $\mathbf{e}$ (or *directional derivative*) at a point $\mathbf{a} \in G$ is $\displaystyle\lim_{t \to 0} \frac{g(\mathbf{a} + t\mathbf{e}) - g(\mathbf{a})}{t}$ whenever this limit exists. It is written either $D_{\mathbf{e}}g(\mathbf{a})$ or $\dfrac{dg}{d\mathbf{e}}(\mathbf{a})$.

(3.2) above shows that if $\mathbf{f}$ is differentiable at $\mathbf{a}$, then each component f_i has directional derivatives in every direction at $\mathbf{a}$. We shall see shortly that the converse of this remark is *false*.

Certain directions are special, namely those of the standard basis elements $\mathbf{e}_i$, the directions of the co-ordinate axes. The directional derivative of a function $g : G \longrightarrow \mathbb{R}$ in the direction $\mathbf{e}_i$ is called the *partial derivative of* g *with respect to* x_i and is often denoted $\frac{\partial g}{\partial x_i}$ or, more precisely, $D_i\, g(\mathbf{a})$. Thus,

$$\frac{\partial g}{\partial x_i} = \lim_{t \to 0} \frac{g(a_1, \cdots, a_{i-1}, a_i+t, a_{i+1}, \cdots, a_n) - g(a_1, \cdots, a_n)}{t}, \tag{3.3}$$

which is just the ordinary derivative of the function considered as a function of x_i only, with all the other components fixed.

In calculating the i-th partial derivative at a point $\mathbf{a} \in \mathbb{R}^n$, we are really considering the function h of one variable defined by $h(x)=g(a_1, \cdots, a_{i-1}, x, a_{i+1}, \cdots, a_n)$ and forming $h'(a_i)$. In practice we calculate partial derivations at a point $\mathbf{x}$ rather than a point $\mathbf{a}$. In forming $\frac{\partial g}{\partial x_i}$ we regard all the other variables x_j, $j \neq i$, as constants. The notation $\frac{\partial g}{\partial x_i}$ has advantages and disadvantages. When greater precision is required we shall employ the notation $D_i\, g$, and regard it as a derivative with respect to the i-th variable (not necessarily called x_i).

Theorem 3.7. *Let G be an open subset of $\mathbb{R}^n$ and let $\mathbf{f} : G \longrightarrow \mathbb{R}^m$ be differentiable at $\mathbf{a} \in G$. Then $D\mathbf{f}(\mathbf{a})$ is represented by the $m \times n$ matrix $A=(a_{ij})$ where $a_{ij} = D_j f_i(\mathbf{a}) = \frac{\partial f_i}{\partial x_j}$, $1 \leq i \leq m$, $1 \leq j \leq n$.*

Proof. Let $\mathbf{L}=D\mathbf{f}(\mathbf{a})$. From (3.2) above, we see that

$$L_i(\mathbf{e}_j) = \lim_{t \to 0} \frac{f_i(\mathbf{a}+t\mathbf{e}_j) - f_i(\mathbf{a})}{t} = D_j f_i(\mathbf{a}).$$

This proves the result since $\mathbf{L}$ is determined by its action on the basis (Theorem 2.82).

The matrix expression is

$$D\mathbf{f}(\mathbf{a}) = \begin{bmatrix} \frac{\partial f_1}{\partial x_1} & \frac{\partial f_1}{\partial x_2} & \cdots & \frac{\partial f_1}{\partial x_n} \\ \vdots & \vdots & & \vdots \\ \frac{\partial f_m}{\partial x_1} & \frac{\partial f_m}{\partial x_2} & \cdots & \frac{\partial f_m}{\partial x_n} \end{bmatrix} \tag{3.4}$$

where all the partial derivatives are to be calculated at $\mathbf{a}$. This matrix is often called the *Jacobi matrix* of $\mathbf{f}$ at $\mathbf{a}$. When $\mathbf{f} : \mathbb{R}^n \longrightarrow \mathbb{R}^n$

this is a square matrix and its determinant is then defined. This determinant is called the *Jacobian* of $\mathbf{f}$ at $\mathbf{a}$ and is denoted $J_{\mathbf{f}}(\mathbf{a})$. Thus, $J_{\mathbf{f}}(\mathbf{a}) = \det D\mathbf{f}(\mathbf{a}) = \det [D_j f_i(\mathbf{a})]$. Other common notations are $\dfrac{\partial(f_1,f_2,\cdots,f_n)}{\partial(x_1,x_2,\cdots,x_n)}$ or $\dfrac{\partial(y_1,y_2,\cdots,y_n)}{\partial(x_1,x_2,\cdots,x_n)}$ where $\mathbf{y}=\mathbf{f}(\mathbf{x})$. If $J_{\mathbf{f}}(\mathbf{a})=0$ then $D\mathbf{f}(\mathbf{a})$ is not invertible and this suggests that $\mathbf{f}$ may exhibit irregular behaviour in a neighbourhood of $\mathbf{a}$. The hypothesis $J_{\mathbf{f}}(\mathbf{a})\neq 0$ will prevent this. The Jacobian often plays the same role for functions from $\mathbb{R}^n$ to $\mathbb{R}^n$ that the derivative f' plays for $f:\mathbb{R}\longrightarrow\mathbb{R}$, for example in the Inverse Function Theorem (chapter 4) and in changing variables in multiple integrals (chapter 6).

WARNING. $\mathbf{f}$ differentiable $\Rightarrow$ all partial derivatives exist but *not* conversely. A *partial* derivative is what it says.

Thus we see that $D\mathbf{f}(\mathbf{a})$ is uniquely determined; it is (represented by) the matrix printed above in (3.4). However, differentiability does not only say that the matrix exists,

the basic affine approximation property must also hold.

Computation of the matrix in (3.4) is a relatively simple matter : it is obtained by applying the methods of calculus of functions of one variable.

The important question to be answered is : if the matrix (3.4) exists, under what additional conditions is $\mathbf{f}$ differentiable at $\mathbf{a}$? It will turn out that *continuity* of the partial derivatives *suffices*; but it is *not necessary*. Before proving this we give some examples to illustrate that existence of the partial derivatives is not enough on its own. This should not be too surprising. Consider a function $f:\mathbb{R}^2\longrightarrow\mathbb{R}$; its graph is a surface. The two partial derivatives can be written $\dfrac{\partial f}{\partial x}, \dfrac{\partial f}{\partial y}$. The derivative $\dfrac{\partial f}{\partial x}$ gives local information on f in the direction of the x-axis. Indeed $f(x, a_2)$ is well approximated by the tangent line with slope $\dfrac{\partial f}{\partial x}(\mathbf{a})$. Similarly $f(a_1, y)$ is well approximated by the tangent with slope $\partial f/\partial y$. There is no reason to believe that we can say anything about the local behaviour of f in any other direction. However, when the partial derivatives are continuous, $\dfrac{\partial f}{\partial x}(\mathbf{b})$ is close to $\dfrac{\partial f}{\partial x}(\mathbf{a})$ for $\mathbf{b}$ near $\mathbf{a}$, so $\dfrac{\partial f}{\partial x}(\mathbf{a})$ gives local information on $f(x,b_2)$ also. Similarly for $\partial f/\partial y$. In all, this gives information on $f(x,y)$ in a small rectangle around (a_1,a_2), so on a neighbourhood of $\mathbf{a}$, and does in fact give differentiability. The details will follow the promised examples.

Example 3.8. Let $f: \mathbb{R}^2 \longrightarrow \mathbb{R}$ be given by $f(x,y) = xy$. Show that f is differentiable at all points of $\mathbb{R}^2$ and find the equation of the tangent plane at a point (a,b).

Solution. We have $\dfrac{\partial f}{\partial x}=y$, $\dfrac{\partial f}{\partial y}=x$, so the candidate for $Df(a,b)$ is (b,a). We have to show that, if $R=xy-[ab+(b,a)\begin{pmatrix}x-a\\ y-b\end{pmatrix}]$ then $\dfrac{|R|}{\|\mathbf{x}-\mathbf{a}\|}\longrightarrow 0$ as $\|\mathbf{x}-\mathbf{a}\|\longrightarrow 0$. Here, $R = xy-ab-b(x-a)-a(y-b) = xy+ab-bx-ay = (x-a)(y-b)$. By the elementary inequality $2uv \leq u^2+v^2$, (or by writing $x=a+r\cos\theta$, $y=b+r\sin\theta$, so that $\|\mathbf{x}-\mathbf{a}\|=r$, and then letting $r\longrightarrow 0$) we have

$$|R| \leq [(x-a)^2+(y-b)^2]/2 = \|\mathbf{x}-\mathbf{a}\|^2/2,$$

so that $\dfrac{|R|}{\|\mathbf{x}-\mathbf{a}\|} \leq \|\mathbf{x}-\mathbf{a}\|/2 \longrightarrow 0$ as $\|\mathbf{x}-\mathbf{a}\|\longrightarrow 0$ as required and the differentiability is shown. The tangent plane at (a,b) has equation $z = f(\mathbf{a})+d_1(x-a)+d_2(y-b)$ where $(d_1,d_2) = Df(a,b) = (b,a)$, that is, $z = ab+b(x-a)+a(y-b)$, or $z = bx+ay-ab$.

Example 3.9. Let $f: \mathbb{R}^2 \longrightarrow \mathbb{R}$ be given by

$$f(x,y) = \begin{cases} xy/(x^2+y^2), & \text{if } (x,y)\neq(0,0) \\ 0\ , & \text{if } (x,y)=(0,0) \end{cases}.$$

Then f is not differentiable at (0,0).

Solution. By elementary calculus (quotient rule), for $(x,y)\neq(0,0)$,

$$\frac{\partial f}{\partial x}=\frac{(x^2+y^2)\,y-xy(2x)}{(x^2+y^2)^2}=\frac{y^3-x^2y}{(x^2+y^2)^2}, \quad\text{and}\quad \frac{\partial f}{\partial y}=\frac{(x^2+y^2)\,x-xy(2y)}{(x^2+y^2)^2}=\frac{x^3-y^2x}{(x^2+y^2)^2}.$$

Observe that we can write $\partial f/\partial y$ down at once by interchanging symbols in the expression for $\partial f/\partial x$ since f is symmetrical in x and y: $f(x,y)=f(y,x)$. This calculation is not valid at (0,0) so we have to compute $\partial f/\partial x$ and $\partial f/\partial y$ at (0,0) directly from the definition. Indeed we have $\dfrac{\partial f}{\partial x}(0,0)= \lim\limits_{h\to 0}\dfrac{f(h,0)-f(0,0)}{h} = \lim\limits_{h\to 0} 0=0$, and, similarly, $\dfrac{\partial f}{\partial y}=0$.
Therefore the candidate for $Df(0,0)$ is the matrix $(0,0)$. Let $R(x,y)=f(x,y)-[f(0,0)+(0,0)\begin{pmatrix}x\\ y\end{pmatrix}]$, $(x,y)\neq(0,0)$. Then $R(x,y)=\dfrac{xy}{x^2+y^2}$. We will show that $\dfrac{|R|}{\sqrt{(x^2+y^2)}} \not\longrightarrow 0$ as $x^2+y^2\longrightarrow 0$. Indeed, taking $x=y=t$,

$\dfrac{|R|}{\sqrt{(x^2+y^2)}} = \dfrac{t^2/2t^2}{\sqrt{2}|t|} = \dfrac{1}{2\sqrt{2}|t|} \longrightarrow \infty$ as $t \longrightarrow 0$. This proves that f is not differentiable at (0,0) even though all partial derivatives exist there.

Since f is not continuous at (0,0) (Example 2.13), Theorem 3.4 shows that f is not differentiable at (0,0), so the direct argument is not needed here. In this example, f does not have a derivative in directions other than those of the co-ordinate axes. For, if $\mathbf{e}=(\cos\theta, \sin\theta)$ is any direction,

$$D_{\mathbf{e}}f(0,0) = \lim_{t \to 0} \frac{f(t\cos\theta,\ t\sin\theta)-f(0,0)}{t} = \lim_{t\to 0} \frac{\cos\theta\sin\theta}{t} = \lim_{t\to 0} \frac{\sin 2\theta}{2t},$$

which limit exists if, and only if, $\sin 2\theta=0$, that is along the co-ordinate axes only.

Example 3.10. Let $f(x,y)= \begin{cases} x^2y/(x^2+y^2), & (x,y)\neq(0,0) \\ 0 & , (x,y)=(0,0) \end{cases}$.

Then f has directional derivatives in every direction at (0,0) but is not differentiable at (0,0).

Solution. For $t\neq 0$, $\dfrac{f(t\cos\theta,\ t\sin\theta)-f(0,0)}{t}=\cos^2\theta\sin\theta$, so, for each fixed θ, the directional derivative in the direction $(\cos\theta, \sin\theta)$ exists and equals $\cos^2\theta\sin\theta$. Taking $\theta=0$ and $\theta=\pi/2$ gives $\dfrac{\partial f}{\partial x}=0$ and $\dfrac{\partial f}{\partial y}=0$. Let $R(x,y)=f(x,y)-[f(0,0)+(0,0)\begin{pmatrix} x \\ y \end{pmatrix}]=\dfrac{x^2y}{x^2+y^2}$. We show that $\dfrac{|R|}{\sqrt{(x^2+y^2)}}$ does not tend to 0 as $x^2+y^2 \longrightarrow 0$. Indeed, when $x=r\cos\theta$, and $y=r\sin\theta$, $\dfrac{|R|}{\sqrt{(x^2+y^2)}}=\cos^2\theta\sin\theta$ for all $r\neq 0$. The limit as $r\longrightarrow 0$ exists for θ fixed but depends on θ so the limit does not exist. In particular, for $\theta=\pi/4$, the limit is $1/2\sqrt{2}\neq 0$. Note that f is continuous at all points.

We now prove the result alluded to above which makes the task of checking differentiability much easier.

Theorem 3.11. *Let G be open in $\mathbb{R}^n$ and let $\mathbf{f}: G\longrightarrow\mathbb{R}^m$. Suppose that all the partial derivatives D_jf_i exist on a neighbourhood of $\mathbf{a}\in G$ and are continuous at $\mathbf{a}$. Then $\mathbf{f}$ is differentiable at $\mathbf{a}$.*

Proof. Since convergence in $\mathbb{R}^n$ is equivalent to the convergence of each component, it suffices to prove the result for $f: G\longrightarrow\mathbb{R}$. Let

$R(\mathbf{h})=f(\mathbf{a}+\mathbf{h})-f(\mathbf{a})-\sum_{j=1}^{n}\frac{\partial f}{\partial x_j}(\mathbf{a})h_j$. This is defined for $\|\mathbf{h}\|<\delta_1$ (say) since G is open. We have to show that $|R(\mathbf{h})|/\|\mathbf{h}\|\longrightarrow 0$ as $\|\mathbf{h}\|\longrightarrow 0$. We write $f(\mathbf{a}+\mathbf{h})-f(\mathbf{a})$ as a telescoping sum where only one co-ordinate changes at each step, that is, we write

$$\begin{aligned} f(a_1+h_1, a_2+h_2,\cdots,a_n+h_n)-f(a_1,a_2,\cdots,a_n)\\ = f(a_1+h_1, a_2+h_2,\cdots,a_n+h_n)-f(a_1,a_2+h_2,\cdots,a_n+h_n)\\ + f(a_1,a_2+h_2,\cdots,a_n+h_n)-f(a_1,a_2,a_3+h_3,\cdots,a_n+h_n)+\cdots\\ + f(a_1,a_2,\cdots,a_{n-1},a_n+h_n)-f(a_1,a_2,\cdots,a_{n-1},a_n). \end{aligned}$$

Let $\delta_2\le\delta_1$ be such that all the partial derivatives of f exist on the δ_2-neighbourhood of $\mathbf{a}$. Let $g(x)=f(x,a_2+h_2,\cdots,a_n+h_n)$, where $\mathbf{h}$ is fixed with $\|\mathbf{h}\|<\delta_2/2$. Then $g'(x)$ exists on the δ_3-neighbourhood of a_1 with $g'(x)=\frac{\partial f}{\partial x_1}(x,a_2+h_2,\cdots,a_n+h_n)$, for $\delta_3<\frac{\sqrt{3}}{2}\delta_2$. (This ensures that the point at which $\partial f/\partial x_1$ is evaluated lies in the δ_2-neighbourhood of $\mathbf{a}$.) By the Mean Value Theorem (*MVT*) of elementary analysis,

$$g(a_1+h_1)-g(a_1)=h_1g'(a_1+\theta_1h_1)\quad\text{where } 0<\theta_1<1.$$

That is, $f(a_1+h_1,a_2+h_2,\cdots,a_n+h_n)-f(a_1,a_2+h_2,\cdots,a_n+h_n)=\frac{\partial f}{\partial x_1}(\mathbf{c}_1)h_1$ where $\mathbf{c}_1=(a_1+\theta_1h_1,a_2+h_2,\cdots,a_n+h_n)$, and therefore $\|\mathbf{c}_1-\mathbf{a}\|^2=\theta_1^2h_1^2+h_2^2+\cdots+h_n^2<\|\mathbf{h}\|^2$. The other pairs of terms are treated in exactly the same way so that we can write

$$f(\mathbf{a}+\mathbf{h})-f(\mathbf{a})=\frac{\partial f}{\partial x_1}(\mathbf{c}_1)h_1+\frac{\partial f}{\partial x_2}(\mathbf{c}_2)h_2+\cdots+\frac{\partial f}{\partial x_n}(\mathbf{c}_n)h_n$$

where $\mathbf{c}_2=(a_1,a_2+\theta_2h_2,a_3+h_3,\cdots,a_n+h_n)$, $0<\theta_2<1$, and this pattern is followed until we reach $\mathbf{c}_n=(a_1,a_2,\cdots,a_{n-1},a_n+\theta_nh_n)$.

Thus we have $R(\mathbf{h})=\sum_{j=1}^{n}\left[\frac{\partial f}{\partial x_j}(\mathbf{c}_j)-\frac{\partial f}{\partial x_j}(\mathbf{a})\right]h_j$.

By the triangle inequality, since $|h_j|\le\|\mathbf{h}\|$ for every j, we get,

$$|R(\mathbf{h})|\le\sum_{j=1}^{n}\left|\frac{\partial f}{\partial x_j}(\mathbf{c}_j)-\frac{\partial f}{\partial x_j}(\mathbf{a})\right|\|\mathbf{h}\|.$$

Since the partial derivatives are continuous at $\mathbf{a}$, for any $\varepsilon>0$, there exists $\delta>0$ such that $\left|\frac{\partial f}{\partial x_j}(\mathbf{c})-\frac{\partial f}{\partial x_j}(\mathbf{a})\right|<\varepsilon/n$ for $\|\mathbf{c}-\mathbf{a}\|<\delta$, $j=1,2,\cdots,n$.

We can always suppose $\delta<\delta_2/2$ and hence we obtain $\|\mathbf{R}(\mathbf{h})\|\le\varepsilon\|\mathbf{h}\|$ for $\|\mathbf{h}\|<\delta$, and this proves the result.

Definition 3.12. If $\mathbf{f}: G\subseteq\mathbb{R}^n\longrightarrow\mathbb{R}^m$ has all partial derivatives $\dfrac{\partial f_i}{\partial x_j}=D_jf_i$ continuous on G we say $\mathbf{f}$ is *continuously differentiable* on G. We write $\mathbf{f}\in C^1(G,\mathbb{R}^m)$, or simply $C^1(G)$ when $m=1$, and say $\mathbf{f}$ *is of class* C^1 on G.

Here the superscript 1 refers to the order of the partial derivatives considered and C denotes continuity.

We have seen that

$$\boxed{\mathbf{f}\in C^1(G,\mathbb{R}^m)}\ \Rightarrow\ \boxed{\mathbf{f}\text{ differentiable on }G}$$
$$\Rightarrow\ \boxed{\text{all partial derivatives exist on }G},$$

but the converse implications may fail. Many results actually need $\mathbf{f}$ to be of class C^1 rather than differentiable.

3.2. THE CHAIN RULE

We now investigate when combinations of differentiable functions are differentiable. Our first result deals with sums and multiplication by scalars.

Theorem 3.13. *Let* $\mathbf{f},\mathbf{g}: G\longrightarrow\mathbb{R}^m$ *be differentiable at* $\mathbf{a}\in G$. *Then* $\mathbf{f}+\mathbf{g}$ *and* $\alpha\mathbf{f}$ $(\alpha\in\mathbb{R})$ *are differentiable at* $\mathbf{a}$ *with* $D(\mathbf{f}+\mathbf{g})(\mathbf{a})=D\mathbf{f}(\mathbf{a})+D\mathbf{g}(\mathbf{a})$, *and* $D(\alpha\mathbf{f})(\mathbf{a})=\alpha D\mathbf{f}(\mathbf{a})$.

Proof. Since $\mathbf{f},\mathbf{g}$ are differentiable at $\mathbf{a}$,

$$\mathbf{f}(\mathbf{x})=\mathbf{f}(\mathbf{a})+D\mathbf{f}(\mathbf{a})(\mathbf{x}-\mathbf{a})+\mathbf{R}(\mathbf{x},\mathbf{a})\ \text{and}\ \mathbf{g}(\mathbf{x})=\mathbf{g}(\mathbf{a})+D\mathbf{g}(\mathbf{a})(\mathbf{x}-\mathbf{a})+\mathbf{S}(\mathbf{x},\mathbf{a}),$$

where $\|\mathbf{R}(\mathbf{x},\mathbf{a})\|/\|\mathbf{x}-\mathbf{a}\|\longrightarrow 0$ as $\|\mathbf{x}-\mathbf{a}\|\longrightarrow 0$, (same for $\mathbf{S}$). Then we have $\mathbf{f}(\mathbf{x})+\mathbf{g}(\mathbf{x})=\mathbf{f}(\mathbf{a})+\mathbf{g}(\mathbf{a})+[D\mathbf{f}(\mathbf{a})+D\mathbf{g}(\mathbf{a})](\mathbf{x}-\mathbf{a})+\mathbf{R}+\mathbf{S}$ and

$$\|\mathbf{R}+\mathbf{S}\|/\|\mathbf{x}-\mathbf{a}\|\le\|\mathbf{R}\|/\|\mathbf{x}-\mathbf{a}\|+\|\mathbf{S}\|/\|\mathbf{x}-\mathbf{a}\|\longrightarrow 0 \text{ as } \|\mathbf{x}-\mathbf{a}\|\longrightarrow 0.$$

Since $D\mathbf{f}(\mathbf{a})+D\mathbf{g}(\mathbf{a})$ is a linear operator, this proves the result for $\mathbf{f}+\mathbf{g}$. The case $\alpha\mathbf{f}$ is simpler and is left to the reader.

We now deal with compositions of functions. Let G be an open subset of $\mathbb{R}^n$, let E be an open subset of $\mathbb{R}^p$ and suppose $\mathbf{f}: E\longrightarrow\mathbb{R}^m$, $\mathbf{g}: G\longrightarrow\mathbb{R}^p$ satisfy $\mathbf{g}(G)\subseteq E$. Then $\mathbf{h}: G\subseteq\mathbb{R}^n\longrightarrow\mathbb{R}^m$ defined by $\mathbf{h}(\mathbf{x})=\mathbf{f}(\mathbf{g}(\mathbf{x}))$ is the composition of $\mathbf{f}$ with $\mathbf{g}$ written $\mathbf{h}=\mathbf{f}\circ\mathbf{g}$, (first apply **g** then **f**).

Theorem 3.14. (*The chain rule.*) *With* **f**,**g** *as above, if* **g** *is differentiable at* $\mathbf{x}\in G$ *and* **f** *is differentiable at* $\mathbf{y}=\mathbf{g}(\mathbf{x})\in E$, *then the composite* $\mathbf{h}=\mathbf{f}\circ\mathbf{g}$ *is differentiable at* $\mathbf{x}\in G$ *and* $D\mathbf{h}(\mathbf{x})=D\mathbf{f}(\mathbf{y})D\mathbf{g}(\mathbf{x})$.

Note. $D\mathbf{f}(\mathbf{y})D\mathbf{g}(\mathbf{x})$ can be thought of either as a composition of linear operators or as a product of matrices.

Proof. Write $\mathbf{L}=D\mathbf{g}(\mathbf{x})$, $\mathbf{M}=D\mathbf{f}(\mathbf{y})$, where $\mathbf{y}=\mathbf{g}(\mathbf{x})$. We want to show that $\lim_{\|\mathbf{k}\|\to 0}\dfrac{\|\mathbf{h}(\mathbf{x}+\mathbf{k})-\mathbf{h}(\mathbf{x})-\mathbf{MLk}\|}{\|\mathbf{k}\|}=0$. We will show that, for any $\varepsilon>0$, there is $\delta>0$ so that $\|\mathbf{h}(\mathbf{x}+\mathbf{k})-\mathbf{h}(\mathbf{x})-\mathbf{MLk}\|<C\varepsilon\|\mathbf{k}\|$, for $\|\mathbf{k}\|<\delta$, where C is a constant, from which the result is immediate. We write

$$\mathbf{h}(\mathbf{x}+\mathbf{k})-\mathbf{h}(\mathbf{x})-\mathbf{MLk}=\mathbf{f}(\mathbf{g}(\mathbf{x}+\mathbf{k}))-\mathbf{f}(\mathbf{g}(\mathbf{x}))-\mathbf{M}(\mathbf{g}(\mathbf{x}+\mathbf{k})-\mathbf{g}(\mathbf{x}))+\mathbf{M}(\mathbf{g}(\mathbf{x}+\mathbf{k})-\mathbf{g}(\mathbf{x})-\mathbf{Lk}).$$

By the triangle inequality we obtain

$$\|\mathbf{h}(\mathbf{x}+\mathbf{k})-\mathbf{h}(\mathbf{x})-\mathbf{MLk}\|\le\|\mathbf{f}(\mathbf{g}(\mathbf{x}+\mathbf{k}))-\mathbf{f}(\mathbf{g}(\mathbf{x}))-\mathbf{M}(\mathbf{g}(\mathbf{x}+\mathbf{k})-\mathbf{g}(\mathbf{x}))\| + \|\mathbf{M}\|\,\|\mathbf{g}(\mathbf{x}+\mathbf{k})-\mathbf{g}(\mathbf{x})-\mathbf{Lk}\|.$$

Since **g** is differentiable at **x** for any $\varepsilon>0$ there is $\delta_1>0$ such that the second term can be made less than $\varepsilon\|\mathbf{M}\|\,\|\mathbf{k}\|$ for $\|\mathbf{k}\|<\delta_1$. We turn to the first term. By Corollary 3.5, there is $\delta_2>0$ such that

$$\|\mathbf{g}(\mathbf{x}+\mathbf{k})-\mathbf{g}(\mathbf{x})\|\le(\|\mathbf{L}\|+1)\|\mathbf{k}\| \quad \text{whenever } \|\mathbf{k}\|<\delta_2$$

As **f** is differentiable at **y**, there exists $\delta_3>0$ such that

$$\|\mathbf{f}(\mathbf{z})-\mathbf{f}(\mathbf{y})-\mathbf{M}(\mathbf{z}-\mathbf{y})\|<\varepsilon\|\mathbf{z}-\mathbf{y}\| \text{ for } \|\mathbf{z}-\mathbf{y}\|<\delta_3.$$

Choosing δ_2 so that $(\|\mathbf{L}\|+1)\delta_2<\delta_3$ we see that we may take $\mathbf{z}=\mathbf{g}(\mathbf{x}+\mathbf{k})$ and $\mathbf{y}=\mathbf{g}(\mathbf{x})$ to get

$\|\mathbf{f}(\mathbf{g}(\mathbf{x}+\mathbf{k}))-\mathbf{f}(\mathbf{g}(\mathbf{x}))-\mathbf{M}(\mathbf{g}(\mathbf{x}+\mathbf{k})-\mathbf{g}(\mathbf{x}))\|<\varepsilon\|\mathbf{g}(\mathbf{x}+\mathbf{k})-\mathbf{g}(\mathbf{x})\|<\varepsilon(\|\mathbf{L}\|+1)\|\mathbf{k}\|$, for $\|k\|<\delta_2$. Taking $\delta=\min(\delta_1,\delta_2)$ proves the result with $C=\|\mathbf{M}\|+\|\mathbf{L}\|+1$.

Example 3.14. Let $\mathbf{g}:\mathbb{R}^3\longrightarrow\mathbb{R}^3$, $f:\mathbb{R}^3\longrightarrow\mathbb{R}$ be differentiable and suppose $h=f\circ\mathbf{g}:\mathbb{R}^3\longrightarrow\mathbb{R}$ is defined. Then $Dh(\mathbf{x})=Df(\mathbf{g}(\mathbf{x}))\,D\mathbf{g}(\mathbf{x})$ is the 1×3 matrix obtained by multiplying together the 1×3 matrix $Df(\mathbf{g}(\mathbf{x}))$ and the 3×3 matrix $D\mathbf{g}(\mathbf{x})$. In terms of components, the partial derivatives therefore satisfy

$(D_1h(\mathbf{x}), D_2h(\mathbf{x}), D_3h(\mathbf{x}))=$

$$(D_1f(\mathbf{y}), D_2f(\mathbf{y}), D_3f(\mathbf{y}))\begin{bmatrix} D_1g_1(\mathbf{x}) & D_2g_1(\mathbf{x}) & D_3g_1(\mathbf{x}) \\ D_1g_2(\mathbf{x}) & D_2g_2(\mathbf{x}) & D_3g_2(\mathbf{x}) \\ D_1g_3(\mathbf{x}) & D_2g_3(\mathbf{x}) & D_3g_3(\mathbf{x}) \end{bmatrix}$$

Therefore $D_ih(\mathbf{x})=\sum_{j=1}^{3} D_jf(\mathbf{y})\cdot D_ig_j(\mathbf{x})$, where $\mathbf{y}=\mathbf{g}(\mathbf{x})$, $i=1,2,3$.

The alternative notation is more commonly used in calculations and reads

$$\frac{\partial h}{\partial x_i} = \frac{\partial f}{\partial y_1}\frac{\partial g_1}{\partial x_i} + \frac{\partial f}{\partial y_2}\frac{\partial g_2}{\partial x_i} + \frac{\partial f}{\partial y_3}\frac{\partial g_3}{\partial x_i},\ i=1,2,3. \tag{3.5}$$

Note that we have to write $\partial f/\partial y_1$ etc. rather than $\partial f/\partial x_1$ to denote $D_1 f(\mathbf{y})$ to avoid possible confusion (a weakness of the notation $\partial f/\partial x$). (3.5) is more easily remembered in its imprecise form. Write $u=h(\mathbf{x})$, $\mathbf{y}=\mathbf{g}(\mathbf{x})$. Then we can put $u=f(\mathbf{y})$ and write

$$\frac{\partial u}{\partial x_i} = \frac{\partial u}{\partial y_1}\frac{\partial y_1}{\partial x_i} + \frac{\partial u}{\partial y_2}\frac{\partial y_2}{\partial x_i} + \frac{\partial u}{\partial y_3}\frac{\partial y_3}{\partial x_i}. \tag{3.6}$$

The common terms ∂y_1, ∂y_2, ∂y_3 can be imagined to be links in a chain; hence the *chain rule*. This last form is often simplest to use when calculating partial derivatives but may lead to error if care is not taken to distinguish between u on the left (a function of $\mathbf{x}$) with u on the right (a different function of $\mathbf{y}$).

Example 3.15. Let $\mathbf{g}(x,y)=(x^2-y^2+xy,\ y^2-1)$, let $\mathbf{f}(u,v)=(u+v,\ 2u,\ v^2)$. Show that $\mathbf{g},\mathbf{f}$ are differentiable, that $\mathbf{f}\circ\mathbf{g}$ exists, and compute the derivative of $\mathbf{f}\circ\mathbf{g}$ at the point (1,1) (*a*) directly, (*b*) using the chain rule.

Solution. We have $\mathbf{g}(x,y)=(g_1(x,y),\ g_2(x,y))$ where $g_1(x,y)=x^2-y^2+xy$, $g_2(x,y)=y^2-1$. Then $\dfrac{\partial g_1}{\partial x}=2x+y$, $\dfrac{\partial g_1}{\partial y}=-2y+x$, $\dfrac{\partial g_2}{\partial x}=0$, $\dfrac{\partial g_2}{\partial y}=2y$.

Each partial derivative is evidently continuous so $\mathbf{g}\in C^1(\mathbb{R}^2,\mathbb{R}^2)$ and is therefore differentiable. Similarly $\mathbf{f}\in C^1(\mathbb{R}^2,\mathbb{R}^3)$. Then $\mathbf{h}:=\mathbf{f}\circ\mathbf{g}$ is defined as a map from $\mathbb{R}^2$ into $\mathbb{R}^3$ since the image set of $\mathbf{g}$ is contained in the domain of $\mathbf{f}$ (all of $\mathbb{R}^2$).

Indeed, $(\mathbf{f}\circ\mathbf{g})(x,y)=\mathbf{f}(g_1(x,y),\ g_2(x,y))=(g_1+g_2,\ 2g_1,\ g_2^2)$

$$=(x^2+xy-1,\ 2x^2-2y^2+2xy,\ y^4-2y^2+1) := (h_1,\ h_2,\ h_3).$$

Then $D(\mathbf{f}\circ\mathbf{g})(x,y) = D\mathbf{h}(x,y) = \begin{bmatrix} \partial h_1/\partial x & \partial h_1/\partial y \\ \partial h_2/\partial x & \partial h_2/\partial y \\ \partial h_3/\partial x & \partial h_3/\partial y \end{bmatrix} = \begin{bmatrix} 2x+y & x \\ 4x+2y & 2x-4y \\ 0 & 4y^3-4y \end{bmatrix}$

so that $D\mathbf{h}(1,1) = \begin{bmatrix} 3 & 1 \\ 6 & -2 \\ 0 & 0 \end{bmatrix}$. We now use the chain rule.

As $\mathbf{g}(1,1)=(1,0)$ we have $D(\mathbf{f}\circ\mathbf{g})(1,1)=D\mathbf{f}(1,0)D\mathbf{g}(1,1)$ which gives

$$\begin{bmatrix} \partial f_1/\partial u & \partial f_1/\partial v \\ \partial f_2/\partial u & \partial f_2/\partial v \\ \partial f_3/\partial u & \partial f_3/\partial v \end{bmatrix} \begin{bmatrix} \partial g_1/\partial x & \partial g_1/\partial y \\ \partial g_2/\partial x & \partial g_2/\partial y \end{bmatrix} = \begin{bmatrix} 1 & 1 \\ 2 & 0 \\ 0 & 0 \end{bmatrix} \begin{bmatrix} 3 & -1 \\ 0 & 2 \end{bmatrix} = \begin{bmatrix} 3 & 1 \\ 6 & -2 \\ 0 & 0 \end{bmatrix} \text{ as before.}$$

The partial derivatives of $\mathbf{h}=f\circ\mathbf{g}$ can be obtained more quickly using the notation of Example 3.6. For example, for $u=g_1(x,y)$, $v=g_2(x,y)$

$$\frac{\partial h_1}{\partial x} = \frac{\partial f_1}{\partial u}\frac{\partial u}{\partial x} + \frac{\partial f_1}{\partial v}\frac{\partial v}{\partial x} = 1\cdot\frac{\partial g_1}{\partial x} + 1\cdot\frac{\partial g_2}{\partial x} = 3+0 = 3\text{ , etc.}$$

Example 3.16. Let $w=f(x,y,z)$ and suppose $z=\phi(x,y)$. Calculate $\dfrac{\partial w}{\partial x}$ and $\dfrac{\partial w}{\partial y}$

Solution. There is some ambiguity in what is being asked here. Since $z=\phi(x,y)$, $w=f(x,y,\phi(x,y)) := h(x,y)$, we are asking for $\dfrac{\partial h}{\partial x}$ and $\dfrac{\partial h}{\partial y}$ (probably). If we write $\mathbf{g}(x,y)=(x,y,\phi(x,y))$ then $h(x,y)=(f\circ\mathbf{g})(x,y)$ and the chain rule gives $Dh(x,y)=Df(g_1,g_2,g_3)\cdot D\mathbf{g}(x,y)$, which yields

$$Dh(x,y)= [D_1f \;\; D_2f \;\; D_3f\,] \begin{bmatrix} D_1g_1 & D_2g_1 \\ D_1g_2 & D_2g_2 \\ D_1g_3 & D_2g_3 \end{bmatrix} = \begin{bmatrix} \dfrac{\partial f}{\partial x} & \dfrac{\partial f}{\partial y} & \dfrac{\partial f}{\partial z} \end{bmatrix} \begin{bmatrix} 1 & 0 \\ 0 & 1 \\ \dfrac{\partial \phi}{\partial x} & \dfrac{\partial \phi}{\partial y} \end{bmatrix}$$

Therefore, $\dfrac{\partial h}{\partial x} = \dfrac{\partial f}{\partial x} + \dfrac{\partial f}{\partial z}\dfrac{\partial \phi}{\partial x}$ and $\dfrac{\partial h}{\partial y} = \dfrac{\partial f}{\partial y} + \dfrac{\partial f}{\partial z}\dfrac{\partial \phi}{\partial y}$.

This can be achieved (more simply?) by using the analogue of the version of the chain rule illustrated in Example 3.14. Indeed, we write $w=f(x,y,\phi(x,y))$ and

$$\frac{\partial w}{\partial x} = \frac{\partial f}{\partial x}\frac{\partial x}{\partial x} + \frac{\partial f}{\partial y}\frac{\partial y}{\partial x} + \frac{\partial f}{\partial z}\frac{\partial \phi}{\partial x} = \frac{\partial f}{\partial x}\cdot 1 + \frac{\partial f}{\partial y}\cdot 0 + \frac{\partial f}{\partial z}\frac{\partial \phi}{\partial x}.$$

If we write w rather than f on the right hand side here we would then have $\dfrac{\partial w}{\partial x}=\dfrac{\partial w}{\partial x}+\dfrac{\partial w}{\partial z}\dfrac{\partial \phi}{\partial x}$ which might be confusing : we could wrongly infer that $\dfrac{\partial w}{\partial z}\dfrac{\partial \phi}{\partial x} = 0$. In fact, w refers to different functions on each side of this equality so $\dfrac{\partial w}{\partial x}$ on the left is not equal to $\dfrac{\partial w}{\partial x}$ on the right.

Example 3.17. Let $w=x+y$ where $y=2x$. What is $\partial w/\partial x$?

Solution. It is tempting to say $\partial w/\partial x=1$ and indeed this is $\partial f/\partial x$ where $f(x,y)=x+y$. But perhaps $\dfrac{\partial w}{\partial x}$ means $\dfrac{\partial}{\partial x}f(x,2x)$? Then $\dfrac{\partial w}{\partial x}=\dfrac{\partial}{\partial x}(3x)=3$. This latter should be written $\dfrac{dw}{dx}$, an ordinary derivative since w is now a function of x alone. Let us see how the chain rule applies here. We have $\frac{d}{dx}f(x,2x) = \frac{\partial f}{\partial x}\cdot 1+\frac{\partial f}{\partial y}\cdot 2 = 1+2 = 3.$

This example illustrates how an ill-specified problem can easily lead to error.

3.3. THE GRADIENT VECTOR

Real-valued functions $\phi : \mathbb{R}^n \longrightarrow \mathbb{R}$ are often called *scalar fields*: to each point $\mathbf{x}$ one associates a *scalar* $\phi(\mathbf{x})$. Functions $\mathbf{F} : \mathbb{R}^n \longrightarrow \mathbb{R}^m$ may be called *vector fields*; this is particularly so for functions $\mathbf{F} : \mathbb{R}^3 \longrightarrow \mathbb{R}^3$ because elements in $\mathbb{R}^3$ can be identified with vectors with their usual meaning. More details may be found in Chapter 7.

Let $\phi : \mathbb{R}^n \longrightarrow \mathbb{R}$ be differentiable. The derivative at a point $\mathbf{x}$, $D\phi(\mathbf{x})$, is a linear operator so can be identified with a $1\times n$ matrix, an element of $\mathbb{R}^n$, or a vector (a vector is usually considered as a column, an $n\times 1$ matrix). This vector is called the *gradient* of ϕ at $\mathbf{x}$ and is written grad$\phi(\mathbf{x})$ or $\nabla\phi(\mathbf{x})$. The symbol ∇ may be read "del". More precisely, $\nabla\phi(\mathbf{x})$ is defined as an element of $\mathbb{R}^n$ by

$$<\nabla\phi(\mathbf{x}),\mathbf{h}> = D\phi(\mathbf{x})\mathbf{h} \text{ for all } \mathbf{h}\in\mathbb{R}^n. \tag{3.7}$$

It is essential here that we use the Euclidean inner product as we are identifying the components of a vector with the action of a linear operator on a basis element.

One use of the gradient of a scalar field is in determining the directions of greatest and least rates of increase of ϕ.

Proposition 3.18. *Let $\phi : G\subseteq\mathbb{R}^n \longrightarrow \mathbb{R}$ be differentiable at $\mathbf{a}\in G$. Then for any direction $\mathbf{e}$, $D_{\mathbf{e}}\phi(\mathbf{a})$ exists and moreover*

$$D_{\mathbf{e}}\phi(\mathbf{a}) = D\phi(\mathbf{a})\mathbf{e} = <\nabla\phi(\mathbf{a}),\mathbf{e}>. \tag{3.8}$$

Proof. $\frac{\phi(\mathbf{a}+t\mathbf{e})-\phi(\mathbf{a})}{t} = D\phi(\mathbf{a})\mathbf{e}+\frac{R(t)}{t}$, see (3.2) above. Therefore the left-hand side does have a limit as $t \longrightarrow 0$ and this gives

$$D_{\mathbf{e}}\phi(\mathbf{a}) = D\phi(\mathbf{a})\mathbf{e} = <\nabla\phi(\mathbf{a}),\mathbf{e}>.$$

It should not be necessary to remind you that directional derivatives can exist even when ϕ is not differentiable, but Proposition 3.18 gives the simplest method of computing directional derivatives, when applicable.

The directional derivative $D_{\mathbf{e}}\phi$ measures the rate of increase of ϕ in the direction $\mathbf{e}$. What direction $\mathbf{e}$ gives the greatest rate of increase? We have $<\nabla\phi(\mathbf{a}),\mathbf{e}> \le \|\nabla\phi(\mathbf{a})\|\,\|\mathbf{e}\|$ with equality only if $\mathbf{e}$ is a scalar multiple of $\nabla\phi(\mathbf{a})$, by Schwarz's inequality. Therefore $D_{\mathbf{e}}\phi(\mathbf{a})$ is a maximum, namely $\|\nabla\phi(\mathbf{a})\|$ when $\mathbf{e}=\nabla\phi(\mathbf{a})/\|\nabla\phi(\mathbf{a})\|$ (recall that $\|\mathbf{e}\|=1$). In other words, the direction of fastest increase is that determined by $\nabla\phi(\mathbf{a})$. Similarly the direction of fastest decrease is that of $-\nabla\phi$.

Another interpretation of $\nabla\phi$ arises when we consider the points $\{\mathbf{x}\in\mathbb{R}^n : \phi(\mathbf{x})=\alpha\}$, where α is a constant. These sets are called *level sets* of ϕ (level curves in $\mathbb{R}^2$). If $\phi:\mathbb{R}^2\longrightarrow\mathbb{R}$, its graph is the surface defined by $z=\phi(x,y)$. The level curves are curves along which the height of the surface is constant. If we image the surface as a landscape with mountains and valleys, then the base plane is a map (scale of one to one) and the level curves are the contour lines on the map. To increase ϕ as quickly as possible from a given value α, it is "obvious" that one should move off at right angles to a level curve. In other words, $\nabla\phi$ is *orthogonal* to the level curve of ϕ. Here, two vectors $\mathbf{u},\mathbf{v}$ are said to be *orthogonal* if $<\mathbf{u},\mathbf{v}>=0$. To say $\nabla\phi$ is orthogonal to a level set of ϕ is to say that $<\nabla\phi,\mathbf{T}>=0$ for every vector $\mathbf{T}$ that is *tangent* to the level set.

We define the notion of tangent direction via curves in the surface. A *path* is a function $\mathbf{c}$ from an interval $[t_1,t_2]$ in $\mathbb{R}$ into $\mathbb{R}^n$, the image set of $\mathbf{c}$ is called a *curve*. This may be visualized by thinking of the parameter t as *time* and $\mathbf{c}(t)$ as the position vector of a moving point. As t varies the point traces out the curve. Let $S=\{\mathbf{x}\in\mathbb{R}^n : \phi(\mathbf{x})=\alpha\}$ be a level set of ϕ. A curve in S is defined by $\mathbf{c}:[t_1,t_2]\longrightarrow S$. A vector $\mathbf{T}$ is *tangent* to S at a point $\mathbf{z}\in S$ if there is a curve in S with $\mathbf{c}(\tau)=\mathbf{z}$ for some $\tau\in(t_1,t_2)$ and $\dfrac{d\mathbf{c}(\tau)}{dt}=\mathbf{T}$. Here $\dfrac{d\mathbf{c}(\tau)}{dt}=\left(\dfrac{dc_1}{dt},\dfrac{dc_2}{dt},\ldots,\dfrac{dc_n}{dt}\right)=D\mathbf{c}(\tau)$. Figure 3.3 indicates why $\dfrac{d\mathbf{c}}{dt}$ is a tangent vector.

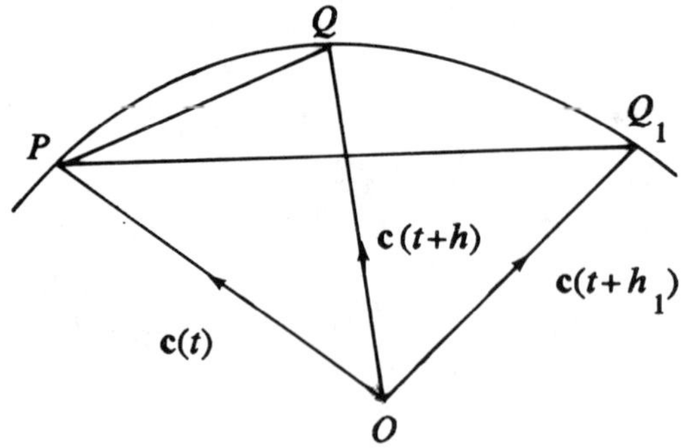

Figure 3.3

Let P be the point with position vector $\mathbf{c}(t)$, Q be the neighbouring point with position vector $\mathbf{c}(t+h)$. Then the chord PQ is given by the vector $\overrightarrow{PQ}=\mathbf{c}(t+h)-\mathbf{c}(t)$. This has the same direction as $\dfrac{\mathbf{c}(t+h)-\mathbf{c}(t)}{h}$, for $h>0$, and this approaches $\dfrac{d\mathbf{c}}{dt}$ as $h\longrightarrow 0$. Since the tangent at P should be a limiting case of the chord as $h\longrightarrow 0$, it is clear that the direction of the tangent should be $\dfrac{d\mathbf{c}}{dt}$. This vector points in the direction of the curve corresponding to increasing t.

Proposition 3.19. *Let $\phi:\mathbb{R}^n\longrightarrow\mathbb{R}$ be differentiable and for $\alpha\in\mathbb{R}$ let $S=\{\mathbf{x}\in\mathbb{R}^n:\phi(\mathbf{x})=\alpha\}$ be a level set of ϕ. Then, for $\mathbf{z}\in S$, $\nabla\phi(\mathbf{z})$ is orthogonal to S.*

Proof. Let $\mathbf{c}$ be a curve in S and let $\mathbf{z}=\mathbf{c}(\tau)$. Since $\mathbf{c}(t)$ lies in S, $\phi(\mathbf{c}(t))=\alpha$, for t in an interval containing τ. By the chain rule, $D(\phi\circ\mathbf{c})(\tau)=D\phi(\mathbf{c}(\tau))\dfrac{d\mathbf{c}(\tau)}{dt}=<\nabla\phi(\mathbf{c}(\tau)),\dfrac{d\mathbf{c}(\tau)}{dt}>$. As the left-hand term is 0 we have $<\nabla\phi(\mathbf{z}),\mathbf{T}>=0$, and this holds for every tangent vector $\mathbf{T}$.

3.4. MEAN VALUE THEOREMS

An important result for functions $f:\mathbb{R}\longrightarrow\mathbb{R}$ is the Mean Value Theorem which we have already employed in the proof of Theorem 3.11. Its importance is that it enables properties of f to be deduced from properties of f'. One of its consequences is that $f'(x)=0$ on an interval implies that $f(x)$ is constant on that interval. We wish to extend this and other results.

Theorem 3.20. *(Mean Value Theorem, MVT.)*
Let $f:G\subseteq\mathbb{R}^n\longrightarrow\mathbb{R}$ be continuous on the line segment $\{\mathbf{a}+t\mathbf{h},\ 0\le t\le 1\}$ and differentiable at every point of the segment $\{\mathbf{a}+t\mathbf{h},\ 0<t<1\}$. Then there exists θ, $0<\theta<1$, such that $f(\mathbf{a}+\mathbf{h})-f(\mathbf{a})=Df(\mathbf{a}+\theta\mathbf{h})\mathbf{h}=<\nabla f(\mathbf{a}+\theta\mathbf{h}),\mathbf{h}>$.

Proof. Let $\phi:[0,1]\longrightarrow\mathbb{R}$ be defined by $\phi(t)=f(\mathbf{a}+t\mathbf{h})$. Then $\phi=f\circ\mathbf{g}$ where $\mathbf{g}:[0,1]\longrightarrow\mathbb{R}^n$ is given by $\mathbf{g}(t)=\mathbf{a}+t\mathbf{h}$. Here, $\mathbf{g}$ is differentiable on $(0,1)$ with $D\mathbf{g}(t)=\mathbf{h}$. By the chain rule, ϕ is differentiable and

$$\phi'(t)=\frac{d\phi}{dt}=Df(\mathbf{g}(t))D\mathbf{g}(t)=Df(\mathbf{a}+t\mathbf{h})\mathbf{h}=<\nabla f(\mathbf{a}+t\mathbf{h}),\mathbf{h}>.$$

The mean value theorem in $\mathbb{R}$ may now be applied to ϕ to give $\phi(1)-\phi(0)=\phi'(\theta)$ for some $\theta\in(0,1)$, that is, $f(\mathbf{a}+\mathbf{h})-f(\mathbf{a})=Df(\mathbf{a}+\theta\mathbf{h})\mathbf{h}$.

Recall from Chapter 2 that a set C in $\mathbb{R}^n$ is said to be *convex* if whenever $\mathbf{x},\mathbf{y}$ are two points of C, the line segment $t\mathbf{x}+(1-t)\mathbf{y}$, $0\le t\le 1$, also lies in C.

Theorem 3.21. *Let C be a closed convex set in $\mathbb{R}^n$ with nonempty interior and let $f : C \longrightarrow \mathbb{R}$ be continuous on C and differentiable on* int(C). *Suppose that $\|Df(\mathbf{x})\| \le M$ for all $\mathbf{x} \in \text{int}(C)$. Then, for all $\mathbf{x},\mathbf{y} \in C$, $|f(\mathbf{x}) - f(\mathbf{y})| \le M\|\mathbf{x} - \mathbf{y}\|$. In particular, if $Df(\mathbf{x})=\mathbf{0}$ for all $\mathbf{x}$ in* int(C) *then f is constant on C.*

Proof. For any $\mathbf{y} \in \text{int}(C)$ and $\mathbf{x} \in C$, the closed line segment $t\mathbf{x}+(1-t)\mathbf{y}$, $0 \le t \le 1$, lies in C and the segment with $0<t<1$ lies in int(C), by problem 2.31. Therefore, by Theorem 3.20, there is $\theta \in (0,1)$ such that

$$f(\mathbf{x}) - f(\mathbf{y}) = Df(\mathbf{y}+\theta(\mathbf{x}-\mathbf{y}))(\mathbf{x}-\mathbf{y}) = \langle \nabla f(\mathbf{y}+\theta(\mathbf{x}-\mathbf{y})), \mathbf{x}-\mathbf{y} \rangle .$$

Using the definition of norm of a linear operator (or the Schwarz inequality) gives $|f(\mathbf{x}) - f(\mathbf{y})| \le \|Df(\mathbf{y}+\theta(\mathbf{x}-\mathbf{y}))\| \, \|\mathbf{x}-\mathbf{y}\| \le M\|\mathbf{x}-\mathbf{y}\|$.

For $\mathbf{y} \in \text{bdy}(C)$ there is a sequence $\mathbf{y}^{(m)}$ of points of int(C) with $\mathbf{y}^{(m)} \to \mathbf{y}$. The inequality holds for each m and passing to the limit proves that it also holds for $\mathbf{y}$. When $Df(\mathbf{c})=\mathbf{0}$ for all $\mathbf{c} \in \text{int}(C)$, we have $M=0$ and $|f(\mathbf{x}) - f(\mathbf{y})| = 0$ for all $\mathbf{y}$ in C so f is constant on C.

There is no Mean Value Theorem for functions $\mathbf{f} : G \subseteq \mathbb{R}^n \longrightarrow \mathbb{R}^m$ when $m>1$ (problem 3.25) but there is an inequality which is useful.

Theorem 3.22. *Let C be a convex set in $\mathbb{R}^n$ with nonempty interior and let $\mathbf{f} : C \longrightarrow \mathbb{R}^m$ be continuous on C and differentiable on* int(C). *Then for $\mathbf{x} \in \text{int}(C)$, and $\mathbf{y} \in C$ there exists $\theta \in (0,1)$ such that $\|\mathbf{f}(\mathbf{y})-\mathbf{f}(\mathbf{x})\| \le \|D\mathbf{f}(\mathbf{c})(\mathbf{y}-\mathbf{x})\|$ where $\mathbf{c}=(1-\theta)\mathbf{x}+\theta\mathbf{y}$. In particular, if $D\mathbf{f}(\mathbf{c})=\mathbf{0}$ for all $\mathbf{c} \in \text{int}(C)$, then $\mathbf{f}$ is constant* on C.

Proof. For $\mathbf{x} \in \text{int}(C)$, and $\mathbf{y} \in C$ write $\mathbf{h}=\mathbf{y}-\mathbf{x}$ and let $\mathbf{u}=\mathbf{f}(\mathbf{y})-\mathbf{f}(\mathbf{x})$. Then $\mathbf{x}+t\mathbf{h} \in \text{int}(C)$ for $0<t<1$ (problem 2.31). If $\mathbf{u}=\mathbf{0}$ the inequality is trivial. For $\mathbf{u} \ne \mathbf{0}$, consider the function $\phi : [0,1] \longrightarrow \mathbb{R}$ defined by $\phi(t) := \langle \mathbf{u},\mathbf{f}(\mathbf{x}+t\mathbf{h}) \rangle$. Then ϕ is differentiable and $\phi'(t) = \langle \mathbf{u}, D\mathbf{f}(\mathbf{x}+t\mathbf{h})\mathbf{h} \rangle$. By the MVT for real-valued functions, there is θ, $0<\theta<1$, such that $\phi(1) - \phi(0) = \phi'(\theta)$. This gives $\langle \mathbf{u},\mathbf{f}(\mathbf{y})-\mathbf{f}(\mathbf{x}) \rangle = \langle \mathbf{u}, D\mathbf{f}(\mathbf{x}+\theta\mathbf{h})\mathbf{h} \rangle$. Hence,

$$\|\mathbf{f}(\mathbf{y})-\mathbf{f}(\mathbf{x})\|^2 = \langle \mathbf{u}, D\mathbf{f}(\mathbf{x}+\theta\mathbf{h})\mathbf{h} \rangle \le \|\mathbf{u}\| \, \|D\mathbf{f}(\mathbf{x}+\theta\mathbf{h})\mathbf{h}\|,$$

by Schwarz's inequality. Cancelling $\|\mathbf{u}\| = \|\mathbf{f}(\mathbf{y})-\mathbf{f}(\mathbf{x})\|$ gives the result.

A simple consequence of this result is that if $\|D\mathbf{f}(\mathbf{c})\| \le M$ for all $\mathbf{c} \in \text{int}(C)$ then $\|\mathbf{f}(\mathbf{y})-\mathbf{f}(\mathbf{x})\| \le M\|\mathbf{y}-\mathbf{x}\|$ for all $\mathbf{x},\mathbf{y} \in C$. Indeed, for $\mathbf{x} \in \text{int}(C)$, $\mathbf{y} \in C$, $\|\mathbf{f}(\mathbf{y})-\mathbf{f}(\mathbf{x})\| \le \|D\mathbf{f}(\mathbf{c})(\mathbf{y}-\mathbf{x})\| \le \|D\mathbf{f}(\mathbf{c})\| \, \|\mathbf{y}-\mathbf{x}\| \le M\|\mathbf{y}-\mathbf{x}\|$. As above, a limiting argument proves that this holds also for $\mathbf{x} \in C$.

In fact, the correct setting for the last part of Theorems 3.21 and 3.22 is on a *connected* set C, as we will now show. Refer back to Chapter 2 and recall that a convex set is pathwise connected, since every two points can be joined by a straight line segment, so is connected.

Theorem 3.23. *Let $G\subseteq\mathbb{R}^n$ be an open, connected set and let $\mathbf{f}: G \longrightarrow \mathbb{R}^m$ be differentiable on G with $D\mathbf{f}(\mathbf{x})=\mathbf{0}$ for all $\mathbf{x}\in G$. Then* **f** *is constant.*

Proof. Let $\mathbf{a}$ be any point of G, and let $U:=\{\mathbf{x}\in G : \mathbf{f}(\mathbf{x})=\mathbf{f}(\mathbf{a})\}$, $V:=\{\mathbf{x}\in G : \mathbf{f}(\mathbf{x})\neq\mathbf{f}(\mathbf{a})\}$. As **f** is differentiable on G, it is continuous and therefore V is an open set. In fact, by Problem 2.5, if $\mathbf{x}\in V$ then $\mathbf{f}(\mathbf{x}) - \mathbf{f}(\mathbf{a})\neq 0$ and there is a δ-neighbourhood of **x** with $\mathbf{f}(\mathbf{y}) - \mathbf{f}(\mathbf{a})\neq 0$ for all **y** in the δ-neighbourhood. We show that U is also open. Indeed let $\mathbf{x}\in U$. Then there is a neighbourhood of **x** lying entirely in the open set G. This neighbourhood is a convex set so, by Theorem 3.22, **f** is constant on that neighbourhood, and the neighbourhood therefore lies in U. We now have $G=U\cup V$ where U and V are disjoint open sets. Since G is connected this is only possible if one of the sets is empty. As $\mathbf{a}\in U$, V must be empty and $G=U$.

It follows readily from this result that the only functions which have constant derivatives are the affine functions (problem 3.26). It is essential in Theorem 3.23 that G has interior points. There is an example, due to H.Whitney [17], of a connected set A in $\mathbb{R}^2$ with $\text{int}(A)=\emptyset$ and of a nonconstant differentiable function with derivative 0 on A.

The above results enable us to give for C^1 mappings a uniform version of the approximation of **f** by its derivative; compare Corollary 3.5. This is extremely useful in obtaining local properties of **f** from properties of $D\mathbf{f}(\mathbf{a})$ and will be used in Chapter 4.

Proposition 3.24. *Let G be an open subset of $\mathbb{R}^n$ and $\mathbf{f}\in C^1(G,\mathbb{R}^m)$. For $\mathbf{a}\in G$ and $\varepsilon>0$ there is $\delta>0$ such that $B(\mathbf{a},\delta)\subseteq G$ and*

$$\|\mathbf{f}(\mathbf{x}) - \mathbf{f}(\mathbf{y}) - D\mathbf{f}(\mathbf{a})[\mathbf{x} - \mathbf{y}]\| \le \varepsilon\|\mathbf{x} - \mathbf{y}\|, \text{ for all } \mathbf{x},\mathbf{y}\in B(\mathbf{a},\delta).$$

Hence $\|\mathbf{f}(\mathbf{x}) - \mathbf{f}(\mathbf{y})\| \le (\|D\mathbf{f}(\mathbf{a})\| + \varepsilon)\|\mathbf{x} - \mathbf{y}\|$, *for all* $\mathbf{x},\mathbf{y}\in B(\mathbf{a},\delta)$.

Proof. Let $\mathbf{L}=D\mathbf{f}(\mathbf{a})$ and define $\mathbf{g}: G\longrightarrow\mathbb{R}^m$ by $\mathbf{g}(\mathbf{x})=\mathbf{f}(\mathbf{x}) - \mathbf{L}\mathbf{x}$. Then $D\mathbf{g}(\mathbf{a})=D\mathbf{f}(\mathbf{a}) - \mathbf{L}=\mathbf{0}$, the zero operator (matrix). By the remark following Theorem 2.86, $\|D\mathbf{g}(\mathbf{x})\|^2\le \sum_{i=1}^{m}\sum_{j=1}^{n}\left|\frac{\partial g_i}{\partial x_j}\right|^2$. Since $\frac{\partial g_i}{\partial x_j}=0$ and each partial derivative is continuous on G, there is a δ-neighbourhood of **a**, contained in G, so that $\|D\mathbf{g}(\mathbf{x})\|\le\varepsilon$ for $\mathbf{x}\in B(\mathbf{a},\delta)$. By the remark following Theorem 3.22, it follows that $\|\mathbf{g}(\mathbf{x})-\mathbf{g}(\mathbf{y})\|\le\varepsilon\|\mathbf{x}-\mathbf{y}\|$, that is, $\|\mathbf{f}(\mathbf{x})-\mathbf{f}(\mathbf{y})- D\mathbf{f}(\mathbf{a})[\mathbf{x}-\mathbf{y}]\|\le\varepsilon\|\mathbf{x}-\mathbf{y}\|$. Hence, for $\mathbf{x},\mathbf{y}\in B(\mathbf{a},\delta)$,

$$\|\mathbf{f}(\mathbf{x})-\mathbf{f}(\mathbf{y})\| \le \|\mathbf{f}(\mathbf{x})-\mathbf{L}\mathbf{x}-(\mathbf{f}(\mathbf{y})-\mathbf{L}\mathbf{y})\| + \|\mathbf{L}\mathbf{x}-\mathbf{L}\mathbf{y}\| \le (\varepsilon+\|\mathbf{L}\|)\|\mathbf{x}-\mathbf{y}\|.$$

3.5. HIGHER ORDER PARTIAL DERIVATIVES

If the first order partial derivatives of a function

$f: G \subseteq \mathbb{R}^n \longrightarrow \mathbb{R}$ exist on G, then the functions $\partial f/\partial x_i$ may also have partial derivatives. We call these *second order partial derivatives* of f and we write

$$\frac{\partial}{\partial x_j}\left(\frac{\partial f}{\partial x_i}\right) \text{ as } \frac{\partial^2 f}{\partial x_j \partial x_i} \text{ or } D_j D_i f \text{ . When } i=j \text{ we write } \frac{\partial^2 f}{\partial x_i^2}.$$

It may be possible to continue this process and obtain third (and higher) order partial derivatives. Of course, when $n=2$ (or $n=3$) we write x,y (or x,y,z) rather than use subscripts.

Example 3.25. Let $f(x,y)=x\,y^2+y\,e^x$.

Then $\partial f/\partial x=y^2+ye^x$, $\partial f/\partial y=2xy+e^x$, $\partial^2 f/\partial x^2=ye^x$, $\partial^2 f/\partial y^2=2x$,

and $$\frac{\partial^2 f}{\partial x\partial y}=\frac{\partial}{\partial x}\left(\frac{\partial f}{\partial y}\right)=\frac{\partial}{\partial x}\left[2xy+e^x\right]=2y+e^x,\ \frac{\partial^2 f}{\partial y\partial x}=\frac{\partial}{\partial y}\left[y^2+ye^x\right]=2y+e^x.$$

Note that the mixed partial derivatives $\partial^2 f/\partial x\partial y$ and $\partial^2 f/\partial y\partial x$ are equal in the above example. This may be either surprising or perhaps intuitively obvious; $\partial^2 f/\partial x\partial y$ first leaves x alone and differentiates with respect to y, then leaves y alone and differentiates with respect to x, whereas $\partial^2 f/\partial y\partial x$ does the same but in the opposite order. Of course, changing the order of operations is not allowed willy-nilly and so we may suspect that different answers are possible.

Example 3.26. Let $$f(x,y)=\begin{cases}\dfrac{xy(x^2-y^2)}{x^2+y^2}, & \text{if } (x,y)\neq(0,0)\\ 0, & \text{at } (0,0)\end{cases}.$$

Then $f\in C^1(\mathbb{R}^2,\mathbb{R})$ but $\dfrac{\partial^2 f}{\partial y\partial x}\neq\dfrac{\partial^2 f}{\partial x\partial y}$ at $(0,0)$.

Solution. For points other than (0,0),

$$\frac{\partial f}{\partial x}=\frac{\partial}{\partial x}\left[y(x^3-y^2x)(x^2+y^2)^{-1}\right]=y(3x^2-y^2)(x^2+y^2)^{-1}-2x^2y(x^2-y^2)(x^2+y^2)^{-2},$$

$$\frac{\partial f}{\partial y}=\frac{\partial}{\partial y}\left[y(x^3-y^2x)(x^2+y^2)^{-1}\right]=(x^3-3xy^2)(x^2+y^2)^{-1}-2xy^2(x^2-y^2)(x^2+y^2)^{-2}.$$

At (0,0), $\dfrac{f(h,0)-f(0,0)}{h}=0$ for all nonzero h so $\partial f/\partial x=0$. Similarly, $\dfrac{\partial f}{\partial y}=0$. Now $$\left|\frac{\partial f}{\partial x}\right|\leq\frac{|y|(|3x^2|+|y^2|)}{x^2+y^2}+\frac{2|x^2||y|(|x|^2+|y|^2)}{(x^2+y^2)^2},$$
and writing (polar co-ordinates) $x=r\cos\theta$, $y=r\sin\theta$ where $r^2=x^2+y^2$, we see that $|\partial f/\partial x|\longrightarrow 0$ as $r\longrightarrow 0$. This shows that $\partial f/\partial x$ is continuous at (0,0). Continuity at other points is a simple consequence of the

rules for sums, products and quotients. In exactly the same way we see that $\partial f/\partial y$ is continuous. Now, at points other than (0,0) we have

$$\frac{\partial^2 f}{\partial y \partial x} = (3x^2-3y^2)(x^2+y^2)^{-1} - 2y^2(3x^2-y^2)(x^2+y^2)^{-2} - (2x^4-6x^2y^2)(x^2+y^2)^{-2} + 8x^2y^2(x^2-y^2)(x^2+y^2)^{-3}$$

and

$$\frac{\partial^2 f}{\partial x \partial y} = (3x^2-3y^2)(x^2+y^2)^{-1} - 2x^2(x^2-3y^2)(x^2+y^2)^{-2} - 2y^2(3x^2-y^2)(x^2+y^2)^{-2} + 8x^2y^2(x^2-y^2)(x^2+y^2)^{-3},$$

and these expressions are equal. However, at (0,0),

$$\frac{\partial^2 f}{\partial y \partial x} = \lim_{k \to 0} \frac{1}{k}\left[\frac{\partial f}{\partial x}(0,k) - \frac{\partial f}{\partial x}(0,0)\right] = \lim_{k \to 0} \frac{1}{k}[-k-0] = -1,$$

while $$\frac{\partial^2 f}{\partial x \partial y} = \lim_{h \to 0} \frac{1}{h}\left[\frac{\partial f}{\partial y}(h,0) - \frac{\partial f}{\partial y}(0,0)\right] = \lim_{h \to 0} \frac{1}{h}[h-0] = 1.$$

In spite of this example (and others) this situation is the exception rather than the rule. We shall give two sets of sufficient conditions under which equality holds. The first result contains the stronger hypotheses but these are usually met in practical examples.

We shall write $f \in C^2(G)$ if $f: G \subseteq \mathbb{R}^n \longrightarrow \mathbb{R}$ and all second order partial derivatives of f are continuous on G and say f is of class C^2. Note that $f \in C^2(G) \Rightarrow f \in C^1(G)$. For, if $f \in C^2$ then $D_i f \in C^1$ (for each i) so by Theorem 3.11, $D_i f$ is differentiable and hence continuous.

Theorem 3.27. *Let $f \in C^2(G)$. Then for all $\mathbf{a} \in G$, and any i, j with $1 \le i, j \le n$, $D_i D_j f(\mathbf{a}) = D_j D_i f(\mathbf{a})$.*

Proof. Let $g(x,y) = f(\hat{\mathbf{a}}(x,y))$, where $\hat{\mathbf{a}}$ denotes $\mathbf{a}$ modified by replacing a_i by x, a_j by y. Clearly g has continuous second order partial derivatives and we have to show that

$$D_1 D_2\, g(x_0, y_0) = D_2 D_1\, g(x_0, y_0), \text{ where } (x_0, y_0) = (a_i, a_j)$$

Consider the expression

$$E(h,k) = g(x_0+h, y_0+k) - g(x_0+h, y_0) - g(x_0, y_0+k) + g(x_0, y_0).$$

Let $\phi(x) = g(x, y_0+k) - g(x, y_0)$, and $\psi(y) = g(x_0+h, y) - g(x_0, y)$.

Then $$E(h,k) = \phi(x_0+h) - \phi(x_0) = \psi(y_0+k) - \psi(y_0).$$

By the mean value theorem,

$$\phi(x_0+h) - \phi(x_0) = h\phi'(x_0+\theta_1 h), \text{ where } 0 < \theta_1 < 1,$$
$$= h[D_1 g(x_0+\theta_1 h, y_0+k) - D_1 g(x_0+\theta_1 h, y_0)]$$

$$= hk\, D_2D_1g(x_0+\theta_1h,\ y_0+\theta_2k), \quad 0<\theta_2<1.$$

Similarly, $\psi(y_0+k) - \psi(y_0) = kh\, D_1D_2g(x_0+\theta_4h,\ y_0+\theta_3k), \quad 0<\theta_3,\theta_4<1.$

Therefore, $D_1D_2g(x_0+\theta_4h,\ y_0+\theta_3k) = D_2D_1g(x_0+\theta_1h,\ y_0+\theta_2k).$

As D_1D_2g and D_2D_1g are continuous, for any $\varepsilon>0$ there is $\delta>0$ so that

$$|D_1D_2g(x_0+\theta_4h,\ y_0+\theta_3k) - D_1D_2g(x_0,y_0)| < \varepsilon/2 \quad \text{for } h^2+k^2<\delta^2$$

and

$$|D_2D_1g(x_0+\theta_1h,\ y_0+\theta_2k) - D_2D_1g(x_0,y_0)| < \varepsilon/2 \quad \text{for } h^2+k^2<\delta^2.$$

Hence, by the triangle inequality, $|D_1D_2g(x_0,y_0) - D_2D_1g(x_0,y_0)| < \varepsilon$. As $\varepsilon>0$ is arbitrary this proves the desired equality.

This result extends to higher order partial derivatives. For example, if f has continuous third order partial derivatives then $D_iD_jD_kf = D_jD_iD_kf = D_jD_kD_if$. The first equality follows by applying the theorem to D_kf, the second is D_j of two equal objects, etc.

The following result is of less practical use but has weaker hypotheses. We only give the version for $n=2$.

Theorem 3.28. *Let $f: G\subseteq\mathbb{R}^2 \longrightarrow \mathbb{R}$ and suppose D_1f, D_2f and D_1D_2f exist and are continuous on the open set G containing (x_0,y_0). Then $D_2D_1f(x_0,y_0)$ exists and equals $D_1D_2f(x_0,y_0)$.*

Proof. Let h, k be chosen so that $(x_0+h,\ y_0+k)$ lies in G.

We have to prove that $\dfrac{D_1f(x_0,\ y_0+k) - D_1f(x_0,\ y_0)}{k} := \dfrac{F(k)}{k}$ has a limit as $k\longrightarrow 0$ and that this limit is $D_1D_2f(x_0,y_0)$.

Since $D_1f(x,y) = \lim\limits_{h\to 0} \dfrac{f(x+h,y)-f(x,y)}{h}$ exists on G, we have

$$\frac{F(k)}{k} = \frac{1}{k}\left[\lim_{h\to 0} \frac{f(x_0+h,y_0+k)-f(x_0+h,y_0) - f(x_0,y_0+k)+f(x_0,y_0)}{h}\right]$$

By the mean value theorem, for any fixed k,

$$\frac{F(k)}{k} = \frac{1}{k}\left[\lim_{h\to 0} \frac{k}{h}\,(D_2f(x_0+h,\ y_0+\theta k) - D_2f(x_0,\ y_0+\theta k))\right]$$

$$= \lim_{h\to 0} D_1D_2f(x_0+\phi h,\ y_0+\theta k)$$

where $0<\theta,\phi<1$, but θ,ϕ depend on x_0, y_0, h and k. Since the behaviour of θ,ϕ as $h\longrightarrow 0$ is unknown we have to proceed carefully. For any $\varepsilon>0$, as $F(k)/k$ is the above limit, there is $\delta_1>0$ (depending on k) so that

$$\left|\frac{F(k)}{k} - D_1D_2f(x_0+\phi h,\ y_0+\theta k)\right| < \varepsilon/2, \text{ for all } |h|<\delta_1\ . \tag{3.9}$$

As D_1D_2f is continuous, there is a δ-neighbourhood V of (x_0,y_0) such that $|D_1D_2f(x,y) - D_1D_2f(x_0,y_0)| < \varepsilon/2$ for all $(x,y)\in V$.

We then have

$$|D_1D_2f(x_0+\phi h,y_0+\theta k) - D_1D_2f(x_0,y_0)| < \varepsilon/2, \tag{3.10}$$

for all h,k with $h^2+k^2<\delta^2$. Take any k with $|k|<\delta/2$. Then, taking h with $|h|<\min(\delta/2,\ \delta_1)$ and applying the triangle inequality to (3.9) and (3.10) we obtain

$$\left|\frac{F(k)}{k} - D_1D_2\,f(x_0,y_0)\right| < \varepsilon \quad \text{for every } |k|<\delta/2.$$

This proves that $\lim_{k\to 0}\frac{F(k)}{k} = D_1D_2f(x_0,y_0)$ as required.

Problems.

1. Show directly that derivatives are unique, that is, if **L** and **M** are linear operators that both satisfy the conditions in Definition 3.1 then **Lx**=**Mx** for every **x**.

2. Show that if $\mathbf{L}:\mathbb{R}^n\longrightarrow\mathbb{R}^m$ is linear and $\mathbf{f}:\mathbb{R}^p\longrightarrow\mathbb{R}^n$ is differentiable at $\mathbf{a}\in\mathbb{R}^p$ then **L**o**f** is differentiable at **a** and $D(\mathbf{L}o\mathbf{f})(\mathbf{a})=\mathbf{L}\,D\mathbf{f}(\mathbf{a})$.

3. Let $\mathbf{a},\mathbf{e}\in\mathbb{R}^n$ and let $\mathbf{g}:\mathbb{R}\longrightarrow\mathbb{R}^n$ be defined by $\mathbf{g}(t)=\mathbf{a}+t\mathbf{e}$. Show that **g** is differentiable and that $D\mathbf{g}(t)=\mathbf{e}$ for all $t\in\mathbb{R}$.

4. Let $f:\mathbb{R}^2\longrightarrow\mathbb{R}$ be given by $f(x,y)=\begin{cases}\dfrac{x^3-y^3}{x^2+y^2} & \text{if } (x,y)\neq(0,0)\\[2mm] 0 & \text{when } (x,y)=(0,0)\end{cases}$.

Calculate $\frac{\partial f}{\partial x}$ and $\frac{\partial f}{\partial y}$ at (0,0). Is f differentiable there?

5. Let $f:\mathbb{R}^2\longrightarrow\mathbb{R}$ be given by $f(x,y)=|x\,y|^{1/2}$. Show that $\partial f/\partial x$ and $\partial f/\partial y$ both exist at (0,0) but f is not differentiable there.

6. Let $f:\mathbb{R}^3\longrightarrow\mathbb{R}$ be defined by $f(x,y,z)=\begin{cases}\dfrac{x\,yz}{(x^2+y^2+z^2)^\alpha}, & (x,y,z)\neq(0,0,0)\\[2mm] 0, & (x,y,z)=(0,0,0)\end{cases}$

where α is a real constant. Show that f is differentiable at (0,0,0) if and only if $\alpha<1$.

7. Let $f:\mathbb{R}^2\longrightarrow\mathbb{R}$ be given by $f(x,y)=(x\,y)^{1/3}$. Show that $\partial f/\partial x$ and $\partial f/\partial y$ both exist at (0,0). Let $\phi(x)=x^2$. Calculate $\frac{d}{dx}f(x,\phi(x))$ at $x=0$ directly and then by the chain rule. Why do the answers differ?

8. Let $w=f(u,v)=u^2+v^2$ and suppose that $u=2x+3y$, $v=4x-5y$. Calculate $\partial w/\partial x$ and $\partial w/\partial y$ (*a*) directly by writing w in terms of x,y ; (*b*) using the chain rule.

9. Let $f(x,y)=\begin{cases}\dfrac{x^3-3y^2x}{x^2+y^2} & \text{if } (x,y)\neq(0,0)\\ 0 & \text{at } (0,0)\end{cases}$. Show that f is not differentiable at (0,0). Now let $x=r\cos\theta$, $y=r\sin\theta$. Show that $f(r\cos\theta,\ r\sin\theta)=r\cos(3\theta)$ when $r\neq0$. Show that $g(r,\theta)=\begin{cases}r\cos3\theta & \text{if } r\neq0\\ 0 & \text{if } r=0\end{cases}$ is differentiable at $r=0$. Why does this not contradict the chain rule?

10. Let $f,\,g:\mathbb{R}\longrightarrow\mathbb{R}$ be differentiable on $\mathbb{R}$ and let $h(x,y)=f(x)+g(y)$. Show that h is differentiable on $\mathbb{R}^2$. Use this to give an example of a function $h:\mathbb{R}^2\longrightarrow\mathbb{R}$ that is differentiable but not of class C^1.

11. Give an example of a function $f:\mathbb{R}^2\longrightarrow\mathbb{R}$ that is differentiable at (0,0) but is discontinuous at all other points.

12. Let $\mathbf{f}:G\subseteq\mathbb{R}^n\longrightarrow\mathbb{R}^m$ be differentiable at $\mathbf{a}\in G$ and let $\mathbf{M}:\mathbb{R}^n\longrightarrow\mathbb{R}^m$ be a linear operator. Prove that, if for some direction $\mathbf{e}$, $D\mathbf{f}(\mathbf{a})\mathbf{e}\neq\mathbf{M}\mathbf{e}$, then there is $\delta>0$ such that for $|t|<\delta$

$$\|\mathbf{f}(\mathbf{a}+t\mathbf{e})-\mathbf{f}(\mathbf{a})-tD\mathbf{f}(\mathbf{a})\mathbf{e}\|<\|\mathbf{f}(\mathbf{a}+t\mathbf{e})-\mathbf{f}(\mathbf{a})-t\mathbf{M}\mathbf{e}\|\ .$$

[The best local affine approximation in the direction $\mathbf{e}$ is $\mathbf{f}(\mathbf{a})+tD\mathbf{f}(\mathbf{a})\mathbf{e}$.]

13. Find the equation of the tangent plane to the surface $z=3-2x^2-y^2$ at the point (1,–1,0).

14. A function $f:G\subseteq\mathbb{R}^n\longrightarrow\mathbb{R}$ is said to be *positively homogeneous of degree p* if $f(t\,\mathbf{x})=t^pf(\mathbf{x})$ for all $\mathbf{x}\in G$ and all $t>0$. Prove that, if f is differentiable on G, then $<\nabla f(\mathbf{x}),\mathbf{x}>=p\,f(\mathbf{x})$. (Euler's theorem.) Prove also the converse as follows. For a fixed $\mathbf{y}\in G$, $\mathbf{y}\neq\mathbf{0}$, let $g(t):=f(t\mathbf{y})$, $t>0$. Show from $<\nabla f(\mathbf{x}),\mathbf{x}>=p\,f(\mathbf{x})$ for $\mathbf{x}\in G$ that $t\,g'(t)=p\,g(t)$ and hence that $f(t\,\mathbf{y})=t^pf(\mathbf{y})$.

15. Let $\mathbf{f}:\mathbb{R}^n \longrightarrow \mathbb{R}^m$ be differentiable at $\mathbf{a}$ and let $\mathbf{h}$ be defined by $\mathbf{h}(\mathbf{x})=\mathbf{f}(\mathbf{x}+\mathbf{a})$. Show that $\mathbf{h}$ is differentiable at $\mathbf{0}$ and that $D\mathbf{h}(\mathbf{0})=D\mathbf{f}(\mathbf{a})$.

16. For functions $f, g:\mathbb{R}^n \longrightarrow \mathbb{R}$ let $\mathbf{h}:\mathbb{R}^n \longrightarrow \mathbb{R}^2$ be defined by $\mathbf{h}(\mathbf{x})=(f(\mathbf{x}), g(\mathbf{x}))$. Prove that $\mathbf{h}$ is differentiable if and only if f and g are. Show that a function $\mathbf{f}:\mathbb{R}^n \longrightarrow \mathbb{R}^m$ is differentiable if and only if each component function is differentiable.

17. Let $H:\mathbb{R}^2 \longrightarrow \mathbb{R}$ be defined by $H(x,y)=xy$. Show that H is differentiable on $\mathbb{R}^2$. Deduce, using problem 16, that the pointwise product fg of differentiable functions $f,g:\mathbb{R}^n \longrightarrow \mathbb{R}$ is differentiable and $D(fg)(\mathbf{a}) = g(\mathbf{a})\,Df(\mathbf{a})+f(\mathbf{a})\,Dg(\mathbf{a})$.

18. Let $\mathbf{f},\mathbf{g}: G\subseteq\mathbb{R}^n \longrightarrow \mathbb{R}^m$ be differentiable and let $h:A \longrightarrow \mathbb{R}$ be defined by $h(\mathbf{x})=<\mathbf{f}(\mathbf{x}),\mathbf{g}(\mathbf{x})>$. Prove that h is differentiable and that for $\mathbf{a}\in A$, $D\,h(\mathbf{a})\mathbf{x} = <D\mathbf{f}(\mathbf{a})\mathbf{x}, \mathbf{g}(\mathbf{a})> + <\mathbf{f}(\mathbf{a}), D\mathbf{g}(\mathbf{a})\mathbf{x}>$.

19. Let $\phi(t)=\begin{cases}\sin(t)/t & \text{if } t\neq 0\\ 1 & \text{at } t=0\end{cases}$. Show that ϕ is differentiable at all points. Let $f:\mathbb{R}^2 \longrightarrow \mathbb{R}$ be defined by $f(x,y)=\dfrac{\cos x-\cos y}{x-y}$ if $x\neq y$ and $f(x,x)=-\sin x$. By writing f in terms of ϕ, show that f is differentiable.

20. Let $f: G\subseteq\mathbb{R}^2 \longrightarrow \mathbb{R}$ be such that $\dfrac{\partial f}{\partial y}$ exists and is continuous on G and that $\dfrac{\partial f}{\partial x}$ exists at $(a,b)\in G$. Prove that f is differentiable at (a,b).

21. Find an expression for the directional derivative of $\phi(x,y)=x^2y+y^3x$ at the point $(1,1)$ in an arbitrary direction $\mathbf{u}=(\cos\theta, \sin\theta)$ and determine the direction which makes this a maximum.

22. Let $f(x,y)=\begin{cases}y^3/(x^2+y^2) & \text{if } (x,y)\neq(0,0)\\ 0 & \text{at } (0,0)\end{cases}$. Show that, for every direction $\mathbf{u}$, the directional derivative of f exists at $(0,0)$ but, for certain directions, $D_{\mathbf{u}}f\neq <\nabla f, \mathbf{u}>$.

23. Let C be an open convex subset of $\mathbb{R}^n$ and let $\mathbf{f}: C \longrightarrow \mathbb{R}^m$ be differentiable on C. Suppose that $\mathbf{x},\mathbf{y}\in C$ and $\mathbf{u}\in\mathbb{R}^n$. Write $\mathbf{h}=\mathbf{y}-\mathbf{x}$ and for $t\in[0,1]$ let $\phi(t) := <\mathbf{u},\mathbf{f}(\mathbf{x}+t\mathbf{h})>$. Show that ϕ is differentiable on $(0,1)$ and that $\phi'(t)=<\mathbf{u}, D\mathbf{f}(\mathbf{x}+t\mathbf{h})\mathbf{h}>$.

24. Let C be an open convex subset of $\mathbb{R}^n$ and let $\mathbf{f} : C \longrightarrow \mathbb{R}^n$ be differentiable on C. Suppose the matrix $D\mathbf{f}(\mathbf{x})$ is positive definite on C, that is, $\sum_{i,j=1}^{n} \frac{\partial f_i}{\partial x_j} h_i h_j > 0$ for all $\mathbf{x} \in C$ and all $\mathbf{h} \neq \mathbf{0}$. Show that $\mathbf{f}$ is one-to-one on C.

25. Let $\mathbf{f} : [0,2\pi] \longrightarrow \mathbb{R}^2$ be defined by $\mathbf{f}(x)=(\sin x, 1-\cos x)$. Show that $\mathbf{f}(2\pi)-\mathbf{f}(0)$ never equals $2\pi D\mathbf{f}(\theta)$ for any $\theta \in (0,2\pi)$ (there can be no *MVT*). Apply Theorem 3.22 to obtain an inequality for $\|\mathbf{f}(t)-\mathbf{f}(0)\|$. Deduce that $\cos t \geq 1 - t^2/2$.

26. Let G be an open connected subset of $\mathbb{R}^n$ and let $\mathbf{f} : G \longrightarrow \mathbb{R}^m$ have a constant derivative on G, that is, $D\mathbf{f}(\mathbf{x})=\mathbf{L}$ for all $\mathbf{x} \in G$. Prove that $\mathbf{f}$ is affine.

27. Let C be a convex subset of $\mathbb{R}^2$ and let $f: C \longrightarrow \mathbb{R}$ be differentiable on C. Suppose that $\partial f/\partial y=0$ on C. Prove that f does not depend on y, that is $f(x,y_1)=f(x,y_2)$ for all (x,y_1), (x,y_2) in C. Is the same true if C is replaced by an open connected set G?

28. For $\alpha > 0$ let $f(x, y)=\dfrac{x y^{\alpha}}{x^2+y^2}$ if $(x, y) \neq (0,0)$ and $f(0,0)=0$. Show that f is of class C^1 if $\alpha > 2$. Show also that for $1 < \alpha < 3$, $\dfrac{\partial^2 f}{\partial x \partial y} \neq \dfrac{\partial^2 f}{\partial y \partial x}$ at $(0,0)$. Include here the possibility that only one of these exists. [Assume α is such that y^{α} is well defined.]

29. Let $g : \mathbb{R} \longrightarrow \mathbb{R}$ be differentiable at all points of $[a,b]$ (but g' is not necessarily continuous). Show that if $g'(a) < 0$ and $g'(b) > 0$ then there exists $c \in (a,b)$ such that $g'(c) = 0$. Deduce that if f is differentiable on $[a,b]$ and $f'(a) = \alpha$ and $f'(b) = \beta$ then f' assumes every value between α and β. (A theorem of Darboux.)

Hints on the problems

1. Prove that for any $\varepsilon > 0$ there is $\delta > 0$ such that $\|\mathbf{Lh}-\mathbf{Mh}\| < 2\varepsilon\|\mathbf{h}\|$ if $\|\mathbf{h}\| < \delta$. For any $\mathbf{x} \neq \mathbf{0}$ consider $\delta\mathbf{x}/\|\mathbf{x}\|$.

9. $\cos 3\theta = 4\cos^3\theta - 3\cos\theta$ is a standard formula.

10. See the first part of 19 for an example.

11. Use some simple formula on irrationals, zero on rationals.

12. *L.*(*R.*)*H.S.* denotes *L*eft (*R*ight) *H*and *S*ide. $\dfrac{\|L.H.S.\|}{|t|} = \dfrac{\|\mathbf{R}(t)\|}{|t|} \longrightarrow 0$

while $\dfrac{\|R.H.S.\|}{|t|} = \|[D\mathbf{f}(\mathbf{a})-\mathbf{M}]\mathbf{e}+\dfrac{\mathbf{R}(t)}{t}\| \geq \|[D\mathbf{f}(\mathbf{a})-\mathbf{M}]\mathbf{e}\| - \dfrac{\|\mathbf{R}(t)\|}{|t|}$.

14. Let $\phi(t)=f(t\mathbf{x})$ and calculate $\phi'(1)$ in two ways.

18. The product is $H\circ\mathbf{h}$.

19. Note that $f(x,y) = -\sin(\frac{x+y}{2})\phi(\frac{x-y}{2})$. Use 17 and the chain rule.

20. Write $f(a+h, b+k) - f(a+h, b)+f(a+h, b) - f(a, b)$, use the *MVT* plus the continuity of $\partial f/\partial y$ on the first pair, use existence of $\partial f/\partial x$ on the other pair.

24. For $\mathbf{a}\neq\mathbf{b}$ write $\mathbf{h}=\mathbf{b}-\mathbf{a}$. Apply the *MVT* to $g(\mathbf{x})=\langle\mathbf{f}(\mathbf{x})-\mathbf{f}(\mathbf{a}),\mathbf{h}\rangle$.

27. Last part, NO. Imagine a domain that bends back on itself.

29. g attains its minimum on $[a,b]$.

Chapter 4

The implicit and inverse function theorems

4.1. EXISTENCE THEOREMS

In this chapter we shall prove two very important results : the inverse function theorem and the implicit function theorem. The first of these studies the invertibility of a function **f**. A function **f** assigns to each **x** in its domain a unique point **y** (the image of **x** under **f**). Often we are interested in the inverse process, that is given **y** determine the **x** from which it came. For example, if $y=(x^2+1)e^x$, it is easy to compute y given x, but it is less easy to determine x so that, for example, $(x^2+1)e^x=1$. Indeed we need to know that there is exactly one x satisfying this equation : the function is one-to-one. Then an inverse function will exist. When **f** is one-to-one on a set S, $\mathbf{f}^{-1}$ is defined from $\mathbf{f}(S)$ into S. Since $\mathbf{f}(S)$ need have no interior point we cannot, in general, discuss differentiability of the inverse. If **f** is a C^1 function it is well approximated by the derivative and we may expect properties of the derivative to carry over into local properties of the function. This is indeed true. The inverse function theorem says that invertibility of $D\mathbf{f}(\mathbf{a})$ implies local invertibility of **f**. Moreover, $\mathbf{f}^{-1}$ is differentiable with derivative $[D\mathbf{f}(\mathbf{a})]^{-1}$.

The second result deals with a situation which is familiar in elementary calculus. Here we meet examples where some relation $F(x,y)=0$ is said to define y implicitly as a function of x. The idea behind this statement is that once x is given then y is also known : loosely the equation $F(x,y)=0$ can be "solved" for y in terms of x. This may be plausible if we think in terms of degrees of freedom. The rough idea is that a set has one degree of freedom if, at any point, it is possible to move in only one direction; two degrees of freedom correspond to it being possible to move in two independent directions, etc. For example in three dimensions the maximum number of degrees of freedom is 3, an open set in $\mathbb{R}^3$ would have exactly 3, but a surface would have 2 and a curve 1.

Consider an equation relating two variables. To begin with there are 2 degrees of freedom but an equation such as $F(x,y)=0$ imposes one constraint and will reduce the number of degrees of freedom by one; we may consider x to be free but y is not, it is determined by x. Although this may be an aid to intuition, it is far from the truth of the matter.

What does it mean to say that $F(x,y)=0$ defines y implicitly as a function of x? It means that there is a function f such that $F(x,y)=0$ is equivalent to $y=f(x)$. Of course, we should not expect to be able to write down a formula for f in general, but, for a given x, there must be a uniquely determined y. The set of points which satisfy $F(x,y)=0$, usually a curve in $\mathbb{R}^2$, must be the graph of some function f. The example $x^2+y^2-1=0$, the circle centred at (0,0) and of radius 1, shows that this may well be impossible : for any x, $-1<x<1$, there are two y's which satisfy the relation. The curve $x^2+y^2=1$ is made up of the graph of two separate functions $f_1(x)=\sqrt{(1-x^2)}$, $f_2(x)=-\sqrt{(1-x^2)}$, $-1\leq x\leq 1$. However, if we take some point on the circle and concentrate on a neighbourhood of that point, that portion of the circle is definitely either part of the graph of f_1 or of f_2, except for the points (-1,0) and (1,0) where the circle bends back on itself. Therefore, any general theorem can only be *local* and must exclude points where there is a vertical tangent, which points can be decided by looking at the derivative. The implicit function theorem is such a general result which deals with many variables not just two. It gives appropriate conditions so that an equation $\mathbf{F(x,y)=0}$ is locally equivalent to $\mathbf{y=f(x)}$ for some function $\mathbf{f}$, and that differentiability properties of $\mathbf{F}$ carry over to $\mathbf{f}$.

The theorems are important ones with many applications some of which we cover in this chapter.

The two theorems are closely related. Indeed the inverse function theorem is a special case of the implicit function theorem : take $\mathbf{F(u,v)=f(u)-v}$. Solving for $\mathbf{u}$ in terms of $\mathbf{v}$ is determining the inverse of $\mathbf{f}$. It is also possible to deduce the implicit function theorem from the inverse function theorem : we have adopted the latter approach.

We have several reasons for tackling the theorems as we do. Firstly, the inverse theorem splits fairly naturally into three parts :

(1) The injectivivity theorem.

For $\mathbf{f}\in C^1$, $D\mathbf{f(a)}$ one-to-one $\Rightarrow$ $\mathbf{f}$ is locally one-to-one.

(2) The surjectivity theorem.

For $\mathbf{f}\in C^1$, $D\mathbf{f(a)}$ onto $\Rightarrow$ $\mathbf{f}$ is locally onto.

(3) Differentiability of the inverse.

(1) is an easy consequence of the approximation of **f** by $D\mathbf{f}(\mathbf{a})$ as embodied in Proposition 3.24.
(2) also uses this approximation together with an argument to show that the equation $\mathbf{f}(\mathbf{x})=\mathbf{z}$ has a solution.
(3) involves showing that the appropriate limit exists.

Secondly, (1) and (2) have independent interest.

Thirdly, it is at least as easy to deduce the differentiability of the function **f** in the implicit function theorem via the inverse function theorem as it is to prove it directly.

Fourthly, it is a commonly held belief by students encountering these theorems for the first time that they are impossibly difficult. We hope to disprove this. To help in this, we have found an easy proof of (2), which seems to be new in this context; the ideas are familiar in more advanced courses but are particularly simple here.

The key step in the proof is to prove that some equation has a unique solution. How can we prove the existence of some quantity? There are basically two methods : one is to exhibit it explicitly (via a formula), the other is to obtain it as a limit. The latter method may well be disguised and called an existence theorem and indeed we shall do so here. We consistently use the existence theorem known as Banach's fixed point theorem, or the contraction mapping principle. We make no apologies for using this theorem as it has many other important applications and should be regarded as essential knowledge. It is a crucial ingredient in many existence theorems, for example in proving the existence of solutions of differential equations; see §4.4. Its use allows us to give the simplified proof of (2) intimated above.

We begin to discuss the existence theorem.

Definition 4.1. Let X be a subset of $\mathbb{R}^n$. A mapping $\mathbf{f}: X \longrightarrow \mathbb{R}^n$ is said to be a *contraction* if there is a constant $k<1$ such that $\|\mathbf{f}(\mathbf{x})-\mathbf{f}(\mathbf{y})\| \leq k\|\mathbf{x}-\mathbf{y}\|$ for all $\mathbf{x},\mathbf{y}$ in X.

Thus, the distance between images is less than the distance between the original points and we expect the size of sets to be contracted. It is obvious that a contraction is uniformly continuous. Proving that a mapping is a contraction is often done using a mean value theorem. The following example illustrates this.

Example 4.2. Let $\mathbf{f}: \overline{\mathrm{B}}(0,r) \subseteq \mathbb{R}^n \longrightarrow \mathbb{R}^n$ be of class C^1 and suppose $|D_i f_j(\mathbf{x})| \leq K$ for all $\mathbf{x} \in \overline{\mathrm{B}}(0,r)$, $1 \leq i,j \leq n$. Then **f** is a contraction provided $K<1/n$.

Solution. By the mean value theorem (3.20) applied to each component, $|f_i(\mathbf{x})-f_i(\mathbf{y})| \leq \|Df_i(\mathbf{y}+\theta(\mathbf{x}-\mathbf{y})\| \, \|\mathbf{x}-\mathbf{y}\|$ where $\|Df_i(\mathbf{z})\|$ is the norm of the linear operator $Df_i(\mathbf{z})$ (or $1\times n$ matrix). Thus,

$$\begin{aligned}\|Df_i(\mathbf{z})\| &= \sup_{\|\mathbf{h}\|=1} \|Df_i(\mathbf{z})\mathbf{h}\| \\ &= \sup_{\|\mathbf{h}\|=1} |<\nabla f_i(\mathbf{z}),\mathbf{h}>| \leq \|\nabla f_i(\mathbf{z})\| \text{ , by Schwarz's inequality,} \\ &= \left[\sum_{j=1}^{n} |D_j f_i(\mathbf{z})|^2\right]^{1/2} \leq \sqrt{n}K .\end{aligned}$$

Therefore we have $|f_i(\mathbf{x})-f_i(\mathbf{y})| \leq K\sqrt{n}\|\mathbf{x}-\mathbf{y}\|$ and this yields

$$\|\mathbf{f}(\mathbf{x})-\mathbf{f}(\mathbf{y})\| = \left[\sum_{i=1}^{n} |f_i(\mathbf{x})-f_i(\mathbf{y})|^2\right]^{1/2} \leq Kn\|\mathbf{x}-\mathbf{y}\|.$$

In fact, a more precise result is given following Theorem 3.22.

Now we prove the basic result : Banach's fixed point theorem.

Theorem 4.3. *Let X be a closed subset of $\mathbb{R}^n$ and let $\mathbf{f}: X \longrightarrow X$ be a contraction. Then $\mathbf{f}$ has a unique fixed point in X, that is, there is precisely one point $\mathbf{z}$ in X with $\mathbf{f}(\mathbf{z})=\mathbf{z}$.*

Proof. Let $\mathbf{x}^{(0)}$ be any point of X and define a sequence $\{\mathbf{x}^{(i)}\}$ recursively by $\mathbf{x}^{(i+1)}=\mathbf{f}(\mathbf{x}^{(i)})$, $i\geq 0$. We shall prove that $\{\mathbf{x}^{(i)}\}$ is a Cauchy sequence and therefore convergent since X is complete (Lemma 1.32). Indeed we have

$$\|\mathbf{x}^{(i+1)}-\mathbf{x}^{(i)}\| = \|\mathbf{f}(\mathbf{x}^{(i)}) - \mathbf{f}(\mathbf{x}^{(i-1)})\| \leq k\|\mathbf{x}^{(i)}-\mathbf{x}^{(i-1)}\|, \; i\geq 1.$$

By an easy induction argument, this gives

$$\|\mathbf{x}^{(i+1)}-\mathbf{x}^{(i)}\| \leq k^i\|\mathbf{x}^{(1)}-\mathbf{x}^{(0)}\|, \; i\geq 1.$$

Now, by the triangle inequality, for $p>0$,

$$\begin{aligned}\|\mathbf{x}^{(i)}-\mathbf{x}^{(i+p)}\| &\leq \|\mathbf{x}^{(i)}-\mathbf{x}^{(i+1)}\| + \|\mathbf{x}^{(i+1)}-\mathbf{x}^{(i+2)}\| + \cdots + \|\mathbf{x}^{(i+p-1)}-\mathbf{x}^{(i+p)}\| \\ &\leq (k^i+k^{i+1}+\cdots+k^{i+p-1})\|\mathbf{x}^{(1)}-\mathbf{x}^{(0)}\| \leq \frac{k^i}{1-k}\|\mathbf{x}^{(1)}-\mathbf{x}^{(0)}\|\end{aligned}$$

since $k^i+k^{i+1}+\cdots+k^{i+p-1}$ is part of the convergent geometric series $\sum_{r=i}^{\infty} k^r$. Since $k<1$, $k^i \longrightarrow 0$ as $i \longrightarrow \infty$, so for any $\varepsilon>0$ there exists $N>0$ such that $\|\mathbf{x}^{(i)}-\mathbf{x}^{(i+p)}\| < \varepsilon$ for all $i\geq N$, $p>0$. This proves that $\{\mathbf{x}^{(i)}\}$ is a Cauchy sequence so $\mathbf{x}^{(i)} \longrightarrow \mathbf{z}$ (say); $\mathbf{z}\in X$ since X is complete. As $\mathbf{f}$ is continuous, $\mathbf{f}(\mathbf{x}^{(i)}) \longrightarrow \mathbf{f}(\mathbf{z})$. But $\mathbf{f}(\mathbf{x}^{(i)})=\mathbf{x}^{(i+1)} \longrightarrow \mathbf{z}$, so we have $\mathbf{z}=\mathbf{f}(\mathbf{z})$. This proves existence. For uniqueness, suppose $\mathbf{f}(\mathbf{z})=\mathbf{z}$ and

$\mathbf{f}(\mathbf{y})=\mathbf{y}$. Then $\|\mathbf{z}-\mathbf{y}\|=\|\mathbf{f}(\mathbf{z})-\mathbf{f}(\mathbf{y})\|\le k\|\mathbf{z}-\mathbf{y}\|$ so that $(1-k)\|\mathbf{z}-\mathbf{y}\|\le 0$ and this is only possible if $\mathbf{z}=\mathbf{y}$.

The sequence $\{\mathbf{x}^{(m)}\}$ used in the proof is often called the *sequence of successive approximations*. The proof shows that this sequence converges irrespective of the starting point $\mathbf{x}^{(0)}$. The process may also be called simple iteration. For, $\mathbf{x}^{(1)}=\mathbf{f}(\mathbf{x}^{(0)})$, $\mathbf{x}^{(2)}=\mathbf{f}(\mathbf{x}^{(1)})=\mathbf{f}(\mathbf{f}(\mathbf{x}^{(0)}))$, $\mathbf{x}^{(3)}=\mathbf{f}(\mathbf{f}(\mathbf{f}(\mathbf{x}^{(0)})))$, $\cdots$, which we may write as $\mathbf{x}^{(m)}=\mathbf{f}^{(m)}(\mathbf{x}^{(0)})$ where $\mathbf{f}^{(m)}$ are *iterates* of $\mathbf{f}$.

A perusal of the proof reveals that it is valid for any norm on $\mathbb{R}^n$. In fact, the proof works in much greater generality : the norm may be replaced by a metric d, and X may be any complete metric space. The proof is as above with $d(\mathbf{x},\mathbf{y})$ replacing $\|\mathbf{x}-\mathbf{y}\|$; linearity of the space is not required. It is in this form that it is known as Banach's fixed point theorem and is particularly useful. It can be useful to use other norms since a function may be a contraction with respect to one norm and not another.

Corollary 4.4. *Let X be a closed linear subspace of $\mathbb{R}^n$ and let $\mathbf{f}: X\longrightarrow X$ be a strict contraction. Then* $\mathbf{I}-\mathbf{f}$ *is a homeomorphism, that is,* $\mathbf{I}-\mathbf{f}: X\longrightarrow X$ *is continuous and has a continuous inverse.*

Proof. For $\mathbf{y}\in X$ let $\mathbf{g}(\mathbf{x})=\mathbf{f}(\mathbf{x})+\mathbf{y}$; $\mathbf{y}$ is kept fixed. Then $\mathbf{g}$ is a contraction of X into X (since X is closed under addition) so has a unique fixed point $\mathbf{z}$ (say) : $\mathbf{z}=\mathbf{f}(\mathbf{z})+\mathbf{y}$, $\mathbf{z}\in X$. This proves that $(\mathbf{I}-\mathbf{f})(\mathbf{z})=\mathbf{y}$ so $\mathbf{I}-\mathbf{f}$ is surjective and it is injective by the uniqueness argument. Therefore $(\mathbf{I}-\mathbf{f})^{-1}$ exists. To establish the continuity write $(\mathbf{I}-\mathbf{f})^{-1}(\mathbf{y})=\mathbf{x}$ and $(\mathbf{I}-\mathbf{f})^{-1}(\mathbf{b})=\mathbf{a}$. Then $\mathbf{y}=\mathbf{x}-\mathbf{f}(\mathbf{x})$ and $\mathbf{b}=\mathbf{a}-\mathbf{f}(\mathbf{a})$, and we have

$$\|\mathbf{x}-\mathbf{a}\|=\|\mathbf{f}(\mathbf{x})+\mathbf{y}-(\mathbf{f}(\mathbf{a})+\mathbf{b})\|\le\|\mathbf{f}(\mathbf{x})-\mathbf{f}(\mathbf{a})\|+\|\mathbf{y}-\mathbf{b}\|\le k\|\mathbf{x}-\mathbf{a}\|+\|\mathbf{y}-\mathbf{b}\|$$

That is, $\|(\mathbf{I}-\mathbf{f})^{-1}(\mathbf{y})-(\mathbf{I}-\mathbf{f})^{-1}(\mathbf{b})\|\le\frac{1}{1-k}\|\mathbf{y}-\mathbf{b}\|$, and this proves that $(\mathbf{I}-\mathbf{f})^{-1}$ is uniformly continuous.

Corollary 4.5. *Let* $\mathbf{L}:\mathbb{R}^n\longrightarrow\mathbb{R}^n$ *be a linear operator with* $\|\mathbf{L}\|<1$. *Then* $\mathbf{I}-\mathbf{L}$ *is invertible and* $\|(\mathbf{I}-\mathbf{L})^{-1}\|\le\dfrac{1}{1-\|\mathbf{L}\|}$.

Proof. Corollary 4.4 applies since $\|\mathbf{L}\mathbf{x}-\mathbf{L}\mathbf{y}\|\le\|\mathbf{L}\|\,\|\mathbf{x}-\mathbf{y}\|$.

Corollary 4.6. *Let* $\mathbf{L},\mathbf{M}:\mathbb{R}^n\longrightarrow\mathbb{R}^n$ *be linear operators and suppose that* $\mathbf{L}^{-1}$ *exists and that* $\|\mathbf{L}-\mathbf{M}\|<\dfrac{1}{\|\mathbf{L}^{-1}\|}$. *Then* $\mathbf{M}^{-1}$ *also exists.*

Proof. We can write $\mathbf{M}=\mathbf{L}+\mathbf{M}-\mathbf{L}=\mathbf{L}(\mathbf{I}+\mathbf{L}^{-1}(\mathbf{M}-\mathbf{L}))$, so that $\mathbf{M}^{-1}$ will exist provided that $\mathbf{I}+\mathbf{L}^{-1}(\mathbf{M}-\mathbf{L})$ has an inverse (Problem 2.44). Since $\|\mathbf{L}^{-1}(\mathbf{M}-\mathbf{L})\| \le \|\mathbf{L}^{-1}\| \, \|\mathbf{M}-\mathbf{L}\| < 1$, the result follows from Corollary 4.5.

The above Corollaries are valid in more general spaces and may be found in courses on Functional Analysis. The proof given applies generally but often another proof, based on geometric series of operators is used, which has the advantage that it is less dependent on the norm. Corollary 4.6 says that if an invertible linear operator $\mathbf{L}$ is given a small perturbation, the perturbed operator is also invertible. If particular, if $\mathbf{L}(\mathbf{x})$ are linear operators that depend continuously on $\mathbf{x}$ and if $\mathbf{L}(\mathbf{x}^{(0)})$ is invertible then so is $\mathbf{L}(\mathbf{x})$ for $\mathbf{x}$ near $\mathbf{x}^{(0)}$.

Using some simple facts from linear algebra this can also be seen as follows. If A is an $n\times n$ matrix, A is invertible if and only if $\det A \neq 0$ and we have $A^{-1}=(\det A)^{-1}\operatorname{adj}(A)$. Here det stands for *determinant* and adj for *adjutant* (sometimes adjoint). The determinant of A is a sum of products of elements of A, a scalar, while $\operatorname{adj}(A)$ is a matrix whose elements are cofactors of A, each cofactor being, up to sign, a determinant of a submatrix of A (see §2.8). Therefore small changes in the elements of A will give small changes in $\det A$. In the above situation if $\det \mathbf{L}(\mathbf{x}^{(0)}) \neq 0$ then $\det \mathbf{L}(\mathbf{x}) \neq 0$ for $\mathbf{x}$ near $\mathbf{x}^{(0)}$.

We shall also use the following version of Theorem 4.3 in the proof of the inverse function theorem.

Theorem 4.7. *Let* $\mathbf{f} : \overline{B}(\mathbf{0},r) \longrightarrow \mathbb{R}^n$ *be a strict contraction with constant* k *and suppose that* $\|\mathbf{f}(\mathbf{0})\| \le r(1-k)$. *Then* $(\mathbf{I}-\mathbf{f})(\overline{B}(\mathbf{0},r)) \supseteq \overline{B}(\mathbf{0},s)$ *for* $s=(1-k)r-\|\mathbf{f}(\mathbf{0})\|$, *in particular, taking* $s=0$, $\mathbf{f}$ *has a unique fixed point in* $\overline{B}(\mathbf{0},r)$.

Proof. We have to show that for $\|\mathbf{y}\| \le s$, there is a unique $\mathbf{x}\in\overline{B}(\mathbf{0},r)$ such that $\mathbf{x}-\mathbf{f}(\mathbf{x})=\mathbf{y}$. Fix such a $\mathbf{y}$ and let $\mathbf{g} : \overline{B}(\mathbf{0},r) \longrightarrow \mathbb{R}^n$ be given by $\mathbf{g}(\mathbf{x})=\mathbf{y}+\mathbf{f}(\mathbf{x})$. Then $\|\mathbf{g}(\mathbf{x}) - \mathbf{g}(\mathbf{z})\| = \|\mathbf{f}(\mathbf{x}) - \mathbf{f}(\mathbf{z})\| \le k\|\mathbf{x} - \mathbf{z}\|$ so $\mathbf{g}$ is a contraction. Moreover, $\mathbf{g}$ maps $\overline{B}(\mathbf{0},r)$ into itself. Indeed, we have

$$\|\mathbf{g}(\mathbf{x})\| \le \|\mathbf{y}\| + \|\mathbf{f}(\mathbf{x})\| \le \|\mathbf{y}\| + \|\mathbf{f}(\mathbf{x}) - \mathbf{f}(\mathbf{0})\| + \|\mathbf{f}(\mathbf{0})\| \le \|\mathbf{y}\| + k\|\mathbf{x}\| + \|\mathbf{f}(\mathbf{0})\|$$
$$\le s + kr + \|\mathbf{f}(\mathbf{0})\| \le r, \quad \text{by the choice of } s.$$

Thus $\mathbf{g}$ has a unique fixed point $\mathbf{z}$ in $\overline{B}(\mathbf{0},r)$, that is, $(\mathbf{I}-\mathbf{f})(\mathbf{z})=\mathbf{y}$ has a unique solution $\mathbf{z}$ in $\overline{B}(\mathbf{0},r)$ for any $\mathbf{y}\in\overline{B}(\mathbf{0},s)$.

4.2. THE INVERSE FUNCTION THEOREM

Let $A\subseteq\mathbb{R}^n$ and let $\mathbf{f} : A \longrightarrow \mathbb{R}^n$ be some function. Recall that $\mathbf{f}$ is said to be *injective* (or *one-to-one*) if $\mathbf{x}\neq\mathbf{y}$ implies $\mathbf{f}(\mathbf{x})\neq\mathbf{f}(\mathbf{y})$:

distinct points have distinct images. This is equivalent to (the contrapositive) $\mathbf{f}(\mathbf{x})=\mathbf{f}(\mathbf{y}) \Rightarrow \mathbf{x}=\mathbf{y}$. When $\mathbf{f}$ is injective, the inverse function $\mathbf{f}^{-1}$ is defined as the function $\mathbf{f}^{-1}:\mathbf{f}(A)\longrightarrow A$ given by $\mathbf{f}^{-1}(\mathbf{y})=\mathbf{x} \Leftrightarrow \mathbf{f}(\mathbf{x})=\mathbf{y}$. This is well defined since $\mathbf{y}$ uniquely determines $\mathbf{x}$.

Example 4.8. Let $f:\mathbb{R}\longrightarrow\mathbb{R}$ be defined by $f(x)=(x^2+1)e^x$. Then $f'(x)=(x^2+1+2x)e^x=(x+1)^2e^x>0$ for all x. Therefore f is strictly increasing and hence injective. Now $f(x)\longrightarrow 0$ as $x\longrightarrow-\infty$, and $f(x)\longrightarrow\infty$ as $x\longrightarrow\infty$ so f maps $\mathbb{R}$ onto $(0,\infty)$. This shows that $f^{-1}:(0,\infty)\longrightarrow\mathbb{R}$ exists, but it is not possible to write down a formula for f^{-1}.

For a continuous function $f:\mathbb{R}\longrightarrow\mathbb{R}$ it is a relatively simple matter to check when f is one-to-one (injective) on an interval : f must be either strictly increasing or strictly decreasing. This can usually be decided from investigation of the sign of f'. If $f'\neq 0$ on an interval then it is either positive or negative everywhere on that interval. Note that this holds without assuming that f' is continuous (a theorem of Darboux, problem 3.29).

The inverse function theorem similarly makes the assumption that its Jacobian be nonzero. The theorem assures the existence of the inverse function and tells us how to differentiate that inverse. However the function is required to be of class C^1 and not merely differentiable; this is essential. The theorem says roughly that if $D\mathbf{f}(\mathbf{a})$ is invertible then $\mathbf{f}$ is invertible on a neighbourhood of $\mathbf{a}$. Occasionally it is possible to determine $\mathbf{f}^{-1}$ explicitly but, more often, this process is either impossible or not sensible and then the theorem is important.

We adopt the three-point procedure outlined in §4.1.

Theorem 4.9. (*The injectivity theorem.*)
Let $V\subseteq\mathbb{R}^n$ be open and let $\mathbf{f}\in C^1(V,\mathbb{R}^m)$ with $m\geq n$. Suppose that $D\mathbf{f}(\mathbf{a})$ is one-to-one for some $\mathbf{a}\in V$. Then there exists $\delta>0$ such that $\mathbf{f}$ is one-to-one on $\mathrm{B}(\mathbf{a},\delta)$. Moreover $\mathbf{f}^{-1}:\mathbf{f}(\mathrm{B}(\mathbf{a},\delta))\longrightarrow\mathrm{B}(\mathbf{a},\delta)$ is continuous.

Note. $D\mathbf{f}(\mathbf{a})$ one-to-one forces $m\geq n$ by the rank-nullity theorem.

Proof. Write $\mathbf{L}=D\mathbf{f}(\mathbf{a})$. Since $\mathbf{L}$ is one-to-one, there exists $k>0$ such that $\|\mathbf{Lx}\|\geq k\|\mathbf{x}\|$ for all $\mathbf{x}\in\mathbb{R}^n$ (problem 2.41). By Proposition 3.24, there is $\delta>0$ such that $\|\mathbf{f}(\mathbf{x})-\mathbf{f}(\mathbf{y})-\mathbf{L}(\mathbf{x}-\mathbf{y})\|\leq k\|\mathbf{x}-\mathbf{y}\|/2$, for $\mathbf{x},\mathbf{y}\in\mathrm{B}(\mathbf{a},\delta)$. Hence, by the Corollary to the triangle inequality,

$$\|\mathbf{L}(\mathbf{x}-\mathbf{y})\|-\|\mathbf{f}(\mathbf{x})-\mathbf{f}(\mathbf{y})\|\leq k\|\mathbf{x}-\mathbf{y}\|/2.$$

On using $\|\mathbf{L}(\mathbf{x}-\mathbf{y})\|\geq k\|\mathbf{x}-\mathbf{y}\|$, this yields $\|\mathbf{f}(\mathbf{x})-\mathbf{f}(\mathbf{y})\|\geq k\|\mathbf{x}-\mathbf{y}\|/2$ for

$\mathbf{x},\mathbf{y}\in B(\mathbf{a},\delta)$ and shows that $\mathbf{f}$ is one-to-one on $B(\mathbf{a},\delta)$. Therefore $\mathbf{f}^{-1} : \mathbf{f}(B(\mathbf{a},\delta)) \longrightarrow B(\mathbf{a},\delta)$ exists and for $\mathbf{u},\mathbf{v}\in\mathbf{f}(B(\mathbf{a},\delta))$, $\mathbf{u}=\mathbf{f}(\mathbf{x})$, $\mathbf{v}=\mathbf{f}(\mathbf{y})$ say, the above can be written

$$\|\mathbf{u}-\mathbf{v}\| \geq k\|\mathbf{f}^{-1}(\mathbf{u})-\mathbf{f}^{-1}(\mathbf{v})\|/2 \quad \text{or} \quad \|\mathbf{f}^{-1}(\mathbf{u})-\mathbf{f}^{-1}(\mathbf{v})\| \leq 2\|\mathbf{u}-\mathbf{v}\|/k,$$

which shows that $\mathbf{f}^{-1}$ is uniformly continuous.

Theorem 4.10. (*The surjectivity theorem.*)
Let V be an open subset of $\mathbb{R}^n$ and let $\mathbf{f}\in C^1(V,\mathbb{R}^p)$ where $p\leq n$. Suppose that $\mathbf{a}\in V$ and $D\mathbf{f}(\mathbf{a})$ is surjective, that is, $D\mathbf{f}(\mathbf{a})$ maps $\mathbb{R}^n$ onto $\mathbb{R}^p$. Then $\mathbf{f}$ is locally surjective, that is, there exist $\delta>0$ and $\varepsilon>0$ such that for every $\mathbf{z}\in\overline{B}(\mathbf{f}(\mathbf{a}),\delta)\subseteq\mathbb{R}^p$ there is $\mathbf{x}\in\overline{B}(\mathbf{a},\varepsilon)\subseteq\mathbb{R}^n$ with $\mathbf{f}(\mathbf{x})=\mathbf{z}$.

Note. $D\mathbf{f}(\mathbf{a})$ surjective forces $p\leq n$ by the rank-nullity theorem.

Proof. There is no loss in taking $\mathbf{a}=\mathbf{0}$ for otherwise we consider $\mathbf{h}(\mathbf{x}) := \mathbf{f}(\mathbf{x}+\mathbf{a})$ defined on the translate $V^t := \{\mathbf{x} : \mathbf{x}+\mathbf{a}\in V\}$. Now write $\mathbf{L}=D\mathbf{f}(\mathbf{0})$. We first show that there is a linear operator $\mathbf{M} : \mathbb{R}^p \longrightarrow \mathbb{R}^n$ such that $\mathbf{LMv}=\mathbf{v}$ for all $\mathbf{v}\in\mathbb{R}^p$. Indeed, let $\{\mathbf{e}_1,\mathbf{e}_2,\cdots,\mathbf{e}_p\}$ be the standard basis of $\mathbb{R}^p$. As $\mathbf{L}$ is surjective, there exist $\mathbf{u}_1,\mathbf{u}_2,\cdots,\mathbf{u}_p\in\mathbb{R}^n$ such that $\mathbf{L}(\mathbf{u}_i)=\mathbf{e}_i$. Define $\mathbf{M}$ on all of $\mathbb{R}^p$ by

$$\mathbf{Mv} = \mathbf{M}\left(\sum_{i=1}^{p} v_i\mathbf{e}_i\right) = \sum_{i=1}^{p} v_i\mathbf{u}_i .$$

Then $\mathbf{M}$ is linear and $\mathbf{LMv} = \mathbf{L}\left(\sum_{i=1}^{p} v_i\mathbf{u}_i\right) = \sum_{i=1}^{p} v_i\mathbf{L}(\mathbf{u}_i) = \mathbf{v}$ for each $\mathbf{v}\in\mathbb{R}^p$, as claimed. Now let $m=\|\mathbf{M}\|$, the operator norm of $\mathbf{M}$. Take $\varepsilon>0$ so that the ε-nbd of $\mathbf{0}$ is contained in V and such that, by Proposition 3.24,

$$\|\mathbf{f}(\mathbf{x})-\mathbf{f}(\mathbf{y})-\mathbf{L}(\mathbf{x}-\mathbf{y})\| \leq \|\mathbf{x}-\mathbf{y}\|/2m, \text{ for all } \mathbf{x},\mathbf{y}\in\overline{B}(\mathbf{0},\varepsilon).$$

Let $\delta=\varepsilon/2m$ and let $\mathbf{z}\in\overline{B}(\mathbf{f}(\mathbf{0}),\delta)$. We now want to show that there is $\mathbf{x}\in\overline{B}(\mathbf{0},\varepsilon)$ with $\mathbf{f}(\mathbf{x})=\mathbf{z}$. To this end define $\mathbf{g} : \overline{B}(\mathbf{0},\varepsilon) \longrightarrow \mathbb{R}^n$ by $\mathbf{g}(\mathbf{x})=\mathbf{MLx}-\mathbf{M}[\mathbf{f}(\mathbf{x})-\mathbf{z}]$. We claim that $\mathbf{g}$ is a strict contraction with constant 1/2 and that $\|\mathbf{g}(\mathbf{0})\|\leq\varepsilon/2$ so that Theorem 4.7 applies. Indeed, for $\mathbf{x},\mathbf{y}\in\overline{B}(\mathbf{0},\varepsilon)$,

$$\begin{aligned}\|\mathbf{g}(\mathbf{x})-\mathbf{g}(\mathbf{y}\| &= \|\mathbf{ML}(\mathbf{x}-\mathbf{y})-\mathbf{M}[\mathbf{f}(\mathbf{x})-\mathbf{f}(\mathbf{y})]\| \\ &\leq m\|\mathbf{L}(\mathbf{x}-\mathbf{y})-[\mathbf{f}(\mathbf{x})-\mathbf{f}(\mathbf{y})]\| \\ &\leq \|\mathbf{x}-\mathbf{y}\|/2.\end{aligned}$$

Furthermore, $\|\mathbf{g}(\mathbf{0})\| = \|\mathbf{M}[\mathbf{f}(\mathbf{0})-\mathbf{z}]\| \leq m\delta=\varepsilon/2$. Thus $\mathbf{g}$ has a fixed point in $\overline{B}(\mathbf{0},\varepsilon)$, that is, there exist $\mathbf{x}$ with $\mathbf{x}=\mathbf{MLx}-\mathbf{M}[\mathbf{f}(\mathbf{x})-\mathbf{z}]$. Then, applying $\mathbf{L}$, we see that $\mathbf{Lx}=\mathbf{Lx}-[\mathbf{f}(\mathbf{x})-\mathbf{z}]$ since $\mathbf{LM}$ is the identity on $\mathbb{R}^p$. This proves $\mathbf{f}(\mathbf{x})=\mathbf{z}$.

Corollary 4.11. (*The open mapping theorem.*)
Let V be an open subset of $\mathbb{R}^n$ and let $\mathbf{f} \in C^1(V, \mathbb{R}^p)$, where $p \leq n$. Suppose that $D\mathbf{f}(\mathbf{x})$ is surjective for every $\mathbf{x} \in V$. Then $\mathbf{f}$ is an open mapping, that is, the image of each open set in V is open in $\mathbb{R}^p$.

Proof. Let U be open in V and let $\mathbf{y} \in \mathbf{f}(U)$, say $\mathbf{y} = \mathbf{f}(\mathbf{x})$ for $\mathbf{x} \in U$. By Theorem 4.10 applied to $\mathbf{f}|_U$ ($\mathbf{f}$ restricted to U), there is $\delta > 0$ so that $B(\mathbf{y}, \delta) \subseteq \mathbf{f}(U)$. This shows that $\mathbf{y}$ is an interior point of $\mathbf{f}(U)$ and therefore $\mathbf{f}(U)$ is open.

Recall from chapter 2 that for a C^1 function $\mathbf{f} : V \subseteq \mathbb{R}^n \longrightarrow \mathbb{R}^n$, the *Jacobian* of $\mathbf{f}$ is defined by $J_{\mathbf{f}}(\mathbf{x}) := \det D\mathbf{f}(\mathbf{x})$.

Theorem 4.12. (*The inverse function theorem.*)
Let V be an open subset of $\mathbb{R}^n$ and let $\mathbf{f} \in C^1(V, \mathbb{R}^n)$. Let $\mathbf{a} \in V$ and suppose the Jacobian $J_{\mathbf{f}}(\mathbf{a})$ is nonzero. Then there is a neighbourhood $U \subseteq V$ of $\mathbf{a}$ such that $\mathbf{f}|_U$ is invertible. The inverse $\mathbf{g} := (\mathbf{f}|_U)^{-1}$ is of class C^1 on an open set $W \subseteq \mathbf{f}(U)$ and $D\mathbf{g}(\mathbf{y}) = [D\mathbf{f}(\mathbf{x})]^{-1}$, where $\mathbf{y} = \mathbf{f}(\mathbf{x})$.

Proof. As $J_{\mathbf{f}}(\mathbf{a}) \neq 0$, $D\mathbf{f}(\mathbf{a})$ is a bijection. Therefore by Theorems 4.9 and 4.10, $\mathbf{f}$ is locally bijective, that is , there is a neighbourhood U of $\mathbf{a}$ such that $\mathbf{f} : U \longrightarrow \mathbf{f}(U)$ is invertible and the inverse function $\mathbf{g}$ is continuous. Moreover, $\mathbf{f}(U)$ is open by Corollary 4.11 . We now prove that $\mathbf{g}$ is differentiable. Since $D\mathbf{f}(\mathbf{a})$ is invertible and $\mathbf{f}$ is of class C^1, there is a neighbourhood of $\mathbf{a}$ such that $D\mathbf{f}(\mathbf{x})$ is invertible for $\mathbf{x}$ in that neighbourhood : see the comments following Corollary 4.6. We can always suppose that this nbd is contained in U. Let $\mathbf{y} = \mathbf{f}(\mathbf{x})$ (equivalently $\mathbf{x} = \mathbf{g}(\mathbf{y})$). We shall show that $D\mathbf{g}(\mathbf{y}) = [D\mathbf{f}(\mathbf{x})]^{-1}$. For $\mathbf{y}$ and $\mathbf{x}$ fixed, let $C = \|[D\mathbf{f}(\mathbf{x})]^{-1}\|$. We want to show that

$$\|\mathbf{g}(\mathbf{y}+\mathbf{k}) - \mathbf{g}(\mathbf{y}) - [D\mathbf{f}(\mathbf{x})]^{-1}\mathbf{k}\| / \|\mathbf{k}\| \longrightarrow 0 \text{ as } \|\mathbf{k}\| \longrightarrow 0.$$

For $\|\mathbf{k}\|$ sufficiently small we can write

$$\mathbf{g}(\mathbf{y}+\mathbf{k}) - \mathbf{g}(\mathbf{y}) - [D\mathbf{f}(\mathbf{x})]^{-1}\mathbf{k} = [D\mathbf{f}(\mathbf{x})]^{-1}\{D\mathbf{f}(\mathbf{x})[\mathbf{g}(\mathbf{y}+\mathbf{k}) - \mathbf{g}(\mathbf{y})] - \mathbf{k}\}$$

and obtain

$$\|\mathbf{g}(\mathbf{y}+\mathbf{k}) - \mathbf{g}(\mathbf{y}) - [D\mathbf{f}(\mathbf{x})]^{-1}\mathbf{k}\| \leq C\|D\mathbf{f}(\mathbf{x})[\mathbf{g}(\mathbf{y}+\mathbf{k}) - \mathbf{g}(\mathbf{y})] - \mathbf{k}\|. \tag{4.1}$$

Write $\mathbf{z} = \mathbf{g}(\mathbf{y}+\mathbf{k})$. Then $\mathbf{y}+\mathbf{k} = \mathbf{f}(\mathbf{z})$ and the right hand side of (4.1) may be written $C\|D\mathbf{f}(\mathbf{x})[\mathbf{z} - \mathbf{x}] - (\mathbf{f}(\mathbf{z}) - \mathbf{f}(\mathbf{x}))\|$. As $\mathbf{f}$ is differentiable at $\mathbf{x}$, for any $\varepsilon > 0$ (with $\varepsilon < 1$) there exists $\delta > 0$ such that

$$\|D\mathbf{f}(\mathbf{x})(\mathbf{z} - \mathbf{x}) - [\mathbf{f}(\mathbf{z}) - \mathbf{f}(\mathbf{x})\| < \frac{\varepsilon}{C}\|\mathbf{z} - \mathbf{x}\| \quad \text{for } \|\mathbf{z} - \mathbf{x}\| < \delta.$$

Then (4.1) gives

$$\|\mathbf{g}(\mathbf{y}+\mathbf{k}) - \mathbf{g}(\mathbf{y}) - [D\mathbf{f}(\mathbf{x})]^{-1}\mathbf{k}\| \le \varepsilon\|\mathbf{z}-\mathbf{x}\| = \varepsilon\|\mathbf{g}(\mathbf{y}+\mathbf{k}) - \mathbf{g}(\mathbf{y})\| . \qquad (4.2)$$

By the Corollary to the triangle inequality,

$$\begin{aligned}\|\mathbf{g}(\mathbf{y}+\mathbf{k}) - \mathbf{g}(\mathbf{y}) - [D\mathbf{f}(\mathbf{x})]^{-1}\mathbf{k}\| &\ge \|\mathbf{g}(\mathbf{y}+\mathbf{k}) - \mathbf{g}(\mathbf{y})\| - \|[D\mathbf{f}(\mathbf{x})]^{-1}\mathbf{k}\| \\ &\ge \|\mathbf{g}(\mathbf{y}+\mathbf{k}) - \mathbf{g}(\mathbf{y})\| - C\|\mathbf{k}\|.\end{aligned}$$

Putting this in (4.2) gives $(1-\varepsilon)\|\mathbf{g}(\mathbf{y}+\mathbf{k}) - \mathbf{g}(\mathbf{y})\| \le C\|\mathbf{k}\|$ for $\|\mathbf{z}-\mathbf{x}\| < \delta$. Using this in (4.2) now yields

$$\|\mathbf{g}(\mathbf{y}+\mathbf{k}) - \mathbf{g}(\mathbf{y}) - [D\mathbf{f}(\mathbf{x})]^{-1}\mathbf{k}\| < \frac{C\varepsilon}{1-\varepsilon}\|\mathbf{k}\| \text{ for } \|\mathbf{z}-\mathbf{x}\| < \delta.$$

Since $\mathbf{g}$ is continuous, there exists $\eta > 0$ such that $\|\mathbf{z}-\mathbf{x}\| = \|\mathbf{g}(\mathbf{y}+\mathbf{k}) - \mathbf{g}(\mathbf{y})\| < \delta$ for $\|\mathbf{k}\| < \eta$, and this proves that

$$\|\mathbf{g}(\mathbf{y}+\mathbf{k}) - \mathbf{g}(\mathbf{y}) - [D\mathbf{f}(\mathbf{x})]^{-1}\mathbf{k}\| \,/\, \|\mathbf{k}\| < \frac{C\varepsilon}{1-\varepsilon} \quad \text{for } \|\mathbf{k}\| < \eta,$$

which shows that $\mathbf{g}$ is differentiable with $D\mathbf{g}(\mathbf{y}) = [D\mathbf{f}(\mathbf{x})]^{-1}$. That $\mathbf{g} \in C^1$ is a consequence of this since for a matrix A, $A^{-1} = \text{adj}(A)/\det(A)$ and $\det(A)$ and the entries of adj(A) are finite sums of products of entries of A. Hence, because the Jacobian is nonzero, continuity of the entries of $D\mathbf{f}(\mathbf{x})$ implies the same for $D\mathbf{g}(\mathbf{y})$.

From the theorem and properties of determinant, the Jacobian of the local inverse $\mathbf{g}$ is given by $J_{\mathbf{g}}(\mathbf{y}) = 1/J_{\mathbf{f}}(\mathbf{x})$. Also it can be shown (by an induction argument) that $\mathbf{g}$ inherits all the differentiability properties of $\mathbf{f}$, for example, if $\mathbf{f}$ is of class C^2 so is $\mathbf{g}$.

Example 4.13. Show that the equations $u = x^3y + y^2$, $v = \ln(x+y)$ have a unique solution $x = x(u,v)$, $y = (u,v)$ on a neighbourhood of $(6, \ln 3)$ with $x(6, \ln 3) = 1$, $y(6, \ln 3) = 2$. Find $\partial x/\partial u$, $\partial x/\partial v$, $\partial y/\partial u$, $\partial y/\partial v$ at $(6, \ln 3)$.

Solution. Let $\mathbf{f}(x,y) = (u,v) = (x^3y + y^2, \ln(x+y))$. Then, to solve for x,y, in terms of u,v is to show that $\mathbf{f}^{-1}$ exists for then we have $(x,y) = \mathbf{f}^{-1}(u,v)$. Since $\partial f_1/\partial x = 3x^2y$, $\partial f_1/\partial y = x^3 + 2y$, $\partial f_2/\partial x = 1/(x+y)$, $\partial f_2/\partial y = 1/(x+y)$ all exist and are continuous for $x+y \neq 0$, $D\mathbf{f}(\mathbf{x})$ exists on a neighbourhood of $(1,2)$ and $D\mathbf{f}(1,2) = \begin{bmatrix} 6 & 5 \\ 1/3 & 1/3 \end{bmatrix}$ so $J_{\mathbf{f}} = 1/3 \neq 0$. Therefore $\mathbf{f}^{-1}$ exists on a neighbourhood of $\mathbf{f}(1,2) = (6, \ln 3)$. Moreover, $D\mathbf{f}^{-1}(\mathbf{y}) = [D\mathbf{f}(\mathbf{x})]^{-1}$ so that

$$D\mathbf{f}^{-1}(6, \ln 3) = \begin{bmatrix} 6 & 5 \\ 1/3 & 1/3 \end{bmatrix}^{-1} = 3\begin{bmatrix} 1/3 & -5 \\ -1/3 & 6 \end{bmatrix} = \begin{bmatrix} 1 & -15 \\ -1 & 18 \end{bmatrix} = \begin{bmatrix} \partial x/\partial u & \partial x/\partial v \\ \partial y/\partial u & \partial y/\partial v \end{bmatrix}$$

We used the fact that for a 2×2 matrix, $\begin{bmatrix} a & b \\ c & d \end{bmatrix}^{-1} = \dfrac{1}{ad-bc}\begin{bmatrix} d & -b \\ -c & a \end{bmatrix}$.

Example 4.14. Let $\mathbf{f}(x,y)=(e^x\cos y, e^x\sin y)$. Show that **f** has a local inverse in a neighbourhood of any point, but **f** is not injective on $\mathbb{R}^2$.

Solution. $D\mathbf{f}(x,y)=\begin{bmatrix} e^x\cos y & -e^x\sin y \\ e^x\sin y & e^x\cos y \end{bmatrix}$ and the Jacobian is $J_{\mathbf{f}}(x,y)=e^x(\cos^2 y+\sin^2 y)=e^x>0$ for all (x,y). Clearly **f** is of class C^1 so **f** has a local inverse by the theorem. However, **f** is not one-to-one on $\mathbb{R}^2$ since, for example, $\mathbf{f}(x, y)=\mathbf{f}(x, y+2m\pi)$ for any integer m.

Example 4.14 illustrates that there may be no global inverse. We shall discuss global invertibility in §4.5.

4.3. THE IMPLICIT FUNCTION THEOREM

The implicit function theorem treats the problem as to when a relationship between variables allows some of them to be "solved" in terms of the others. We begin by giving the intuitive reasons for the hypotheses that will be made.

We consider the space $\mathbb{R}^{k+n}$ as the Cartesian product $\mathbb{R}^k\times\mathbb{R}^n$ and write elements of $\mathbb{R}^{k+n}$ as $\mathbf{z}=(\mathbf{x},\mathbf{y})$ where $\mathbf{x}=(x_1, x_2,\cdots, x_k)$, $\mathbf{y}=(y_1, y_2,\cdots, y_n)$. Let $\mathbf{F}:\mathbb{R}^{k+n}\longrightarrow\mathbb{R}^n$ be a function. We are interested in the relation $\mathbf{F}(\mathbf{x},\mathbf{y})=\mathbf{0}$, or the set of points $(\mathbf{x},\mathbf{y})$ which satisfy this equation. The relation $\mathbf{F}(\mathbf{x},\mathbf{y})=\mathbf{0}$ is a system of n equations in $(k+n)$ unknowns:

$$F_j(x_1,\cdots,x_k, y_1,\cdots, y_n) = 0, \quad 1\le j\le n,$$

where F_j are the component functions of **F**.

We may think of elements of $\mathbb{R}^{n+k}$ as having $(n+k)$ degrees of freedom. Each equation $F_j(\mathbf{z})=0$ reduces the degrees of freedom by one in general; it might not if, for example, $F_3(\mathbf{z})=F_1(\mathbf{z})+F_2(\mathbf{z})$ for then the equation $F_3(\mathbf{z})=0$ is not an extra constraint; it gives no extra information. Of course it is not always obvious when an equation is independent of the earlier ones. Independent equations will each reduce the degrees of freedom by one so, if all n equations are independent, we are left with k degrees of freedom. This suggests that $x_1,\cdots,x_k$ are free but $y_1,\cdots,y_n$ are determined in terms of these, that is to say, we can "solve" for **y** in terms of **x**. Therefore we must impose some condition that ensures that the equations $F_j(\mathbf{z})=0$ are independent.

As mentioned in §4.1, any theorem must be local in nature. We can only establish some result on a neighbourhood of a point $(\mathbf{a},\mathbf{b})$ where $\mathbf{F}(\mathbf{a},\mathbf{b})=\mathbf{0}$.

We shall impose a condition on the derivative $D\mathbf{F}(\mathbf{a},\mathbf{b})$. This is most easily understood in terms of matrices. Recall that $D\mathbf{F}(\mathbf{z})$ is the $n\times(k+n)$ matrix $\left[\dfrac{\partial F_i}{\partial z_j}\right]$, that is,

$$D\mathbf{F}(\mathbf{z}) = \begin{bmatrix} \partial F_1/\partial x_1 & \cdots & \partial F_1/\partial x_k & \partial F_1/\partial y_1 & \cdots & \partial F_1/\partial y_n \\ \vdots & & \vdots & \vdots & & \vdots \\ \partial F_n/\partial x_1 & \cdots & \partial F_n/\partial x_k & \partial F_n/\partial y_1 & \cdots & \partial F_n/\partial y_n \end{bmatrix}.$$

We define $D_{\mathbf{y}}\mathbf{F}(\mathbf{z})$ to be the $n\times n$ matrix $\left[\dfrac{\partial F_i}{\partial y_j}\right]$, that is,

$$D_{\mathbf{y}}\mathbf{F}(\mathbf{z}) = \begin{bmatrix} \partial F_1/\partial y_1 & \cdots & \partial F_1/\partial y_n \\ \vdots & & \vdots \\ \partial F_n/\partial y_1 & \cdots & \partial F_n/\partial y_n \end{bmatrix}.$$

In terms of linear operators, at a point $\mathbf{c}=(\mathbf{a},\mathbf{b})\in\mathbb{R}^k\times\mathbb{R}^n$, $D_{\mathbf{y}}\mathbf{F}(\mathbf{c})$ is the linear operator from $\mathbb{R}^n$ to $\mathbb{R}^n$ defined by

$$[D_{\mathbf{y}}\mathbf{F}(\mathbf{c})](\mathbf{y})=[D\mathbf{F}(\mathbf{c})](\mathbf{0},\mathbf{y}),\ \mathbf{y}\in\mathbb{R}^n. \tag{4.3}$$

Similarly we can define $D_{\mathbf{x}}\mathbf{F}(\mathbf{c})$ as the linear operator from $\mathbb{R}^k$ to $\mathbb{R}^n$

$$[D_{\mathbf{x}}\mathbf{F}(\mathbf{c})](\mathbf{x})=[D\mathbf{F}(\mathbf{c})](\mathbf{x},\mathbf{0}),\ \mathbf{x}\in\mathbb{R}^k. \tag{4.4}$$

A more precise notation would be to write $D_{(1)}\mathbf{F}$ and $D_{(2)}\mathbf{F}$ in place of $D_{\mathbf{x}}\mathbf{F}$ and $D_{\mathbf{y}}\mathbf{F}$ respectively. These are called *block partial derivatives* of $\mathbf{F}$. Since $(\mathbf{x},\mathbf{y})\in\mathbb{R}^k\times\mathbb{R}^n$ can be written $(\mathbf{x},\mathbf{y})=(\mathbf{x},\mathbf{0})+(\mathbf{0},\mathbf{y})$ we see that

$$[D\mathbf{F}(\mathbf{c})](\mathbf{x},\mathbf{y})=[D_{\mathbf{x}}\mathbf{F}(\mathbf{c})]\mathbf{x}+[D_{\mathbf{y}}\mathbf{F}(\mathbf{c})]\mathbf{y}. \tag{4.5}$$

The condition to be imposed in the implicit function theorem is that $\det(D_{\mathbf{y}}\mathbf{F}(\mathbf{c}))\neq 0$, where det denotes the determinant of a matrix. This is equivalent to saying that $D_{\mathbf{y}}\mathbf{F}(\mathbf{c})$ is an invertible linear operator (or $n\times n$ matrix). It is also the same as saying that the rows equations $F_i(\mathbf{z})=0$ are independent. We assume here that the reader has some knowledge of linear algebra, sufficient to understand the concept of linear independence and a little about determinants (see §2.8).

Why do we assume $D_{\mathbf{y}}\mathbf{F}(\mathbf{c})$ to be invertible? We aim to prove a general result so it must hold in the special case when $\mathbf{F}$ has the form $\mathbf{F}(\mathbf{x},\mathbf{y})=\mathbf{h}(\mathbf{x})+\mathbf{L}(\mathbf{y})$, where $\mathbf{L}$ is linear. Then $\mathbf{F}(\mathbf{x},\mathbf{y})=\mathbf{0}$ is the system of equations $\mathbf{L}\mathbf{y}=-\mathbf{h}(\mathbf{x})$, or in full

$$\begin{array}{cccc} a_{11}y_1+a_{12}y_2+\cdots+a_{1n}y_n=-h_1(\mathbf{x}) \\ \vdots \quad\quad \vdots \quad\quad \vdots \quad\quad \vdots \\ a_{n1}y_1+a_{n2}y_2+\cdots+a_{nn}y_n=-h_n(\mathbf{x}) \end{array}$$

For each $\mathbf{x}$, this is a system of linear equations for $\mathbf{y}$ and has a unique solution if, and only if, $\det \mathbf{L}=\det(a_{ij})\neq 0$, and $\mathbf{y}=\mathbf{L}^{-1}(-\mathbf{h}(\mathbf{x}))$ in this case. Since $\mathbf{L}=D_{\mathbf{y}}\mathbf{F}$ here, we see that the condition $\det D_{\mathbf{y}}\mathbf{F}(\mathbf{c})\neq 0$ is indeed appropriate. Of course, when $n=k=1$, the condition on $F(x,y)$ is $\partial F/\partial y\neq 0$ which corresponds to the curve $F(x,y)=0$ having a non-vertical tangent.

Theorem 4.16. (*The implicit function theorem.*)
Let V be an open set in $\mathbb{R}^{k+n}$ and let $\mathbf{F}\in C^1(V,\mathbb{R}^n)$. Suppose that

(*i*) $\mathbf{F}(\mathbf{a},\mathbf{b})=\mathbf{0}$ *for some* $(\mathbf{a},\mathbf{b})\in V$;
(*ii*) $\det[D_{\mathbf{y}}\mathbf{F}(\mathbf{a},\mathbf{b})]\neq 0$ (*equivalently* $D_{\mathbf{y}}\mathbf{F}(\mathbf{a},\mathbf{b})$ *is invertible*).

Then there is a neighbourhood $U\subseteq\mathbb{R}^k$ of $\mathbf{a}$, *a neighbourhood W of* $\mathbf{b}$ *in* $\mathbb{R}^n$ *and a unique function* $\mathbf{f}: U\longrightarrow W$ *such that*

(*a*) $\mathbf{f}$ *is of class* C^1 on U,
(*b*) $\mathbf{f}(\mathbf{a})=\mathbf{b}$,
(*c*) *for all* $\mathbf{x}\in U$, $(\mathbf{x},\mathbf{f}(\mathbf{x}))\in V$, *and* $\mathbf{F}(\mathbf{x},\mathbf{f}(\mathbf{x}))=\mathbf{0}$.

Proof Let $\mathbf{H}(\mathbf{x},\mathbf{y})=(\mathbf{x},\mathbf{F}(\mathbf{x},\mathbf{y}))$ a map from a neighbourhood of $(\mathbf{a},\mathbf{b})$ in $\mathbb{R}^{k+n}$ into $\mathbb{R}^{k+n}$. We will show that $\mathbf{H}$ satisfies the hypotheses of the inverse function theorem. Note that $H_i(\mathbf{x},\mathbf{y})=\begin{cases} x_i & \text{for } 1\le i\le k \\ F_{i-k}(\mathbf{x},\mathbf{y}) & \text{for } k<i\le k+n \end{cases}$,
so that $\mathbf{H}\in C^1$. We first give the arguments using matrices. $D\mathbf{H}(\mathbf{z})$ is the $(k+n)\times(k+n)$ matrix $(\partial H_i/\partial x_j)$, $1\le i,j\le k+n$, and can be written in

block matrix notation as $D\mathbf{H}(\mathbf{z})=\left[\begin{array}{c:c} \mathbf{I}_{k\times k} & \mathbf{0}_{k\times n} \\ \hdashline D_{\mathbf{x}}\mathbf{F} & D_{\mathbf{y}}\mathbf{F} \end{array}\right]$ where $\mathbf{I}_{k\times k}$ is the $k\times k$ identity matrix and $\mathbf{0}_{k\times n}$ is the $k\times n$ zero matrix. Note that $D_{\mathbf{x}}\mathbf{F}$ is an $n\times k$ matrix, $D_{\mathbf{y}}\mathbf{F}$ is $n\times n$. Therefore, $J_{\mathbf{H}}(\mathbf{z})=\det D\mathbf{H}(\mathbf{z})=\det D_{\mathbf{y}}\mathbf{F}(\mathbf{z})\neq 0$, using properties of determinant (expanding by rows). Thus, by the inverse function theorem, there is a neighbourhood of $(\mathbf{a},\mathbf{b})$ on which $\mathbf{H}^{-1}$ exists and is of class C^1. Write $\mathbf{H}^{-1}(\mathbf{x},\mathbf{0}):=(\mathbf{g}(\mathbf{x}),\mathbf{f}(\mathbf{x}))$. Then we have

$$\mathbf{H}^{-1}(\mathbf{u},\mathbf{v})=(\mathbf{x},\mathbf{y}) \Leftrightarrow (\mathbf{u},\mathbf{v})=\mathbf{H}(\mathbf{x},\mathbf{y})=(\mathbf{x},\mathbf{F}(\mathbf{x},\mathbf{y})) \Leftrightarrow \mathbf{u}=\mathbf{x} \text{ and } \mathbf{v}=\mathbf{F}(\mathbf{x},\mathbf{y}).$$

Therefore $\mathbf{H}^{-1}(\mathbf{x},\mathbf{0})=(\mathbf{x},\mathbf{y})$ and $\mathbf{F}(\mathbf{x},\mathbf{y})=\mathbf{0}$. Thus $\mathbf{g}(\mathbf{x})=\mathbf{x}$ and $\mathbf{F}(\mathbf{x},\mathbf{f}(\mathbf{x}))=\mathbf{0}$. Now set $\mathbf{G}(\mathbf{x})=(\mathbf{x},\mathbf{f}(\mathbf{x}))=\mathbf{H}^{-1}(\mathbf{x},\mathbf{0})$. Then we see that $\mathbf{G}$ is of class C^1 since $\mathbf{H}^{-1}$ is, by the inverse function theorem. Moreover $\mathbf{G}$ is C^1 if and only if the same is true of $\mathbf{f}$ as is seen by computing partial derivatives of G_i. This proves that $\mathbf{f}$ is a C^1 function.

Now we give the arguments using linear operators. We have to show that $D\mathbf{H}(\mathbf{a},\mathbf{b})$ is invertible. It is easily shown that $\mathbf{H}$ is differentiable and that $D\mathbf{H}(\mathbf{a},\mathbf{b})(\mathbf{x},\mathbf{y})=(\mathbf{x},[D\mathbf{F}(\mathbf{a},\mathbf{b})](\mathbf{x},\mathbf{y}))$; the argument is given in full in Lemma 4.22. Let $\mathbf{L}_1:=D_{\mathbf{x}}\mathbf{F}(\mathbf{a},\mathbf{b})$, $\mathbf{L}_2:=D_{\mathbf{y}}\mathbf{F}(\mathbf{a},\mathbf{b})$ be the block partial derivatives as in (4.3) and (4.4) so that $D\mathbf{F}(\mathbf{a},\mathbf{b})(\mathbf{x},\mathbf{y})=\mathbf{L}_1\mathbf{x}+\mathbf{L}_2\mathbf{y}$. Then $[D\mathbf{H}(\mathbf{a},\mathbf{b})]^{-1}$ is the linear operator $\mathbf{M}(\mathbf{u},\mathbf{v}):=(\mathbf{u},\mathbf{L}_2^{-1}(\mathbf{v}-\mathbf{L}_1\mathbf{u}))$ since

$$[D\mathbf{H}(\mathbf{a},\mathbf{b})\circ\mathbf{M}](\mathbf{u},\mathbf{v})=(\mathbf{u}, D\mathbf{F}(\mathbf{a},\mathbf{b})\circ\mathbf{M}(\mathbf{u},\mathbf{v}))=(\mathbf{u}, \mathbf{L}_1\mathbf{u}+\mathbf{L}_2[\mathbf{L}_2^{-1}(\mathbf{v}-\mathbf{L}_1\mathbf{u})])=(\mathbf{u},\mathbf{v}).$$

It is possible to give a direct proof of the existence of a unique continuous function $\mathbf{f}$ defined implicitly by a relation of the form $\mathbf{F}(\mathbf{x},\mathbf{y})=\mathbf{0}$ which only requires $\mathbf{F}$ to be a C^1 function with respect to the $\mathbf{y}$ variables. Again the basic Banach fixed point theorem is used (problem 4.15).

We have for simplicity always supposed that the variables of interest have been ordered appropriately. It is possible to state a version of the theorem where this has not been done but we prefer to relabel the variables, if necessary.

Of course, once it is known that $\mathbf{f}$ is differentiable we may compute $D\mathbf{f}$ by applying the chain rule to the relation $\mathbf{F}(\mathbf{x},\mathbf{f}(\mathbf{x}))=\mathbf{0}$; see Lemma 4.22 and Proposition 4.23 *infra*.

First we see some examples. In these we shall only have to compute the determinant of a 2×2 matrix but, in general, it is necessary to be able to handle larger matrices.

Example 4.17. Let $F(x,y)=x^2+y^2-1$. Then $\frac{\partial F}{\partial y}=2y$ which is continuous. The point (0,1) satisfies $F(0,1)=0$ and $\frac{\partial F}{\partial y}(0,1)=2\neq 0$ so by the implicit function theorem there is a neighbourhood U of 0, a neighbourhood W of the point 1 and a function $f: U \longrightarrow W$ such that $y=f(x)$ on U. Of course $f(x)=\sqrt{(1-x^2)}$ in this case. Notice that (1,0) also satisfies $F(1,0)=0$ but $\frac{\partial F}{\partial y}=0$ there so the theorem fails, as it must. However at (1,0), $\partial F/\partial x\neq 0$ so $x=g(y)$ on some nbd; indeed $x=\sqrt{(1-y^2)}$.

Example 4.18. Let $F(x,y)=x^3+y^3-1$. Then $\frac{\partial F}{\partial y}=3y^2$ is continuous. The point (1,0) satisfies $F(1,0)=0$ but $\partial F/\partial y=0$ there, so the hypotheses of the theorem are not satisfied. However, $x^3+y^3-1=0 \Leftrightarrow y=(1-x^3)^{1/3}$ which exists for all x. So the equation $x^3+y^3=1$ does define y implicitly as a function of x.

Example 4.18 illustrates the fact that the hypotheses in the Implicit Function Theorem are *sufficient* but not *necessary* conditions. Nevertheless, as intimated earlier, one cannot expect to prove a general result under weaker assumptions.

Example 4.19. Show that the equations $x^3u+yv^3=2$, $xu^3+y^3v=2$ define u,v implicitly as functions of x,y near the point $(x,y,u,v)=(1,1,1,1)$.

Solution. Let $\mathbf{F}(x,y,u,v)=(x^3u+yv^3-2,\ xu^3+y^3v-2)$, a map from $\mathbb{R}^4$ to $\mathbb{R}^2$. Then $\mathbf{F}(1,1,1,1)=(0,0)$ and clearly the partial derivatives of F_1 and F_2 are continuous so $\mathbf{F}$ is of class C^1.

We want $\det\begin{bmatrix}\partial F_1/\partial u & \partial F_1/\partial v\\ \partial F_2/\partial u & \partial F_2/\partial v\end{bmatrix}\neq 0$ at (1,1,1,1).

In fact, $\det\begin{bmatrix}\partial F_1/\partial u & \partial F_1/\partial v\\ \partial F_2/\partial u & \partial F_2/\partial v\end{bmatrix}=\det\begin{bmatrix}x^3 & 3yv^2\\ 3u^2x & y^3\end{bmatrix}=\det\begin{bmatrix}1 & 3\\ 3 & 1\end{bmatrix}$

$$=1-9=-8 \quad \text{at } (1,1,1,1).$$

Therefore, the implicit function theorem applies, and $(u,v)=\mathbf{f}(x,y)$ on a neighbourhood of $(x,y,u,v)=(1,1,1,1)$.

The implicit function theorem can give much information on so-called *bifurcation* problems. These involve problems of the form $\mathbf{F}(\mathbf{x},\lambda)=\mathbf{0}$ where $\mathbf{F}: \mathbb{R}^n\times\mathbb{R}\longrightarrow\mathbb{R}^n$ and λ is regarded as a parameter. Let

$S_\lambda = \{\mathbf{x} : \mathbf{F}(\mathbf{x},\lambda)=\mathbf{0}\}$ be the solutions corresponding to a fixed value of the parameter. The parameter is often one of the controls of the problem and its value affects the state of the system. It is important to know that a small change in λ does not affect the qualitative behaviour of the corresponding solution $\mathbf{x}\in S_\lambda$. Values of the parameter for which this fails are called *bifurcation values*. Typically, there is some trivial solution which exists for all λ; for simplicity we suppose this is the zero solution: $\mathbf{F}(\mathbf{0},\lambda)=\mathbf{0}$ for all λ. Then a bifurcation value μ is such that every small neighbourhood of $(\mathbf{0},\mu)$ contains points $(\mathbf{x},\lambda)$ with $\mathbf{F}(\mathbf{x},\lambda)=\mathbf{0}$ and $\mathbf{x}\neq\mathbf{0}$. When $n=1$, solutions of $F(x,\lambda)=0$ can be plotted in the (x,λ)-plane. The line $x=0$ is included and μ is a bifurcation value roughly means that there is another curve of solutions (x,λ) which intersects $(0,\mu)$. When $\mathbf{F}(\mathbf{x},\lambda)=\lambda\mathbf{x}-\mathbf{L}\mathbf{x}$, where $\mathbf{L} : \mathbb{R}^n \longrightarrow \mathbb{R}^n$ is linear (a matrix), μ is a bifurcation value precisely when μ is an *eigenvalue* of $\mathbf{L}$; then if $\mathbf{x}$ is a corresponding eigenvector there are solutions $(t\mathbf{x},\mu)$ for all scalars t. When $n=1$ this is a straight line in the (x,λ) plane.

Example 4.20. A necessary condition for bifurcation.
Let $\mathbf{F} : \mathbb{R}^n\times\mathbb{R} \longrightarrow \mathbb{R}^m$ satisfy $\mathbf{F}(\mathbf{0},\lambda)=\mathbf{0}$ for all $\lambda\in\mathbb{R}$ and suppose that μ is a bifurcation value. If $\mathbf{F}$ is continuous and is of class C^1 in $\mathbf{x}$ then $\det(D_{\mathbf{x}}\mathbf{F})=0$, that is, $D_{\mathbf{x}}\mathbf{F}$ is not invertible at $(\mathbf{0},\mu)$.

Solution. If $\det(D_{\mathbf{x}}\mathbf{F})\neq 0$ at $(\mathbf{0},\mu)$, we may invoke the implicit function theorem to show that $\mathbf{F}(\mathbf{x},\lambda)=\mathbf{0}$ has a unique solution $\mathbf{x}=\mathbf{f}(\lambda)$ on a neighbourhood of $(\mathbf{0},\mu)$. Since $(\mathbf{0},\lambda)$ is a solution on this neighbourhood there can be no solution with $\mathbf{x}\neq\mathbf{0}$ so μ is not a bifurcation value.

The above conditions are certainly not sufficient, in general. It would take us too far afield to discuss this topic in more detail but the implicit function theorem is a most useful tool in extracting information about the nature of solutions. We give one easy positive result for linear operators ($n\times n$ matrices).

Given a linear operator from $\mathbb{R}^n$ to $\mathbb{R}^n$, equivalently a $n\times n$ matrix A, a real number λ is called an *eigenvalue* of A if there is a nonzero $\mathbf{x}\in\mathbb{R}^n$ such that $A\mathbf{x}=\lambda\mathbf{x}$; $\mathbf{x}$ is called an *eigenvector* corresponding to λ. [We restrict attention to real eigenvalues and eigenvectors but complex ones are important.] Thus, if λ is an eigenvalue of A, the nullspace $N(A-\lambda I)\neq\{\mathbf{0}\}$. An eigenvalue λ is said to be *simple* if $N(A-\lambda I)$ is 1-dimensional, that is, $N(A-\lambda I)=\{c\mathbf{x}^{(0)} : c \text{ a scalar}\}$ and $N(A-\lambda I)=N(A-\lambda I)^2$. [The algebraic and geometric multiplicities of λ are both 1.] Then $\mathbb{R}^n$ can be written $\mathbb{R}^n=N(A-\lambda I)\oplus R(A-\lambda I)$ where $R(A-\lambda I)$

stands for the range of $A-\lambda I$ and $\oplus$ denotes *direct sum*. This means that $\mathbf{x}\in\mathbb{R}^n$ can be written $\mathbf{x}=\mathbf{n}+\mathbf{r}$ for $\mathbf{n}\in N(A-\lambda I)$ and $\mathbf{r}\in R(A-\lambda I)$ where $N(A-\lambda I)\cap R(A-\lambda I)=\{\mathbf{0}\}$ [this means that $\mathbf{x}$ uniqely determines $\mathbf{n}$ and $\mathbf{r}$]. To see this, recall that the rank-nullity theorem shows that the space of all elements of the form $\mathbf{n}+\mathbf{r}$ has dimension n so equals $\mathbb{R}^n$. For the other part, suppose that $\mathbf{w}\in N(A-\lambda I)\cap R(A-\lambda I)$. Then $\mathbf{w}=(A-\lambda I)\mathbf{z}$ for some $\mathbf{z}$ and $(A-\lambda I)\mathbf{w}=\mathbf{0}$. Thus, $(A-\lambda I)^2\mathbf{z}=\mathbf{0}$, that is, $\mathbf{z}\in N(A-\lambda I)^2=N(A-\lambda I)$ by hypothesis. This shows that $(A-\lambda I)\mathbf{z}=\mathbf{0}$ that is $\mathbf{w}=\mathbf{0}$.

Example 4.21. Let $A(t)$ be a family of linear operators for t in some open interval containing 0 with all entries of class C^1. Suppose that $\lambda^{(0)}$ is a simple eigenvalue of $A(0)$ with corresponding eigenvector $\mathbf{x}^{(0)}$. Then there is a neighbourhood U of $t=0$ such that, for every $t\in U$, $A(t)$ has an eigenvalue λ.

Solution. The assertion is that the equation $A(t)\mathbf{x}=\lambda\mathbf{x}$ has a nonzero solution $\mathbf{x}$. Let $\mathbf{x}^{(0)}$ be an eigenvector of $A(0)$ corresponding to $\lambda^{(0)}$. Define $\mathbf{F}$ by

$$\mathbf{F}(\mathbf{x},\lambda,t)=(A(t)\mathbf{x}-\lambda\mathbf{x},\ <\mathbf{x},\mathbf{x}^{(0)}>-1).$$

$\mathbf{F}$ is defined on a neighbourhood of $(\mathbf{x}^{(0)},\lambda^{(0)},0)$ in $\mathbb{R}^{n+2}$ with values in $\mathbb{R}^{n+1}$, and $\mathbf{F}(\mathbf{x}^{(0)},\lambda^{(0)},0)=\mathbf{0}$. We may write the components of $\mathbf{F}$ as

$$F_i=\sum_{j=1}^n a_{ij}x_j-\lambda x_i,\ i=1,2,\cdots,n,\qquad F_{n+1}=\sum_{j=1}^n x_jx_j^{(0)}-1.$$

We will use the implicit function theorem to "solve" for $\mathbf{x},\lambda$ as functions of t. We need to show that $\mathbf{L}:=D_{(\mathbf{x},\lambda)}\mathbf{F}(\mathbf{x}^{(0)},\lambda^{(0)},0)$ is invertible. We have $\mathbf{L}(\mathbf{x},\mu)=((A(0)-\lambda^{(0)})\mathbf{x}-\mu\mathbf{x}^{(0)},<\mathbf{x},\mathbf{x}^{(0)}>)$; for example, the matrix of partial derivatives is, in block notation

$$([\cdot]^{\mathrm{T}}\text{ denotes a column})\quad\left[\begin{array}{c|c} A(0)-\lambda^{(0)}I & [\mathbf{x}^{(0)}]^{\mathrm{T}} \\ \hline -\mathbf{x}^{(0)} & 0\end{array}\right].$$

To prove that $\mathbf{L}$ is invertible it suffices, by the rank-nullity theorem to show that $\mathbf{L}$ is one-to-one. For this we have to show that $\mathbf{L}(\mathbf{x},\mu)=\mathbf{0}$ implies that $\mathbf{x}=\mathbf{0}$ and $\mu=0$. When $\mathbf{L}(\mathbf{x},\mu)=\mathbf{0}$ we have

$$(A(0)-\lambda^{(0)})\mathbf{x}-\mu\mathbf{x}^{(0)}=\mathbf{0}\ \text{ and }\ <\mathbf{x},\mathbf{x}^{(0)}>=0. \tag{4.6}$$

The first equation in (4.6) says that $\mu\mathbf{x}^{(0)}\in R(A(0)-\lambda^{(0)}I)$. But also $\mu\mathbf{x}^{(0)}\in N(A(0)-\lambda^{(0)}I)$ so $\mu\mathbf{x}^{(0)}=\mathbf{0}$ and therefore $\mu=0$. Then, the first equation gives $(A(0)-\lambda^{(0)})\mathbf{x}=\mathbf{0}$, that is $\mathbf{x}=c\mathbf{x}^{(0)}$, and then the second equation yields $c=0$. Hence, by the implicit function theorem, $\mathbf{x}=\mathbf{x}(t)$,

$\lambda=\lambda(t)$ on a nbd and so $A(t)\mathbf{x}(t)-\lambda(t)\mathbf{x}(t)=\mathbf{0}$, and $<\mathbf{x}(t),\mathbf{x}^{(0)}> = 1$, whence $\mathbf{x}(t)\neq\mathbf{0}$.

Since we know from the implicit function theorem when implicitly defined functions are differentiable we can calculate the derivative by use of the chain rule. The result will follow from the next lemma.

Lemma 4.22. *Suppose* $\mathbf{F}:\mathbb{R}^{k+n}\longrightarrow\mathbb{R}^{n}$ *and* $\mathbf{f}:\mathbb{R}^{k}\longrightarrow\mathbb{R}^{n}$ *are differentiable and let* $\mathbf{H}(\mathbf{x})=\mathbf{F}(\mathbf{x},\mathbf{f}(\mathbf{x}))$, $\mathbf{x}\in\mathbb{R}^{k}$. *Then* $\mathbf{H}$ *is differentiable and*

$$D\mathbf{H}(\mathbf{x})=D_{\mathbf{x}}\mathbf{F}(\mathbf{x},\mathbf{f}(\mathbf{x}))+D_{\mathbf{y}}\mathbf{F}(\mathbf{x},\mathbf{f}(\mathbf{x}))\,D\mathbf{f}(\mathbf{x}).$$

Proof. Let $\mathbf{G}(\mathbf{x})=(\mathbf{x},\mathbf{f}(\mathbf{x}))$. Then $\mathbf{G}:\mathbb{R}^{k}\longrightarrow\mathbb{R}^{n+k}$ and $\mathbf{H}=\mathbf{F}\circ\mathbf{G}$. Now $\mathbf{G}$ is differentiable and $D\mathbf{G}(\mathbf{x})\mathbf{h}=(\mathbf{h},D\mathbf{f}(\mathbf{x})\mathbf{h})$, for we have

$$\mathbf{G}(\mathbf{x}+\mathbf{h})-\mathbf{G}(\mathbf{x})-(\mathbf{h},D\mathbf{f}(\mathbf{x})\mathbf{h})=(\mathbf{0},\mathbf{f}(\mathbf{x}+\mathbf{h})-\mathbf{f}(\mathbf{x})-D\mathbf{f}(\mathbf{x})\mathbf{h})=(\mathbf{0},\mathbf{R}(\mathbf{x},\mathbf{h}))$$

and $\|(\mathbf{0},\mathbf{R}(\mathbf{x},\mathbf{h}))\|/\|\mathbf{h}\|=\|\mathbf{R}(\mathbf{x},\mathbf{h})\|/\|\mathbf{h}\|\longrightarrow 0$ as $\|\mathbf{h}\|\longrightarrow 0$.

In terms of matrices, $D\mathbf{G}(\mathbf{x})=\begin{bmatrix}\mathbf{I}\\ \hdashline D\mathbf{f}(\mathbf{x})\end{bmatrix}$, ($\mathbf{I}$ denotes the $k\times k$ identity matrix) and this is also evident from computing the partial derivatives (but that does not prove differentiability). Now, by the chain rule, $\mathbf{H}$ is differentiable and $D\mathbf{H}(\mathbf{x})=D\mathbf{F}(\mathbf{x},\mathbf{f}(\mathbf{x}))\,D\mathbf{G}(\mathbf{x})$ so that, writing $\mathbf{z}=(\mathbf{x},\mathbf{f}(\mathbf{x}))$ and using the notion of block partial derivative

$$D\mathbf{H}(\mathbf{x})\mathbf{h} = \mathrm{D}\mathbf{F}(\mathbf{z})\,(\mathbf{h},D\mathbf{f}(\mathbf{x})\mathbf{h}) = [D_{\mathbf{x}}\mathbf{F}(\mathbf{z})]\mathbf{h}+[D_{\mathbf{y}}\mathbf{F}(\mathbf{z})]D\mathbf{f}(\mathbf{x})\mathbf{h}.$$

In terms of matrices this may be written

$$D\mathbf{H}(\mathbf{x}) = [D_{\mathbf{x}}\mathbf{F}(\mathbf{z})\,\vdots\,D_{\mathbf{y}}\mathbf{F}(\mathbf{z})]\begin{bmatrix}\mathbf{I}\\ \hdashline D\mathbf{f}(\mathbf{x})\end{bmatrix} = [D_{\mathbf{x}}\mathbf{F}(\mathbf{z})]+[D_{\mathbf{y}}\mathbf{F}(\mathbf{z})]\,D\mathbf{f}(\mathbf{x}).$$

Proposition 4.23. *Under the assumptions of the implicit function theorem, the implicitly defined function* $\mathbf{f}$ *has derivative given by*

$$D\mathbf{f}(\mathbf{x}) = -[D_{\mathbf{y}}\mathbf{F}(\mathbf{x},\mathbf{f}(\mathbf{x}))]^{-1}D_{\mathbf{x}}\mathbf{F}(\mathbf{x},\mathbf{f}(\mathbf{x})).$$

Proof. Write $\mathbf{z}=(\mathbf{x},\mathbf{f}(\mathbf{x}))$. By applying Lemma 4.22, because $\mathbf{H}(\mathbf{x})=\mathbf{0}$, we have $\mathbf{0}=D_{\mathbf{x}}\mathbf{F}(\mathbf{z})+D_{\mathbf{y}}\mathbf{F}(\mathbf{z})\,D\mathbf{f}(\mathbf{x})$ and hence $D\mathbf{f}(\mathbf{x})=-[D_{\mathbf{y}}\mathbf{F}(\mathbf{z})]^{-1}D_{\mathbf{x}}\mathbf{F}(\mathbf{z})$, since $D_{\mathbf{y}}\mathbf{F}(\mathbf{z})$ is invertible.

This may seem complicated but in practice one can often avoid calculating inverses of matrices by solving equations for partial derivatives instead; the procedure is, of course, equivalent. The equations for the partial derivatives are obtained by the usual elementary calculus method of "implicit differentiation", which means using the chain rule.

In the special case when $F:\mathbb{R}^2 \longrightarrow \mathbb{R}$ and $F(x,y)=0$ implicitly defines a function $f(x)$ this may be written :

$$0 = \frac{\partial}{\partial x}[F(x,f(x))] = \frac{\partial F}{\partial x} + \frac{\partial F}{\partial y}f'(x) \text{ so that } f'(x) = -\frac{\partial F/\partial x}{\partial F/\partial y}.$$

Note that the mnemonic of "cancelling ∂F" fails here.

Example 4.24. Show that the equations $x+y+t=0$, $xyt+\sin(xyt)=0$ define x,y implicitly as functions of t in a neighbourhood of the point $(t,x,y)=(-1,0,1)$. Calculate $\dfrac{dx}{dt}$ and $\dfrac{dy}{dt}$ at this point.

Solution. Let $\mathbf{F}(t,x,y)=(x+y+t, xyt+\sin(xyt))$. Then $\mathbf{F}(-1,0,1)=(0,0)$ (note that this has to be checked). Now t plays the role of $\mathbf{x}$ and (x,y) the role of $\mathbf{y}$ in the theorem. Then we have

$$D_{\mathbf{y}}\mathbf{F}=\begin{bmatrix} \partial F_1/\partial x & \partial F_1/\partial y \\ \partial F_2/\partial x & \partial F_2/\partial y \end{bmatrix} = \begin{bmatrix} 1 & 1 \\ yt+yt\cos(xyt) & xt+xt\cos(xyt) \end{bmatrix}$$

and we see that $\mathbf{F}$ is of class C^1. At $(-1,0,1)$,

$\det D_{\mathbf{y}}\mathbf{F}=\det\begin{bmatrix} 1 & 1 \\ -1-\cos(0) & 0 \end{bmatrix}=2.$ Therefore, by the implicit function theorem, $(x,y)=\mathbf{f}(t)$ on some neighbourhood, and $\mathbf{f}$ is differentiable. Write $x(t)$, $y(t)$ for the components of $\mathbf{f}(t)$. Then we have,

$$x(t)+y(t)+t=0, \quad \text{and} \quad x(t)y(t)t+\sin[x(t)y(t)t]=0$$

Differentiating with respect to t gives $x'(t)+y'(t)+1=0$, and

$$[x'(t)y(t)t+x(t)y'(t)t+x(t)y(t)][1+\cos(x(t)y(t)t)]=0.$$

At the point $(-1,0,1)$ this gives the two equations

$$x'+y'+1=0, \qquad -x'(1+\cos(0))=0$$

Therefore $x'=0$ and $y'=-1$, at that point.

Example 4.25. Show that the equations $-2xy+u^2+v^2=0$, $x^3-y^3+u^3+v^3=0$ define x,y implicitly as functions of u,v in a neighbourhood of $(x,y,u,v)=(1,1,1,-1)$. If $x=f(u,v)$, $y=g(u,v)$ on that neighbourhood, calculate $\partial f/\partial u$, $\partial f/\partial v$ and $\partial g/\partial u$, $\partial g/\partial v$ at $(u,v)=(1,-1)$ and write down the affine approximation to $(f(u,v), g(u,v))$ at that point.

Solution. Let $\mathbf{F}(u,v,x,y)=(u^2+v^2-2xy, u^3+v^3+x^3-y^3)$.
Then $\mathbf{F}(1,-1,1,1)=(0,0)$ and clearly all partial derivatives are continuous, so $\mathbf{F}$ is a C^1 function. Also

$$\det\begin{bmatrix} \partial F_1/\partial x & \partial F_1/\partial y \\ \partial F_2/\partial x & \partial F_2/\partial y \end{bmatrix} = \det\begin{bmatrix} -2y & -2x \\ 3x^2 & -3y^2 \end{bmatrix} = 6y^3+6x^3=12$$

at the given point, so indeed $(x,y)=\mathbf{f}(u,v)=(f(u,v), g(u,v))$ on some neighbourhood. Then, on this neighbourhood

$$-2f(u,v)g(u,v)+u^2+v^2=0, \quad \text{and} \quad f^3(u,v)-g^3(u,v)+u^3+v^3=0.$$

Differentiate these equations with respect to u keeping v constant to obtain

$$-2f\cdot\partial g/\partial u - 2g\cdot\partial f/\partial u + 2u = 0, \quad 3f^2\cdot\partial f/\partial u - 3g^2\cdot\partial g/\partial u + 3u^2 = 0$$

At $(u,v)=(1,-1)$, $f=1$, $g=1$ so we have

$$-\partial g/\partial u - \partial f/\partial u + 1 = 0, \quad \partial f/\partial u - \partial g/\partial u + 1 = 0$$

and therefore $\partial f/\partial u=0$, $\partial g/\partial u=1$, at that point. Exactly similarly, by differentiating with respect to v we obtain $\partial f/\partial v=-1$, $\partial g/\partial v=0$.
The affine approximation to $\mathrm{f}(u,v)$ at $(1,-1)$ is

$$\mathrm{f}(1,-1)+D\mathrm{f}(1,-1)\begin{pmatrix}u-1\\v+1\end{pmatrix}=\begin{pmatrix}1\\1\end{pmatrix}+\begin{pmatrix}0&-1\\1&0\end{pmatrix}\begin{pmatrix}u-1\\v+1\end{pmatrix}=\begin{pmatrix}1\\1\end{pmatrix}+\begin{pmatrix}-v-1\\u-1\end{pmatrix}=\begin{pmatrix}-v\\u\end{pmatrix},$$

that is $f(u,v)=-v+R_1$, $g(u,v)=u+R_2$ in a neighbourhood of $(1,-1)$ where R_1, R_2 are "small".

4.4. AN EXISTENCE THEOREM FOR DIFFERENTIAL EQUATIONS

An ordinary differential equation (O.D.E) of order n in *standard form* is an equation of the type (writing $y^{(n)}$ for the n-th derivative of a function y)

$$y^{(n)}=f(t, y, y', \cdots, y^{(n-1)}). \tag{4.7}$$

A *solution* of (4.7) on an interval I is a function $y\in C^{(n)}(I, \mathbb{R})$ which satisfies (4.7) identically on I. For simple first and second-order equations it is often easy to "solve" them explicitly. The methods of separation of variables and use of an integrating factor are probably familiar to the reader. However, many equations cannot be "solved" and any solution can only be found approximately by some numerical method on a computer. However, use of a numerical method requires a licence, namely one needs to know before one starts that there is a solution. Otherwise, what does the computer output mean? Also the more we know in advance about any solution the easier it will be to compute accurate approximations. In other words we require an *existence theorem* plus whatever other information can be gleaned from the given equation.

Differential equations usually have many solutions. For example, "solving" a first-order O.D.E. requires one integration, so an arbitrary constant (or parameter) arises. Similarly a second-order O.D.E. involves two arbitrary constants, or two parameters. These parameters are determined by some given initial conditions. Unfortunately, these well-known "facts" fall short of the truth as the following example illustrates.

Example 4.26. Solve the differential equation $\dfrac{dy}{dt} = y(1-y)$.

Solution. We can write this as (separation of variables method) $\dfrac{1}{y(1-y)}\dfrac{dy}{dt} = 1$, that is, $d/dt\,[\ln(y/(1-y)] = 1$. Therefore, $\ln(y/(1-y))=t+c$, c an arbitrary constant, and hence $y/(1-y)=ae^t$, or $y=ae^t/(1+ae^t)$, a an arbitrary (nonzero) constant. The meaning of a is seen by setting $t=0$; $a=y(0)/(1-y(0))$. However, we have "lost" a solution. In the first step we divided by $y(1-y)$. This is only valid if $y\neq 0$ and $y\neq 1$. Now $y(t)\equiv 0$ and $y(t)\equiv 1$ are seen to be solutions. The former is included in the general expression $y(t)=ae^t/(1+ae^t)$ when $a=0$, but no value of a gives the latter solution. However, formally replacing a gives the expression $y(t)=y(0)e^t/(1-y(0)+y(0)e^t)$ and this is valid even for $y(0)=1$.

More complicated examples show that a second-order equation can have solutions which consist of a 2-parameter family plus a 1-parameter family plus isolated solutions. The well-known "fact" that a second-order equation has a 2-parameter family of solutions is true for *linear* differential equations but not generally.

We may regard a first-order equation $dy/dt=f(t,y)$ as specifying the velocity of a moving particle whose position at time t is measured by $y(t)$. If we observe the particle at some instant, say at $t=0$, we expect the motion to be predicted by the O.D.E. and $y(t)$ to be determined uniquely. In other words, we may believe that an

$$\textit{initial value problem (I.V.P)}\quad \frac{dy}{dt} = f(t,y),\ y(0) = \alpha,$$

should have a unique solution on an interval containing 0. This is expecting too much in general so we must restrict the class of functions f to make this true. Also the example above with initial condition $y(0)=-1/\varepsilon$ becomes infinite when $e^t=1+\varepsilon$. Therefore, one can take $\varepsilon>0$ so small that the solution which starts at $t=0$ becomes infinite in an arbitrarily small time, namely $t=\ln(1+\varepsilon)$. Thus, even for a function $f(t,y)$ as simple as $y-y^2$, one can only expect solutions to exist on small intervals containing the initial point, in other words, existence can only be *local* in general.

We shall give an existence and uniqueness theorem for I.V.P's for *systems* of first-order equations, that is, we have n equations for n functions, $y_1,\cdots,y_n$, of the form

$$y_i=f_i(t, y_1, y_2,\cdots, y_n),\ 1\le i\le n.$$

We can conveniently write these as

$$\mathbf{y}'(t)=\mathbf{f}(t,\mathbf{y}). \tag{4.8}$$

The initial condition takes the form

$$\mathbf{y}(t_0)=\mathbf{z}. \tag{4.9}$$

Every equation of order n in standard form as in (4.7) can be written as such a system. To see this write $y_1 = y$, $y_2 = y'$, $\cdots$, $y_n = y^{(n-1)}$. We see that

$$\begin{aligned} y_1' &= y_2 := f_1(t,\mathbf{y}) \\ y_2' &= y_3 := f_2(t,\mathbf{y}) \\ &\vdots \\ y_{n-1}' &= y_n := f_{n-1}(t,\mathbf{y}) \end{aligned}$$

and
$$y_n' = f(t,\mathbf{y}) := f_n(t,\mathbf{y}) .$$

Initial conditions on $y_1, y_2, \cdots, y_n$ are initial conditions on $y, y', \cdots, y^{(n-1)}$.

Suppose f in (4.7) is infinitely differentiable (C^∞) and that (4.7) has a C^∞ solution y on some interval I containing the initial point 0. Since $y, y', \cdots, y^{(n-1)}$ are prescribed at 0, the equation (4.7) can be used to calculate $y^{(n)}(0)$. By differentiating (4.7), $y^{(n+1)}$ can be written in terms of lower order derivatives of y so that $y^{(n+1)}(0)$ can be calculated. Proceeding in this way all derivatives of y at 0 can be computed and so we could obtain y via its Taylor series expansion. This is not suggested as a practical method but illustrates that prescribing the initial values of $y, y', \cdots, y^{(n-1)}$ is an appropriate condition to assure existence of a solution. Here again the series may only converge near 0, so the solution may only be a local one.

We turn to the study of (4.8). We assume that f is continuous on $I \times G$ where I is an open interval in $\mathbb{R}$ and G is an open set in $\mathbb{R}^n$. We shall make a further restriction later. We note that the IVP (4.8), (4.9) is equivalent to (by the fundamental theorem of the calculus)

$$\mathbf{y}(t) = \mathbf{y}(t_0) + \int_{t_0}^{t} \mathbf{f}(s,\mathbf{y}(s)). \tag{4.10}$$

Here the integral of function from I to $\mathbb{R}^n$ is defined by integrating each component : $\int_{t_0}^{t} \mathbf{g}(s)\, ds$ has components $\int_{t_0}^{t} g_i(s)\, ds$.

We require a preliminary lemma which says that an inequality for sums becomes an inequality for integrals. The Lemma actually holds for any norm and a proof may be given using approximating sums and the triangle inequality, but we shall give the proof only for the Euclidean norm. A proof for the norm $\|\cdot\|_\infty$ is even simpler.

Lemma 4.27. *Let* $\mathbf{g} : [a,b] \longrightarrow \mathbb{R}^n$ *be continuous on* $[a,b]$. *Then*

$$\left\| \int_a^b \mathbf{g}(t)\, dt \right\| \le \int_a^b \|\mathbf{g}(t)\|\, dt.$$

Proof. Since $\big|\, \|\mathbf{g}(t)\| - \|\mathbf{g}(s)\| \,\big| \le \|\mathbf{g}(t) - \mathbf{g}(s)\|$, by the Corollary to the triangle inequality, $t \longmapsto \|\mathbf{g}(t)\|$ is continuous and therefore integrable. Now write $\mathbf{G} = \int_a^b \mathbf{g}$ and denote the components of $\mathbf{G}$ and $\mathbf{g}$ by G_i and g_i respectively. Then $\|\mathbf{G}\|^2 = \langle \mathbf{G}, \mathbf{G} \rangle = \sum_{i=1}^n G_i \left(\int_a^b g_i \right) = \int_a^b \left(\sum_{i=1}^n G_i g_i \right)$.

By Schwarz's inequality, $\sum_{i=1}^n G_i g_i(t) \le \|\mathbf{G}\|\, \|\mathbf{g}(t)\|$ and therefore

$$\|\mathbf{G}\|^2 \le \int_a^b \|\mathbf{G}\|\, \|\mathbf{g}(t)\|\, dt = \|\mathbf{G}\| \int_a^b \|\mathbf{g}(t)\|\, dt.$$

If $\|\mathbf{G}\| \ne 0$ we may divide by $\|\mathbf{G}\|$ to obtain the conclusion, whereas if $\|\mathbf{G}\| = 0$ the result holds trivially.

In order to obtain a local existence and uniqueness result we shall require the function $\mathbf{f}$ in (4.7) to satisfy a *local Lipschitz condition.*

Definition 4.28. Let G be an open subset of $\mathbb{R}^{n+1}$ and let $\mathbf{f} : G \longrightarrow \mathbb{R}^n$. $\mathbf{f}$ is said to satisfy a *local Lipschitz condition* (in $\mathbb{R}^n$ variables) if for each $(t_0, \mathbf{y}^{(0)}) \in G$ there is a neighbourhood U of $(t_0, \mathbf{y}^{(0)})$ and a constant $L = L(t_0, \mathbf{y}^{(0)})$ such that $\|\mathbf{f}(t,\mathbf{x}) - \mathbf{f}(t,\mathbf{y})\| \le L \|\mathbf{x} - \mathbf{y}\|$ for all $(t,\mathbf{x}),\ (t,\mathbf{y}) \in U$.

It is readily seen that if $\mathbf{f}$ satisfies such a local Lipschitz condition then for any compact subset K of G there exists L_K such that

$$\|\mathbf{f}(t,\mathbf{x}) - \mathbf{f}(t,\mathbf{y})\| \le L_K \|\mathbf{x} - \mathbf{y}\| \text{ for all } (t,\mathbf{x}),\ (t,\mathbf{y}) \in K.$$

The simplest criterion that ensures $\mathbf{f}$ satisfies a local Lipschitz condition is :

Proposition 4.29. *If* $\mathbf{f} \in C^1(G, \mathbb{R}^n)$ *then* $\mathbf{f}$ *satisfies a local Lipschitz condition on* G.

Proof. This is an immediate consequence of the approximation result of Proposition 3.24.

Theorem 4.30. *Let G be an open connected set in $\mathbb{R}^{n+1}$ and suppose $\mathbf{f}: G \longrightarrow \mathbb{R}^n$ satisfies a local Lipschitz condition on G. Suppose that $(t_0, \mathbf{z}) \in G$, let $E = \{|t-t_0| \le a, \|\mathbf{y}-\mathbf{z}\| \le b\}$ be an $(n+1)$-dimensional "cylinder" that lies in G, and let L be the corresponding Lipschitz constant for E. Write $M = \max_E \|\mathbf{f}(t,\mathbf{y})\|$, and take $\delta < \min\{a, b/M, 1/L\}$. Then, the IVP $\mathbf{y}' = \mathbf{f}(t,\mathbf{y})$, $\mathbf{y}(t_0) = \mathbf{z}$ has a unique solution on the interval $I = [t_0 - \delta, t_0 + \delta]$.*

Proof. We apply the contraction mapping theorem to a subset of the normed linear space $C(I, \mathbb{R}^n)$ whose norm is the uniform norm. Let $S := \{\mathbf{u} \in C(I, \mathbb{R}^n) : \mathbf{u}(t_0) = \mathbf{z}, \ \|\mathbf{u}(t) - \mathbf{z}\| \le b \text{ for all } t \in I\}$. S endowed with the uniform norm is certainly a metric space. In fact, S is complete, for if $\{\mathbf{u}^{(m)}\}$ is a Cauchy sequence in S, $\{\mathbf{u}^{(m)}\}$ is Cauchy in $C(I, \mathbb{R}^n)$ hence convergent to $\mathbf{u} \in C(I, \mathbb{R}^n)$. Since $\|\mathbf{u}^{(m)}(t) - \mathbf{z}\| \le b$ for all $t \in I$ and $\mathbf{u}^{(m)}(t)$ converges to $\mathbf{u}(t)$ it follows that $\|\mathbf{u}(t) - \mathbf{z}\| \le b$ so $\mathbf{u} \in S$ (problem 1.27). For $\mathbf{u} \in S$ let $\mathbf{w}$ be defined by

$$\mathbf{w}(t) = \mathbf{z} + \int_{t_0}^{t} \mathbf{f}(s, \mathbf{u}(s))\, ds. \tag{4.11}$$

Then $\mathbf{w}$ is continuous on I. Moreover, $\mathbf{w} \in S$. For by Lemma 4.27,

$$\|\mathbf{w}(t) - \mathbf{z}\| \le \int_{t_0}^{t} \|\mathbf{f}(s, \mathbf{u}(s))\|\, ds \le |t - t_0| M \le M\delta \le b.$$

We use (4.11) to define a function $F : S \longrightarrow S$ by

$$F\,\mathbf{y} = \mathbf{w}, \quad \mathbf{w} \text{ as in (4.11).}$$

We intend to prove that F has a unique fixed point in S which, in view of (4.11), is the conclusion of the theorem. Now let $\mathbf{u}, \mathbf{v} \in S$. Then

$$\begin{aligned}\|F\,\mathbf{u}(t) - F\,\mathbf{v}(t)\| &\le \int_{t_0}^{t} \|\mathbf{f}(s,\mathbf{u}(s)) - \mathbf{f}(s,\mathbf{v}(s))\|\, ds, \text{ by Lemma 4.27.} \\ &\le \int_{t_0}^{t} L\|\mathbf{u}(s) - \mathbf{v}(s)\|\, ds, \text{ since } \mathbf{u}(s), \mathbf{v}(s) \in E, \\ &\le L|t - t_0|\, \|\mathbf{u} - \mathbf{v}\|_u .\end{aligned}$$

Therefore, $\|F\,\mathbf{u} - F\,\mathbf{v}\|_u \le L\,\delta\, \|\mathbf{u} - \mathbf{v}\|_u$. Since $L\,\delta < 1$, F is a contraction on the complete metric space S so has a unique fixed point by Theorem 4.3. This proves the result.

It is possible to prove that under the same assumptions there is a unique solution on the interval $[t-\delta,t_0-\delta]$ for any $\delta\leq\min(a,b/M)$; the restriction concerning L can be removed. An elegant way of doing this is by considering an equivalent norm on S: for simplicity of notation consider $t_0=0$. For $t\geq 0$, let $\|\mathbf{u}\|_\alpha:=\sup\{\|\mathbf{u}(t)\|e^{-\alpha t}:0\leq t\leq\delta\}$, with α to be chosen. This does define a norm equivalent to the uniform norm. Now

$$\|F\,\mathbf{u}(t)-F\,\mathbf{v}(t)\|\leq\int_0^t L\|\mathbf{u}(s)-\mathbf{v}(s)\|e^{-\alpha s}e^{\alpha s}\,ds$$

$$\leq L\|\mathbf{u}-\mathbf{v}\|_\alpha\int_0^t e^{\alpha s}\,ds\leq L\|\mathbf{u}-\mathbf{v}\|_\alpha\,[e^{\alpha t}-1]/\alpha.$$

Hence $\|F\,\mathbf{u}(t)-F\,\mathbf{v}(t)\|_\alpha=\sup\|F\,\mathbf{u}(t)-F\,\mathbf{v}(t)\|e^{-\alpha t}\leq(L/\alpha)\|\mathbf{u}-\mathbf{v}\|_\alpha$ and so F is a contraction in the metric space $(S,\|\cdot\|_\alpha)$ for α chosen larger than L and hence F has a unique fixed point. The solution for $t<0$ can be obtained by "reversing time", that is, making a change of variable.

Example.4.31 The IVP $\dfrac{dy}{dt}=t^2+y^2$, $y(0)=1$ has a unique solution on $[0,\delta]$. Estimate δ.

Solution. $f(t,y)=t^2+y^2$ is certainly locally Lipschitz since f is C^1. In this case we see directly that

$$|f(t,y)-f(t,x)|=|y^2-x^2|=|y+x|\,|y-x|$$

and the local Lipschitz constant is a bound for $|y+x|$. To estimate δ we take $\delta=\min(a,b/M)$ with the notation of the theorem. Now

$$M=\max\{t^2+y^2:|t|\leq a,\ |y-1|\leq b\}=a^2+(1+b)^2$$

so $\delta=\min\{a,\ b/(a^2+(1+b)^2)\}$. To obtain an estimate as large as possible we should maximize $b/(a^2+(1+b)^2)$ for a given a. The maximum is when $b^2=1+a^2$ and is $1/(2b+2)$. Since $b>1$ here, we see that certainly $\delta<1/4$. Taking $a=1/4$ we obtain $\delta=0.246$. It is not worthwhile to attempt to improve this further, other ideas are more appropriate and are studied in courses on differential equations.

This estimate for δ does not give a maximal interval of existence but does assure that the solution exists on that particular interval. In the above example it may be shown by using the inequalities $y^2\leq t^2+y^2\leq 1+y^2$ for $0\leq t\leq 1$ that the solution of the IVP becomes infinite somewhere between $t=\pi/4$ and $t=1$, and so ceases to exist.

The above existence result applies to an ODE of order n in standard form. Occasionally one meets an ODE of order n in the form

$$g(t,y,y',\cdots,y^{(n)})=0. \tag{4.12}$$

We denote the $(n+2)$ variables on which g depends by $(t,\mathbf{u},\upsilon)$, where $\mathbf{u}\in\mathbb{R}^n$.

Proposition 4.32. *Let D be an open subset of $\mathbb{R}^{n+2}$ and let $g\in C^1(D,\mathbb{R})$. Suppose that $(t_0,\mathbf{u}_0^{(0)},\upsilon_0)\in D$ satisfies $g(t_0,\mathbf{u}_0^{(0)},\upsilon_0) = 0$ and that*

$\frac{\partial g}{\partial \upsilon}(t_0,\mathbf{u}_0^{(0)},\upsilon_0)\neq 0$. *Then the IVP* : $g(t, y, y', \cdots, y^{(n)}) = 0$,

$$y(t_0)=u_1^{(0)},\quad y'(t_0)=u_2^{(0)},\cdots,\quad y^{(n-1)}(t_0)=u_n^{(0)},\quad y^{(n)}(t_0)=\upsilon_0$$

has a unique local solution.

Proof. By the implicit function theorem $g(t,\mathbf{u},\upsilon)=0$ is equivalent to $\upsilon=f(t,\mathbf{u})$ on a neighbourhood of $(t_0,\mathbf{u}^{(0)},\upsilon_0)$, where $f\in C^1$. The local solution of the corresponding ODE in standard form gives a solution of the problem as posed.

4.5. GLOBAL HOMEOMORPHISMS

Topological properties of a space are those that are preserved under homeomorphisms. A homeomorphism is a continuous bijection f which maps open sets into open sets and so has a continuous inverse. The Inverse Function theorem asserts that if **f** is of class C^1 on an open set V and the Jacobian $J_{\mathbf{f}}(\mathbf{x})$ is non-zero on V, then **f** is a local homeomorphism (actually it says even more than this). It is of interest to know when **f** has an inverse on all of V – a global homeomorphism. For this we need **f** to be one-to-one on V. We shall give one result in this direction which at least highlights where things may go wrong. More advanced methods seem to be needed to obtain stronger results.

A physical interpretation arises in the theory of elasticity. Suppose some body made of elastic material occupies some region $\overline{\Omega}$ in $\mathbb{R}^3$ and is then deformed so that it occupies a new region $\overline{D}$. The deformation can be described by saying where points in $\overline{\Omega}$ move, in other words, by a function $\mathbf{f}:\overline{\Omega}\to\overline{D}$. The hypothesis $J_{\mathbf{f}}(\mathbf{x})\neq 0$ means that small volumes of material cannot be compressed to a single point. The function **f** will fail to be one-to-one when two points in $\overline{\Omega}$ are moved to the same point. This may happen by points of the boundary being brought into contact. If we prevent this possibility then it is reasonable to believe that the only other possibility is for points on the boundary to be pushed into the interior. The theorem below says that this is indeed so.

Theorem 4.33. *Let Ω be a bounded open subset of $\mathbb{R}^n$ and suppose there is a bounded, open set V with $\Omega \subseteq \mathrm{cl}(\Omega) \subseteq V$. Suppose that $\mathbf{f} \in C^1(V,\mathbb{R}^n)$ and that $J_{\mathbf{f}}(\mathbf{x}) \neq 0$ for all $\mathbf{x} \in V$. Further suppose that $\mathbf{f}$ is one-to-one on* $\mathrm{bdy}(\Omega)$. *Then*

$\mathbf{f}$ *is one-to-one on* $\mathrm{cl}(\Omega)$ *if and only if* $\mathbf{f}(\mathrm{bdy}(\Omega)) = \mathrm{bdy}(\mathbf{f}(\Omega))$.

Before giving the proof we make a few observations.

(1). In general, there is very little connection between $\mathbf{f}(\mathrm{bdy}(\Omega))$ and $\mathrm{bdy}(\mathbf{f}(\Omega))$.

(2). Since $\mathbf{f}$ is continuous and $\mathrm{cl}(\Omega)$ is compact, $\mathbf{f}(\mathrm{cl}(\Omega)) = \mathrm{cl}(\mathbf{f}(\Omega))$. Indeed $\mathbf{f}(\Omega) \subseteq \mathbf{f}(\mathrm{cl}(\Omega))$ and $\mathbf{f}(\mathrm{cl}(\Omega))$ is compact, hence closed, so that $\mathrm{cl}(\mathbf{f}(\Omega)) \subseteq \mathbf{f}(\mathrm{cl}(\Omega))$. Conversely, if $\mathbf{y} \in \mathbf{f}(\mathrm{cl}(\Omega))$ then $\mathbf{y} = \mathbf{f}(\mathbf{x})$ for some $\mathbf{x} \in \mathrm{cl}(\Omega)$. Then there exists $\{\mathbf{x}^{(m)}\} \in \Omega$ with $\mathbf{x}^{(m)} \to \mathbf{x}$. By continuity $\mathbf{f}(\mathbf{x}^{(m)}) \to \mathbf{f}(\mathbf{x}) = \mathbf{y}$ so $\mathbf{y} \in \mathrm{cl}(\mathbf{f}(\Omega))$.

(3). Since $\mathbf{f} \in C^1(V,\mathbb{R}^n)$ and $J_{\mathbf{f}}(\mathbf{x}) \neq 0$, $\mathbf{f}$ maps open sets onto open sets (Corollary 4.11). Therefore points of Ω are mapped into the open set $\mathbf{f}(\Omega)$ and so points of $\mathrm{bdy}(\mathbf{f}(\Omega))$ cannot be images of points of Ω, that is, we have $\mathrm{bdy}(\mathbf{f}(\Omega)) \subseteq \mathbf{f}(\mathrm{bdy}(\Omega))$.

Proof of Theorem 4.33. We write $D := \mathbf{f}(\Omega)$ so that D is an open set and by (2) $\mathbf{f}(\mathrm{cl}(\Omega)) = \mathrm{cl}(D)$. First suppose that $\mathbf{f}$ is one-to-one on $\mathrm{cl}(\Omega)$. By (3) we have only to show that $\mathbf{f}(\mathrm{bdy}(\Omega)) \subseteq \mathrm{bdy}(D)$. Thus, let $\mathbf{y} \in \mathbf{f}(\mathrm{bdy}(\Omega))$, $\mathbf{y} = \mathbf{f}(\mathbf{x})$ for some $\mathbf{x} \in \mathrm{bdy}(\Omega)$. If $\mathbf{y} \notin \mathrm{bdy}(D)$ then $\mathbf{y} \in D = \mathbf{f}(\Omega)$ so $\mathbf{y} = \mathbf{f}(\mathbf{z})$ for some $\mathbf{z} \in \Omega$. This contradicts $\mathbf{f}$ being one-to-one on $\mathrm{cl}(\Omega)$.
We turn to the converse so we suppose that $\mathbf{f}(\mathrm{bdy}(\Omega)) = \mathrm{bdy}(\mathbf{f}(\Omega))$. Let $M := \{\mathbf{y} \in \mathrm{cl}(D) : \mathbf{f}^{-1}(\mathbf{y})$ contains at least two points of $\mathrm{cl}(\Omega)\}$.
Our aim is to prove that M is empty. So suppose $M \neq \emptyset$. Let $r := \inf\{\, \mathrm{dist}(\mathbf{y},\mathrm{bdy}(D)) : \mathbf{y} \in M\}$. We claim that there is a point $\mathbf{y} \in M$ such that $\mathrm{dist}(\mathbf{y},\mathrm{bdy}(D)) = r$. Indeed, by definition of infimum, there exists $\mathbf{y}^{(k)} \in M$ such that $r \le \mathrm{dist}(\mathbf{y}^{(k)},\mathrm{bdy}(D)) \le r + 1/k$. By compactness of $\mathrm{cl}(D)$ we can suppose that $\mathbf{y}^{(k)} \to \mathbf{y}$ and hence $\mathrm{dist}(\mathbf{y},\mathrm{bdy}(D)) = r$. Since $\mathbf{y}^{(k)} \in M$, there exist $\mathbf{x}^{(k)}, \mathbf{z}^{(k)} \in \mathrm{cl}(\Omega)$ with $\mathbf{x}^{(k)} \neq \mathbf{z}^{(k)}$ and $\mathbf{f}(\mathbf{x}^{(k)}) = \mathbf{y}^{(k)} = \mathbf{f}(\mathbf{z}^{(k)})$. By compactness, subsequences converge to $\mathbf{x}, \mathbf{z} \in \mathrm{cl}(\Omega)$, say, and by continuity, $\mathbf{f}(\mathbf{x}) = \mathbf{y} = \mathbf{f}(\mathbf{z})$. If $\mathbf{x} = \mathbf{z}$, $\mathbf{f}(\mathbf{x}^{(k)}) = \mathbf{f}(\mathbf{z}^{(k)})$ would contradict the local one-to-one property for k sufficiently large. Therefore $\mathbf{x} \neq \mathbf{z}$ and $\mathbf{y} \in M$. We next claim that $r = 0$. If not, $\mathbf{y} \in \mathrm{int}D = D$ and $\mathbf{y} = \mathbf{f}(\mathbf{x}) = \mathbf{f}(\mathbf{z})$, $\mathbf{x} \neq \mathbf{z}$. Then $\mathbf{x}, \mathbf{z} \in \Omega$ since by assumption $\mathbf{f}(\mathrm{bdy}(\Omega)) = \mathrm{bdy}(D)$. Take $\varepsilon > 0$ so that $\mathrm{B}(\mathbf{x},\varepsilon) \subseteq \Omega$, $\mathrm{B}(\mathbf{z},\varepsilon) \subseteq \Omega$ and $\mathrm{B}(\mathbf{x},\varepsilon) \cap \mathrm{B}(\mathbf{z},\varepsilon) = \emptyset$. By the Inverse Function theorem we can suppose ε is

so small that f restricted to each neighbourhood is a homeomorphism. Then $\mathbf{f}(B(\mathbf{x},\varepsilon))$ and $\mathbf{f}(B(\mathbf{z},\varepsilon))$ are open sets containing $\mathbf{y}$ so contain a δ-neighbourhood of $\mathbf{y}$. This shows that $B(\mathbf{y},\delta)\subseteq M$. This neighbourhood contains points of M closer to the boundary of D than $\mathbf{y}$, a contradiction. Finally $r=0$ means that $\mathbf{y}\in\text{bdy}(D)=\mathbf{f}(\text{bdy}(\Omega))$. Since f is one-to-one on $\text{bdy}(\Omega)$, this is impossible.

We believe that if Ω is a connected set with a smooth boundary then $\text{bdy}\mathbf{f}(\Omega)=\mathbf{f}(\text{bdy}\Omega)$ is a consequence of the other hypotheses. This seems to require some more advanced topological ideas. However, we can prove the following result.

Theorem 4.34. *Under the hypotheses of Theorem 4.33, suppose that* $\text{bdy}(\Omega)$ *and* $\text{bdy}(\mathbf{f}(\Omega))$ *have the same finite number of components (maximal connected subsets). Then* $\text{bdy}(\mathbf{f}(\Omega))=\mathbf{f}(\text{bdy}(\Omega))$. *In particular, this holds if* $\text{bdy}(\Omega)$ *is connected.*

Proof. Write D in place of $\mathbf{f}(\Omega)$. Let Γ be a component of $\text{bdy}(\Omega)$. We first show that $\mathbf{f}(\Gamma)$ cannot intersect both $\text{bdy}(D)$ and D. Indeed, suppose this does occur for some component Γ and let $A:=\{\mathbf{x}\in\Gamma : \mathbf{f}(\mathbf{x})\in D=\mathbf{f}(\Omega)\}$. A is nonempty and open relative to Γ since f is continuous and D is an open set. Therefore $\Gamma\setminus A$ is a closed, nonempty set. Now $\text{cl}(A)$ and $\Gamma\setminus A$ cannot be separated, that is we have $\inf\{\|\mathbf{x}-\mathbf{z}\| : \mathbf{x}\in\text{cl}(A), \mathbf{z}\in\Gamma\setminus A\}=0$, for otherwise $\Gamma=\text{cl}(A)\cup\Gamma\setminus A$ is not connected (problem 2.36). Since $\text{cl}(A)$ and $\Gamma\setminus A$ are compact sets, there exists $\mathbf{x}\in\text{cl}(A)\cap\Gamma\setminus A$. Then there is $\{\mathbf{a}^{(m)}\}\subseteq A$ such that $\mathbf{a}^{(m)}\longrightarrow\mathbf{x}$ so $\mathbf{f}(\mathbf{a}^{(m)})\longrightarrow\mathbf{f}(\mathbf{x})$. Since $\mathbf{a}^{(m)}\in A$, there is $\mathbf{z}^{(m)}\in\Omega$ such that $\mathbf{f}(\mathbf{a}^{(m)})=\mathbf{f}(\mathbf{z}^{(m)})$. By compactness of $\text{cl}(\Omega)$, we can suppose that $\mathbf{z}^{(m)}\longrightarrow\mathbf{z}\in\text{cl}(\Omega)$. Then $\mathbf{f}(\mathbf{z})=\mathbf{f}(\mathbf{x})$, so $\mathbf{z}\in\text{bdy}(\Omega)$ since $\mathbf{x}\notin A$. As f is one-to-one on $\text{bdy}(\Omega)$ this implies that $\mathbf{x}=\mathbf{z}$. However, $\mathbf{a}^{(m)}\longrightarrow\mathbf{x}$ and $\mathbf{z}^{(m)}\longrightarrow\mathbf{x}$ and $\mathbf{f}(\mathbf{a}^{(m)})=\mathbf{f}(\mathbf{z}^{(m})$ then contradicts the fact that f is locally one-to-one.

The above argument shows that if Γ is a component of $\text{bdy}(\Omega)$ then either $\mathbf{f}(\Gamma)\subseteq\text{bdy}(D)$ or else $\mathbf{f}(\Gamma)\subseteq D$. If Γ is a component of $\text{bdy}(\Omega)$ and $\mathbf{f}(\Gamma)\subseteq\text{bdy}(D)$ then $\mathbf{f}(\Gamma)$ belongs to exactly one component of $\text{bdy}(D)$ since $\mathbf{f}(\Gamma)$ is connected and components are maximal. Thus, if there is a component Γ with $\mathbf{f}(\Gamma)\subseteq D$ then since, by hypothesis, the number of components of $\text{bdy}(D)$ equals the number of components of $\text{bdy}(\Omega)$ we see that there would be a component of $\text{bdy}(D)$ containing no points of $\mathbf{f}(\text{bdy}(\Omega))$. Since $\text{bdy}(D)\subseteq\mathbf{f}(\text{bdy}(\Omega))$ this is impossible. The conclusion is that $\mathbf{f}(\Gamma)\subseteq\text{bdy}(D)$ for every component Γ and therefore $\text{bdy}(D)=\mathbf{f}(\text{bdy}(\Omega))$.

Remark. I discussed these results with several of my colleagues. In particular, Ms. Weiyu Zhao provided many of the ideas for the given proofs.

4.6. THE BROUWER FIXED POINT THEOREM

If f is a continuous function and maps a closed interval $[a,b]$ into itself, then f has at least one *fixed point*, a point x with $f(x)=x$. This is a simple consequence of the intermediate value theorem applied to $f(x)-x$. There is an analogous result in $\mathbb{R}^n$, known as Brouwer's fixed point theorem, which is far from obvious. A few years ago a new elementary proof of this was found by Milnor **[6]**. We shall give a proof of the result based on the simplification by Rogers **[8]** of Milnor's work. We shall need to call on a result on the change of variables in a multiple integral which will be shown in Chapter 6. We prefer to present the result here as it fits into the theme of existence theorems.

Theorem.4.35. (*Brouwer's fixed point theorem.*)
Let $\overline{\mathbf{B}} = \{\mathbf{x} : \|\mathbf{x}\| \le 1\}$ *denote the closed unit ball in* $\mathbb{R}^n$ *and let* $\mathbf{f} : \overline{\mathbf{B}} \longrightarrow \overline{\mathbf{B}}$ *be continuous. Then* $\mathbf{f}$ *has a fixed point in* $\overline{\mathbf{B}}$.

The proof of this will be based on the following Lemma.

Lemma 4.36. (*The no-retraction Lemma.*)
It is not possible for a map $\mathbf{f}$ *of class* C^1 *(on an open set containing* $\overline{\mathbf{B}}$*) to map* $\overline{\mathbf{B}}$ *onto the unit sphere* $\mathrm{S}=\{\mathbf{x} : \|\mathbf{x}\|=1\}$ *and to satisfy* $\mathbf{f}(\mathbf{x})=\mathbf{x}$ *for all* $\mathbf{x}\in\mathrm{S}$.

Proof. Suppose such an $\mathbf{f}$ exists, write $\mathbf{g}(\mathbf{x})=\mathbf{f}(\mathbf{x})-\mathbf{x}$ and for $0\le t\le 1$ let $\mathbf{f}_t(\mathbf{x})=(1-t)\mathbf{x}+t\mathbf{f}(\mathbf{x})=\mathbf{x}+t\mathbf{g}(\mathbf{x})$. By convexity of $\overline{\mathbf{B}}$, $\mathbf{f}_t$ maps $\overline{\mathbf{B}}$ into itself. Moreover since $\mathbf{f}\in C^1$, $\mathbf{g}\in C^1$ too and so $\mathbf{g}$ satisfies a Lipschitz condition on the compact set $\overline{\mathbf{B}}$, say $\|\mathbf{g}(\mathbf{x})-\mathbf{g}(\mathbf{y})\| \le M\|\mathbf{x}-\mathbf{y}\|$ for all $\mathbf{x},\mathbf{y}\in\overline{\mathbf{B}}$. Then $\mathbf{f}_t$ is one-to-one on $\overline{\mathbf{B}}$ if $0\le t<1/M$. For if $\mathbf{f}_t(\mathbf{x})=\mathbf{f}_t(\mathbf{y})$ we have $\|\mathbf{x}-\mathbf{y}\|=t\|\mathbf{g}(\mathbf{x})-\mathbf{g}(\mathbf{y})\| \le tM\|\mathbf{x}-\mathbf{y}\|$ and so $\mathbf{x}=\mathbf{y}$. Now $D\mathbf{f}_t(\mathbf{x})=\mathbf{x}+tD\mathbf{g}(\mathbf{x})$ where all the partial derivatives of $\mathbf{g}$ are continuous on $\overline{\mathbf{B}}$ and hence bounded. Therefore $D\mathbf{f}_t(\mathbf{x})$ is invertible for t sufficiently small, say $0\le t\le t_0$, by Theorem 4.xx. We show next that $\mathbf{f}_t$ maps $\overline{\mathbf{B}}$ *onto* $\overline{\mathbf{B}}$. Indeed, by the inverse function theorem, $\mathbf{f}_t$ maps the open ball $\mathbf{B}$ onto an *open* set $G_t\subseteq\mathbf{B}$. Since $\mathbf{f}_t$ is one-to-one on $\overline{\mathbf{B}}$, $\mathbf{f}_t$ maps S onto bdy(G_t), by Theorem

4.33. But $\mathbf{f}_t(\mathbf{x})=\mathbf{x}$ for $\mathbf{x}\in\mathbf{S}$ so we must have $\mathbf{S}=\mathrm{bdy}(G_t)$ and hence $\mathbf{B}\setminus G_t=\mathbf{B}\setminus\mathrm{cl}(G_t)$. To see this note that if $\mathbf{y}\in\mathbf{B}\setminus G_t$ then $\mathbf{y}\in\mathbf{B}$ and $\mathbf{y}\notin G_t$ so $\mathbf{y}\notin\mathbf{S}=\mathrm{bdy}(\mathbf{B})=\mathrm{bdy}(G_t)$ and therefore $\mathbf{y}\notin\mathrm{cl}(G_t)$. The reverse inclusion is trivial. Now, if $\mathbf{B}\neq G_t$ we would have $\mathbf{B}=(\mathbf{B}\setminus G_t)\cup G_t$, a disjoint union of open sets, and, as $\mathbf{B}$ is an open connected set, one of these must be empty. Thus, $\mathbf{B}=G_t$. Now consider the integral $I(t):=\int_{\overline{\mathbf{B}}}|J_{\mathbf{f}_t}(\mathbf{x})|\,d\mathbf{x}$, where J is the Jacobian. By the change of variables theorem, (Chapter 6, which does not depend on this Lemma), when $0\le t\le t_0$, this integral is the n-dimensional volume of $\overline{\mathbf{B}}$, a constant. But from $D\mathbf{f}_t(\mathbf{x})=\mathbf{x}+tD\mathbf{g}(\mathbf{x})$ it is clear that $I(t)$ is a polynomial in t and hence is constant on all of $[0,1]$. However $\mathbf{h}(\mathbf{x}):=<\mathbf{f}_1(\mathbf{x}),\mathbf{f}_1(\mathbf{x})>\ =1$ identically on $\overline{\mathbf{B}}$ and therefore $D\mathbf{h}(\mathbf{x})\mathbf{y}=<D\mathbf{f}_1(\mathbf{x})\mathbf{y}\,,\mathbf{f}_1(\mathbf{x})>\ =0$ (problem 3.18). This shows that $D\mathbf{f}_1(\mathbf{x})$ is singular for all $\mathbf{x}$ and so $J_{\mathbf{f}_1}(\mathbf{x})=0$. This would give $I(1)=0$, the required contradiction.

Proof of Theorem 4.35. First suppose that $\mathbf{f}$ is of class C^1 and suppose that $\mathbf{f}(\mathbf{x})\neq\mathbf{x}$ for all $\mathbf{x}\in\overline{\mathbf{B}}$. Let $t(\mathbf{x})$ be the positive solution of the quadratic equation $\|\mathbf{x}+t(\mathbf{x}-\mathbf{f}(\mathbf{x}))\|^2=1$. By the quadratic formula,

$$\|\mathbf{x}-\mathbf{f}(\mathbf{x})\|^2 t(\mathbf{x})=<\mathbf{x},\mathbf{f}(\mathbf{x})-\mathbf{x}>+\{<\mathbf{x},\mathbf{x}-\mathbf{f}(\mathbf{x})>^2+(1-\|\mathbf{x}\|^2)(\|\mathbf{x}-\mathbf{f}(\mathbf{x})\|^2\}^{1/2}. \quad (4.13)$$

For $\|\mathbf{x}\|\neq 1$ the discriminant, $\{\cdot\}$ above, is positive since $\|\mathbf{x}-\mathbf{f}(\mathbf{x})\|\neq 0$. Furthermore, if $\|\mathbf{x}\|=1$, $<\mathbf{x},\mathbf{x}-\mathbf{f}(\mathbf{x})>\ \neq 0$ for otherwise $<\mathbf{f}(\mathbf{x}),\mathbf{x}>\ =\ <\mathbf{x},\mathbf{x}>\ =1$ and equality would hold in Schwarz's inequality and hence we would have $\mathbf{f}(\mathbf{x})=\mathbf{x}$. Thus, the discriminant is never 0 in $\overline{\mathbf{B}}$. Since the square root function is continuosly differentiable on $(0,\infty)$ and $\|\mathbf{f}(\mathbf{x})-\mathbf{x}\|\neq 0$, (4.13) implies that t is a C^1 function. Now set $\mathbf{g}(\mathbf{x}):=\mathbf{x}+t(\mathbf{x})(\mathbf{x}-\mathbf{f}(\mathbf{x}))$. Then $\mathbf{g}$ is a C^1 map of $\mathbf{B}$ onto $\mathbf{S}$ and $\mathbf{g}(\mathbf{x})=\mathbf{x}$ for $\mathbf{x}\in\mathbf{S}$. This contradicts Lemma 4.36 so proves the theorem for C^1 maps. Next, if $\mathbf{f}$ is continuous on $\overline{\mathbf{B}}$, by an extension of the Weierstrass approximation theorem, $\mathbf{f}$ can be uniformly approximated by C^1 functions $\mathbf{f}^{(m)}:\overline{\mathbf{B}}\to\overline{\mathbf{B}}$ with $\mathbf{f}^{(m)}\to\mathbf{f}$ uniformly on $\overline{\mathbf{B}}$. Each $\mathbf{f}^{(m)}$ has a fixed point in $\overline{\mathbf{B}}$ say $\mathbf{f}^{(m)}(\mathbf{x}^{(m)})=\mathbf{x}^{(m)}$. As $\overline{\mathbf{B}}$ is compact $\{\mathbf{x}^{(m)}\}$ has a convergent subsequence whose limit $\mathbf{x}$ is a fixed point of $\mathbf{f}$ because of the uniform convergence (problem 2.20).

Problems.

1. Show that $F(x,y)=x^3y+y^3x-2=0$ defines y as a function f of x near (1,1). Find the derivative of f at $x=1$.

2. Show that the equation $x^2+2y^2+3z^2=1$ defines z implicitly as a function $f(x,y)$ near the point $(1/\sqrt{6},1/\sqrt{6},1/\sqrt{6})$. Find the equation of the tangent plane to the graph of f at this point.

3. Consider the equation $(x+1)^2y-xy^2=4$. Is y defined implicitly as a function of x in a neighbourhood of the following points: (*i*) (−1,2) ; (*ii*) (2,1) ; (*iii*) (1,2) ?

4. Show that the equations $xz-yt=1$, $xyzt=20$ determine z and t as functions of x,y in a nbd of the point $(x,y,z,t)=(1,2,5,2)$ and determine $\partial z/\partial x$ and $\partial z/\partial y$ at this point. Show that the given equations do not define x and z as functions of y,t in any nbd of (1,2,5,2).

5. Suppose that $F(x,y,z)=0$ implicitly defines each variable as a function of the other two. Prove that $\dfrac{\partial y}{\partial x}\dfrac{\partial z}{\partial y}\dfrac{\partial x}{\partial z}=-1$. The term $\partial y/\partial x$ means $\partial g/\partial x$ where $y=g(x,z)$, similarly for the others.

6. Let $f:\mathbb{R}\longrightarrow\mathbb{R}$ be defined by $f(x)=\begin{cases}x/2+x^2\sin(1/x) & \text{for } x\neq 0\\ 0 & \text{when } x=0\end{cases}$. Show that f is differentiable at all points, that $f'(0)\neq 0$ but f is not one-to-one on any nbd of 0. Why does this not contradict the inverse function theorem ?

7. Solve the equations $x^3-yu=0$, $xy^2+uv=0$ explicitly for u and v as functions of x,y and compare with the conclusions of the implicit function theorem.

8. Under what restrictions can the equations $x=r\cos\theta$, $y=r\sin\theta$ be solved for r,θ as functions of x,y. When is this inverse function differentiable? Compare with the conclusions of the inverse function theorem.

9. Solve the equations $x=r\sin\theta\cos\phi$, $y=r\sin\theta\sin\phi$, $z=r\cos\theta$ for r,θ,ϕ in terms of x,y,z. When is this inverse function differentiable? Compare with the conclusions of the inverse function theorem.

10. Suppose that $\mathbf{f}:\mathbb{R}^n\longrightarrow\mathbb{R}^n$ is differentiable and is invertible on a neighbourhood of a point $\mathbf{a}$ and that $J_{\mathbf{f}}(\mathbf{a})=0$. Show that $\mathbf{f}^{-1}$ is not differentiable at $\mathbf{f}(\mathbf{a})$.

11. Show that $\mathbf{f} : \mathbb{R}^2 \longrightarrow \mathbb{R}^2$ defined by $\mathbf{f}(x,y)=(x^3+2xy+y^2, x^2+y)$ is locally invertible near (1,1). Calculate $D\mathbf{f}^{-1}(4,2)$ and obtain the affine approximation to $\mathbf{f}^{-1}(u,v)$ near (4,2).

12. Let $\mathbf{f} : \mathbb{R}^3 \rightarrow \mathbb{R}^3$ be defined by $\mathbf{f}(x,y,z)=(x+y+z, xy+yz+zx, xyz)$. Let a,b,c be distinct real numbers. Show that $\mathbf{f}$ has an inverse $\mathbf{g}$ in a nbd of $\mathbf{a}=(a,b,c)$ and that $J_{\mathbf{g}}(\mathbf{f}(\mathbf{a}))=[(c-b)(a-c)(b-a)]^{-1}$.

13. Let $\mathbf{f} : \mathbb{R}^3 \rightarrow \mathbb{R}^3$ be defined by $\mathbf{f}(x,y,z)=(x-xy, xy-xyz, xyz)$. Prove that $\mathbf{f} \in C^1(\mathbb{R}^3)$. Let $U := \{\mathbf{x} \in \mathbb{R}^3 : xy \neq 0\}$. Show that $J_{\mathbf{f}}(\mathbf{x}) \neq 0$ for all $\mathbf{x} \in U$. Check that $\mathbf{f}$ is one-to-one on U and determine a formula for $\mathbf{f}^{-1} : \mathbf{f}(U) \rightarrow U$ and calculate the Jacobian of $\mathbf{f}^{-1}$ at a point $(a,b,c) \in \mathbf{f}(U)$.

14. Let $U := \{\mathbf{x} \in \mathbb{R}^2 : x_1 + x_2 \neq 1\}$ and let $\mathbf{f} : U \longrightarrow \mathbb{R}^2$ be given by the formulae $f_i(x_1, x_2) = x_i/(1+x_1+x_2)$, $i=1,2$. Show that $\mathbf{f}$ is one-to-one on U, determine a formula for $\mathbf{f}^{-1}$ and calculate $J_{\mathbf{f}}$.

15. Assume the hypotheses of the implicit function theorem are satisfied with the exception that one only assumes that $\mathbf{F}$ is of class C^1 in $\mathbf{y}$, that is $\partial F_i/\partial y_j$ are continuous. For simplicity of notation take $(\mathbf{a},\mathbf{b})=(\mathbf{0},\mathbf{0})$ and write $\mathbf{L}=D_{\mathbf{y}}\mathbf{F}(\mathbf{0},\mathbf{0})$. Prove that, for each fixed $\mathbf{x}$ in a nbd $B(\mathbf{0},s)$, the function $\mathbf{G}(\mathbf{x},\mathbf{y}) := \mathbf{y}-\mathbf{L}^{-1}\mathbf{F}(\mathbf{x},\mathbf{y})$, regarded as a function of $\mathbf{y}$, has a unique fixed point on a nbd $B(\mathbf{0},r)$, which we denote simply by $\mathbf{y}$. Write $\mathbf{f}(\mathbf{x})=\mathbf{y}$ in this case and prove that $\mathbf{F}(\mathbf{x},\mathbf{f}(\mathbf{x}))=\mathbf{0}$ and that $\mathbf{f}$ is continuous.

16. Let K be a compact subset of an open set G and suppose that $\mathbf{f}$ is locally Lipschitz on G. Prove that $\mathbf{f}$ satisfies a (uniform) Lipschitz condition on K.

17. Reduce the differential equation

$$\frac{d^2y}{dx^2} = \left\{1+\left(\frac{dy}{dx}\right)^2\right\}^{1/2}$$

to a system of first-order equations. Estimate an interval of existence for the solution with $y(0) = 0$, $y'(0) = 0$.

18. By using the inequalities

$$y^2 \le t^2+y^2 \le 1+y^2 \text{ for } 0 \le t \le 1,$$

show that the solution $y(t)$ of the IVP, $\dfrac{dy}{dt} = t^2+y^2$, $y(0) = 1$, satisfies $1/(1-t) \le y(t) \le \tan(t+\pi/4)$ as long as it exists and deduce that $y(t) \longrightarrow \infty$ somewhere between $t = \pi/4$ and $t = 1$.

Hints on the problems.

3. (*iii*) The theorem fails but the equation can be thought of as a quadratic equation for y. Try to solve it for x near 1.

4. If x,z were functions of y,t on a nbd then $xz=1+yt=20/yt$ on that nbd. However this would imply that $yt=4$.

11. If $\mathbf{f}^{-1}$ were differentiable then one could differentiate the equation $\mathbf{x}=\mathbf{f}^{-1}(\mathbf{f}(\mathbf{x}))$ by the chain rule.

15. By Theorem 3.22 there is $r>0$ so that $\|\mathbf{G}(\mathbf{x},\mathbf{y})-\mathbf{G}(\mathbf{x},\mathbf{z})\| < 1/2\|\mathbf{y}-\mathbf{z}\|$ for $\|\mathbf{x}\| \le r$, $\mathbf{y},\mathbf{z} \in \overline{\mathbf{B}}(\mathbf{0},r)$. Also by continuity of $\mathbf{G}$ at $(\mathbf{0},\mathbf{0})$, there is $s>0$, $s \le r$, such that $\|\mathbf{G}(\mathbf{x},\mathbf{0})\| < r/2$ for $\|\mathbf{x}\| \le s$. Apply Theorem 4.7.

18. $y' \ge y^2$ implies $(1/y^2)y' \ge 1$ or $d/dt(-1/y - t) \ge 0$ so $-1/y - t$ is an increasing function of t.

Chapter 5

Maxima and Minima

5.1. EXTREMA AND CRITICAL POINTS

In this chapter we deal exclusively with scalar-valued functions, which may be termed scalar fields. Let A be a nonempty set of $\mathbb{R}^n$ and let $f: A \longrightarrow \mathbb{R}$. Since the elements of the image set $f(A)$ lie in $\mathbb{R}$, the usual ordering on $\mathbb{R}$ may be applied and it makes sense to talk of largest and smallest values. Of course, there need be no largest value (maximum) or smallest value (minimum). To begin with, $f(A)$ has to be bounded above (or below) at the very least. Then $\sup f(A)$ (or $\inf f(A)$) certainly exists but there might be no point of A which actually achieves this value; some extra properties of A and of f are needed to ensure the maximum (or minimum) is attained.

Definition 5.1. Let $f: A \subseteq \mathbb{R}^n \longrightarrow \mathbb{R}$ be some function. We say that f *attains its maximum over* A *at* $\mathbf{a} \in A$ if $f(\mathbf{x}) \leq f(\mathbf{a})$ for all $\mathbf{x} \in A$. $f(\mathbf{a})$ is called the *maximum value* of f. If $f(\mathbf{x}) < f(\mathbf{a})$ for all $\mathbf{x} \in A$ other than $\mathbf{a}$, we say f has a *strict maximum* at $\mathbf{a}$. Minima and strict minima are defined similarly. Maxima and minima are often called *extrema*.

Thus, f has a maximum at $\mathbf{a}$ precisely when $f(\mathbf{a})$ is the largest possible value of $f(\mathbf{x})$ for $\mathbf{x} \in A$. This value may be attained at other points too.

We recall from Theorem 2.23 that if A is *compact* and f is *continuous* on A then f attains its maximum and minimum values on A. However, that theorem gives no clue as to how these extreme values may be found. It is here that differentiation plays an important part.

Theorem 5.2. *Let* $f: G \subseteq \mathbb{R}^n \longrightarrow \mathbb{R}$ *be differentiable on the open set* G *and let* A *be a subset of* G. *If* f *attains its maximum over* A *at an interior point* $\mathbf{a} \in A$ *then* $Df(\mathbf{a}) = \nabla f(\mathbf{a}) = \mathbf{0}$.

Proof. Since $\mathbf{a} \in \text{int}(A)$, there is a δ-neighbourhood of $\mathbf{a}$ lying entirely within A. Then, for any direction $\mathbf{e}$ ($\|\mathbf{e}\| = 1$), $\mathbf{a} + t\mathbf{e} \in A$ for $|t| < \delta$. Since $f(\mathbf{a})$ is the maximum, $f(\mathbf{a}) \geq f(\mathbf{a} + t\mathbf{e})$ for all t, $|t| < \delta$. Therefore,

$D_{\mathbf{e}}f(\mathbf{a}) = \lim_{t\to 0} \dfrac{f(\mathbf{a}+t\mathbf{e})-f(\mathbf{a})}{t}$ must be zero since it exists (because f is differentiable at $\mathbf{a}$) and, for $t<0$ the ratios are positive, for $t>0$, the ratios are negative; hence $D_{\mathbf{e}}f(\mathbf{a})$ must be both ≥ 0 and ≤ 0. [This is nothing but the one-dimensional argument that if g has a maximum on an interval $(a-\delta,\ a+\delta)$ then $g'(a)=0$.] As this is true for all directions, if $\nabla f(\mathbf{a}) \neq \mathbf{0}$ we may take $\mathbf{e} = \nabla f(\mathbf{a}) \,/\, \|\nabla f(\mathbf{a})\|$ (recall $\|\mathbf{e}\|=1$) and obtain

$$D_{\mathbf{e}}f(\mathbf{a}) = \langle \nabla f(\mathbf{a}), \mathbf{e} \rangle = \langle \nabla f(\mathbf{a}), \nabla f(\mathbf{a}) \,/\, \|\nabla f(\mathbf{a})\| \rangle = \|\nabla f(\mathbf{a})\| \neq 0,$$

(Proposition 3.18), a contradiction. [Equally well : all partial derivatives of f are 0 by the above.]

There is obviously a dual result for minima. This is easy to deal with since f has a minimum at $\mathbf{b}$ $\Leftrightarrow$ $-f$ has a maximum at $\mathbf{b}$.

Definition 5.3. A point $\mathbf{a}$ is called a *critical point* of f if $\nabla f(\mathbf{a}) = \mathbf{0}$.

Corollary 5.4. *Under the assumptions made in Theorem* 5.1, *if f has a maximum (or minimum) over A at $\mathbf{a}\in A$ then either $\mathbf{a}$ is a critical point of f or $\mathbf{a}\in$* bdy(A).

Proof. Either $\mathbf{a}\in\text{int}(A)$ and then $\mathbf{a}$ is a critical point, or $\mathbf{a}\in\text{bdy}(A)$, since $A\subseteq\text{cl}(A)=\text{int}(A)\cup\text{bdy}(A)$ (problem 1.15).

Finding critical points of f is essentially a routine matter. We calculate all the partial derivatives of f and set them equal to 0 (simultaneously). The problem is then purely one of algebraic manipulation. However, critical points are not necessarily extreme points. In one dimension they are the points where the slope of the graph is zero, and an example such as $f(x)=x^3$ at $x=0$, shows that this can occur at points other than maxima or minima. For a function of two variables, its graph $z=f(x,y)$ is a surface in $\mathbb{R}^3$ which may be thought of as a terrain. Maxima are mountain peaks, minima are valley bottoms but one can also have mountain passes where the gradient is $\mathbf{0}$ but one direction is a minimum, the other a maximum. Such a point is usually called a *saddle point* because the saddle of a horse is perhaps the most common example of the shape. For more variables, other types of behaviour are possible, but we shall use the term *saddle point* to mean a critical point which is neither a maximum or a minimum.

Continuing with the above analogy we see that there can be several mountain peaks, the maximum being the highest. The other peaks are locally highest. Points that yield maxima or minima are called *extreme points* and their values are termed *extrema*.

Definition 5.5. A point $\mathbf{a}$ is called a *local maximum of f* if there is $\delta>0$ such that $f(\mathbf{a})\geq f(\mathbf{x})$ for all $\mathbf{x}\in B(\mathbf{a},\delta)$. Strict local maxima, local and strict local minima have the obvious definitions.

Corollary 5.6. *Local extrema always occur at critical points.*

Proof. They are, by definition, interior points.

The converse of Corollary 5.6 is false as indicated in the preamble. Moreover, there need be no local extrema on a closed bounded set A even though global extrema exist; consider $f(x,y)=x+y$ on $A=\{(x,y):x^2+y^2\leq 1\}$. However, finding all the critical points gives a complete list of all possible candidates for *local* extrema.

Example 5.7. Let $f(x,y)=x^2-y^2$. Then $(0,0)$ is the only critical point but it is neither a maximum nor a minimum.

Solution. $\partial f/\partial x=2x$, $\partial f/\partial y=2y$, so critical points are those points (x,y) where $2x=0$ and $2y=0$, that is $(0,0)$ only. Now $f(0,0)=0$ but $f(h,0)=h^2>0$ while $f(0,k)=-k^2<0$ so $(0,0)$ is a saddle point. Indeed, the surface $z=x^2-y^2$ resembles a saddle very well.

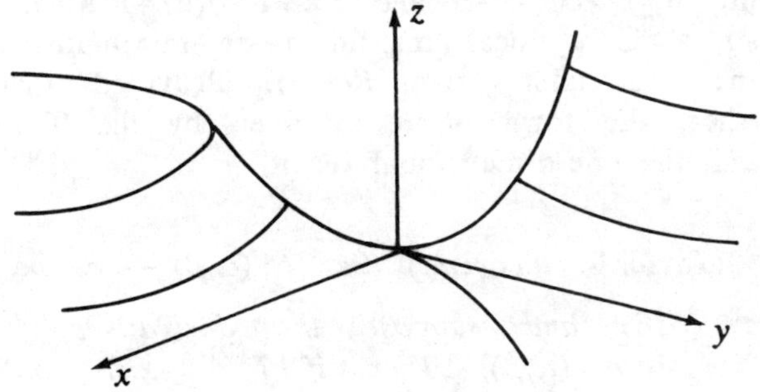

Figure 5.1

Example 5.8. Let $f(x,y,z)=x^2+2y^2+3z^2-xyz$. Show that f has a local minimum at $(0,0,0)$.

Solution. $\partial f/\partial x=2x-yz$, $\partial f/\partial y=4y-xz$, $\partial f/\partial z=6z-xy$ and these are simultaneously 0 at $(0,0,0)$, so $(0,0,0)$ is a critical point. Consider a neighbourhood V of $(0,0,0)$ and for $(h,k,l)\in V$ look at $f(h,k,l)-f(0,0,0)=h^2+2k^2+3l^2-hkl$. It is intuitively obvious that, for h,k, and l small enough, the term hkl, being of third order, is negligible compared with the terms of order two. We shall show this in the example at hand but in future will omit such arguments. Now, for $h^2+k^2+l^2<1$, by the simple inequality $ab\le \frac{a^2+b^2}{2}$ we obtain

$$|hkl| \le \frac{h^2k^2+l^2}{2} < \frac{h^2+l^2}{2} \le \frac{h^2+k^2+l^2}{2}$$

and hence $f(h,k,l)-f(0,0,0) > \dfrac{h^2+3k^2+5l^2}{2} > 0.$

This shows that 0 is the minimum of f on the 1–nbd of (0,0,0).

There are a number of other critical points in the above example but determining whether or not they are maxima or minima by the above method would take some effort. It is convenient to have a test which is easily applied. The test we shall give is analogous to the second derivative test in one variable calculus, but to give it for more variables requires the use of the Taylor series of scalar fields, to which we now turn.

5.2. TAYLOR EXPANSIONS AND THE SECOND DERIVATIVE TEST

Let $f:\mathbb{R}^n \longrightarrow \mathbb{R}$ be differentiable at a point $\mathbf{a}$. Then $f(\mathbf{a}+\mathbf{h})=f(\mathbf{a})+Df(\mathbf{a})\mathbf{h}+R(\mathbf{a},\mathbf{h})$, where $|R(\mathbf{a},\mathbf{h})|/\|\mathbf{h}\| \longrightarrow 0$ as $\|\mathbf{h}\| \longrightarrow 0$. When $\mathbf{a}$ is a critical point of f, $Df(\mathbf{a})=\mathbf{0}$ and $f(\mathbf{a}+\mathbf{h})-f(\mathbf{a})=R(\mathbf{a},\mathbf{h})$. Therefore to determine whether $\mathbf{a}$ is a local maximum or minimum of f requires knowledge of the remainder term $R(\mathbf{a},\mathbf{h})$. Provided that f can be differentiated further, the form of R is given by the Taylor expansion of f. We first recall the one-dimensional result.

Theorem 5.9. (*Taylor's theorem.*) *Let $f:(\alpha,\beta)\longrightarrow\mathbb{R}$ be $(p-1)$ times differentiable on (α,β) and suppose the derivatives $f, f', \cdots, f^{(p-1)}$ are continuous on $[a,b]\subset(\alpha,\beta)$. Then if $f^{(p)}$ exists on (a,b) there exists $\xi\in(a,b)$ such that*

$$f(b)=f(a)+f'(a)(b-a)+\frac{f''(a)(b-a)^2}{2!}+\cdots+\frac{f^{(p-1)}(a)(b-a)^{(p-1)}}{(p-1)!}+E_p$$

where $E_p = \dfrac{f^{(p)}(\xi)(b-a)^p}{p!}$.

The proof may be found in many books on elementary analysis, for example references [12] and [13].

The remainder term written above is usually called the *Lagrange remainder*. There are other forms of the remainder which have advantages in some situations but, the Lagrange form is probably the simplest. Indeed it resembles the first neglected term save that $f^{(p)}$ is to be evaluated at a point ξ, not at a. The point ξ may be written in the form $(1-\theta)a+\theta b$ where $0<\theta<1$. This has the advantage of covering the case when $b<a$ too.

We use this result to prove the n-dimensional version. Recall that $f\in C^p(G)$ means that f has continuous partial derivatives of order p on G.

Theorem 5.10. (*Taylor expansion with remainder.*)
Let $G\subseteq\mathbb{R}^n$ be an open set and let $f\in C^p(G)$. Suppose $\mathbf{a}\in G$ and that $\mathbf{a}+t\mathbf{h}\in G$ for all $t\in[0,1]$. Then, there exists θ, $0<\theta<1$, such that

$$f(\mathbf{a}+\mathbf{h})=f(\mathbf{a})+\sum_{i=1}^{n}\frac{\partial f}{\partial x_i}(\mathbf{a})\,h_i+\frac{1}{2!}\sum_{i,j=1}^{n}\frac{\partial^2 f}{\partial x_i\partial x_j}(\mathbf{a})\,h_ih_j+\cdots$$

$$+\frac{1}{(p-1)!}\sum_{i_1,i_2,\cdots,i_{p-1}}\frac{\partial^{(p-1)}f}{\partial x_{i_1}\partial x_{i_2}\cdots\partial x_{i_{p-1}}}(\mathbf{a})\,h_{i_1}h_{i_2}\cdots h_{i_{p-1}}+E_p(\mathbf{h})$$

where $$E_p=\frac{1}{p!}\sum_{i_1,i_2,\cdots,i_p}\frac{\partial^p f}{\partial x_{i_1}\partial x_{i_2}\cdots\partial x_{i_p}}(\mathbf{a}+\theta\mathbf{h})\,h_{i_1}h_{i_2}\cdots h_{i_p}\,.$$

Note that a sum such as

$$\sum_{i_1,i_2,\cdots,i_q}\frac{\partial^q f}{\partial x_{i_1}\partial x_{i_2}\cdots\partial x_{i_q}}(\mathbf{a}+\theta\mathbf{h})\,h_{i_1}h_{i_2}\cdots h_{i_q}$$

is a multiple sum over all integers $i_1,i_2,\cdots,i_q$ between 1 and n. Because $f\in C^{(p)}(G)$, the order of differentiation is irrelevant; see Theorem 3.27.

Proof. Let ϕ be defined by $\phi(t)=f(\mathbf{a}+t\mathbf{h})$. Then ϕ is defined for all t such that $\mathbf{a}+t\mathbf{h}\in G$. This is open, since G is open (problem 1.18) and contains $[0,1]$ by hypothesis. We shall apply Theorem 5.9 to ϕ on $[0,1]$, so we have to calculate derivatives of ϕ. Since $f\in C^1$, ϕ is differentiable, and, by the chain rule,

$$\phi'(t)=Df(\mathbf{a}+t\mathbf{h})\mathbf{h}=\langle\nabla f(\mathbf{a}+t\mathbf{h}),\mathbf{h}\rangle=\sum_{i=1}^{n}\frac{\partial f}{\partial x_i}(\mathbf{a}+t\mathbf{h})\,h_i\,.$$

Since $f\in C^2$, $\partial f/\partial x_i$ are C^1 functions so are differentiable and we can continue the computation process. Indeed

$$\phi''(t)=\frac{d}{dt}\left[\sum_{i=1}^{n}\frac{\partial f}{\partial x}(\mathbf{a}+t\mathbf{h})h_i\right]=\sum_{i=1}^{n}\left[\sum_{j=1}^{n}\frac{\partial^2 f}{\partial x_i\partial x_j}(\mathbf{a}+t\mathbf{h})h_j\right]h_i\,.$$

This process continues but we write $i_1,i_2,\cdots$ in place of $i,j,\cdots$ and have $\phi^{(q)}(t)=\sum_{i_1}\sum_{i_2}\cdots\sum_{i_q}\frac{\partial^q f}{\partial x_{i_1}\partial x_{i_2}\cdots\partial x_{i_q}}(\mathbf{a}+\theta\mathbf{h})\,h_{i_1}h_{i_2}\cdots h_{i_q}\quad,\ 1\le q\le p.$

By Theorem 5.9, $\phi(1)=\phi(0)+\phi'(0)+\frac{\phi''(0)}{2!}+\cdots+\frac{1}{(p-1)!}\phi^{(p-1)}(0)+E_p(\theta)$

where $0<\theta<1$, and this gives the conclusion of the theorem.

Example 5.11. Find the Taylor expansion of $f(x,y,z) = xyz$ at the point (1,1,1).

Solution. We have

$$f(1+h_1,1+h_2,1+h_3)=f(1,1,1)+h_1\frac{\partial f}{\partial x}(1,1,1)+h_2\frac{\partial f}{\partial y}(1,1,1)+h_3\frac{\partial f}{\partial z}(1,1,1)$$

$$+\frac{1}{2!}(h_1,h_2,h_3)\begin{bmatrix} D_1D_1f & D_1D_2f & D_1D_3f \\ D_2D_1f & D_2D_2f & D_2D_3f \\ D_3D_1f & D_3D_2f & D_3D_3f \end{bmatrix}\begin{bmatrix} h_1\\ h_2\\ h_3\end{bmatrix}+\cdots$$

Thus, $f(1+h_1,1+h_2,1+h_3)=1+h_1+h_2+h_3+\frac{1}{2}(h_1,h_2,h_3)\begin{bmatrix} 0 & z & y\\ z & 0 & x\\ y & x & 0\end{bmatrix}\begin{bmatrix} h_1\\ h_2\\ h_3\end{bmatrix}+\cdots$

where $(x,y,z)=(1,1,1)$. In particular this says that

$$(1+h_1)(1+h_2)(1+h_3)=1+(h_1+h_2+h_3)+h_1h_2+h_2h_3+h_3h_1+E_3.$$

Of course, in this case one can multiply out the left hand side and obtain $E_3=h_1h_2h_3$. Here Taylor's theorem gives the remainder to be

$$E_3=\frac{1}{6}\sum_{i,j,k=1}^{3}\frac{\partial^3 f}{\partial x_i\partial x_j\partial x_k}(1+\theta h_1,1+\theta h_2,1+\theta h_3)\,h_1h_2h_3\,,$$

which looks very complicated. However, $\frac{\partial^3 f}{\partial x_i\partial x_j\partial x_k}=0$ except when i, j, k are all different and one has, for example, $\frac{\partial^3 f}{\partial x\partial y\partial z}=1$. Since $\partial x\partial y\partial z$ can be permuted this term occur six times in the above sum and gives $E_3=h_1h_2h_3$ again. In this example $E_4=0$ which is to be expected since f has only terms of order three.

For the purposes of investigating the nature of extreme points we shall be mainly interested in considering only terms up to order two. Then, at a critical point $\mathbf{a}$, since $Df(\mathbf{a})=\mathbf{0}$, the Taylor expansion is

$$f(\mathbf{a}+\mathbf{h}) = f(\mathbf{a})+\frac{1}{2}\sum_{i,j=1}^{n}\frac{\partial^2 f}{\partial x_i \partial x_j}(\mathbf{a}+\theta\mathbf{h})\, h_i h_j \tag{5.1}$$

if f is of class C^2. If f is of class C^3 we may write

$$f(\mathbf{a}+\mathbf{h}) = f(\mathbf{a})+ \frac{1}{2}\sum_{i,j=1}^{n}\frac{\partial^2 f}{\partial x_i \partial x_j}(\mathbf{a})\, h_i h_j + E_3 \tag{5.2}$$

where E_3 may be thought of as higher order terms. The matrix of second order partial derivatives $(\partial^2 f / \partial x_i \partial x_j)$ is called the *Hessian matrix* of f, and is denoted by H, or $H_f(\mathbf{a})$. Twice the sum of the second order terms $\sum_{i,j=1}^{n}\frac{\partial^2 f}{\partial x_i \partial x_j} h_i h_j$ is a *quadratic form* in the variables $h_1,\cdots,h_n$ and may be written as a matrix product $\mathbf{h}^{\mathrm{T}}H\mathbf{h}$, where $\mathbf{h}$ denotes the column vector with components $h_1, h_2,\cdots, h_n$, $\mathbf{h}^{\mathrm{T}}$ (the transpose of $\mathbf{h}$) is the corresponding row vector. Note that H is a *symmetric* matrix, $H^{\mathrm{T}}=H$, when f is of class C^2. The theory of quadratic forms is usually covered in a course of linear algebra. We recall some definitions.

Definition 5.12. A quadratic form $q(\mathbf{h})=\sum_{i,j=1}^{n} a_{ij}h_i h_j$, with $a_{ij}=a_{ji}$ is said to be *positive definite* (respectively *negative definite*) if $q(\mathbf{h})>0$ for all $\mathbf{h}\neq 0$ (resp. $q(\mathbf{h})<0$ for $\mathbf{h}\neq 0$), *positive semidefinite* (resp. *negative semidefinite*) if $q(\mathbf{h})\geq 0$ for all $\mathbf{h}$ (respectively $q(\mathbf{h})\leq 0$), and *indefinite* if there exist $\mathbf{h}_1,\mathbf{h}_2$ with $q(\mathbf{h}_1)>0$ and $q(\mathbf{h}_2)<0$.

Criteria for these are given in courses of linear algebra and include properties of determinants of submatrices of $A=(a_{ij})$. Also it is possible to give a criterion in terms of the eigenvalues $\lambda_1,\cdots,\lambda_n$ of A. This is:

A is positive definite $\Leftrightarrow$ all the eigenvalues of A are positive.

(The eigenvalues are real because A is symmetric.)

However, in practice, the easiest way to determine whether or not a quadratic form is positive definite is by the process of completing the square. For example, in two variables x,y a quadratic form is $ax^2+bxy+cy^2$ and we can write this as $a(x+by/2a)^2+(c-b^2/4a)y^2$ so this is positive definite if, and only if, $a>0$ and $c-b^2/4a>0$.

We now give the second derivative test for local extrema.

Theorem 5.13. *Let $G \subseteq \mathbb{R}^n$ be open and let $f \in C^2(G)$. Suppose that $\mathbf{a} \in G$ is a critical point of f. Write $Q(\mathbf{x},\mathbf{h})$ for the quadratic form*
$\sum_{i,j=1}^{n} \frac{\partial^2 f}{\partial x_i \partial x_j}(\mathbf{x})\, h_i h_j$. *Then*
(*a*) *$Q(\mathbf{a},\mathbf{h})$ positive definite implies that $\mathbf{a}$ is a strict local minimum,*
(*b*) *$Q(\mathbf{a},\mathbf{h})$ negative definite implies that $\mathbf{a}$ is a strict local maximum,*
(*c*) *$Q(\mathbf{a},\mathbf{h})$ indefinite implies that $\mathbf{a}$ is a saddle point.*
Conversely,
(*a'*) *if $\mathbf{a}$ is a local minimum then $Q(\mathbf{a},\mathbf{h})$ is positive semidefinite,*
(*b'*) *if $\mathbf{a}$ is a local maximum then $Q(\mathbf{a},\mathbf{h})$ is negative semidefinite.*

The proof requires a preliminary lemma.

Lemma 5.14. *Let $q(\mathbf{h}) = \sum_{i,j=1}^{n} a_{ij} h_i h_j$ be a positive definite quadratic form, that is, $q(\mathbf{h}) > 0$ for all $\mathbf{h} \neq \mathbf{0}$. Then there exists $m > 0$ such that $q(\mathbf{h}) \geq m\|\mathbf{h}\|^2$ for all $\mathbf{h}$.*

Proof. As q is a finite sum of products of components of $\mathbf{h}$, it is continuous on $\mathbb{R}^n$. Let m be the minimum of q on the set $S = \{\mathbf{h} \in \mathbb{R}^n : \|\mathbf{h}\| = 1\}$; $m > 0$ since the minimum is attained (problem 2.7). For $\mathbf{h} \neq \mathbf{0}$, $\mathbf{h}/\|\mathbf{h}\|$ lies in S so $q(\mathbf{h}/\|\mathbf{h}\|) \geq m$. Since $q(c\mathbf{h}) = c^2 q(\mathbf{h})$ for any scalar c, this gives the result.

Proof of Theorem 5.13. (*a*) From (5.1) it is evident that $\mathbf{a}$ is a strict local minimum of f if $Q(\mathbf{a}+\theta\mathbf{h},\mathbf{h})$ is positive definite. Since θ, which depends on $\mathbf{h}$, is unknown, to verify this we have to show that $Q(\mathbf{x},\mathbf{h})$ is positive definite for all $\mathbf{x}$ in some neighbourhood of $\mathbf{a}$. Now $|Q(\mathbf{x},\mathbf{h}) - Q(\mathbf{a},\mathbf{h})| \leq \sum_{i,j=1}^{n} |D_i D_j f(\mathbf{x}) - D_i D_j f(\mathbf{a})| \cdot |h_i h_j|$. As $D_i D_j f$ is continuous for each i,j, for any $\varepsilon > 0$, there is a δ-neighbourhood of $\mathbf{a}$ so that $|D_i D_j f(\mathbf{x}) - D_i D_j f(\mathbf{a})| < \varepsilon$ for $\mathbf{x} \in B(\mathbf{a},\delta)$, $1 \leq i,j \leq n$. Then

$$|Q(\mathbf{x},\mathbf{h}) - Q(\mathbf{a},\mathbf{h})| \leq \varepsilon \sum_{i,j=1}^{n} |h_i h_j| = \varepsilon \left[\sum_{i=1}^{n} |h_i| \right]^2 \leq \varepsilon n \|\mathbf{h}\|^2,$$

by Schwarz's inequality (or Cauchy's, Lemma 1.4). By Lemma 5.14 there is $m > 0$ such that $Q(\mathbf{a},\mathbf{h}) \geq m\|\mathbf{h}\|^2$ and taking $\varepsilon < m/n$ then yields $Q(\mathbf{x},\mathbf{h}) \geq Q(\mathbf{a},\mathbf{h}) - |Q(\mathbf{x},\mathbf{h}) - Q(\mathbf{a},\mathbf{h})| > 0$ for every $\mathbf{x}$ in the δ-nbd of $\mathbf{a}$.
To prove (*b*) apply (*a*) to $-f$. Since (*c*) is a consequence of (*a'*) and (*b'*) and (*b'*) is proved by applying (*a'*) to $-f$, it remains to prove (*a'*). There is a δ-neighbourhood of $\mathbf{a}$ in G such that $f(\mathbf{x}) \geq f(\mathbf{a})$ for all $\mathbf{x}$ in the δ-neighbourhood. From (5.1) it follows that $Q(\mathbf{a}+\theta\mathbf{h},\mathbf{h}) \geq 0$ for all $\mathbf{h}$ in $B(\mathbf{0},\delta)$; note that θ depends on $\mathbf{h}$, it is not a universal constant. Suppose, for a contradiction argument, that $Q(\mathbf{a},\mathbf{k}) < 0$ for

some $\mathbf{k}\in\mathbb{R}^n$. Since $f\in C^2$, $Q(\mathbf{x},\mathbf{k})$ depends continuously on $\mathbf{x}$, so there is a δ_1-neighbourhood V of $\mathbf{a}$ with $Q(\mathbf{x},\mathbf{k})<0$ for all $\mathbf{x}\in V$. Obviously we can suppose $\delta_1<\delta$. Now take $\mathbf{h}=\delta_1\mathbf{k}/\|\mathbf{k}\|$. Then there exists θ, $0<\theta<1$, such that $Q(\mathbf{a}+\theta\mathbf{h},\mathbf{h})\geq 0$. Since $Q(\mathbf{x},c\mathbf{h})=c^2Q(\mathbf{x},\mathbf{h})$ for any constant c, this gives $Q(\mathbf{a}+\theta\mathbf{h},\mathbf{k})\geq 0$. However $\mathbf{a}+\theta\mathbf{h}\in V$ so this contradicts $Q(\mathbf{x},\mathbf{k})<0$ for all $\mathbf{x}\in V$, and proves that no such $\mathbf{k}$ exists.

Parts (a)-(c) of Theorem 5.13 give *sufficient conditions* for a critical point to be of the stated type whereas parts (a'), (b') give *necessary conditions*. These conditions are not identical. Indeed (a)-(c) constitute the test to apply to critical points but the test will be inconclusive when these conditions are not met. In particular if the Hessian determinant, $\det H$, is zero at $\mathbf{a}$, $\mathbf{a}$ is said to be *degenerate*; the test works for nondegenerate critical points. A common situation where $\det H=0$ is when all second order terms are 0 at $\mathbf{a}$: in this case it is rather obvious that everything depends on the higher order terms.

Example 5.15. Find the critical points of $f(x,y)=x^4+y^4+6x^2y^2+8x^3$ and determine their nature.

Solution. $\dfrac{\partial f}{\partial x} = 4x^3+12xy^2+24x^2$, and $\dfrac{\partial f}{\partial y} = 4y^3+12x^2y = 4y(y^2+3x^2)$.

Critical points occur when both of these expressions equal 0. Since y^2+3x^2 is zero only at (0,0), the second equation gives $y=0$. Then we must have $4x^2(x+6)=0$ so that $x=0$ or $x=-6$. Thus, there are two critical points (0,0) and (−6,0). We now compute the Hessian matrix. By routine differentiations,

$$H=\begin{bmatrix}\partial^2 f/\partial x^2 & \partial^2 f/\partial x\partial y\\ \partial^2 f/\partial y\partial x & \partial^2 f/\partial y^2\end{bmatrix}=\begin{bmatrix}12x^2+12y^2+48x & 24xy\\ 24xy & 12y^2+12x^2\end{bmatrix}.$$

At (−6,0), $H=\begin{bmatrix}144 & 0\\ 0 & 432\end{bmatrix}$ and the quadratic form is $144h^2+432k^2$ which is positive definite, therefore (−6,0) is a strict local minimum.

At (0,0), $H=\begin{bmatrix}0 & 0\\ 0 & 0\end{bmatrix}$ so (0,0) is degenerate and the test fails. In this case we look at $f(h,k)-f(0,0)$ for h^2+k^2 sufficiently small. Here $f(h,k)-f(0,0)=h^4+k^4+6h^2k^2+8h^3$. The dominant term, for h^2+k^2 small, is the lowest order term, namely $8h^3$. This can take both signs, therefore (0,0) is a saddle point.

A convenient notation in problems involving only 2 or 3 variables is to write f_x, f_y for $\partial f/\partial x$, $\partial f/\partial y$, etc. For two variables the

Hessian matrix is $H=\begin{bmatrix} f_{xx} & f_{xy} \\ f_{yx} & f_{yy} \end{bmatrix}$. Then we have :

H is positive definite $\Leftrightarrow f_{xx}>0$ and $\det H=f_{xx}f_{yy}-(f_{xy})^2>0$
($\Rightarrow$ strict local minimum).

H is negative definite $\Leftrightarrow f_{xx}<0$ and $\det H=f_{xx}f_{yy}-(f_{xy})^2>0$
($\Rightarrow$ strict local maximum).

H is indefinite $\Leftrightarrow \det H<0$
($\Rightarrow$ saddle point).

These conditions can be checked without reference to the quadratic forms and constitute the simple form of the test encountered in a first course on Advanced Calculus. The proofs of these facts are simple consequences of completing the square on the quadratic form.

Example 5.16. Find the critical points of $f(x,y)=x^2y+y^2x-3xy$ and determine their nature.

Solution. $f_x=2xy+y^2-3y$, $f_y=x^2+2yx-3x$. Critical points occur at points (x,y) where $y(y-3+2x)=0$ and $x(x-3+2y)=0$. Solving these equations, either algebraically or by interpreting them as equations of straight lines in the (x,y) plane, gives solutions (0,0), (1,1), (3,0) and (0,3). We compute the second derivatives and obtain $f_{xx}=2y$, $f_{xy}=2x+2y-3$, $f_{yy}=2x$. At (0,0), $f_{xx}=0$ and $\det H=\det\begin{bmatrix} 0 & -3 \\ -3 & 0 \end{bmatrix}=-9$ and therefore (0,0) is a saddle point. In fact, the quadratic form here is $-6hk$ and this can take both signs. At the point (1,1), $f_{xx}=2$ and $f_{xx}f_{yy}-f_{xy}^2=3$ so (1,1) is a local minimum. Here the quadratic form is $2h^2+2hk+2k^2=2(h+k/2)^2+3k^2/2$ and is positive definite. At (3,0), $f_{xx}=0$ and $f_{xx}f_{yy}-f_{xy}^2=-9$, therefore (3,0) is a saddle point. Also the quadratic form here is $6k^2+6hk = 6\,[(k+h/2)^2-h^2/4]$ which can be positive (for example when $h=0$, $k\neq 0$) or negative (when $h=-2k$), so is indeed indefinite.

A simple, yet useful, example of the techniques is to the problem of fitting the "best" straight line to a set of data. Theory may predict that experimental data should lie on a straight line $y=mx+c$, where m and c are to be determined. The actual observations may, and often do, contain measurement errors. Therefore given data $(x_1,y_1),\cdots,(x_p,y_p)$, there is usually no straight line containing all of the data points. Indeed if a line has equation $y=mx+c$, each data point deviates from the line at $x=x_i$ by an amount $d_i=y_i-(mx_i+c)$. A good fit would be when all deviations d_i are small. Since some may be positive and others negative, we should minimize all of the absolute values $|d_i|$. Since $f(x)=|x|$ is not differentiable at $x=0$ it is better to minimize the sum of the squares

$$f(m,c)=\sum_{i=1}^{p} d_i^2=\sum_{i=1}^{p}(y_i-mx_i-c)^2. \tag{5.3}$$

This is known as the *method of least squares* . The critical points of f are when $\partial f/\partial m=0$ and $\partial f/\partial c=0$, that is, when

$$\sum_{i=1}^{p} 2(y_i-mx_i-c)(-x_i)=0 \quad \text{and} \quad \sum_{i=1}^{p} 2(y_i-mx_i-c)(-1)=0. \tag{5.4}$$

The second equation in (5.4) says that, at a critical point, $\sum_{i=1}^{p} d_i=0$, that is, the positive and negative deviations cancel, as is to be expected. Expanding out the equations in (5.4) gives

$$m\sum x_i^2+c\sum x_i-\sum x_iy_i=0 \quad \text{and} \quad m\sum x_i+pc-\sum y_i=0, \tag{5.5}$$

where all sums run from $i=1$ to $i=p$. These equations have a unique solution provided that $\begin{bmatrix}\sum x_i^2 & \sum x_i\\ \sum x_i & p\end{bmatrix}$ is nonsingular. Since

$\sum x_i\le\left[\sum x_i^2\right]^{1/2}\left[\sum 1^2\right]^{1/2}=p^{1/2}\left[\sum x_i^2\right]^{1/2}$ by Schwarz's inequality, with equality only if all the x_i are equal, the determinant of the above matrix is nonzero when x_i are distinct (as they ought to be with sensible data). Moreover, this unique critical point is a local minimum. To see this, apply the second derivative test. Indeed, since $f_{mm}=2\sum x_i^2$, $f_{mc}=2\sum x_i$, and $f_{cc}=2p$ we see that $f_{mm}>0$ and $f_{mm}f_{cc}-(f_{mc})^2>0$ except when all the x_i are equal, exactly as above.

The calculations here are ideally suited to a programmable calculator or a home computer since the solution can be written down explicitly once the data is given.

Example 5.17. Find the equation of the straight line that best fits the data (-3,5), (-2,4), (0,1.5), (1.5,0), (4,-3), (5,-4.5) by the least squares method.

Solution. It is routine to check that $\sum x_i^2=56.25$, $\sum x_i=5.5$, $\sum x_i y_i=-57.5$ and $\sum y_i=3$. Therefore the equations are

$$56.25\,m+5.5\,c+57.5 = 0 \quad \text{and} \quad 5.5\,m+6\,c-3 = 0,$$

with (approximate) solution $m=-1.177$, $c=1.579$, so the best straight line has equation $y = -1\cdot177\,x+1\cdot579$. Note that none of the points actually lie on the line but, nevertheless, they do not miss by much.

When the second derivative test is inconclusive at a point (a,b,c), looking directly at $f(a+h,\, b+k,\, c+l)-f(a,\, b,\, c)$, for all of h, k, and l small, may often be successful.

Example 5.18. Show that (0,0,0) is the only critical point of $f(x,y,z)=\tan^{-1}(x^2+y^2+z^2)-(xy+yz+zx)$ and determine its nature.

Solution. $f_x=\dfrac{2x}{1+(x^2+y^2+z^2)^2}-(y+z)$, $f_y=\dfrac{2y}{1+(x^2+y^2+z^2)^2}-(z+x)$,

$f_z=\dfrac{2z}{1+(x^2+y^2+z^2)^2}-(x+y)$. At a critical point, all three terms are zero. Consider

$$0=f_x-f_y = \frac{2(x-y)}{1+(x^2+y^2+z^2)^2}-(y-x)=(x-y)\left[1+\frac{2}{1+(x^2+y^2+z^2)^2}\right].$$

This shows that $x=y$ at a critical point. By symmetry $y=z$ too. So the critical point is (x,x,x) where $\dfrac{2x}{1+(3x^2)^2}-2x=0$, that is $x=0$. To determine its nature one could try the second derivative test. In fact the test fails since the point is degenerate (check this). We use the Taylor expansion of $\tan^{-1}(u)=u-u^3/3+u^5/5-\cdots$ for $|u|<1$. Then for $(h,\,k,\,l)$ near (0,0,0),

$$f(h,\,k,\,l)-f(0,0,0)=h^2+k^2+l^2-(h^2+k^2+l^2)^3/3-(hk+kl+lh)+E$$

where E consists of higher order terms from the series for $\tan^{-1}(u)$ which we ignore. We need to keep the term $(h^2+k^2+l^2)^3$ since the critical point is degenerate. Now for h small,

$$f(h,\,h,\,0)=2h^2-(2h^2)^3/3-h^2>0 \quad \text{whereas} \quad f(h,\,h,\,h)=-(3h^2)^3/3<0\,.$$

This shows that (0,0,0) is a saddle point. The fact that $f(h,h,h)-f(0,0,0)$ contains no terms of order 2 confirms that (0,0,0) is degenerate.

5.3. PROBLEMS WITH CONSTRAINTS

We frequently encounter extrema problems where constraints are present. For example, the problem of determining the closed cylinder of largest volume for a given surface area (most economically shaped can), the problem of finding the point on a given curve nearest to the origin, problems of maximising output on a farm of fixed size, etc. These all take the form: find the extrema of $f(\mathbf{x})$ subject to $g(\mathbf{x})=0$. Instead of finding the extrema of f with $\mathbf{x}$ free to vary, $\mathbf{x}$ is constrained by being required to satisfy the equation $g(\mathbf{x})=0$. Let $S=\{\mathbf{x}\in\mathbb{R}^n : g(\mathbf{x})=0\}$. This is a level set of g and, when $n=3$, typically is a surface. Extrema of f subject to the constraint $g(\mathbf{x})=0$ are the extrema of f restricted to S. For example, f has a local minimum at $\mathbf{a}$ subject to the constraint $g(\mathbf{x})=0$ if $\mathbf{a}\in S$ and there is a neighbourhood U of $\mathbf{a}$ such that $f(\mathbf{x})\geq f(\mathbf{a})$ for all $\mathbf{x}\in S\cap U$. One approach to these problems is to use the equation $g(\mathbf{x})=0$ directly to write the problem as a simpler one. We illustrate this method by some simple examples.

Example 5.19. Find the extreme values of $f(x,y)=\sqrt{3}x+y$ subject to $x^2+y^2=1$.

Solution. The constraint here is that (x,y) should lie on the circle centred at 0 and with radius 1. All such points can be written $x=\cos\theta, y=\sin\theta$, $0\leq\theta\leq 2\pi$, and then we want extrema of $f(\theta)=\sqrt{3}\cos\theta+\sin\theta$. By elementary calculus, $f'(\theta)=-\sqrt{3}\sin\theta+\cos\theta$, and critical points occur when $\tan\theta=1/\sqrt{3}$ that is, for $\theta=\pi/6$ and $7\pi/6$. At $\theta=\pi/6$, $f=2$, at $\theta=7\pi/6$, $f=-2$. These are the maximum and minimum values (they exist since f is continuous on the circle, a compact set).

Another way to try this problem is to write $y=\sqrt{(1-x^2)}$. However, this only corresponds to the upper semicircle so we must also take $g(x)=-\sqrt{(1-x^2)}$ separately. Writing $y=\sqrt{(1-x^2)}$,$-1\leq x\leq 1$, and considering $f(x)=\sqrt{3}x+\sqrt{(1-x^2)}$, we find $f'(x)=0$ when $x=\sqrt{3}\sqrt{(1-x^2)}$. Thus, $x\geq 0$ and $x^2=3(1-x^2)$, therefore $x=\sqrt{3}/2$. This gives the maximum value 2. The minimum occurs on the lower semicircle.

In this particular example, one can find the extreme points of $f(\theta)$ without any calculus by observing that one can always write $a\cos\theta+b\sin\theta$ in the form $R\cos(\theta-\alpha)$ so that $R\cos\alpha=a$ and $R\sin\alpha=b$ which gives $R=\sqrt{(a^2+b^2)}$ and $\tan\alpha=b/a$. In the above example we get $\sqrt{3}\cos\theta+\sin\theta=2\cos(\theta-\alpha)$ where $\cos\alpha=\sqrt{3}/2$, $\sin\alpha=1/2$. Obviously the maximum of this expression is 2 when $\theta=\alpha$, that is $\theta=\pi/6$.

This example shows that there are several possible methods of solving such problems and it may not be clear which one will be best.

Example 5.20. Find the extrema of $f(x, y)=x^2-2x+y^2$ on the curve $y^2=3x$ (a parabola).

Solution. The problem is equivalent to finding the extrema of $f(x)=x^2+x$. Now, $f'(x)=2x+1$, so there is one critical point at $x=-1/2$. However, $x=-1/2$ does not correspond to a point on the curve $y^2=3x$ since $y^2\geq 0$ always. What has gone wrong? We forgot to say where $f(x)$ was defined. As $y^2=3x$ implies $x\geq 0$ we have $f(x)=x^2+x$ for $x\geq 0$ and then $f'(x)=2x+1\geq 1$ on $[0,\infty)$ so that f is strictly increasing and therefore f takes its minimum value at $x=0$ (an endpoint of the interval) and this minimum value is 0.

In the above examples it has been rather easy to use the equation for the constraint to reduce the problem to one involving fewer variables. However, it is often either impossible or not useful to carry out this process. Fortunately there is another method which avoids this process. It is best described geometrically.

Suppose we wish to maximize $f(\mathbf{x})$ subject to the constraint $g(\mathbf{x})=0$. For the purpose of the geometrical interpretation we work in $\mathbb{R}^3$. Typically $g(\mathbf{x})=0$ is a surface S, a level set of g, and we are trying to maximize $f(\mathbf{x})$ for $\mathbf{x}\in S$. If $\mathbf{a}\in S$ is a local maximum of f restricted to S, we can no longer say that $f(\mathbf{a})$ is a maximum in any fixed direction but only for those directions which correspond to points staying in S. In other words, directional derivatives of f should be 0 for all directions that are tangent to S at $\mathbf{a}$. If $\mathbf{T}$ is any such tangent to S at $\mathbf{a}$ we should have $<\nabla f(\mathbf{a}),\mathbf{T}>=0$. This means that $\nabla f(\mathbf{a})$ should be parallel to $\nabla g(\mathbf{a})$ (see Proposition 3.19), in other words, there is a scalar λ such that $\nabla f(\mathbf{a})=\lambda\nabla g(\mathbf{a})$. Another way of seeing this is to recall that $\nabla f(\mathbf{a})$ points in the direction in which f increases fastest. If $\nabla f(\mathbf{a})$ is not orthogonal to S it will be possible to move in S so as to increase f and $\mathbf{a}$ is not the maximum. Exactly similar considerations apply to minima.

Theorem 5.21. (*The Lagrange multiplier rule.*)
Let $G\subseteq\mathbb{R}^n$ be open and let $f, g: G\longrightarrow\mathbb{R}$ be of class C^1 on G. Let $S=\{\mathbf{x}\in G : g(\mathbf{x})=0\}$ and suppose f has a constrained extremum at $\mathbf{a}\in S$. Then, if $\nabla g(\mathbf{a})\neq\mathbf{0}$, there exists $\lambda\in\mathbb{R}$ such that $\nabla f(\mathbf{a})=\lambda g(\mathbf{a})$.

The intuition behind this result is given above and a proof may be given using these ideas, but requires a discussion of manifolds which is beyond the scope of this text. We will give a proof using the implicit function theorem. As we shall shortly prove a more general result than Theorem 5.21 we will give a proof now in the simpler case $n=2$, which will illustrate the ideas and hopefully make the later proof more easily understood.

Proof for $n=2$. Since $\nabla g(\mathbf{a})\neq\mathbf{0}$ at least one of the partial derivatives g_x, g_y is nonzero, and by a change of notation if necessary, we can, and do, suppose $g_y\neq 0$ at $\mathbf{a}$. Then, by the implicit function theorem, there is a neighbourhood U of a_1 (the first component of $\mathbf{a}$) and a function $h : U \longrightarrow \mathbb{R}$ such that $g(x, y)=0$ is equivalent to $y=h(x)$. Then $g(x, h(x))=0$ on U, so by differentiation, $0=g_x+g_y h'$ on U, in particular at a_1. The function $F(x)=f(x, h(x))$, $x\in U$, has an extremum at $a_1\in U$ and therefore $F'(a_1)=0$. Differentiating F gives $0=f_x+f_y h'$ at a_1. Since $h'=-g_x/g_y$ at a_1 (recall that $g_y\neq 0$) we obtain $0=f_x-g_x(f_y/g_y)$. Write $\lambda=f_y/g_y$ (evaluated at a_1). Then $f_y=\lambda g_y$ by the definition of λ and $f_x=\lambda g_x$ from the previous equation.

Note that the theorem gives *necessary* conditions for an extremum. It says that all possible extrema occur at points $\mathbf{a}$ on S, where $\nabla f(\mathbf{a})=\lambda\nabla g(\mathbf{a})$ or possibly at points $\mathbf{b}$ where $\nabla g(\mathbf{b})=\mathbf{0}$, but not all such points need be extrema. Theorem 5.21 is called the *Lagrange multiplier rule*, λ is the Lagrange multiplier. It is sometimes called an undetermined multiplier since its value need not be determined. To apply the method we attempt to find all points $\mathbf{a}$ which lie on S and satisfy $\nabla f(\mathbf{a})=\lambda\nabla g(\mathbf{a})$ and then check whether they are maxima or minima. Since λ is also unknown we have $(n+1)$ unknowns $a_1,a_2,\cdots,a_n,\lambda$. $\nabla f(\mathbf{a})=\lambda\nabla g(\mathbf{a})$ gives the n equations

$$\partial f/\partial x_i=\lambda\partial g/\partial x_i,\ i=1,2,\cdots,n, \tag{5.6}$$

and the final equation is

$$g(\mathbf{a})=0. \tag{5.7}$$

Although an extra unknown has been introduced, there is usually a considerable gain from the symmetry of the new problem. We look at more examples.

Example 5.22. Find the points on the surface $z^2=xy+1$ that are nearest to the origin (0,0,0).

Solution. We want to minimize the distance of a point (x,y,z) on the surface to the origin. This is equivalent to minimizing the square of the distance namely $f(x,y,z)=x^2+y^2+z^2$, which is simpler to handle without the square root. Let $g(x,y,z)=xy-z^2+1$. Then $\nabla f=\lambda\nabla g$ gives the equations (1) : $2x=\lambda y$, (2) : $2y=\lambda x$, (3) : $2z=-\lambda 2z$, (4) : $xy-z^2=-1$.

From (3) we see that *either* $z=0$ *or* $\lambda=-1$. If $z=0$ then (4) reads $xy=-1$ and (1) then yields $2x^2=\lambda yx=-\lambda$. Similarly (2) yields $2y^2=\lambda xy=-\lambda$ so $x^2=y^2$ or $x=\pm y$. Since $xy=-1$ we obtain two points (1,−1,0) and (−1,1,0). The alternative is that $\lambda=-1$. Then $2x=-y$ and $2y=-x$ so $x=y=0$ and then we must have $z^2=1$. This gives the further two points (0,0,1), (0,0,−1). At these points $f=1$, the *minimum*. At (1,−1,0) and (−1,1,0), $f=2$, but this is *not* the maximum (nor the minimum) for we can take x,y very large and a corresponding z to get f very large: there is no maximum (for example, $y=x=\sqrt{99}$, $z=10$ gives $f=298$). The geometrical interpretation says that a minimum should exist.

Remark. In the above example it was apparent that a minimum existed and a maximum did not, but how might we prove this? There is a test using an extension of the Hessian method. However, we will use a more direct approach. If $\mathbf{a}$ is a point which is a candidate for an extremum, we consider directly $E=f(\mathbf{x})-f(\mathbf{a})$ for $\mathbf{x}$ near $\mathbf{a}$ and $\mathbf{x}\in S$, in other words, those $\mathbf{x}$ near $\mathbf{a}$ which *satisfy the constraint*.

In the above example, consider the point (0,0,1). Then points near (0,0,1) can be written $(h, k, 1+l)$ where $h^2+k^2+l^2$ is small. We cannot consider *all* such points, only those which lie on S, that is satisfy $(1+l)^2=hk+1$. Then we have

$$E=f(h,k,1+l)-f(0,0,1)=h^2+k^2+(1+l)^2-1=h^2+k^2+hk,$$

since the constraint is satisfied. Hence, $E=(h+k/2)^2+3k^2/4>0$ on completing the square so indeed (0,0,1) is a local minimum.

For the point (1,−1,0) we consider nearby points of the form $(1+h, -1+k, l)$ where $l^2=(1+h)(-1+k)+1=-h+k+hk$. Note that we must have $-h+k+hk\geq 0$ for l to exist. Then,

$$E=f(1+h,-1+k,l)-f(1,-1,0)=2h+h^2-2k+k^2+l^2=h^2+k^2+hk+h-k,$$

on using $l^2=-h+k+hk$. Now h,k are not entirely free since we must satisfy $-h+k+hk\geq 0$. However we can take $h=0$, $k>0$ to find $E=k^2-k<0$ for small k. Then we can take $h=k$ and get $E=3h^2>0$. In all (1,−1,0) is neither a local maximum nor a local minimum. See also Example 5.28, *infra*.

We now give an example to show how blind application of the multiplier rule can have awkward consequences.

Example 5.23. Find the point on the curve $y^2=x^3$ that is closest to the point (−1,0).

Solution. The distance from a point (x,y) on the curve to the point (−1,0) is given by $(\text{distance})^2=(x+1)^2+y^2$. The problem is therefore that of minimizing $f(x,y)=(x+1)^2+y^2$ subject to the constraint $y^2-x^3=0$. Let $g(x,y)=y^2-x^3$. We look for the points where $\nabla f=\lambda\nabla g$. These satisfy

$$2(x+1)=-\lambda 3x^2 \quad \text{and} \quad 2y=\lambda 2y$$

The second equation gives: either $y=0$ or $\lambda=1$. If $\lambda=1$ then $2x+2+3x^2=0$ from the first equation and this implies that $x<0$, and the point is not on the curve $y^2=x^3$ ($x\geq 0$ is necessary). Therefore $y=0$ when $y^2=x^3$ gives $x=0$. However, the first equation then reads $2=0$, which is impossible. The reason is that $\nabla g=\begin{pmatrix}-3x^2\\ 2y\end{pmatrix}=\mathbf{0}$ at $(0,0)$ so we cannot assert that λ exists but we do have to consider such points on S when seeking extrema; that is what Theorem 5.21 asserts. In fact $(0,0)$ gives the minimum as is readily seen by considering $f(x)=(x+1)^2+x^3$ for $x\geq 0$. Of course, a sketch makes this plain too.

We now turn to the problem of finding extrema of a scalar function f subject to several constraints $g_1(\mathbf{x})=0,\cdots,g_m(\mathbf{x})=0$.

Theorem 5.24. (*Lagrange multiplier rule, several constraints.*)
Let G be an open subset of $\mathbb{R}^n$ and let $f\in C^1(G)$. Let $m<n$ and suppose $g_i\in C^1(G)$, $i=1,2,\cdots,m$. Let $\mathbf{g}: G\longrightarrow\mathbb{R}^m$ have components g_i and let $S=\{\mathbf{x}\in G:\mathbf{g}(\mathbf{x})=\mathbf{0}\}$. Suppose that f restricted to S has an extremum at $\mathbf{a}\in S\cap G$. For $\mathbf{x}\in G$ write $\mathbf{y}=(x_1,\cdots,x_{n-m})\in\mathbb{R}^{n-m}$, $\mathbf{z}=(x_{n-m+1},\cdots,x_n)\in\mathbb{R}^m$ and suppose that $\det D_{\mathbf{z}}\mathbf{g}(\mathbf{a})\neq 0$, [equivalently : $D\mathbf{g}(\mathbf{a})$ has maximal rank m]. Then there exists $\lambda=(\lambda_1,\lambda_2,\cdots,\lambda_m)\in\mathbb{R}^m$ such that

$$\nabla f(\mathbf{a})=\lambda_1\nabla g_1(\mathbf{a})+\cdots+\lambda_m\nabla g_m(\mathbf{a}). \tag{5.8}$$

The idea of this proof is exactly similar to the special case proved above in Theorem 5.21, but the notation is more difficult to handle.

Proof. Let $\lambda\in\mathbb{R}^m$ be the unique solution of the m equations

$$\sum_{k=1}^{m}\lambda_k D_i g_k(\mathbf{a})=D_i f(\mathbf{a}),\quad i=n-m+1,\cdots,n, \tag{5.9}$$

or, in matrix notation, $\lambda D_{\mathbf{z}}\mathbf{g}(\mathbf{a})=D_{\mathbf{z}}f(\mathbf{a})$. The solution exists since $D_{\mathbf{z}}\mathbf{g}(\mathbf{a})$ is invertible and indeed may be written $\lambda=D_{\mathbf{z}}f(\mathbf{a})[D_{\mathbf{z}}\mathbf{g}(\mathbf{a})]^{-1}$. By the implicit function theorem, $\mathbf{g}(\mathbf{y},\mathbf{z})=\mathbf{0}$ is equivalent to $\mathbf{z}=\mathbf{h}(\mathbf{y})$ defined on a neighbourhood U of $\mathbf{b}=(a_1,\cdots,a_{n-m})$. Then we have $\mathbf{g}(\mathbf{y},\mathbf{h}(\mathbf{y}))=\mathbf{0}$ on U and, by the chain rule, (compare Lemma 4.22),

$\mathbf{0}=D_{\mathbf{y}}\mathbf{g}(\mathbf{y},\mathbf{h}(\mathbf{y}))+D_{\mathbf{z}}\mathbf{g}(\mathbf{y},\mathbf{h}(\mathbf{y}))D\mathbf{h}(\mathbf{y})$, in particular $D_{\mathbf{y}}\mathbf{g}(\mathbf{a})+D_{\mathbf{z}}\mathbf{g}(\mathbf{a})D\mathbf{h}(\mathbf{b})=\mathbf{0}$, or $D\mathbf{h}(\mathbf{b})=-[D_{\mathbf{z}}\mathbf{g}(\mathbf{a})]^{-1}D_{\mathbf{y}}\mathbf{g}(\mathbf{a})$. Also, if we write $F(\mathbf{y})=f(\mathbf{y},\mathbf{h}(\mathbf{y}))$ for $\mathbf{y}\in U$, F has an extremum at $\mathbf{b}\in U$ (since $(\mathbf{y},\mathbf{h}(\mathbf{y}))\in S$) and therefore $DF(\mathbf{b})=\mathbf{0}$. Again, by the chain rule, $DF(\mathbf{b})=D_{\mathbf{y}}f(\mathbf{a})+D_{\mathbf{z}}f(\mathbf{a})\,D\mathbf{h}(\mathbf{b})$. Therefore $\mathbf{0}=D_{\mathbf{y}}f(\mathbf{a})-D_{\mathbf{z}}f(\mathbf{a})[D_{\mathbf{z}}\mathbf{g}(\mathbf{a})]^{-1}D_{\mathbf{y}}\mathbf{g}(\mathbf{a})=D_{\mathbf{y}}f(\mathbf{a})-\lambda D_{\mathbf{y}}\mathbf{g}(\mathbf{a})$.
The long hand version of this last equation is

$$\sum_{k=1}^{m}\lambda_k D_i g_k(\mathbf{a})=D_i f(\mathbf{a}),\quad i=1,2,\cdots,n-m, \tag{5.10}$$

and when combined with (5.9) yields the conclusion of the theorem.

The hypothesis $\det D_{\mathbf{z}}\mathbf{g}(\mathbf{a})\neq 0$ is the requirement that the $m\times m$ matrix is invertible or that the m rows of $D_{\mathbf{g}}(\mathbf{a})$ and also the last m columns are linearly independent vectors. By changing notation this is the same as requiring any m columns to be linearly independent, which condition is usually stated, in courses of linear algebra, as $D_{\mathbf{g}}(\mathbf{a})$ has rank m (its maximal value). The intuitive meaning of this condition is that the constraints are independent : there is no relationship between them.

Example 5.25. Find the shortest distance from the ellipse $x^2+2y^2=6$ to the line $y+x=5$.

Solution. Let (x,y) be any point on the ellipse and let (u,v) be any point on the line. We wish to minimize the square of the distance of (x,y) to (u,v), namely $f(x,y,u,v)=(x-u)^2+(y-v)^2$ subject to the constraints $g(x,y,u,v)=x^2+2y^2-6=0$ and $h(x,y,u,v)=u+v-5=0$. Note that $Dg=\begin{bmatrix}2x & 4y & 0 & 0\\ 0 & 0 & 1 & 1\end{bmatrix}$ has rank 2 unless $x=y=0$. By the Lagrange multiplier rule, we seek solutions of $\nabla f(\mathbf{a})=\lambda\nabla g(\mathbf{a})+\mu\nabla h(\mathbf{a})$ (we write λ, μ rather than use subscripts). These equations are

$$2(x-u)=2\lambda x,\qquad 2(y-v)=4\lambda y,\qquad -2(x-u)=\mu,\qquad -2(y-v)=\mu. \tag{5.11}$$

(5.11(*iii*)) and (5.11(*iv*)) show that $x-u=y-v$ and so, by (5.11(*i*)) and (5.11(*ii*)) we see that either $x=2y$ or $\lambda=0$. However, $\lambda=0$ is impossible since the equations $u+v=5$, $u^2+2v^2=6$ have no solution (the line and the ellipse do not intersect). Thus $x=2y$ and so $y=u-v$ from above. Since $u+v=5$ too, this gives $u=(y+5)/2$, $v=(5-y)/2$. Using the ellipse constraint we see that $6y^2=6$ so that $y=\pm 1$. Therefore there are two pairs of candidate points $(x,y)=(2,1)$, $(u,v)=(3,2)$ and $(x,y)=(-2,-1)$, $(u,v)=(2,3)$. The corresponding values of f are 2 and 32 so the minimum distance is $\sqrt{2}$. Note that it is geometrically evident that the minimum exists but a maximum does not. The pair of points where $f=32$ does not correspond to an extremum of f.

When there are three variables and two constraints say $g(\mathbf{x})=0$, $h(\mathbf{x})=0$, the equations take the form

$$f_x - \lambda g_x - \mu h_x=0,\quad f_y - \lambda g_y - \mu h_y=0,\quad f_z - \lambda g_z - \mu h_z=0,$$

or in terms of matrices $\begin{bmatrix} f_x & g_x & h_x \\ f_y & g_y & h_y \\ f_z & g_z & h_z \end{bmatrix}\begin{bmatrix} 1 \\ -\lambda \\ -\mu \end{bmatrix} = \begin{bmatrix} 0 \\ 0 \\ 0 \end{bmatrix}$, say $A\mathbf{w}=\mathbf{0}$.

Since this equation has a nonzero solution $\mathbf{w}=\begin{bmatrix} 1 \\ -\lambda \\ -\mu \end{bmatrix}$, A must be singular, that is, $\det A=0$. This gives an equation that does not involve λ, μ, and *may* be useful in determining the candidates for extrema.

5.4. EXTREMA OVER CLOSED SETS

Consider the problem of determining the maximum and minimum values of a continuous function on a closed, bounded set A. These values are certainly attained (Theorem 2.23). We may attempt to find them as follows. Firstly, if $\text{int}(A)\neq\emptyset$, find the critical points of f to determine local extrema. Secondly, consider points on $\text{bdy}(A)$. Frequently $\mathbf{x}\in\text{bdy}(A)$ can be written in the form $\mathbf{g}(\mathbf{x})=\mathbf{0}$ so this is a problem with constraints and can be tackled by the Lagrange multiplier rule. Thirdly, pick out the greatest and least values.

Example 5.26. Find the greatest and least values of $x^2-xy+y^2-z^2$ on and within the ellipsoid $3x^2+3y^2+z^2=4$.

Solution. We have to find the maximum and minimum of $f(x,y)=x^2-xy+y^2-z^2$ on the set $A = \{(x,y,z) : 3x^2+3y^2+z^2\le 4\}$. Interior critical points are when

$$2x-y=0,\ -x+2y=0,\ \text{and}\ -2z=0, \tag{5.12}$$

which gives the point (0,0,0) [which is inside A] at which point $f=0$.

Boundary extrema are extrema of f subject to $g(x,y) := 3x^2+3y^2+z^2-4=0$. By the Lagrange multiplier rule these occur among the solutions of

$$2x-y=6\lambda x,\quad -x+2y=6\lambda y,\quad -2z=2\lambda z,\quad 3x^2+3y^2+z^2=4. \tag{5.13}$$

Note that $\nabla g=(6x, 6y, 2z)$ is $\mathbf{0}$ only at (0,0,0), not on the ellipsoid. Then, from (5.13)(*iii*), either $z=0$ or $\lambda=-1$. If $z=0$, then

$$y=(2-6\lambda)x,\quad x=(2-6\lambda)y.$$

Therefore, $x=0 \Leftrightarrow y=0$ and since both cannot be 0 here we see that neither is 0. Then $y=(2-6\lambda)^2 y$ so $2-6\lambda=\pm 1$, that is, $6\lambda=1$ or $6\lambda=3$ which give $y=x$ or $y=-x$. The points $(x,\pm x,0)$ satisfy (5.13)(*iv*) only if $6x^2=4$ and this gives 4 points $(\pm 2/\sqrt{6},\pm 2/\sqrt{6},0)$ and $(\pm 2/\sqrt{6},\mp 2/\sqrt{6},0)$ with corresponding values for f of 2/3 and 2. Alternatively, if $\lambda=-1$, then $y=8x$, and $x=8y$ so that $x=y=0$ and the corresponding points are $(0,0,z)$ where $z^2=4$. This gives 2 points where $f=-4$. In total, the minimum -4 occurs at $(0,0,\pm 2)$, the maximum 2 occurs at $(\pm\sqrt{(2/3)},\mp\sqrt{(2/3)},0)$.

Example 5.27. Find the maximum and minimum of $f(x,y)=x^2+y^2+xy+1$ over the set $A=\{(x,y) : xy\geq -1\}$.

Solution. Interior critical points satisfy $2x+y=0$, $2y+x=0$, that is $(0,0)$ only and $f=1$ there. This is the minimum for we can complete the square on f to obtain $f(x,y)=(x+y/2)^2+3y^2/4+1$. For boundary extrema, let $g(x,y)=1+xy$ and apply the Lagrange multiplier rule. We want points where

$$2x+y=\lambda y, \qquad 2y+x=\lambda x, \qquad xy=-1$$

that is,

$$2x=(\lambda-1)y, \qquad 2y=(\lambda-1)x, \qquad xy=-1 .$$

Multiply the first equation by y the second by x to see that $(\lambda-1)y^2=(\lambda-1)x^2=2xy=-2$. Thus, $x^2=y^2$ (and are nonzero) and since also $xy=-1$ this gives the two points $(1,-1)$ and $(-1,1)$. At these points $f=2$. However, *these points are not maxima.* A is not a bounded region so we cannot assert that f has a maximum. Indeed, it does not, for we can take x,y both very large and positive to make f as large as we please. The two points $(1,-1)$, $(-1,1)$ are not local extrema of f on A either. This is most easily seen by drawing the two branches of the hyperbola $xy=-1$ and noting that f has an extremum on the hyperbola precisely when $h(x,y)=x^2+y^2$ has an extremum on this set (they coincide). This is the square of the distance from $(0,0)$ to the curve and can be increased by moving along the curve and decreased by moving off the curve into the region $xy>-1$. These points are local minima of f restricted to the curve $xy=-1$ but not local minima over A.

Problems.

1. Find the critical points of $f(x,y) = x^4+y^4 - (x - y)^2$ and determine their nature.

2. Show that $(1,1,1)$ is a critical point of $f(x,y,z) = x^4+y^4+z^4 - 4xyz$ and determine its nature.

3. Let G be an open bounded set in $\mathbb{R}^n$ and let $f:\mathrm{cl}(G)\to\mathbb{R}$ be continuous on $\mathrm{cl}(G)$ and differentiable in G. Prove that, if f is constant on $\mathrm{bdy}(G)$ then there exists $\mathbf{x}\in G$ such that $Df(\mathbf{x}) = 0$. (A

generalization of Rolle's theorem).

4. Find the extrema of $f(x,y,z) = 1/x + 1/y + 1/z$, $(x,y,z) \neq (0,0,0)$, subject to the constraint $x+y+z = 3$ and determine their nature.

5. Find the minimum of $4x^3+y^3+z^3$ subject to $x+y+z = 1, x>0, y>0, z>0$.

6. Find the point on the line of intersection of the planes $x - y = 1$, $4x+z = 1$ which is nearest to the origin.

7. Find the points on the curve of intersection of the cylinder $x^2+y^2 = 90$ and the plane $3x+y+z = 40$ that are nearest to and furthest from the origin.

8. Find the shortest distance from a point on the plane $ax+by+cz = 1$ to the origin.

9. (*a*) Show that the maximum of $x^2y^2z^2$ on the sphere $x^2+y^2+z^2 = r^2$ is $(r^2/3)^3$.
(*b*) Show that the minimum of $x^2+y^2+z^2$ on the surface $x^2y^2z^2 = (r^2/3)^3$ is r^2.

10. Find the maximum of $x_1^2x_2^2\cdots x_n^2$ subject to $x_1^2+x_2^2+\cdots+x_n^2 = 1$. Deduce that, for any n positive real numbers a_i,

$$(a_1a_2\cdots a_n)^{1/n} \leq \frac{a_1+a_2+\cdots+a_n}{n},$$

the inequality of the means (geometric mean $\leq$ arithmetic mean).

11. Let $\{a_1,a_2,\cdots,a_n\}$ be positive real numbers. For $p>1$ let q satisfy $\frac{1}{p}+\frac{1}{q} = 1$. Find the maximum of $\sum_{i=1}^{n} a_ix_i$ subject to $\sum_{i=1}^{n} |x_i|^q = 1$.

Deduce that $\sum_{i=1}^{n} a_ix_i \leq \left\{\sum_{i=1}^{n} a_i^p\right\}^{1/p}\left\{\sum_{i=1}^{n} |x_i|^q\right\}^{1/q}$,
(Hölder's inequality for sums). Hence show that

$$\left\{\sum_{i=1}^{n} |x_i+y_i|^p\right\}^{1/p} \leq \left\{\sum_{i=1}^{n} |x_i|^p\right\}^{1/p} + \left\{\sum_{i=1}^{n} |y_i|^p\right\}^{1/p},$$

(Minkowski's inequality for sums).

12. Show that $f(x,y) = (x - y^3)(x - 3y^3)$ has a saddle point at (0,0) but f restricted to any straight line through (0,0) has a minimum at the origin.

13. Find the maximum and minimum of $(x-y)(y-z)(z-x)$ subject to the constraint $x^2+y^2+z^2 = 2$.

14. (*i*) Show that if $f:\mathbb{R}\to\mathbb{R}$ is differentiable on $\mathbb{R}$ and 0 is a local minimum and there are no other critical points, then 0 is a global minimum.
(*ii*) Let $f:\mathbb{R}^2\to\mathbb{R}$ be given by $f(x,y)=x^2+y^2(1-x)^3$. Show that (0,0) is the only critical point, is a strict local minimum but there is no global minimum.

15. A postal service limits the size of parcels that it will deliver according to the rule that length + girth should not exceed 2 metres. For a rectangular box the length is the longest side and the girth is twice the width + twice the depth. Find the dimensions of the box of largest volume that the postal service will accept. Repeat for a cylindrical container where the girth is now the circumference of the circular cross-section.

16. Find the maximum and minimum of $x^3+y^3+z^3$ subject to the constraints $x^2+y^2+z^2 = 1$, $x+y+z = 1$.

17. Find the points on the curve cut on the cylinder $x^2+y^2 = 4$ by the surface $xyz = 2$ which are nearest to the origin.

Hints on the problems.

3. Consider where f attains its maximum and minimum values.

6, 7, 8. Minimize the square of the distance, namely $x^2+y^2+z^2$.

10. Consider $a_i/(a_1+\cdots+a_n)$

11. Show $d/dx(|x|^q)=q|x|^{q-2}x$ starting from $(x^2)^{q/2}$. Last part: $|x_i+y_i|^p = |x_i+y_i|^{p-1}|x_i+y_i| \le |x_i+y_i|^{p-1}|x_i| + |x_i+y_i|^{p-1}|y_i|$; apply Hölder's inequality.

13. Use symmetry to reduce the number of cases. Consider $\lambda=0$ and $\lambda\neq0$.

16. x, y, and z are each a solution of the same quadratic equation.

Chapter 6

Integration

6.1. THE INTEGRAL OVER N-DIMENSIONAL RECTANGLES

For a bounded function $f\colon [a,b] \longrightarrow \mathbb{R}$, the notion of the Riemann integral of f, usually denoted $\int_a^b f(x)\,dx$, is a familiar topic in elementary calculus. The theoretical development is given in a first course of Analysis. We intend to discuss the extension of this notion to allow the integration of functions of several variables.

Before we begin, some comments are in order. The Riemann integral has theoretical drawbacks and for more advanced work the superior theory of the Lebesgue integral is essential. However, this topic deserves a course unto itself so cannot be discussed in the limited space here. The deficiencies of the Riemann integral become apparent when convergence properties of integrals of sequences of functions are discussed and when the negligible sets of the theory are investigated. These difficulties can often be ignored when the practical problem of evaluating integrals is the main concern.

One of our aims is to give a *necessary and sufficient* condition for a bounded function f to be integrable which involves a measure of the size of the set of discontinuities of f. These notions may well be new to the reader even for a function $f\colon [a,b] \longrightarrow \mathbb{R}$.

We start by defining the integral over particularly simple subsets of $\mathbb{R}^n$, the analogues of intervals in $\mathbb{R}$.

Definition 6.1. An *n-dimensional rectangle* (or *n-rectangle*) E is a Cartesian product of n intervals, that is, for $\mathbf{a}=(a_1, a_2,\cdots, a_n)$ and $\mathbf{b}=(b_1, b_2,\cdots, b_n)$ where $a_i<b_i$ for each i,

$$E(\mathbf{a},\mathbf{b}) := \{\mathbf{x}\in\mathbb{R}^n : a_i\leq x_i\leq b_i \text{ for } i=1,2,\cdots,n\}. \tag{6.1}$$

When $n=1$ this set is a closed interval, for $n=2$ it is a rectangle in the usual sense, for $n=3$ the terminology rectangular box is descriptive. We have defined n-rectangles as closed sets but we could equally consider open n-rectangles or even allow some parts of the boundary to be included and other parts excluded. An *n-cube* is an n-rectangle with equal edge lengths $b_i - a_i$.

Intervals have lengths, rectangles have areas, boxes have volumes. To cover the general case we use the terminology *Jordan content*, or simply *content*. The n-dimensional content of an n-dimensional rectangle $E(\mathbf{a},\mathbf{b})$ is defined to be $c(E) := \prod_{i=1}^{n} (b_i - a_i)$.

Note that an interval $[a,b]$ can be considered as a degenerate rectangle $[a,b] \times [c,d]$ with $c=d$. It has 1-dimensional content $b-a$ but 2-dimensional content 0.

A finite union of n-rectangles E_i is said to be *non-overlapping* if no two have common interior points ; they may have common boundary points. If S is such a finite union then $c(S)$ is defined to be the sum of the contents $c(E_i)$. If a set S can be written in two ways as a finite union of non-overlapping n-rectangles, it is easily seen that the sum of the contents is the same so the content of such a set S is well defined. Also for two non-overlapping sets of this type S_1, S_2, it follows that $c(S_1 \cup S_2) = c(S_1) + c(S_2)$. This extends to any finite union, the so-called *finite additivity* property of content.

Later we shall meet the concept of content of sets other than finite unions of n-rectangles.

The definition of the Riemann integral of real-valued functions defined on n-rectangles is exactly similar to that of integration of functions defined on intervals. The main difference is that the length of intervals must be replaced by the content of n-rectangles.

Definition 6.2. A *partition* P of an interval $[a,b]$ (also called *dissection* or *subdivision*) is a finite set of points $\{x_0, x_1, \cdots, x_p\}$ such that $a = x_0 < x_1 < \cdots < x_p = b$. If $E(\mathbf{a},\mathbf{b})$ is an n-rectangle and P_i are partitions of $[a_i, b_i]$ then $P = P_1 \times \cdots \times P_n$ is a partition of E.

Thus, a partition divides E into smaller n-rectangles. If P_i divides $[a_i, b_i]$ into p_i sub-intervals then P divides E into $p_1 p_2 \cdots p_n$ sub-n-rectangles. In $\mathbb{R}^2$ a partition of a rectangle is obtained by drawing lines parallel to the co-ordinate axes forming a grid.

A partition P_2 is *finer* than a partition P_1 if all n-rectangles in P_2 are also in P_1. P_2 may be called a *refinement* of P_1. In $\mathbb{R}^2$ a refinement is obtained by drawing extra grid lines.

Let P be a partition of $E(\mathbf{a},\mathbf{b})$ into p sub-n-rectangles E_i and let $f: E(\mathbf{a},\mathbf{b}) \longrightarrow \mathbb{R}$ be bounded. The *upper* and *lower sums* are defined by

$$U(f, P) := \sum_{i=1}^{p} M_i\, c(E_i)\,, \qquad L(f, P) := \sum_{i=1}^{p} m_i\, c(E_i)\,, \tag{6.2}$$

where $M_i := \sup\{f(\mathbf{x}) : \mathbf{x} \in E_i\}$, and $m_i := \inf\{f(\mathbf{x}) : \mathbf{x} \in E_i\}$.

It is evident that, for every partition P, $L(f, P) \leq U(f, P)$.

When $n=1$ and $f>0$, the upper sum is an approximation from above of the area under the graph of f and the lower sum is an approximation from below. A similar interpretation can be given when f is defined on a rectangle in $\mathbb{R}^2$. Then upper and lower sums approximate the volume under the surface $z=f(x,y)$. It is intuitively clear that a subdivision should lead to better approximations. The next result proves this.

Proposition 6.3. *If P' is finer than P then*

$$L(f, P) \leq L(f, P') \quad \textit{and} \quad U(f, P') \leq U(f, P).$$

Proof. Suppose that an n-rectangle E in P is divided into sub-rectangles E_i, $1 \leq i \leq p$. Then

$$M_i := \sup\{f(\mathbf{x}) : \mathbf{x} \in E_i\} \leq M := \sup\{f(\mathbf{x}) : \mathbf{x} \in E\} \quad \text{for every } i.$$

Therefore, $\sum_{i=1}^{p} M_i\, c(E_i) \leq \sum_{i=1}^{p} M\, c(E_i) = M\, c(E)$. Adding all such terms proves that $U(f, P') \leq U(f, P)$. The proof for the lower sum is exactly similar using the fact that a lower bound for a set is a lower bound for any subset.

Corollary 6.4. *If P and P' are any 2 partitions,*

$$L(f, P) \leq U(f, P'). \tag{6.3}$$

Proof. Let P_1 be a refinement of both P and P'. By Proposition 6.3,

$$L(f, P) \leq L(f, P_1) \leq U(f, P_1) \leq U(f, P').$$

The result of Corollary 6.4 is very plausible when $f>0$ for $n=1$ (or $n=2$) since $L(f, P)$ should be less than or equal to the area (or volume) under the graph of f and $U(f, P')$ should be greater than or equal to the same area (or volume). The Riemann integrable functions are those for which the approximations from above and below lead to the same answer.

Definition 6.5. A bounded function $f: E(\mathbf{a},\mathbf{b}) \longrightarrow \mathbb{R}$ is said to be *Riemann integrable* if, for every $\varepsilon > 0$ there exists a partition P_ε such that $U(f, P_\varepsilon) - L(f, P_\varepsilon) < \varepsilon$.

This is often known as Riemann's ε-condition for integrability. Note that it is equivalent to

$$\sup\{L(f, P) : P \text{ a partition of } E(\mathbf{a},\mathbf{b})\} \\ = \inf\{U(f, P') : P' \text{ a partition of } E(\mathbf{a},\mathbf{b})\}. \quad (6.4)$$

For it is always true that we have $\leq$ by (6.3), and Riemann's ε-condition implies that we have $\geq$. It is usually simpler to check Riemann's condition rather than use (6.4). The common value in (6.4) is called the *integral of f over* $E(\mathbf{a},\mathbf{b})$ and is denoted $\int_E f(\mathbf{x})\, d\mathbf{x}$, or $\int_E f(\mathbf{x})\, dx_1 dx_2 \cdots dx_n$, or more briefly, $\int_E f$. $\int_E f$ is the unique real number S such that $L(f, P) \leq S \leq U(f, P)$ for all partitions P.

There is an alternative definition, equivalent to the above one, which is sometimes useful. Given a partition P of $E(\mathbf{a},\mathbf{b})$ into n-rectangles E_i, $1 \leq i \leq p$, a *Riemann sum* is a sum of the form

$$S(f, P) := \sum_{i=1}^{p} f(\mathbf{x}^{(i)})\, c(E_i)\,, \quad \mathbf{x}^{(i)} \in E_i\,.$$

Definition 6.6. A bounded function $f: E(\mathbf{a},\mathbf{b}) \longrightarrow \mathbb{R}$ is said to be Riemann integrable if there is a number S such that, for every $\varepsilon > 0$, there is a partition P_ε such that $|S(f, P) - S| < \varepsilon$ for every partition P finer than P_ε and every choice of points $\mathbf{x}^{(i)} \in E_i$.

Obviously, for any partition P, $L(f, P) \leq S(f, P) \leq U(f, P)$ so if f is integrable according to Definition 6.5 it is also integrable by Definition 6.6. The converse is also true. For, suppose f is integrable according to Definition 6.6. Then, for any $\varepsilon > 0$ there exists a partition P_ε such that $|S(f, P) - S| < \varepsilon/4$ for every P finer than P_ε and every choice of points $\mathbf{x}^{(i)} \in E_i$. Fix such a refinement. By the definition of supremum, there exists $\mathbf{x}^{(i)} \in E_i$ such that $f(\mathbf{x}^{(i)}) > M_i - \delta$, for $\delta = \varepsilon/(4c(E))$. The corresponding Riemann sum then satisfies $S_1(f, P) > \sum_{i=1}^{p} (M_i - \delta)\, c(E_i) = U(f, P) - \varepsilon/4$. Similarly there is a Riemann sum with $S_2(f,P) < L(f,P) + \varepsilon/4$. Since $|S_1 - S| < \varepsilon/4$ and $|S_2 - S| < \varepsilon/4$, we have

$|S_1 - S_2| < \varepsilon/2$ and hence $U(f, P) - L(f, P) < S_1(f, P) + \varepsilon/4 - S_2(f, P) + \varepsilon/4 < \varepsilon$.

We will usually say simply *integrable* or *R-integrable* rather than Riemann integrable.

From either definition one can prove the well-known properties of integral many of which are familiar when $n=1$. Note that we can now give examples of integrable functions, namely constants : $f(\mathbf{x})=k$ for all $\mathbf{x}$. The integral of f over E exists and is $k\,\mathrm{c}(E)$ since all upper and lower sums have this value.

Theorem 6.7. *Let f,g be integrable over $E(\mathbf{a},\mathbf{b})$ and let $k \in \mathbb{R}$. Then we have :*

(1) *Linearity. $kf+g$ is integrable and* $\int_E (kf+g) = k\int_E f + \int_E g$.

(2) *If $f(\mathbf{x}) \geq 0$ for all $\mathbf{x} \in E$ then* $\int_E f \geq 0$; *more generally, if $f(\mathbf{x}) \geq g(\mathbf{x})$ for all $\mathbf{x} \in E$ then* $\int_E f \geq \int_E g$.

(3) *If $m \leq f(\mathbf{x}) \leq M$ for all $x \in E$ then* $m\,\mathrm{c}(E) \leq \int_E f \leq M\,\mathrm{c}(E)$.

(4) *If f is integrable so is $|f|$ and* $\left|\int_E f\right| \leq \int_E |f|$.

Proof. (1) We give the proof for $k=1$ and leave the general case to the reader. For $\varepsilon > 0$ let P_1, P_2 be such that $U(f, P_1) - L(f, P_1) < \varepsilon/2$ and $U(g, P_2) - L(g, P_2) < \varepsilon/2$ and let P be a partition that is finer than both. Then it follows that

$$U(f+g, P) \leq U(f, P) + U(g, P) \leq U(f, P_1) + U(g, P_2),$$

since $\sup\{f(\mathbf{x}) + g(\mathbf{x}) : \mathbf{x} \in A\} \leq \sup\{f(\mathbf{x}) : \mathbf{x} \in A\} + \sup\{g(\mathbf{x}) : \mathbf{x} \in A\}$ for any set A. Similarly, $L(f, P_1) + L(g, P_2) \leq L(f, P) + L(g, P) \leq L(f+g, P)$.

Hence

$$U(f+g, P) - L(f+g, P) \leq U(f, P_1) + U(g, P_2) - [L(f, P_1) + L(g, P_2)] < \varepsilon,$$

so $f+g$ is integrable. Since

$$L(f, P) + L(g, P) \leq L(f+g, P) \leq U(f+g, P) \leq U(f, P) + U(g, P),$$

it is readily seen that $\int_E (f+g) = \int_E f + \int_E g$.

(2) If $f(\mathbf{x}) \geq 0$ for all $\mathbf{x}$ then every lower sum is non-negative and hence the integral is non-negative too. For the second part, $f(\mathbf{x}) - g(\mathbf{x}) \geq 0$ so by (1), $\int_E f - \int_E g = \int_E (f-g) \geq 0$, by the part just shown.

(3) Since $f(\mathbf{x}) - m \geq 0$ for all $\mathbf{x}$, $\int_E f \geq \int_E m = m\,\mathrm{c}(E)$, using (2). Similarly, $M - f(\mathbf{x}) \geq 0$ gives the other inequality.

(4) For a partition P, $M_j - m_j = \sup\{f(\mathbf{x}) : \mathbf{x} \in E_j\} - \inf\{f(\mathbf{y}) : \mathbf{y} \in E_j\}$

$$= \sup\{f(\mathbf{x}) : \mathbf{x} \in E_j\} + \sup\{-f(\mathbf{y}) : \mathbf{y} \in E_j\}$$
$$= \sup\{f(\mathbf{x}) - f(\mathbf{y}) : \mathbf{x},\mathbf{y} \in E_j\}$$
$$= \sup\{|f(\mathbf{x}) - f(\mathbf{y})| : \mathbf{x},\mathbf{y} \in E_j\},$$

because both $f(\mathbf{x})-f(\mathbf{y})$ and $f(\mathbf{y})-f(\mathbf{x})$ are considered in the preceding line. Since $\big||f(\mathbf{x})| - |f(\mathbf{y})|\big| \leq |f(\mathbf{x}) - f(\mathbf{y})|$, by the corollary to the triangle inequality, we see that $U(|f|, P) - L(|f|, P) \leq U(f, P) - L(f, P)$. This proves that $|f|$ is integrable whenever f is. Since $-|f(\mathbf{x})| \leq f(\mathbf{x}) \leq |f(\mathbf{x})|$, (3) yields $\left|\int_E f\right| \leq \int_E |f|$.

A large class of examples of integrable functions is furnished by the functions that are continuous. We shall not give a direct proof of this fact here but instead we will develop notions which will allow us to prove this as a consequence of a *necessary and sufficient* condition for a function to be R-integrable. This will be in terms of the set of discontinuities of the function. Riemann integrable functions need not be continuous. Also there are functions which are not Riemann integrable as the following example shows.

Example 6.8. The function $f : [0,1] \longrightarrow \mathbb{R}$ defined by

$$f(x) = \begin{cases} 1, & x \text{ irrational} \\ 0, & x \text{ rational} \end{cases},$$

is not Riemann integrable.

Solution. Every upper sum$=1$, every lower sum$=0$ since every interval always contains both rational and irrational points (compare Lemma 1.30). The function f here is discontinuous at every point of $[0,1]$.

We now begin to discuss the set of discontinuities of a function.

Defininition 6.9. The *oscillation* of a bounded function f on a set S is $\operatorname{osc} f(S) := \sup\{f(\mathbf{x}) - f(\mathbf{y}) : \mathbf{x},\mathbf{y} \in S\}$.

Observe that $\operatorname{osc} f(S) = \sup\{f(\mathbf{x}) : \mathbf{x} \in S\} - \inf\{f(\mathbf{y}) : \mathbf{y} \in S\}$; such terms occur in forming upper and lower sums. Also we can write $\operatorname{osc} f(S) = \sup\{|f(\mathbf{x}) - f(\mathbf{y})| : \mathbf{x},\mathbf{y} \in S\} = \operatorname{diam} f(S)$. It is evident that $\operatorname{osc} f(S_1) \leq \operatorname{osc} f(S_2)$ if $S_1 \subseteq S_2$.

Definition 6.10. The *oscillation of f at* $\mathbf{x} \in S$ is the number

$$\omega_f(\mathbf{x}) := \inf\{\operatorname{osc} f(\mathrm{B}(\mathbf{x},\delta) \cap S) : \delta > 0\} = \lim_{\delta \to 0+} \operatorname{osc} f(\mathrm{B}(\mathbf{x},\delta) \cap S). \tag{6.5}$$

The infimum here is the limit as $\delta \longrightarrow 0+$ since $\operatorname{osc} f(\mathrm{B}(\mathbf{x},\delta)\cap S)$ is a decreasing function of δ.

Proposition 6.11. *A bounded function $f: A \subseteq \mathbb{R}^n \longrightarrow \mathbb{R}$ is continuous at $\mathbf{a} \in A$ if, and only if, $\omega_f(\mathbf{a})=0$.*

Proof. If f is continuous at $\mathbf{a}$, then for every $\varepsilon>0$ there exists $\delta>0$ such that $\mathbf{x}\in A$ and $\|\mathbf{x}-\mathbf{a}\|<\delta \Rightarrow |f(\mathbf{x})-f(\mathbf{a})|<\varepsilon/2$. Then $\mathbf{x},\mathbf{y}\in A\cap \mathrm{B}(\mathbf{a},\delta)$ implies that $|f(\mathbf{x})-f(\mathbf{y})| \leq |f(\mathbf{x})-f(\mathbf{a})|+|f(\mathbf{a})-f(\mathbf{y})|<\varepsilon$ and hence $\omega_f(\mathbf{a})<\varepsilon$. As $\varepsilon>0$ is arbitrary, this proves $\omega_f(\mathbf{a})=0$. Conversely, if $\omega_f(\mathbf{a})=0$, for every $\varepsilon>0$ there is $\delta>0$ (δ depends on $\mathbf{a}$ in general) so that $\operatorname{osc} f[\mathrm{B}(\mathbf{a},\delta)\cap A]<\varepsilon$, so that certainly $|f(\mathbf{x})-f(\mathbf{a})|<\varepsilon$ for all $\mathbf{x}\in \mathrm{B}(\mathbf{a},\delta)\cap A$. This proves that f continuous at $\mathbf{a}$.

The set $D:=\{\mathbf{x}\in E : \omega_f(\mathbf{x})>0\}$ is thus the set of discontinuities of f on E. Now f is integrable if there is a partition P such that $U(f,P)-L(f,P)=\sum_{i=1}^{p}(M_i-m_i)\,\mathrm{c}(E_i)$ can be made arbitrarily small. On n-rectangles E_i where f is continuous the terms $M_i-m_i=\operatorname{osc} f(E_i)$ can be made small. Since f is bounded, the whole sum can be made small if the total content of the n-rectangles E_i on which f is not continuous is small enough. This is the idea of Theorem 6.18 *infra*. The appropriate measure of the size of D is made in the following definition.

Definition 6.12. Let A be a bounded subset of $\mathbb{R}^n$. We say that A has *(n-dimensional Jordan) content zero* if for every $\varepsilon>0$ there is a *finite* cover of A by n-rectangles $\{E_k : k=1,2,\cdots,q\}$ with $\sum_{k=1}^{q}\mathrm{c}(E_k)<\varepsilon$. We say that A has *(Lebesgue) measure zero* if for every $\varepsilon>0$ there is a *countable* cover of A by n-rectangles $\{E_k : k=1,2,\cdots\}$ with $\sum_{k=1}^{\infty}\mathrm{c}(E_k)<\varepsilon$.

We could equally well use open (boundary excluded) or closed rectangles in this definition. It is clear that a set of content zero also has measure zero but the converse is false. The concepts coincide for *compact* sets. It is particularly useful to allow countable coverings as was realized by Lebesgue and led to the development of the theory of the Lebesgue integral. Accounts of this may be found in references [4], [5], and [16] and in many other books.

Note that a line segment in $\mathbb{R}^2$ has 2-dimensional measure 0 though it has nonzero 1-dimensional measure (length), for we can cover by a rectangle of finite length and arbitrarily small width. In more advanced texts these matters are discussed further but we shall assume the intuitive fact that "nice" curves in $\mathbb{R}^2$ have measure 0. In particular, if A is a region in $\mathbb{R}^2$ with a "nice" boundary, then this boundary has measure 0. We illustrate this with an example.

Example 6.13. Let $f: [a,b] \longrightarrow \mathbb{R}$ be continuous on $[a,b]$. Then the graph $G = \{ (x, f(x)) \in \mathbb{R}^2 : x \in [a,b]\}$ has 2-dimensional measure zero.

Solution. Let $\varepsilon > 0$ be arbitrary. As f is uniformly continuous on $[a,b]$ (Theorem 2.29), there is $\delta(\varepsilon) > 0$ such that $|f(x) - f(y)| < \varepsilon$ for $|x-y| < \delta$. Let $a = x_0 < x_1 < \cdots < x_m = b$ be a partition of $[a,b]$ into intervals of equal length $\delta_1 \leq \delta$. The rectangles $\{[x_j, x_{j+1}] \times [f(x_j) - \varepsilon, f(x_j) + \varepsilon],\ 0 \leq j \leq m-1\}$ include the graph of f, by the uniform continuity requirement, and their total area is $2m\delta_1\varepsilon = 2\varepsilon(b-a)$. Since ε is arbitrary this proves the result.

There are examples of continuous functions from $[0,1]$ to $\mathbb{R}^2$ whose image is all of $[0,1] \times [0,1]$, known as space-filling curves. These examples show that the continuous image of a set of measure zero need not have measure zero. We shall not consider such functions. Obviously, a curve in $\mathbb{R}^2$ that consists of a finite union of graphs of continuous functions, is a set of 2-dimensional measure 0.

The difference between measure zero and content zero is shown by the following example.

Example 6.14. The set $\mathbb{Q} \cap (0,1)$ of all rational numbers in $(0,1)$ has measure zero but does not have content zero.

Solution. The rationals in $(0,1)$ can be enumerated in a single sequence $\{r_1, r_2, \cdots\}$. Then, for $\varepsilon > 0$,

$$\{ [r_i - \varepsilon/2^{i+1}, r_i + \varepsilon/2^{i+1}] : i = 1, 2, \cdots\}$$

is a countable cover of $\mathbb{Q}$ with total length $\sum_{i=1}^{\infty} \frac{\varepsilon}{2^i} = \varepsilon$. To see that the set does not have content zero, observe that if a finite union of closed intervals covers $\mathbb{Q} \cap (0,1)$ then this union must cover all of $(0,1)$. For, the union is a closed set and if $x \in (0,1)$ is not in the union there would be a neighbourhood of x not in this union which neighbourhood would contain rationals in $(0,1)$. This would contradict the covering property so the total length cannot be less than 1.

The fact that the rationals have measure zero is a special case of the following result since single points have measure zero.

Lemma 6.15. *If* $\{F_n\}$ *is a sequence of sets each of measure zero then* $\bigcup_{n=1}^{\infty} F_n$ *also has measure zero.*

Proof. For an arbitrary $\varepsilon>0$ and for each $n\in\mathbb{N}$, there is a countable cover of F_n by n-rectangles $\{E_k^{(n)} : k=1,2,\cdots\}$ such that $\sum_{k=1}^{\infty} c(E_k^{(n)})<\varepsilon/2^n$. Then $\{E_k^{(n)} : k=1,2,\cdots, n=1,2,\cdots\}$ is a countable cover of $\cup F_n$ and $\sum_{n=1}^{\infty}\sum_{k=1}^{\infty} c(E_k^{(n)}) < \sum_{n=1}^{\infty} \varepsilon/2^n=\varepsilon$.

We intend to prove that f is Riemann integrable $\Leftrightarrow$ the set of discontinuities has measure zero. To do this we study the set $D_\varepsilon :=\{\mathbf{x}\in E : \omega_f(\mathbf{x})\geq\varepsilon\}$.

Lemma 6.16. D_ε *is a closed set.*

Proof. We show that the set $\{\mathbf{x}\in E : \omega_f(\mathbf{x})<\varepsilon\}$ is relatively open in E. So suppose $\omega_f(\mathbf{x})<\varepsilon$. Then, by definition of infimum, there exists $\delta>0$ such that $\operatorname{osc} f(B(\mathbf{x},\delta)\cap E)<\varepsilon$. Then, for $\mathbf{y}\in E$ with $\|\mathbf{y}-\mathbf{x}\|<\delta/2$, $B(\mathbf{y},\delta/2)\subset B(\mathbf{x},\delta)$ and so $\operatorname{osc} f(B(\mathbf{y},\delta/2)\cap E)<\varepsilon$ and hence $\omega_f(\mathbf{y})<\varepsilon$.

This result shows that each D_ε is a compact subset of $E(\mathbf{a},\mathbf{b})$.

We now show that if the oscillation of a function is small at each point of an n-rectangle then the oscillation on sufficiently small sub-n-rectangles is also small. This is very similar to continuity on a compact set implying uniform continuity.

Lemma 6.17. *Suppose that* $E(\mathbf{a},\mathbf{b})$ *is a closed n-rectangle and* $\omega_f(\mathbf{x})<\varepsilon$ *for all* $\mathbf{x}\in E$. *Then there exist a partition P of E into n-rectangles* E_k *such that* $\operatorname{osc} f(E_k)<\varepsilon$.

Proof. For each $\mathbf{x}\in E$ there is $\delta_{\mathbf{x}}>0$ such that $\operatorname{osc} f[B(\mathbf{x},\delta_{\mathbf{x}})\cap E]<\varepsilon$. The open cover $\{B(\mathbf{x},\delta_{\mathbf{x}}/3\sqrt{n}) : \mathbf{x}\in E\}$ of the compact set E has a finite subcover $\{B(\mathbf{x}^{(i)}, \delta_i/3\sqrt{n}) : i=1,2,\cdots,q\}$ say. Take F_i to be an n-cube so that $B(\mathbf{x}^{(i)}, \delta_i/2\sqrt{n})\subseteq F_i\subseteq B(\mathbf{x}^{(i)},\delta_i)$. This is possible since

$\|\mathbf{x}\|_\infty \le \|\mathbf{x}\| \le \sqrt{n}\|\mathbf{x}\|_\infty$ (problem 1.5) and the set $\{\mathbf{x} : \|\mathbf{x}\|_\infty \le r\}$ is an n-cube centred at $\mathbf{0}$ whose edges have length $2r$. The change from $\delta_i/3\sqrt{n}$ to $\delta_i/2\sqrt{n}$ ensures that each F_i contains interior points of E. Then we may take $E_i = F_i \cap E$ and we have

$$\operatorname{osc} f(E_i) = \operatorname{osc} f(F_i \cap E) \le \operatorname{osc} f[B(\mathbf{x}^{(i)}, \delta_i) \cap E] < \varepsilon.$$

We can now characterize the Riemann integrable functions.

Theorem 6.18. *Let $f : E(\mathbf{a},\mathbf{b}) \longrightarrow \mathbb{R}$ be a bounded function. Then f is Riemann integrable on $E \Leftrightarrow$ the set D of discontinuities of f has measure zero.*

Proof. Let $D = \{\mathbf{x} \in E : \omega_f(\mathbf{x}) > 0\}$ and for $\varepsilon > 0$, let $D_\varepsilon = \{\mathbf{x} \in E : \omega_f(\mathbf{x}) \ge \varepsilon\}$. We note that $D = \bigcup_{k=1}^{\infty} D_{1/k}$ for, if $\omega_f(\mathbf{x}) > 0$, there is some $k \in \mathbb{N}$ so that $\omega_f(\mathbf{x}) \ge 1/k$. First suppose f is integrable. By Lemma 6.15 it suffices to show that each compact set $D_{1/k}$ has content zero. We will show that D_ε has content zero for every $\varepsilon > 0$. Suppose that for some $\varepsilon > 0$, D_ε does not have content zero. Then there is $\eta > 0$ such that any finite cover of D_ε has total content $\ge \eta$. Let P be any partition of E into n-rectangles $E_1, E_2, \cdots, E_p$. Then

$$U(f, P) - L(f, P) = \sum_{i=1}^{p} (M_i - m_i)\, c(E_i) \ge \sum{}' \varepsilon\, c(E_i)$$

where $\sum'$ denotes the sum over those i for which E_i contains points of D_ε. Since the corresponding n-rectangles E_i give a covering of D_ε, $\sum' c(E_i) \ge \eta$, and hence $U(f, P) - L(f, P) \ge \varepsilon\eta$. This means that the Riemann ε-condition cannot be satisfied, contradicting the fact that f is integrable.

Conversely, suppose that D has measure zero. Then, for an arbitrary $\varepsilon > 0$, $D_\varepsilon \subseteq D$ and the compact set D_ε has measure zero too and hence content zero. Hence there is a partition of E into n-rectangles $\{E_i : 1 \le i \le p\}$, say, such that the total content of the n-rectangles E_i that contain D_ε does not exceed ε. For i such that $E_i \cap D_\varepsilon = \emptyset$, $\omega_f(\mathbf{x}) < \varepsilon$ for all $\mathbf{x} \in E_i$, so, by Lemma 6.17, E_i can be further divided into n-rectangles on which the oscillation of f is less than ε. Once this has been done for each such i, they can be incorporated into a refined partition P which, for simplicity of notation, we write as $\{E_i : 1 \le i \le q\}$. Then $U(f, P)-L(f, P) = \sum_{i=1}^{q} (M_i-m_i)\, c(E_i) := S_1+S_2$, where S_1 is the sum over those i for which $M_i-m_i = \operatorname{osc} f(E_i) < \varepsilon$ and S_2 is the sum of the remaining terms. Then we have $S_1 < \varepsilon \sum_1 c(E_i) \le \varepsilon\, c(E)$. For the terms in S_2 we have $M_i-m_i \le M-m$ where M (respectively m) is the supremum (respectively infimum) of f on E. Therefore $S_2 < (M-m)\sum_2 c(E_i) < \varepsilon(M-m)$, since the total content of these sets does not exceed ε. Thus $U(f, P) - L(f, P) < \varepsilon(M - m+c(E))$, and since ε is arbitrary, this shows that f is integrable.

This result shows, in particular, that a continuous function is integrable, as is a function that is continuous except at finitely many points. Indeed a set of measure zero is allowed which includes countable sets. The Cantor set (Problem 1.37) is an example of an uncountable set of measure zero but it is rather difficult to find examples of functions whose set of discontinuities is an uncountable set of measure zero.

6.2. MULTIPLE INTEGRALS AS ITERATED INTEGRALS

We now consider methods of calculating integrals over n-rectangles. We concentrate on so-called double integrals, integrals in $\mathbb{R}^2$, since this illustrates the general method. Thus, let $E=[a,b]\times[c,d]$ be a rectangle. If $f : E \longrightarrow \mathbb{R}$ is integrable over E its integral is written $\int_E f$ or more fully $\iint_E f(x,y)\, dx\, dy$. The reason for the second notation will become more apparent when we learn how to compute these integrals. This will be done by the process of "first integrating with respect to x" and "then with respect to y" (or in reverse order), so two integration processes are involved.

Some notation will be useful. If y is fixed we write $f(\cdot,y)$ for the function whose value at x is $f(x,y)$. This assignment may be written $x \longmapsto f(x,y)$. Similarly, for x fixed, we write $f(x,\cdot)$ for the function $y \longmapsto f(x,y)$.

For a rectangle $E=[a,b]\times[c,d]$, let $[a,b]$ be subdivided into intervals $\{I_r : 1 \le r \le p\}$ and let $[c,d]$ be subdivided into intervals $\{J_s : 1 \le s \le q\}$ and let P be the corresponding partition of E into rectangles $E_{rs} := I_r \times J_s$. We denote the length of an interval I by $l(I)$. Then the area (or content) of E_{rs} is $l(I_r)\, l(J_s)$. If $f : E \longrightarrow \mathbb{R}$ is integrable over E then there is a partition P such that the integral is well approximated by a sum of the form $\sum_{r=1}^{p} \sum_{s=1}^{q} f(x_r,y_s)\, l(I_r)\, l(J_s)$, corresponding to the points (x_r,y_s) being regularly arranged in the grid. This sum may be written in two ways as

$$\sum_{s=1}^{q} l(J_s)\left[\sum_{r=1}^{p} f(x_r,y_s)\, l(I_r)\right] = \sum_{r=1}^{p} l(I_r)\left[\sum_{s=1}^{q} f(x_r,y_s)\, l(J_s)\right]. \tag{6.6}$$

The first corresponds to evaluating the sum over all rectangles by first summing on the horizontal strip (with $y=y_1$), then adding the contributions from the second strips, $y=y_2$ etc. The second term first sums over the vertical strips $(x=x_1)$ and adds the contribution over each such vertical strip. Concentrating on the first term in (6.6), the term summed over r is an approximating sum for $\int_a^b f(x,y_s)\, dx$. If we write $G(y) := \int_a^b f(x,y)\, dx$, we see that the first term in (6.6) is an approximating sum for $\int_c^d G(y)\, dy$. A similar interpretation holds for the second term. We therefore expect

$$\int_E f = \int_c^d \left[\int_a^b f(x,y)\, dx\right] dy = \int_a^b \left[\int_c^d f(x,y)\, dy\right] dx. \tag{6.7}$$

(6.7) writes the double integral over E as *iterated integrals*. We may also write, for example,

$$\int_c^d dy \int_a^b f(x,y)\, dx \quad \text{in place of} \quad \int_c^d \left[\int_a^b f(x,y)\, dx\right] dy\,.$$

Iterated integrals may often be evaluated by the elementary methods of integral calculus, essentially use of the fundamental theorem of the calculus. Unfortunately, if $\int_E f$ exists, it does not follow that (6.7) holds. In fact, the iterated integrals may fail to exist. The simplest instance of this is when f is continuous except on a horizontal line $y=c$, $a \le x \le b$. This set has 2-dimensional measure zero so $\int_E f$ exists but $\int_a^b f(x,c)\, dx$ does not exist since the function $f(\cdot,c)$ is discontinuous on a set of positive (1-dimensional) measure.

Of course, this cannot happen if f is continuous on E so we expect (6.7) to hold in this case. We shall, in fact, prove more: (6.7) holds when f is integrable over E and each function $f(\cdot,y)$ (and $f(x,\cdot)$) is integrable over $[a,b]$ (respectively $[c,d]$).

Theorem 6.19. *Let f be Riemann integrable over $E=[a,b]\times[c,d]$ and suppose that, for each fixed $x\in[a,b]$, the function $f(x,\cdot)$ is Riemann integrable over $[c,d]$. Then*

$$\int_E f = \int_a^b F(x)\,dx, \quad \textit{where } F(x) := \int_c^d f(x,y)\,dy\,. \tag{6.8}$$

Similarly if f and $f(\cdot,y)$ are R-integrable then

$$\int_E f = \int_a^b G(y)\,dy, \quad \textit{where } G(y) := \int_a^b f(x,y)\,dx\,. \tag{6.9}$$

Proof. Let $a=x_0<x_1<\cdots<x_p=b$ and $c=y_0<y_1<\cdots<y_q=d$ be partitions of $[a,b]$ and $[c,d]$ into subintervals $\{I_r\}$ and $\{J_s\}$ respectively, and let P denote the corresponding partition of E. For $\varepsilon>0$ we suppose that the partition is so fine that $U(f,P)-L(f,P)<\varepsilon/2$. Here

$$U(f,P)=\sum_{r=1}^{p}\sum_{s=1}^{q} M_{rs}\,l(I_r)l(J_s), \quad \text{where} \quad M_{rs}=\sup\{f(x,y) : x\in I_r,\ y\in J_s\},$$

with analogous definitions for $L(f,P)$ and m_{rs}. By the additivity of the integral of functions of one variable,

$$F(x)=\int_c^d f(x,y)\,dy = \sum_{s=1}^{q}\int_{y_{s-1}}^{y_s} f(x,y)\,dy := \sum_{s=1}^{q} F_s(x)$$

By Theorem 6.7 (3), for any x,

$$m_s(x)\,l(J_s)\le F_s(x)\le M_s(x)\,l(J_s)$$

where $M_s(x)=\sup\{f(x,y) : y\in J_s\}$, $m_s(x)=\inf(f(x,y) : y\in J_s\}$. Therefore, for every $x\in I_r$,

$$m_{rs}\,l(J_s)\le F_s(x)\le M_{rs}\,l(J_s)\,.$$

Let x_r be a point of I_r. Then the preceding line holds with $x=x_r$. Multiply by $l(I_r)$ and sum over s and then over r to get

$$L(f,P)\le \sum_{r=1}^{p} F(x_r)\,l(I_r)\le U(f,P)\,.$$

Since the inner sum is an approximating sum for $\int_a^b F$, this proves that F is integrable and $\left|\int_E f - \int_a^b F(x)\,dx\right|<\varepsilon$. Since ε is arbitrary, (6.8) is proved. The proof for (6.9) is exactly similar.

We may conclude that (6.7) holds when f has at most finitely many discontinuities in E. However, more importantly, (6.7) holds if the set of discontinuities of f has 2-dimensional measure zero and the set of discontinuities of $f(x,\cdot)$ (and $f(\cdot,y)$) has 1-dimensional measure zero. The use of this result is often termed "changing the order of integration".

There is an exactly similar result for multiple integrals. We state the result for triple integrals only, as this is an important case.

Theorem 6.20. *Let f be integrable over the 3-rectangle $E(\mathbf{a},\mathbf{b})$, that is, $E=[a_1,b_1]\times[a_2,b_2]\times[a_3,b_3]$ and suppose that, for each fixed z, $f(\cdot,\cdot,z)$ is integrable over $E_{12}:=[a_1,b_1]\times[a_2,b_2]$. Then*

$$\int_E f = \int_{a_3}^{b_3} dz \iint_{E_{12}} f(x,y,z)\,dx\,dy\,. \tag{6.10}$$

Similarly, if for each (x,y) the function $f(x,y,\cdot)$ is integrable over $[a_3,b_3]$ then

$$\int_E f = \iint_{E_{12}} dx\,dy \int_{a_3}^{b_3} f(x,y,z)\,dz\,. \tag{6.11}$$

The proof of Theorem 6.20 is similar to that of Theorem 6.19 and is omitted.

There are other versions obtained by interchanging the roles of the variables. Note that (6.11) is the version most commonly used as it says : first integrate with respect to z and reduce the problem to that of evaluating a double integral. Combining this with Theorem 6.19 gives, when all the integrals exist,

$$\iiint_{E(\mathbf{a},\mathbf{b})} f = \int_{a_3}^{b_3} dz \int_{a_2}^{b_2} dy \int_{a_1}^{b_1} f(x,y,z)\,dx,$$

and these iterated integrals may also be taken in other orders provided that the relevant integrals exist.

In order to evaluate iterated integrals it is necessary to be able to calculate single integrals. The fundamental theorem of the calculus is the important result needed together with a result on change of variable. We give these by way of reminder.

Theorem 6.21. (*a*) *The fundamental theorem of the calculus (first form). If f is integrable on $[a,b]$ and $F(x):=\int_a^b f$ then f is uniformly continuous on $[a,b]$. If also f is continuous at $c\in[a,b]$, then F is differentiable at c and $F'(c)=f(c)$.*

(*b*) *The fundamental theorem of the calculus (second form).*
Let $F:[a,b]\longrightarrow\mathbb{R}$ be continuous on $[a,b]$, differentiable on (a,b) and suppose that F' is R-integrable. Then $\int_a^b F'(x)\,dx=F(b)-F(a)$.

[*Note : F' need not be continuous*].

(c) Change of variable.
Let $\phi : [\alpha,\beta] \longrightarrow [a,b]$ and $f : [a,b] \longrightarrow \mathbb{R}$ be such that f, ϕ and ϕ' are continuous. Then $\int_{\phi(\alpha)}^{\phi(\beta)} f(x)\,dx = \int_\alpha^\beta f(\phi(t))\phi'(t)\,dt$.

Proof (*a*). By additivity of the integral and Theorem 6.7 (4) we have $|F(x)-F(y)| = |\int_x^y f| \le \int_x^y |f| \le M|y-x|$, where M is a bound for $|f|$, which proves the uniform continuity. For $h>0$, we write

$$m_h = \inf\{f(x) : x \in [c,c+h]\}, \quad \text{and} \quad M_h = \sup\{f(x) : x \in [c,c+h]\}.$$

By Theorem 6.7 (3), it follows that
$m_h \le \dfrac{F(c+h)-F(c)}{h} = \dfrac{1}{h}\int_c^{c+h} f \le M_h$. As $h \to 0+$, $m_h \to f(c)$ and $M_h \to f(c)$ by continuity, and hence $\dfrac{F(c+h)-F(c)}{h} \longrightarrow f(c)$ as $h \longrightarrow 0+$. The case $h<0$ is exactly similar and therefore $F'(c)=f(c)$.
(*b*) Let $P=\{x_0, x_1, \cdots, x_m\}$ be a partition of $[a,b]$. By the Mean Value Theorem there exists $c_i \in (x_{i-1}, x_i)$ such that

$$F(x_i) - F(x_{i-1}) = F'(c_i)(x_i - x_{i-1}).$$

Then $L(F',P) \le \sum_{i=1}^m F'(c_i)(x_i - x_{i-1}) = \sum_{i=1}^m F(x_i) - F(x_{i-1}) \le U(F',P)$. Since the sum is telescoping we get $L(F',P) \le F(b)-F(a) \le U(F',P)$. As this holds for any partition P and F' is integrable it follows that $\int_a^b F' = F(b)-F(a)$.

(*c*) Let $F(x) = \int_{\phi(\alpha)}^x f$. Then $F'=f$ and $\int_{\phi(\alpha)}^{\phi(\beta)} f(x)dx = F(\phi(\beta)) - F(\phi(\alpha))$.
Also, by the chain rule, $d/dt[F(\phi(t))] = f(\phi(t))\,\phi'(t)$ and so

$$\int_\alpha^\beta f(\phi(t))\phi'(t)\,dt = F(\phi(\beta)) - F(\phi(\alpha)) = \int_{\phi(\alpha)}^{\phi(\beta)} f(x)\,dx\ .$$

Part (*b*) shows that integration is often the opposite of differentiation. If f is continuous, (*b*) can be deduced from (*a*). The notation $[\,F(x)\,]_a^b := F(b) - F(a)$ is a convenient one for calculation purposes. We illustrate the use of this theorem in the computation of a double integral.

Example 6.22. Evaluate $\iint_E x \sin(y)\,dx\,dy$ over the rectangle $E=[0,1]\times[0,\pi/2]$.

Solution. This may be written

$$\int_0^1 dx \int_0^{\pi/2} x \sin y\,dy = \int_0^1 [-x \cos y]_0^{\pi/2}\,dx = \int_0^1 x\,dx = [x^2/2]_0^1 = 1/2.$$

6.3. INTEGRALS OVER MORE GENERAL SETS

It is necesary to extend the definition of integration so as to be able to integrate over more general regions. We shall consider the integral of *bounded functions* over certain *bounded* sets. Given a bounded region A and a bounded function $f: A \longrightarrow \mathbb{R}$ the method is to extend f to a rectangle E enclosing A by defining the extension to be zero outside A. Intuitively, this ought not to contribute to the value of the integral. We show that this procedure works.

Let A be a bounded set in $\mathbb{R}^n$. Given a bounded function $f: A \longrightarrow \mathbb{R}$ and a rectangle E containing A we define $f_A : E \longrightarrow \mathbb{R}$ by $f_A(x) = \begin{cases} f(x), & x \in A \\ 0, & x \in E \setminus A \end{cases}$, that is, f_A is the extension by 0 of f to A.

Definition 6.23. A bounded function $f: A \longrightarrow \mathbb{R}$ is said to be Riemann integrable over A if there is a rectangle E containing A such that $\int_E f_A$ exists.

It is easily seen, by considering approximating sums, that this definition is independent of the choice of the rectangle E; we may therefore omit reference to E and write $\int_A f$.

Given a set A, the *characteristic function* of A is the function $\chi_A : \mathbb{R}^n \longrightarrow \mathbb{R}$ defined by $\chi_A(\mathbf{x}) = \begin{cases} 1, & \text{if } \mathbf{x} \in A \\ 0, & \text{otherwise} \end{cases}$.

Note that $f_A(\mathbf{x}) = f(\mathbf{x})\, \chi_A(\mathbf{x})$ and that f is integrable over A if and only if $\int_E f \chi_A$ is integrable over a rectangle E containing A and $\int_A f = \int_E f \chi_A$.

In $\mathbb{R}^2$, if we imagine the graph $z = \chi_A(\mathbf{x})$ as a surface of height one sitting above A, then its volume should be (area of A)$\times 1$ so we have area of $A = \int_A 1 = \int_E \chi_A$ since upper (and lower) sums approximate this from above (and below). In fact this is used to define the area of a set or, more generally, its content.

Definition 6.24. We say that a bounded set A is *Jordan measurable* if χ_A is integrable and we then define the *content* of A, c(A) to be

$$c(A)=\int_A 1 = \int_E \chi_A .$$

An alternative terminology is "*A has content*" when A is Jordan measurable. For sets in $\mathbb{R}^1$, content is usually termed length, in $\mathbb{R}^2$ it is area, in $\mathbb{R}^3$ volume. Content may be called n-dimensional volume.

Since bdy(A) is the set of discontinuities of χ_A we see that

$$A \text{ is Jordan measurable} \Leftrightarrow \text{bdy}(A) \text{ has measure } 0. \tag{6.12}$$

For a rectangle $E\,[\mathbf{a},\mathbf{b}]=[a_1,b_1]\times[a_2,b_2]\times\cdots\times[a_n,b_n]$, its content as defined here coincides with the earlier definition, as is readily seen by carrying out the iterated integrals.

Lemma 6.25. *A set A has content* 0 *according to Definition* 6.24 *if and only if A has content* 0 *according to Definition* 6.12.

Proof. If A has content 0 according to Definition 6.24, then χ_A is integrable over a rectangle $E\supseteq A$ with integral 0. Thus, for any $\varepsilon>0$, there is a partition P of E with $U(\chi_A, P)<\varepsilon$. If E_j is a sub-rectangle of E that contains at least one point of A then $M_j = 1$. Other rectangles contribute 0. Hence the sum of the contents of those E_j that contain points of A is less than ε. For the converse, if A is covered by n-rectangles with total content less than ε, these can be incorporated into a partition P of a containing n-rectangle E and only these contribute to the upper sum $U(\chi_A,P)$ so this is less than ε.

As noted previously, the concepts content 0 and measure 0 are distinct although they coincide for compact sets. Thus, if A is a bounded set, A is Jordan measurable $\Leftrightarrow$ c(bdy$A$)=0,
since bdy$A$ is closed. We know that $\mathbb{Q}\cap[0,1]$ has measure 0 since it is a countable set, but it is not Jordan measurable because $\chi_{\mathbb{Q}\cap[0,1]}$ is not integrable (it is discontinuous at all points of [0,1]); its boundary is all of [0,1]. It is easy to see that if E_1, E_2 have content 0 then so does $E_1\cup E_2$ and if $F\subseteq E$ with c(E)=0, then c(F)=0 too.

We can now give a necessary and sufficient condition for a function to be Riemann integrable over a Jordan measurable set.

Theorem 6.26. *Let A be bounded and Jordan measurable and let* $f:A\longrightarrow\mathbb{R}$ *be bounded. Then f is integrable over A if and only if the set* D_A *of discontinuities of f in A has measure* 0.

Proof. Let f_A be the usual extension of f to a rectangle E. Then f_A is integrable $\Leftrightarrow$ the set D_E of discontinuities of f_A in E has measure 0. Now $D_E \subseteq D_A \cup \mathrm{bdy}(A)$ and $\mathrm{bdy}(A)$ has measure 0 hence so does D_E.

We now intend to show that sets of content 0 are negligible as far as Riemann integration is concerned.

Proposition 6.27. *Let N be a set of content 0 and let $f : N \longrightarrow \mathbb{R}$ be bounded. Then f is integrable over N and $\int_N f = 0$.*

Proof. f is integrable over N since the set of discontinuities of f in N certainly has content 0. Let E be a rectangle containing N and suppose that $m \le f(\mathbf{x}) \le M$ for $\mathbf{x} \in N$. Then

$$m\,\chi_N(\mathbf{x}) \le f_N(\mathbf{x}) \le M\,\chi_N(\mathbf{x}) \text{ for all } \mathbf{x} \in E,$$

and by Theorem 6.7 (3), this gives

$$m\,\mathrm{c}(N) = \int_E m\chi_N \le \int_E f_N = \int_N f \le \int_E M\chi_N = M\,\mathrm{c}(N), \text{ so } \int_N f = 0.$$

Theorem 6.28. *Let $f,g : A \longrightarrow \mathbb{R}$ be bounded and suppose f is integrable over A and that $f(\mathbf{x}) = g(\mathbf{x})$ except on a set N of content 0. Then g is integrable over A and $\int_A g = \int_A f$.*

Proof. Let E be a rectangle containing A and let f_A, g_A be the extensions by 0 of f,g to E. Let $h(\mathbf{x}) = f_A(\mathbf{x}) - g_A(\mathbf{x})$, so that $h(\mathbf{x}) = 0$ for $\mathbf{x} \in E \setminus N$. Therefore $h(\mathbf{x}) = h(\mathbf{x})\,\chi_N(\mathbf{x})$ for all $\mathbf{x} \in E$ and

$$\int_E h = \int_E h\,\chi_N = \int_N h = 0,$$

by Proposition 6.27. Finally, since $g_A(\mathbf{x}) = f_A(\mathbf{x}) - h(\mathbf{x})$, we obtain

$$\int_A g = \int_E g_A = \int_E f_A - \int_E h = \int_A f, \text{ by Theorem 6.7 (1).}$$

Theorem 6.28 is *false* if we replace "content 0" by "measure 0" as may be seen by taking $f(x) = 1$ for $x \in [0,1]$ and $g(x) = \begin{cases} 0\,, & x \in \mathbb{Q} \cap [0,1] \\ 1\,, & x \in [0,1] \setminus \mathbb{Q} \end{cases}$; g is not Riemann integrable on $[0,1]$ but differs from f only on a set of measure 0. This is a drawback of the Riemann integral.

It is important in calculating integrals to be able to decompose the region into simpler pieces. The additivity property we need is as follows.

Theorem 6.29. (*Region decomposition, additivity of the integral.*)
Let A, A_1 and A_2 be Jordan measurable sets. Suppose f is integrable over $A = A_1 \cup A_2$ where $c(A_1 \cap A_2) = 0$. Then f is integrable over each of A_1 and A_2 and $\int_A f = \int_{A_1} f + \int_{A_2} f$. The same holds if f is given to be integrable over each of A_1 and A_2 rather than A.

Proof. Take an n-rectangle E containing $A = A_1 \cup A_2$. Then

$$f_A = f\chi_{A_1} + f\chi_{A_2} - f\chi_{A_1 \cap A_2}.$$

If f is integrable over A, $f\chi_{A_1}$ is integrable over E since its discontinuities are those of f together with part of $\mathrm{bdy}(A_1)$, which has zero content. Similarly the other terms are integrable. Since $f\chi_{A_1 \cap A_2}$ is zero except on a set of content 0, its integral is 0. The result then follows from Theorem 6.7 (1). If f is given as integrable over A_1 and A_2 the result follows similarly.

Corollary 6.30. *Let A be Jordan measurable and let S be a Jordan measurable subset of A. Then if f is integrable over A, we have*

$$\int_A f = \int_{A \setminus S} f + \int_S f.$$

Proof. Recall that $A \setminus S = \{\mathbf{x} \in A : \mathbf{x} \notin S\}$. It is easy to check that $\mathrm{bdy}(A \setminus S) \subseteq \mathrm{bdy}(A) \cup \mathrm{bdy}(S)$ so that $\mathrm{bdy}(A \setminus S)$ has content 0 and $A \setminus S$ is Jordan measurable. Theorem 6.29 is now applicable.

Integrals satisfy a mean value property which we now prove.

Theorem 6.31. (*MVT for integrals.*)
Suppose that f is integrable over a Jordan measurable set A and let $m := \inf\{f(\mathbf{x}) : \mathbf{x} \in A\}$, $M := \sup\{f(\mathbf{x}) : \mathbf{x} \in A\}$. Then there exists $\alpha \in \mathbb{R}$ with $m \le \alpha \le M$ such that

$$\int_A f = \alpha\, c(A).$$

If, in addition, f is continuous on A and A is connected then $\alpha = f(\mathbf{a})$ for some $\mathbf{a} \in A$.

Proof. Let E be an n-rectangle containing A. Then

$$m\,\chi_A(\mathbf{x}) \le f_A(\mathbf{x}) \le M\,\chi_A(\mathbf{x}) \quad \text{for all } \mathbf{x} \in E,$$

so that

$$m\,\mathrm{c}(A) \le \int_A f \le M\,\mathrm{c}(A). \tag{6.13}$$

If $\mathrm{c}(A) = 0$ we may take any α. If $\mathrm{c}(A) > 0$ let $\alpha = \int_A f \,/\, \mathrm{c}(A)$. When f is continuous on A and A is connected, if $\mathrm{c}(A) = 0$ then any $\mathbf{a} \in A$ will do. When $\mathrm{c}(A) > 0$, if both inequalities in (6.13) are strict, the result follows from Corollary 2.68. If one inequality in (6.13) is an equality, for example the right-hand one, we claim that $\mathbf{a}$ exists such that $f(\mathbf{a}) = M$. Suppose that $f(\mathbf{x}) < M$ for all $\mathbf{x} \in A$. Since $\mathrm{c}(A) > 0$ it follows that $\mathrm{int}A \ne \emptyset$ because $\mathrm{int}A = \emptyset$ gives $A = \mathrm{bdy}(A)$ and $\mathrm{bdy}(A)$ has content 0. Therefore A contains a neighbourhood and hence contains a closed rectangle S, say, and $\mathrm{c}(S) > 0$. Since f is continuous on the compact set S, there is $\varepsilon > 0$ so that $f(\mathbf{x}) \le M - \varepsilon$ for all $\mathbf{x} \in S$. Then, by Corollary 6.30, and the finite additivity of content,

$$\int_A f = \int_S f + \int_{A \setminus S} f \le (M-\varepsilon)\mathrm{c}(S) + M\mathrm{c}(A \setminus S) < M\mathrm{c}(A),$$

a contradiction.

6.4. EVALUATION OF MULTIPLE INTEGRALS

We consider the process of actually computing multiple integrals. We shall concentrate on double integrals, integrals of functions defined on regions in $\mathbb{R}^2$, since this case reveals the general method. We suppose that the region can be described in a simple fashion.

Theorem 6.32. *Let α, β be continuous real functions defined on an interval $[a,b]$ such that $\alpha(x) \le \beta(x)$ for every $x \in [a,b]$, and let $A \subseteq \mathbb{R}^2$ be defined by $A := \{(x,y) : a \le x \le b,\ \alpha(x) \le y \le \beta(x)\}$. Suppose $f : A \longrightarrow \mathbb{R}$ is bounded and is continuous except possibly on the boundary of A and for at most finitely many points in the interior of A. Then*

$$\int_A f = \iint_A f(x,y)\,dxdy = \int_a^b \left[\int_{\alpha(x)}^{\beta(x)} f(x,y)\,dy\right] dx. \tag{6.14}$$

Proof. Let $E = [a,b] \times [c,d]$ contain A and apply Theorem 6.19 to f_A on E noting that $\mathrm{c}(\mathrm{bdy}(A)) = 0$ and that

$$\int_c^d f_A(x,y)\,dy = \int_c^{\alpha(x)} f_A(x,y)\,dy + \int_{\alpha(x)}^{\beta(x)} f_A(x,y)\,dy + \int_{\beta(x)}^d f_A(x,y)\,dy$$

where the outer two integrals are zero.

An exactly similar result holds for a region that can be written in the form $A = \{(x,y) : \phi(y) \le x \le \psi(y),\ c \le y \le d\}$ for continuous functions ϕ, ψ with $\phi(y) \le \psi(y)$ on $[c,d]$. We refer to these types of region as *elementary regions*. They are immediately recognizable once sketched, as they can be filled by lines parallel to the y-axis (or x-axis), one line for each x between a and b (y between c and d) such that the endpoints lie on single curves, in practice curves defined by a single formula. See Figure 6.1.

A region may be both types simultaneously, for example, the circle $A = \{(x,y) : x^2+y^2 = 1\}$ can be written as

$$A_1 = \{-1 \leq x \leq 1, -\sqrt{(1-x^2)} \leq y \leq \sqrt{(1-x^2)}\},$$

or as

$$A_2 = \{-\sqrt{(1-y^2)} \leq x \leq \sqrt{(1-y^2)}, -1 \leq y \leq 1\}.$$

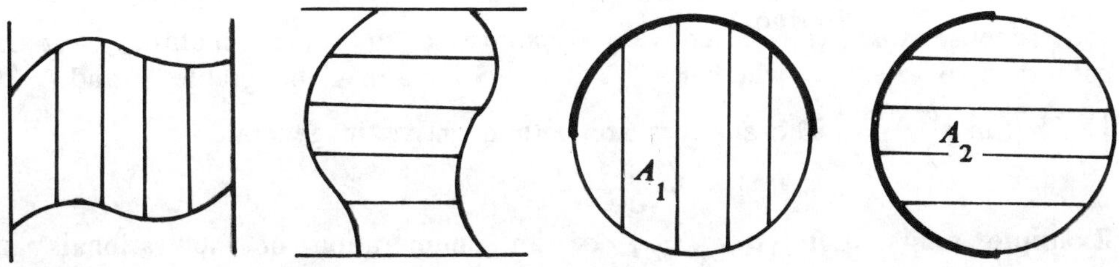

Figure 6.1

For such regions, it is sometimes easy to calculate one of the iterated integrals by use of the fundamental theorem of the calculus, but not the other. More complicated regions can often be decomposed into a finite number of elementary regions and the above method plus additivity enables integrals over such regions to be calculated.

Example 6.33. Let A be the triangle bounded by the straight lines $y = 0$, $x = 1$ and $y = x$. Evaluate $\iint_A \exp(x^2)\, dx\, dy$.

Solution.

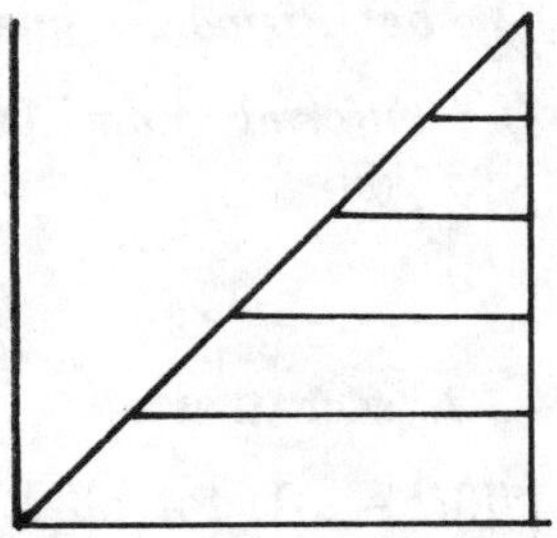

Figure 6.2

By Theorem 6.32 we can write

$$\iint_A \exp(x^2)\, dx\, dy = \int_0^1 \left[\int_y^1 \exp(x^2)\, dx\right] dy = \int_0^1 \left[\int_0^x \exp(x^2)\, dy\right] dx.$$

The first iterated integral cannot be evaluated by the fundamental theorem of calculus since the only function $F(x)$ whose derivative is $\exp(x^2)$ has to be written as an integral. However,

$$\int_0^x \exp(x^2)dy = x\exp(x^2) \text{ and } \int_0^1 x\exp(x^2)\,dx = \Big[\exp(x^2)/2\Big]_0^1 = (e-1)/2,$$

gives the required answer.

6.5. INTEGRATION AND UNIFORM CONVERGENCE

Let $f^{(m)}: [a,b] \to \mathbb{R}$, $m=1,2,\cdots$ be Riemann integrable over $[a,b]$ and suppose that $\{f^{(m)}\}$ converges pointwise to some function f. It is natural to ask whether f is Riemann integrable and if $\int_a^b f = \lim \int_a^b f^{(m)}$. The answers are both negative in general.

Example 6.34. Let $\{r_1, r_2, \cdots\}$ be an enumeration of the rationals in $[0,1]$ and let $f^{(m)}(x) = \begin{cases} 1, & \text{if } x=r_i \text{ for } i\in\{1,2,\cdots,m\} \\ 0, & \text{otherwise} \end{cases}$.
Then $f^{(m)}$ is Riemann integrable over $[0,1]$ for each m but the limit function is $f(x) = \chi_{\mathbb{Q}\cap[0,1]}$ and is not Riemann integrable.

Example 6.35. Let $f^{(m)}(x) = mx\exp(-mx^2)$, $x\in[0,1]$. Then $f^{(m)}(x) \to 0$ for each x but $\int_0^1 f^{(m)}(x) = [-\exp(-mx^2)/2]_0^1 \to 1/2$ while $\int_0^1 0 = 0$.

However, as might be guessed, uniform convergence is a sufficient condition.

Theorem 6.36. *Let $\{f^{(m)}\}$ be Riemann integrable functions on $[a,b]$ and suppose that $f^{(m)} \to f$ uniformly on $[a,b]$. Then f is Riemann integrable and $\int_a^b f = \lim_{m\to\infty} \int_a^b f^{(m)}$.*

Proof. For $\varepsilon>0$ there exists N such that

$$|f^{(m)}(x) - f(x)| < \varepsilon' := \varepsilon/[3(b-a)], \text{ for all } x\in[a,b] \text{ and } m\geq N. \quad (6.15)$$

Let P be a partition of $[a,b]$ with $U(f^{(N)}, P) - L(f^{(N)}, P) < \varepsilon/3$, for this N. Since $f(x) \leq f^{(N)}(x) + \varepsilon'$ for all $x\in[a,b]$, we have for any sub-interval I, $\sup\{f(x) : x\in I\} \leq \sup\{f^{(N)}(x) : x\in I\} + \varepsilon'$. Therefore

$$U(f,P) \le U(f^{(N)},P) + \varepsilon'(b-a).$$

Similarly, $f(x) \ge f^{(N)}(x) - \varepsilon'$ yields $L(f,P) \ge L(f^{(N)},P) - \varepsilon'(b-a)$.
Hence, $U(f,P) - L(f,P) \le U(f^{(N)},P) + \varepsilon/3 - L(f^{(N)},P) + \varepsilon/3 < \varepsilon$.
This proves that f is integrable.
Now, by (6.15) and properties of the integral, for $m \ge N$,

$$\left| \int_a^b f - \int_a^b f^{(m)} \right| \le \int_a^b |f - f^{(m)}| < \varepsilon'(b-a) = \varepsilon/3.$$

This shows that $\lim_{m\to\infty} \int_a^b f^{(m)}$ exists and equals $\int_a^b f$.

As a corollary we have :

Theorem 6.37. *Let $f^{(m)}$ be R-integrable functions on $[a,b]$. If $\sum f^{(m)}(x)$ converges uniformly to $F(x)$ on $[a,b]$ then F is integrable and $\int_a^b F = \sum_{k=1}^{\infty} \int_a^b f^{(k)}$ (integration "term-by-term" is correct).*

Proof. Write $F^{(m)} = \sum_{k=1}^{m} f^{(k)}$, so that $F^{(m)} \to F$ uniformly on $[a,b]$. Each $F^{(m)}$ is integrable being a finite sum of integrable functions. Therefore by Theorem 6.36,

$$\int_a^b F = \lim_{m\to\infty} \int_a^b F^{(m)} = \lim_{m\to\infty} \int_a^b \sum_{k=1}^{m} f^{(k)} = \lim_{m\to\infty} \sum_{k=1}^{m} \int_a^b f^{(k)} = \sum_{k=1}^{\infty} \int_a^b f^{(k)}.$$

Example 6.38. Show that $\ln(2) = 1 - 1/2 + 1/3 - 1/4 + \cdots$.

Solution. For $0 \le t < 1$, $\frac{1}{t+1} = \sum_{k=0}^{\infty} (-t)^k$ being a geometric series with ratio $(-t)$ and the series converges uniformly on $[0,1-\delta]$ $(0<\delta<1/2)$. Integrating term by term from 0 to x gives $\ln(1+x) = \sum_{k=1}^{\infty} \frac{(-1)^{k+1} x^k}{k}$ for $\delta \le x \le 1-\delta$. Since δ may be arbitrarily small, this holds for $0 \le x < 1$. We must show that

$$\lim_{x\to 1-} \sum_{k=1}^{\infty} \frac{(-1)^{k+1} x^k}{k} = \sum_{k=1}^{\infty} \frac{(-1)^{k+1}}{k}.$$

Write $f^{(k)}(x) := (-1)^{k+1} x^k$, $g^{(k)} := 1/k$. Then

$$\left| \sum_{k=1}^{m} f^{(k)}(x) \right| = |x - x^2 + x^3 - \cdots + (-1)^{m-1} x^m| \le 1 \text{ for all } m \text{ and } x \in [0,1].$$

By Dirichlet's test, $\sum_{k=1}^{\infty} g^{(k)} f^{(k)}(x)$ is uniformly convergent on [0,1] and hence $F(x) := \sum_{k=1}^{\infty} \frac{(-1)^{k+1} x^k}{k}$ is continuous on [0,1]. Since $\ln(1+x)$ is continuous and coincides with $F(x)$ on (0,1) we have $F(x)=\ln(1+x)$ on [0,1], in particular at $x=1$.

We have deferred until now a result on differentiation. This is because it is possible to prove one result rather easily via the Fundamental Theorem of the Calculus. The question we ask is if $f^{(m)}(x)$ converges uniformly to a function $f(x)$ and if each $f^{(m)}$ is differentiable, is f differentiable and is $f'(x) = \lim_{m\to\infty} f^{(m)\prime}(x)$? The answer is negative in general.

Example 6.39. Let $f^{(m)}(x)=\sin(m^2 x)/m$, $0\le x\le 1$. Then $f^{(m)}$ converges uniformly to the zero function but $f^{(m)\prime}(x)=m\cos(m^2 x)$ does not have a limit for any x.

Theorem 6.40. *Let $f^{(m)}$ be continuously differentiable on $[a,b]$ and suppose that $f^{(m)}(x)\to f(x)$ pointwise and $f^{(m)\prime}\to g$ uniformly on $[a,b]$. Then f is differentiable and $f'=g$.*

Proof. By Theorem 6.36 applied to the interval $[a,x]$

$$\int_a^x g = \lim_{m\to\infty} \int_a^x f^{(m)\prime} = \lim_{m\to\infty} [f^{(m)}(x) - f^{(m)}(a)] = f(x) - f(a).$$

Since g is continuous, the Fundamental theorem of the Calculus proves that $f' = g$.

Theorem 6.41. *Let $f^{(m)}\in C^1([a,b])$. If $\sum_{m=1}^{\infty} f^{(m)}(x)$ converges pointwise to $F(x)$ on $[a,b]$ and $\sum_{m=1}^{\infty} f^{(m)\prime}$ converges uniformly to G on $[a,b]$, then $F'(x) = G(x)$, that is, differentiation "term-by-term" is correct.*

Proof. A direct consequence of Theorem 6.40.

6.6. CHANGE OF VARIABLE IN A MULTIPLE INTEGRAL

The formula for changing variables for integrals in $\mathbb{R}$ was proved above as a consequence of the chain rule. It asserts that if $g : [\alpha,\beta] \to \mathbb{R}$ is of class C^1 and if f is continuous on $g([\alpha,\beta])$ then

$$\int_{g(\alpha)}^{g(\beta)} f = \int_{\alpha}^{\beta} (f \circ g)\, g' . \tag{6.16}$$

We intend to prove the following extension to multiple integrals.

Theorem 6.42. *Let Ω be a bounded domain in $\mathbb{R}^n$. Suppose $\mathbf{g} \in C^1(\Omega,\mathbb{R})$ is one-to-one on Ω and that $J_{\mathbf{g}}$ is never 0 on Ω. Let $f : \mathbf{g}(\Omega) \to \mathbb{R}$ be continuous. Then, for every Jordan measurable compact subdomain A of Ω, we have*

$$\int_{\mathbf{g}(A)} f = \int_A f \circ \mathbf{g}\, |J_{\mathbf{g}}| . \tag{6.17}$$

The appearance of the absolute value sign in (6.17) is due to the fact that integrals in $\mathbb{R}^n$ are not oriented. In (6.16) if $g' > 0$ on $[\alpha,\beta]$ then $g(\beta) > g(\alpha)$ whereas if $g' < 0$, $g(\beta) < g(\alpha)$. These two cases can be written as $\int_{g([\alpha,\beta])} f = \int_{\alpha}^{\beta} (f \circ g)\,|g'|$, where the integral on the left means $\int_a^b f$ where $[a,b] = g([\alpha,\beta])$.

The idea of the proof of Theorem 6.42 is as follows.

$\int_{\mathbf{g}(A)} f$ can be approximated by sums $\sum_{i=1}^{m} f(\mathbf{x}^{(i)})\, c(D_i)$ where $\{D_i\}$ is a subdivision of $\mathbf{g}(A)$. This sum may be written $\sum_{i=1}^{m} f(\mathbf{g}(\mathbf{y}^{(i)}))\, c(\mathbf{g}(A_i))$, where A_i is a subdivision of A. It will be shown that $c(\mathbf{g}(A_i))$ is very close to $|J_{\mathbf{g}}(\mathbf{y}^{(i)})|\, c(A_i)$ for suitably fine subdivisions so the latter sum is approximately $\sum_{i=1}^{m} f(\mathbf{g}(\mathbf{y}^{(i)}))\, |J_{\mathbf{g}}(\mathbf{y}^{(i)})|\, c(A_i)$ which approximates $\int_A (f \circ \mathbf{g})\,|J_{\mathbf{g}}|$.

The proof given here is based on that of J.Schwartz **[10]**; different types of proof are given by Apostol **[1]**, and Rudin **[9]**.

We first show that $\mathbf{g}$ and $\mathbf{g}^{-1}$ preserve Jordan measurable sets. This will be a consequence of the first lemma.

Lemma 6.43. *Let Ω be an open set in $\mathbb{R}^n$ and let $\mathbf{g} : \Omega \to \mathbb{R}^n$ be locally Lipschitz in Ω. Let X be a subset of Ω with $\mathrm{cl}(X) \subseteq \Omega$. Then, if X has content 0 so too does $\mathbf{g}(X)$.*

Proof. Since $c(X) = 0$, for an arbitrary $\varepsilon > 0$, there are finitely many closed n-cubes K_i which cover X and with $\sum c(K_i) < \varepsilon$. Since $\mathrm{cl}(X)$ is compact and disjoint from $\mathrm{bdy}(\Omega)$, ε can be taken so small that $K_0 = \cup K_i$ lies in Ω. Then K_0 is a compact subset of Ω so $\mathbf{g}$ satisfies a Lipschitz condition on K_0, say $\|\mathbf{g}(\mathbf{x}) - \mathbf{g}(\mathbf{y})\| \leq M\|\mathbf{x} - \mathbf{y}\|$. Let K be one of these n-cubes and let its centre be $\mathbf{a}$ and let the edge length be $2l$. Then, for $\mathbf{x} \in K$, $\|\mathbf{x} - \mathbf{a}\|_\infty \leq l$, where $\|\mathbf{y}\|_\infty = \max|y_i|$; see problem 1.4. Then, using problem 1.5,

$$\|\mathbf{g}(\mathbf{x}) - \mathbf{g}(\mathbf{a})\|_\infty \leq \|\mathbf{g}(\mathbf{x}) - \mathbf{g}(\mathbf{a})\| \leq M\|\mathbf{x} - \mathbf{a}\| \leq M\sqrt{n}\|\mathbf{x} - \mathbf{a}\|_\infty,$$

so that $\mathbf{g}(\mathbf{x})$ belongs to the n-cube centred at $\mathbf{g}(\mathbf{a})$ and with edge length $2M\sqrt{n}l$. Hence $c(\mathbf{g}(K)) \leq (2M\sqrt{n}l)^n = (M\sqrt{n})^n c(K)$. Adding the finitely many such terms proves that $c(\mathbf{g}(X)) \leq (M\sqrt{n})^n \varepsilon$. Since ε is arbitrary, the lemma is proved.

Corollary 6.44. *Let $V \subseteq \mathbb{R}^p$ be open and let $\mathbf{g} \in C^1(V, \mathbb{R}^n)$ where $p < n$. If A is a compact subset of V then $\mathbf{g}(A)$ has content 0.*

Proof. Let $\Omega := V \times \mathbb{R}^{n-p}$, an open subset of $\mathbb{R}^n$ and let

$$\mathbf{h}(x_1, \cdots, x_p, x_{p+1}, \cdots, x_n) = \mathbf{g}(x_1, \cdots, x_p).$$

By computing partial derivatives we see that $\mathbf{h}$ is of class C^1. Let $A_0 = A \times \{\mathbf{0}\}$, so that A_0 has content 0 in $\mathbb{R}^n$. Therefore $\mathbf{h}(A_0) = \mathbf{g}(A)$ has content 0 by the preceding lemma.

Proposition 6.45. *Let Ω be a bounded domain in R^n and let $\mathbf{g} \in C^1(\Omega, \mathbb{R}^n)$ satisfy $J_\mathbf{g} \neq 0$ on Ω and suppose $\mathbf{g}$ is one-to-one on Ω. Then, if X is a Jordan measurable subset of $\mathbf{g}(\Omega)$, $\mathbf{g}^{-1}(X)$ is also Jordan measurable.*

Proof. We have to show that $\mathrm{bdy}(\mathbf{g}^{-1}(X))$ has content 0. We assert that $\mathrm{bdy}(\mathbf{g}^{-1}(X)) \subseteq \mathbf{g}^{-1}(\mathrm{bdy}(X))$. Indeed, if $\mathrm{int}(X) \neq \emptyset$ then equality holds since $\mathbf{g}^{-1}$ is of class C^1 and one-to-one, using (the easy part of) Theorem 4.33. On the other hand, if $\mathrm{int}(X) = \emptyset$ and $\mathbf{y} \in \mathrm{bdy}(\mathbf{g}^{-1}(X))$ then there is a sequence $\{\mathbf{x}^{(m)}\} \subseteq X$ with $\mathbf{g}^{-1}(\mathbf{x}^{(m)}) \to \mathbf{y}$ and hence $\mathbf{x}^{(m)} \to \mathbf{g}(\mathbf{y})$. Thus, $\mathbf{g}(\mathbf{y}) \in \mathrm{cl}(X) = \mathrm{bdy}(X)$ (since $\mathrm{int}(X) = \emptyset$) and this proves that $\mathbf{y} \in \mathbf{g}^{-1}(\mathrm{bdy}(X))$. Since X is Jordan measurable, $c(\mathrm{bdy}(X)) = 0$ and the same holds for $\mathbf{g}^{-1}(\mathrm{bdy}(X))$ by Lemma 6.43.

Now we want to show that for small sets A_i in Ω, $c(\mathbf{g}(A_i))$ is well approximated by $|J_\mathbf{g}(\mathbf{y}^{(i)})|\, c(A_i)$, for some point $\mathbf{y}^{(i)} \in A_i$. The idea is that $\mathbf{g}$ is well approximated by the affine approximation determined by $D_{\mathbf{g}}$ so we first investigate the behaviour of the content of images of sets under linear operators.

Proposition 6.46. *Let* $\mathbf{L}:\mathbb{R}^n\to\mathbb{R}^n$ *be a linear operator and let* A *be a Jordan measurable set. Then* $c(\mathbf{L}(A)) = |\det \mathbf{L}|\, c(A)$.

Proof. We identify $\mathbf{L}$ with the matrix that represents it. If $\mathbf{L}$ is singular, that is, $\det \mathbf{L} = 0$, then $\mathbf{L}$ maps $\mathbb{R}^n$ into a proper linear subspace of $\mathbb{R}^n$. This subspace can also be obtained as the image of some linear operator $\mathbf{M}:\mathbb{R}^p\to\mathbb{R}^n$ with $p<n$. By Corollary 6.44, the content is then 0 and so the result holds for $\mathbf{L}$ singular.

Now suppose $\mathbf{L}$ is nonsingular. Then $\mathbf{L}(A)$ is Jordan measurable by Proposition 6.45. We use the well-known fact from linear algebra that every nonsingular matrix can be written as a product of elementary matrices, that is $\mathbf{L}$ is the composition of linear operators of the following three types :

$$\mathbf{L}_1(x_1,\cdots,x_n) = (kx_1, x_2,\cdots,x_n),\ k\neq 0,$$
$$\mathbf{L}_2(x_1,\cdots,x_n) = (x_1+x_2, x_2,\cdots,x_n),$$
$$\mathbf{L}_3(x_1,\cdots,x_i, x_{i+1},\cdots,x_n) = (x_1,\cdots,x_{i+1}, x_i,\cdots,x_n).$$

By properties of determinant, $\det \mathbf{L}_1 = k$, $\det \mathbf{L}_2 = 1$, $\det \mathbf{L}_3 = -1$.

First consider a rectangle $R = [a_1,b_1]\times[a_2,b_2]\times\cdots\times[a_n,b_n]$. Let $S = \mathbf{L}(R)$, where $\mathbf{L}$ is to be one of the three types. We wish to show that $c(\mathbf{L}(R)) = |\det \mathbf{L}|\, c(R)$ or $\int_{\mathbf{L}(R)} 1 = |\det \mathbf{L}| \int_R 1$, that is,

$\int_E \chi_S = |\det \mathbf{L}| \int_R \chi_R$, where E is an n-rectangle containing $S=\mathbf{L}(R)$ and χ_S is, as usual, the characteristic function of S : $\chi_S(\mathbf{x}) = 1$ for $\mathbf{x}\in S$, 0 for $\mathbf{x}\notin S$. We note that $\chi_S\circ\mathbf{L} = \chi_R$ since $\mathbf{L}$ is invertible, hence one-to-one.

Now $\mathbf{L}_1(R) = [ka_1,kb_1]\times[a_2,b_2]\times\cdots\times[a_n,b_n] := S$.

Then we have $c(R) = \int_R \chi_R = \int_R \chi_S\circ\mathbf{L}_1$

$$= \int_{a_n}^{b_n} dx_n \cdots \int_{a_2}^{b_2} dx_2 \int_{a_1}^{b_1} \chi_S(kx_1,x_2,\cdots,x_n)\, dx_1$$

$$= \int_{a_n}^{b_n} dx_n \cdots \int_{a_2}^{b_2} dx_2 \int_{ka_1}^{kb_1} \chi_S(t_1,x_2,\cdots,x_n)\frac{dt_1}{k},$$

by the change of variables for a single integral. Re-naming the dummy variable t_1 as x_1, gives $c(R) = \frac{1}{|k|}\int_S \chi_S = \frac{1}{|k|}c(S)$.

This proves the result for linear operators of type $\mathbf{L}_1$.

Now for $\mathbf{L}_2$ we have $S \subseteq E := [a_1+a_2, b_1+b_2] \times [a_2,b_2] \times \cdots \times [a_n,b_n]$ and

$$\int_R \chi_S \circ \mathbf{L}_2 = \int_{a_n}^{b_n} dx_n \cdots \int_{a_2}^{b_2} dx_2 \int_{a_1}^{b_1} \chi_S(x_1+x_2, x_2, \cdots, x_n)\, dx_1$$
$$= \int_{a_n}^{b_n} dx_n \cdots \int_{a_2}^{b_2} dx_2 \int_{a_1+x_2}^{b_1+x_2} \chi_S(x_1, x_2, \cdots, x_n)\, dx_1$$
$$= (b_n-a_n)\cdots(b_3-a_3)\int_{a_2}^{b_2}\left[\int_{a_1+x_2}^{b_1+x_2} 1\, dx_1\right] dx_2.$$

Now the iterated integral is calculated as follows,

$$\int_{a_2}^{b_2}\left[\int_{a_1+x_2}^{b_1+x_2} 1\, dx_1\right] dx_2 = \int_{a_2}^{b_2} [x_1]_{a_1+x_2}^{b_1+x_2} = (b_2-a_2)(b_1-a_1),$$

and so $c(R) = c(S)$.

Similarly, using the fact that an iterated integral over a rectangle can be written in any order the result follows for $\mathbf{L}_3$. Therefore, given any nonsingular $\mathbf{L}$, as $\mathbf{L}$ is the composition of operators of the above type and since the determinant of a product is the product of the determinants the result holds for $\mathbf{L}$.

Now we consider a Jordan measurable set A and let R be a rectangle containing A. For $\varepsilon > 0$ there is a partition P of R into subrectangles $R_1, \cdots, R_m$ such that

$$U(\chi_A, P) - c(A) < \varepsilon \quad \text{and} \quad c(A) - L(\chi_A, P) < \varepsilon.$$

Since $\chi_A = 1$ for points in A and 0 otherwise,

$U(\chi_A, P) = \sum_{\mathrm{I}} c(R_i)$ where I denotes these i for which $R_i \cap A \neq \emptyset$ and

$L(\chi_A, P) = \sum_{\mathrm{II}} c(R_i)$, where II denotes these i for which $R_i \subseteq A$. By the result for rectangles $c(\mathbf{L}(R_i)) = |\det \mathbf{L}|\, c(R_i)$.

By the additivity of content on non-overlapping Jordan measurable sets noting that $\cup_{\mathrm{II}} \mathbf{L}(R_i) \subseteq \mathbf{L}(A) \subseteq \cup_{\mathrm{I}} \mathbf{L}(R_i)$, we see that the Jordan measurable set $\mathbf{L}(A)$ lies between two sets of contents

$$|\det \mathbf{L}| \sum_{\mathrm{II}} c(R_i) \text{ and } |\det \mathbf{L}| \sum_{\mathrm{I}} c(R_i).$$

By the above selection these differ by at most

$$|\det \mathbf{L}|\, [U(\chi_A, P) - L(\chi_A, P)].$$

Hence, as ε is arbitrary, we have $c(\mathbf{L}(A)) = |\det \mathbf{L}|\, c(A)$.

Our next aim is to show that if A is a sufficiently small n-cube centred at $\mathbf{x}$ and $\mathbf{g}$ is a C^1 mapping as in the statement of Theorem 6.42, then $c(\mathbf{g}(A))$ is approximately equal to $|J_{\mathbf{g}}(\mathbf{x})|\, c(A)$. It is convenient to prove a preliminary lemma.

Lemma 6.47. *Let $K\subseteq\mathbb{R}^n$ be a closed n-cube with edge length 2a and centre $\mathbf{0}$. Suppose Ω is an open set containing K and $\mathbf{h}\in C^1(\Omega,\mathbb{R}^n)$. Suppose $\mathbf{h}$ is one-to-one on Ω, that $J_{\mathbf{h}}\neq 0$ on Ω, and that there is $\varepsilon>0$ with $\varepsilon<1/\sqrt{n}$ such that $\|\mathbf{h}(\mathbf{x})-\mathbf{x}\|<\varepsilon\|\mathbf{x}\|$ for all $\mathbf{x}\in K$.* Then

$$[1-\varepsilon\sqrt{n}]^n \le c(\mathbf{h}(K))\,/\,c(K) \le [1+\varepsilon\sqrt{n}]^n.$$

Proof We note that $\|\mathbf{h}(\mathbf{x})-\mathbf{x}\|_\infty<\delta\|\mathbf{x}\|_\infty$ for all $x\in K$, where $\delta=\varepsilon\sqrt{n}$. For $\mathbf{x}\in\mathrm{bdy}(K)$, $\|\mathbf{h}(\mathbf{x})-\mathbf{x}\|_\infty<\delta a$, and therefore $(1-\delta)a<\|\mathbf{h}(\mathbf{x})\|_\infty<(1+\delta)a$, by the triangle inequality for $\|\cdot\|_\infty$. Thus, $\mathbf{h}(\mathrm{bdy}(K))=\mathrm{bdy}(\mathbf{h}(K))$ does not intersect the boundary of the n-cube R with edges of length $2(1+\delta)a$. Since $\mathbf{h}(K)$ is connected it must either lie entirely within R or entirely outside R. Since $\mathbf{h}(\mathbf{0})$ lies in R so does $\mathbf{h}(K)$. Similarly, the smaller n-cube of edge length $2(1-\delta)a$ lies inside $\mathbf{h}(K)$. Therefore the Jordan measurable set $\mathbf{h}(K)$ has content between the content of these two n-cubes, that is, $(1-\delta)^n \le c(\mathbf{h}(K))/c(K) \le (1+\delta)^n$.

Proposition 6.48. *Let Ω be a bounded domain in $\mathbb{R}^n$ and let $\mathbf{g}\in C^1(\Omega,\mathbb{R}^n)$ be one-to-one on Ω and satisfy $J_{\mathbf{g}}\neq 0$ on Ω. Suppose that A is a Jordan measurable set with $\mathrm{cl}(A)\subseteq\Omega$. For $\varepsilon>0$ there exists $\delta>0$ such that for any closed n-cube K with centre $\mathbf{x}\in A$ and edge length less than 2δ,*

$$|J_{\mathbf{g}}(\mathbf{x})|(1-\varepsilon)^n \le c(\mathbf{g}(K))\,/\,c(K) \le |J_{\mathbf{g}}(\mathbf{x})|(1+\varepsilon)^n.$$

Proof. Fix $\mathbf{x}\in A$ and write $\mathbf{g}(\mathbf{y})=\mathbf{g}(\mathbf{x})+D\mathbf{g}(\mathbf{x})(\mathbf{y}-\mathbf{x})+\mathbf{R}(\mathbf{y},\mathbf{x})$. Since $J_{\mathbf{g}}(\mathbf{x})\neq 0$, $D\mathbf{g}(\mathbf{x})$ is invertible. Let $\mathbf{M}=[D\mathbf{g}(\mathbf{x})]^{-1}$. Then we can write $\mathbf{M}\,[\mathbf{g}(\mathbf{y})-\mathbf{g}(\mathbf{x})]-(\mathbf{y}-\mathbf{x})=\mathbf{M}\mathbf{R}(\mathbf{y},\mathbf{x})$. Let $\delta(\varepsilon)$ be such that
$\|\mathbf{M}\,[\mathbf{g}(\mathbf{y})-\mathbf{g}(\mathbf{x})]-(\mathbf{y}-\mathbf{x})\|\le\|\mathbf{M}\|\,\|\mathbf{R}(\mathbf{y},\mathbf{x})\|<\varepsilon\|\mathbf{y}-\mathbf{x}\|\,/\,\sqrt{n}$, for $\|\mathbf{y}-\mathbf{x}\|<\delta$.
Write $\mathbf{z}=\mathbf{y}-\mathbf{x}$ and $\mathbf{h}(\mathbf{z})=\mathbf{M}\,[\mathbf{g}(\mathbf{x}+\mathbf{z})-\mathbf{g}(\mathbf{x})]$. Then

$$\|\mathbf{h}(\mathbf{z})-\mathbf{z}\|\le\varepsilon\|\mathbf{z}\|\,/\,\sqrt{n},\ \text{for } \|\mathbf{z}\|<\delta.$$

Therefore if K_0 is the n-rectangle centred at $\mathbf{0}$ and with edge length less that $2\delta/\sqrt{n}$, K_0 lies in $B(\mathbf{0},\delta)$ and applying Lemma 6.47 yields

$$(1-\varepsilon)^n c(K_0)/\le c(\mathbf{h}(K_0)) \le (1+\varepsilon)^n c(K_0).$$

Note that the same ε can be used for any smaller δ too.
But, by definition of $\mathbf{h}$,

$$c(\mathbf{h}(K_0)) = |\det \mathbf{M}|\ c[\mathbf{g}(K_0+\mathbf{x}) - \mathbf{g}(\mathbf{x})] = |\det \mathbf{M}|\ c(\mathbf{g}(K)),$$

where $K = K_0+\mathbf{x}$, is the corresponding n-cube centred at $\mathbf{x}$, and we observe that content is invariant under translation. Since

$$\det \mathbf{M} = \det[D_{\mathbf{g}}(\mathbf{x})]^{-1} = 1 / \det D_{\mathbf{g}}(\mathbf{x}) = 1 / J_{\mathbf{g}}(\mathbf{x}),$$

by properties of determinant, the proof is complete.

Lemma 6.49. *Let* $D = \bigcup_{i=1}^{m} D_i$ *where* D_i *are non-overlapping Jordan measurable sets and let* h *be uniformly continuous on* D. *Then, for any* $\varepsilon > 0$ *there exists* $\delta(\varepsilon) > 0$ *such that if* $\operatorname{diam}(D_i) < \delta$ *we have*

$$\left| \int_D h - \sum_{i=1}^{m} h(\mathbf{x}^{(i)})\, c(D_i) \right| < \varepsilon\, c(D), \textit{ for any choice of points } \mathbf{x}^{(i)} \in D_i .$$

Proof. Since h is uniformly continuous on D, given $\varepsilon > 0$ there is $\delta > 0$ such that $|h(\mathbf{x}) - h(\mathbf{y})| < \varepsilon$ whenever $\mathbf{x},\mathbf{y} \in D$ and $\|\mathbf{x} - \mathbf{y}\| < \delta$.

Now $\int_D h = \sum_{i=1}^{m} \int_{D_i} h$, by additivity of the integral. For $\mathbf{x}^{(i)} \in D_i$,

$$\left| \int_D h - \sum_{i=1}^{m} h(\mathbf{x}^{(i)})\, c(D_i) \right| = \left| \sum_{i=1}^{m} \int_{D_i} [h(\mathbf{x}) - h(\mathbf{x}^{(i)})] \right|,$$

since $h(\mathbf{x}^{(i)})$ is a constant. Therefore,

$$\left| \int_D h - \sum_{i=1}^{m} h(\mathbf{x}^{(i)})\, c(D_i) \right| \le \sum_{i=1}^{m} \int_{D_i} |h(\mathbf{x}) - h(\mathbf{x}^{(i)})| < \sum_{i=1}^{m} \int_{D_i} \varepsilon = \varepsilon c(D).$$

After these long preparations we are ready to prove the change of variables theorem.

Proof of Theorem 6.42. We first suppose that $f(\mathbf{x}) \ge 0$ at all points. Let R be an n-rectangle containing A. Fix an arbitrary ε with $0 < \varepsilon < 1$. Let P be a partition of R into n-cubes $\{K_i\}$ of edge length $2a$. Let $\{K_1, \cdots, K_p\}$ denote those n-cubes that intersect A and write $K_0 = \bigcup_{i=1}^{p} K_i$. Now K_0 is a compact set and we can suppose a is so small that $K_0 \subseteq \Omega$. Then $\mathbf{g}$ is of class C^1 so satisfies a Lipschitz condition on K_0, say $\|\mathbf{g}(\mathbf{u}) - \mathbf{g}(\mathbf{v})\| \le M \|\mathbf{u} - \mathbf{v}\|$, where we can always take $M \ge 1$. In particular, sets of diameter d in K_0 are mapped into sets of diameter at most Md in $\mathbf{g}(K_0)$. Let K be the union of the sets K_i which lie entirely within A, say $K = \bigcup_{i=1}^{m} K_i$. Since A is Jordan measurable we can suppose that a is chosen so small that

$$c(A) \le \bigcup_{i=1}^{m} c(K_i) + \varepsilon \quad \text{and} \quad \sum_{i=m+1}^{p} c(K_i) < \varepsilon.$$

We also suppose a is so small that $M\,\mathrm{diam}(K_i) < \delta$, δ as in Lemma 6.49 and as in Proposition 6.48. By Lemma 6.49 applied to $f\circ\mathbf{g}\,|J_{\mathbf{g}}|$,

$$\left| \int_K f\circ\mathbf{g}\,|J_{\mathbf{g}}| - \sum_{i=1}^{m} (f\circ\mathbf{g})(\mathbf{u}^{(i)})\,|J_{\mathbf{g}}(\mathbf{u}^{(i)})|\; c(K_i) \right| < \varepsilon c(K), \tag{6.18}$$

where we choose $\mathbf{u}^{(i)}$ to be the centre of K_i. Similarly, since $\mathbf{g}(K_i)$ are Jordan measurable non-overlapping sets we have

$$\left| \int_{\mathbf{g}(K)} f - \sum_{i=1}^{m} f(\mathbf{x}^{(i)}) c(\mathbf{g}(K_i)) \right| < \varepsilon c(\mathbf{g}(K)), \tag{6.19}$$

where we choose $\mathbf{x}^{(i)} = \mathbf{g}(\mathbf{u}^{(i)})$. By Proposition 6.48,

$$|J_{\mathbf{g}}(\mathbf{u}^{(i)})|\; c(K_i)\,(1-\varepsilon)^n \le c(\mathbf{g}(K_i)) \le |J_{\mathbf{g}}(\mathbf{u}^{(i)})|\; c(K_i)\,(1+\varepsilon)^n,$$

and because $f(\mathbf{x}) \ge 0$, we have

$$\sum_{i=1}^{m} f(\mathbf{g}(\mathbf{u}^{(i)}))\,|J_{\mathbf{g}}(\mathbf{u}^{(i)})|\; c(K_i)\,(1-\varepsilon)^n \le \sum_{i=1}^{m} f(\mathbf{x}^{(i)})\, c(\mathbf{g}(K_i)) \le$$

$$\le \sum_{i=1}^{m} f(\mathbf{g}(\mathbf{u}^{(i)}))\,|J_{\mathbf{g}}(\mathbf{u}^{(i)})|\; c(K_i)\,(1+\varepsilon)^n. \tag{6.20}$$

Combining (6.18), (6.19) and (6.20) and noting that $(1+\varepsilon)^n \le 1+2^n\varepsilon$

$$\left| \int_K f\circ\mathbf{g}\,|J_{\mathbf{g}}| - \int_{\mathbf{g}(K)} f \right| \le M_1\varepsilon \quad \text{for a constant } M_1.$$

Since $c(A \setminus K) = c(A) - c(K) < \varepsilon$ we also have

$$\left| \int_A (f\circ\mathbf{g})\,|J_{\mathbf{g}}| - \int_K (f\circ\mathbf{g})\,|J_{\mathbf{g}}| \right| = \left| \int_{A\setminus K} (f\circ\mathbf{g})\,|J_{\mathbf{g}}| \right|$$

$$\le M_2 c(A \setminus K) < M_2\varepsilon,$$

where M_2 is a bound for the bounded function $(f\circ\mathbf{g})\,|J_{\mathbf{g}}|$.

Similarly, since by Lemma 6.47, $c(\mathbf{g}(A \setminus K)) \le (M_3\sqrt{n})^n\varepsilon$, we have

$$\left| \int_{\mathbf{g}(A)} f - \int_{\mathbf{g}(K)} f \right| \le M_4\varepsilon.$$

Combining these results proves that

$$\left| \int_A f\circ\mathbf{g}\,|J_{\mathbf{g}}| - \int_{\mathbf{g}(A)} f \right| \le M_5\varepsilon,$$

and, as ε is arbitrary, we have equality. To remove the assumption that $f(\mathbf{x}) \geq 0$ note that since f is bounded we can add a positive constant α so that $f_\alpha(\mathbf{x}) := f(\mathbf{x}) + \alpha \geq 0$. The theorem applies to f_α and also to the constant function α. By additivity of the integral this gives the result for $f_\alpha - \alpha$.

For applications of this result it is important to allow a weakening of the hypotheses. Since sets of content 0 are negligible as far as Riemann integration is concerned it is possible to weaken the hypotheses by allowing $J_{\mathbf{g}}$ to be 0 on a set of content 0. We shall not prove this version but we see below examples where its use is needed.

The application of the change of variables formula for multiple integrals is very different from that of single integrals. In the latter case the aim is to find a transformation ϕ that makes the new integrand $(f \circ \phi)\phi'$ recognizable as easily integrable. The integration is again over an interval. But for multiple integrals, the domain of integration is all important, and the aim is to transform to a very simple domain such as a rectangle or an elementary region.

The next example illustrates this. The change of variable is chosen so that the integration is taken over a rectangle. Without changing variables the given region would have to be divided into several elementary regions and the computations would be harder.

Example 6.50. By making a suitable change of variables evaluate $\iint_D xy\,dx\,dy$, where D is the region in the first quadrant bounded by the curves $xy = 1$, $xy = 2$, $y = x$, $y = 4x$.

Solution. Setting $u = xy$, $v = y/x$ we see that D is the image of the rectangle R bounded by $u = 1$, $u = 2$, $v = 1$, $v = 4$.
Now $\iint_D xy\,dx\,dy = \iint_R u\,|J|\,du\,dv$ where $J = \dfrac{\partial(x,y)}{\partial(u,v)}$. It is easier to calculate $J^{-1} = \dfrac{\partial(u,v)}{\partial(x,y)} = \begin{vmatrix} \dfrac{\partial u}{\partial x} & \dfrac{\partial u}{\partial y} \\ \dfrac{\partial v}{\partial x} & \dfrac{\partial v}{\partial y} \end{vmatrix} = \begin{vmatrix} y & x \\ -y/x^2 & 1/x \end{vmatrix} = 2y/x = 2v$, which is positive in R. Therefore $\iint_D xy\,dx\,dy = \iint_R u\,(1/2v)\,du\,dv$

$$= \frac{1}{2}\int_0^1 u\,du \int_1^4 (1/v)\,dv = \frac{1}{4}\ln(4).$$

We now examine two important changes of variable, the use of polar coordinates in $\mathbb{R}^2$ and spherical polar coordinates in $\mathbb{R}^3$. Given a point $(x,y) \in \mathbb{R}^2$ we may write $x = r\cos\theta$, $y = r\sin\theta$ where $r^2 = x^2 + y^2$ and $\tan\theta = y/x$. Given x and y, θ is only determined modulo multiples of 2π so we must restrict attention to $r \geq 0$ and $0 \leq \theta \leq 2\pi$ (or $-\pi \leq \theta \leq \pi$). For $(x,y) = (0,0)$, θ cannot be determined from these equations. If we consider $x = r\cos\theta$, $y = r\sin\theta$ as a function $\mathbf{g}$ from the (r,θ)-plane to the (x,y)-plane we see that

$$J_{\mathbf{g}} = \frac{\partial(x,y)}{\partial(r,\theta)} = \begin{vmatrix} \partial x/\partial r & \partial x/\partial\theta \\ \partial y/\partial r & \partial y/\partial\theta \end{vmatrix} = \begin{vmatrix} \cos\theta & -r\sin\theta \\ \sin\theta & r\cos\theta \end{vmatrix} = r$$

and this vanishes when $r = 0$. Therefore, the version of the change of variables theorem proved above is inapplicable on a region containing the origin. For such a region we need the stronger form that was intimated after the proof.

Suppose we wish to calculate $\iint_D f(x,y)\,dx\,dy$ when D is the disk $\{x^2 + y^2 \leq a^2\}$. This is the image under the polar coordinate map $\mathbf{g}$ of the rectangle $R = \{(r,\theta) : 0 \leq \theta < 2\pi, 0 \leq r \leq a\}$.

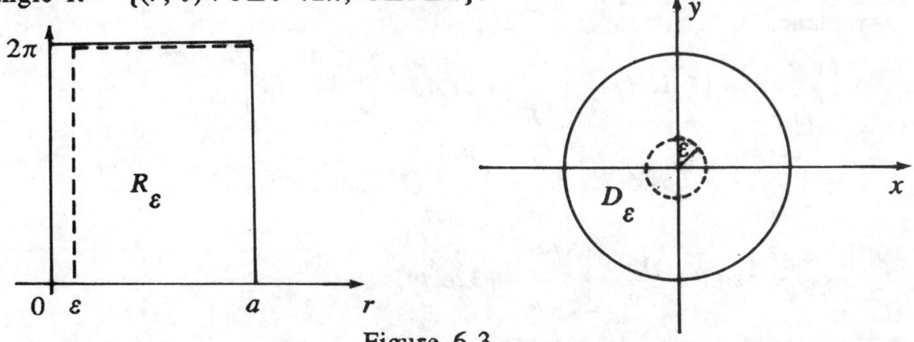

Figure 6.3

$\mathbf{g}$ is not one-to-one on R since all of the boundary $0 \leq \theta < 2\pi$, $r = 0$ is mapped onto the origin. However this segment is of content 0 and so the stronger version of the theorem is applicable. One could also give a direct argument in this case by considering the rectangle $R_\varepsilon := \{\varepsilon \leq r \leq a, 0 \leq \theta < 2\pi\}$ whose image is the annulus $D_\varepsilon := \{\varepsilon^2 \leq x^2 + y^2 \leq a^2\}$. Then $\iint_{D_\varepsilon} f\,dx\,dy = \iint_{R_\varepsilon} (f \circ g)\, r\,dr\,d\theta$ and the integrals over $D \backslash D_\varepsilon$ and $R \backslash R_\varepsilon$ can be made arbitrarily small, since the function $rf(r\cos\theta, r\sin\theta)$ is clearly integrable over all of R and these sets have content less than $C\varepsilon$ (C a constant). Letting $\varepsilon \to 0$ proves the result.

Similar arguments apply to the change from (x,y,z) to spherical polar coordinates (r,θ,ϕ) where $x = r\sin\theta\cos\phi$, $y = r\sin\theta\sin\phi$, $z = r\cos\theta$, $0 \leq r < \infty$, $0 \leq \theta \leq \pi$, $0 \leq \phi < 2\pi$.

We find

$$J = \frac{\partial(x,y,z)}{\partial(r,\theta,\phi)} = \begin{vmatrix} \sin\theta\cos\phi & r\cos\theta\cos\phi & -r\sin\theta\cos\phi \\ \sin\theta\cos\phi & r\cos\theta\cos\phi & -r\sin\theta\cos\phi \\ \cos\theta & -r\sin\theta & 0 \end{vmatrix} = r^2\sin\theta$$

which again vanishes when $r = 0$ and also for $\theta = 0$ and π. Either the direct type of argument outlined above or the stronger form of the change of variables theorem shows the validity of the method.

We give one example of each change of variables here. Others may be found in the examples in Chapter 7 and the problems provide some practice for the reader.

Example 6.51. Evaluate $\iint_D \sqrt{x^2+y^2}\,dxdy$ where D is interior of the circle $x^2+y^2 = 2ax$ $(a>0)$.

Solution. D is described by the polar equation $r^2 = 2ar\cos\theta$ and we can obtain the new limits of integration either by sketching the region in the $r-\theta$ plane or by considering rays $\theta =$ constant in the $x-y$ plane.

$$\iint_D \sqrt{x^2+y^2}\,dx\,dy = \iint_R r\,r\,dr\,d\theta = \int_{-\pi/2}^{\pi/2} d\theta \int_0^{2a\cos\theta} r^2\,dr$$

$$= \int_{-\pi/2}^{\pi/2} \frac{8a^3}{3}\cos^3\theta\,d\theta = \frac{8a^3}{3}\int_{-\pi/2}^{\pi/2} (1-\sin^2\theta)\cos\theta\,d\theta$$

$$= \frac{8a^3}{3}\left[\sin\theta - \frac{\sin^3\theta}{3}\right]_{-\pi/2}^{\pi/2} = 32a^3/9.$$

Example 6.52. Evaluate $\iiint_V z^2\,dx\,dy\,dz$ where V is the interior of the sphere $x^2+y^2+z^2 = 1$.

Solution. Using spherical polar coordinates we have

$$\iiint_V z^2\,dx\,dy\,dz = \int_0^{2\pi} d\phi \int_0^{\pi} d\theta \int_0^1 (r^2\cos^2\theta)\,r^2\sin\theta\,dr = 2\pi\,\frac{2}{3}\,\frac{1}{5} = \frac{4\pi}{15}.$$

Problems.

1. Let $f(x) = \begin{cases} 0, & 0\le x<1 \\ 1, & 1\le x\le 2 \end{cases}$. Show, using upper and lower sums, that f is Riemann integrable over $[0,2]$. Show also that f is not the derivative

of any function.

2. Use Theorem 6.18 to prove that if f,g are integrable on $[a,b]$ then so are $|f|$, fg and f^2.

3. Let $F(x) = \begin{cases} x^2\sin(1/x^2), & 0 \le x < 1 \\ 0, & x = 0 \end{cases}$. Show that F' exists at all points of $[0,1]$ and is continuous except at 0 but F' is not integrable on $[0,1]$. Why does this not contradict Theorem 6.18 ?

4. Prove that if f is integrable over $[0,1]$ then

$$\lim_{n\to\infty} \frac{1}{n} \sum_{k=0}^{n} f(k/n) = \int_0^1 f.$$

5. Let f be integrable over an n-rectangle E (of positive content) and suppose that $f(\mathbf{x}) > 0$ for all $\mathbf{x} \in E$. Show that $\int_E f > 0$.

6. Show that (*a*) $\int_0^1 dx \int_0^1 \frac{(x-y)}{(x+y)^3}\, dy = \frac{1}{2}$, and

(*b*) $\int_0^1 dy \int_0^1 \frac{(x-y)}{(x+y)^3}\, dx = -\frac{1}{2}$. Why do the answers differ?

7. Prove that the set of discontinuities of the characteristic function χ_A of a set A is bdy(A).

8. Show that if $\{\mathbf{x}^{(m)}\}$ is a convergent sequence, then its content satisfies $c\{\mathbf{x}^{(m)}\} = 0$. Give an example of a sequence which does not have zero content.

9. Show that if A,B are Jordan measurable sets then so are

$A \cup B$, $A \cap B$ and $A \setminus B$.

10. Prove that, if f is integrable over a Jordan measurable set A and $k \in \mathbb{R}$, then kf is integrable over A and $\int_A kf = k\int_A f$.

11. Let $\alpha : [a,b] \to \mathbb{R}$ be continuous and satisfy $\alpha(x) > 0$ on $[a,b]$. Let $A := \{ (x,y) \in \mathbb{R}^2 : a \le x \le b,\ 0 \le y \le \alpha(x) \}$. Show that A is Jordan measurable and that $c(A) = \int_a^b \left[\int_0^{\alpha(x)} 1\, dy \right] dx = \int_a^b \alpha(x)\, dx$.

12. Evaluate the integral of xy over the region in the first quadrant bounded by the parabolas $y^2=x$ and $x^2=y$.

13. Show that $\iint e^{x-y}\,dx\,dy=\cosh(k)-1$, where the region of integration is the triangular region bounded by the lines $x=0$, $y=0$, $x+y=k$, where $k>0$.

14. Show that $\displaystyle\int_0^4 dx \int_{\sqrt{x}}^2 \frac{dy}{(x+y)^2} = 4(\sqrt{2}-1)$.

15. Show that $\iint xy^2\,dx\,dy = \frac{\pi}{8} - \frac{1}{5}$, where the area of integration is that area containing the point (1/2,0) and bounded by the circle $(x-1)^2+y^2=1$ and the hyperbola $2x^2-y^2=1$.

16. Evaluate $\displaystyle\int_0^{\pi/2}\int_x^{\pi/2} \frac{\sin y}{x+y}\,dy$.

17. Evaluate $\displaystyle\iint_A \frac{1}{y^2+4ax}\,dx\,dy$ $(a>0)$, where the field of integration is the region bounded by the y-axis, the line $y=1$ and the parabola $y^2=4ax$.

18. Let $A=\{(x,y): 0\le x\le 1,\ 0\le y\le 1\}$. Integrate $\iint_A x^y\,dx\,dy$ in 2 ways and hence show that $\displaystyle\int_0^1 \frac{x-1}{\ln(x)}\,dx = \ln(2)$.

19. Prove that $\iiint e^{x+2y+3z}\,dx\,dy\,dz = \frac{1}{6}(e-1)^3$, where integration is throughout the region bounded by the planes $x=0$, $y=0$, $z=0$ and $x+y+z=1$.

20. Let $f^{(m)}: [0,1]\to\mathbb{R}$ be continuous on [0,1] and suppose that $f^{(m)}\to f$ uniformly on [0,1]. Show that

$$\lim_{m\to\infty}\int_0^{1-1/m} f^{(m)}(x)\,dx = \int_0^1 f(x)\,dx$$

21. Suppose that $\{a^{(m)}\}$ is a sequence of real numbers such that either $\sum ma^{(m)}$ is absolutely convergent or $ma^{(m)}$ decreases to 0. Let $C(x):=\sum a^{(m)}\cos(mx)$. Show that, under either hypothesis, $C'(x)=-\sum ma^{(m)}\sin(mx)$ for $x\in(0,2\pi)$.

22. By making a suitable change of variable show that the area bounded by the circles $x^2+y^2 = 2x$, $x^2+y^2 = 4x$ and the lines $y = x$, $y = 0$ is $3(\pi+2)/4$.

23. Show, by using the substitution $u = y^2/x$, $v = x^2/y$, that $\iint_A xy\,dx\,dy = \frac{5}{12}$, where the field of integration is the area bounded by the loop of the curve $x^3+y^3 = xy$ and the arcs of the parabolas $y^2 = x$, $x^2 = y$ joining the origin to the point (1,1).

24. Integrate the function z^4 over the ball in $\mathbb{R}^3$ with centre the origin and radius a.

25. Show that $\iiint z\,dx\,dy\,dz = \dfrac{9\pi a^4}{8}$, where integration is throughout the volume inside the sphere $x^2+y^2+z^2 = 2az$ $(a>0)$ but outside the sphere $x^2+y^2+z^2 = a^2$.

26. Integrate $(x^2+y^2+z^2)^{-3/2}$ over the region bounded by the spheres $x^2+y^2+z^2 = a^2$ and $x^2+y^2+z^2 = b^2$, where $a>b>0$.

27. Evaluate $\int_V r\,dV$ where $r^2 = x^2+y^2+z^2$ and V is the region bounded by the cone $x^2+y^2 = z^2$ and the plane $z = 3$.

Hints on the problems

1. $P_\varepsilon=\{0,1-\varepsilon,1+\varepsilon,2\}$. Last part see problem 3.29 (Darboux' theorem).

4. Consider the partition P_n consisting of $n+1$ equally spaced points. For $\varepsilon>0$, there is P_ε such that $U(f,P_\varepsilon)-\varepsilon\leq$ Integral $\leq U(f,P_\varepsilon)$. Consider the partition P' consisting of all points of $P_n \cup P_\varepsilon$. Estimate $U(f,P')$.

5. E has positive content and the set of discontinuities of f has content 0 so there must be a point of E at which f is continuous.

9. Show $\text{bdy}(A\cup B)\subseteq\text{bdy}(A)\cup\text{bdy}(B)$. All follows from this.

10. Replace by an n-rectangle by considering $f\chi_A$.

21. Theorem 6.41 (see also Example 2.58).

22. Polars.

Chapter 7.

Vector Calculus

7.1. VECTORS

In this chapter we study particularly the space $\mathbb{R}^3$ and functions whose domains and image sets are subsets of $\mathbb{R}^3$. Elements of $\mathbb{R}^3$ are to be thought of as *vectors*, objects possessing magnitudes and directions. Vectors are often written as columns but we shall write them as before, as a row, along the line.

Consider a rectangular Cartesian coordinate system with origin O. The coordinates of a generic point may be written (x_1, x_2, x_3) or (x,y,z). Given a point X with coordinates (x,y,z) we call $\mathbf{x} := (x,y,z)$ the *position vector* of X. We think of $\mathbf{x}$ as the *directed line segment* from O to X, written $\overrightarrow{OX}$, that is, the set of points $\{t\mathbf{x} : 0 \le t \le 1\}$, taken in the order of increasing t. This is interpreted geometrically by drawing the line segment from O to X and adding an arrow to indicate the direction.

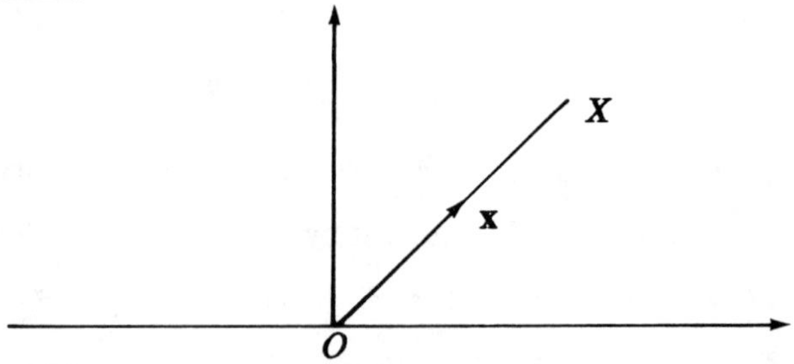

Figure 7.1.

If P,Q are two points with associated position vectors $\mathbf{p},\mathbf{q}$, the directed line segment $\overrightarrow{PQ}$ is similarly defined via the set of points $\{(1-t)\mathbf{p}+t\mathbf{q} : 0 \le t \le 1\}$.

Given a vector, that is, an element $\mathbf{a}$ of $\mathbb{R}^3$, one can identify $\mathbf{a}$ with the directed line segment $\overrightarrow{OA}$, where A is the point with position vector $\mathbf{a}$. However, it is sometimes convenient to identify $\mathbf{a}$ with a line segment $\overrightarrow{PQ}$ where P is any point of $\mathbb{R}^3$ and $\mathbf{q} = \mathbf{p}+\mathbf{a}$. $\overrightarrow{OA}$ and $\overrightarrow{PQ}$ are then parallel line segments with equal lengths, their lengths being $\|\mathbf{a}\|$, usually called the *magnitude* of $\mathbf{a}$. One reason for this identification is the rule for addition. If **a**,**b** are two vectors, $\mathbf{a}+\mathbf{b} = (a_1+b_1, a_2+b_2, a_3+b_3)$. The corresponding line segment is obtained by writing $\mathbf{a} = \overrightarrow{OA}$, $\mathbf{b} = \overrightarrow{AC}$, then $\mathbf{a}+\mathbf{b} = \overrightarrow{OC}$. This is the "nose-to-tail" rule for adding vectors. If we represent **b** as $\overrightarrow{OB}$ then **a**+**b** is the diagonal of a parallelogram determined by O, A and B.

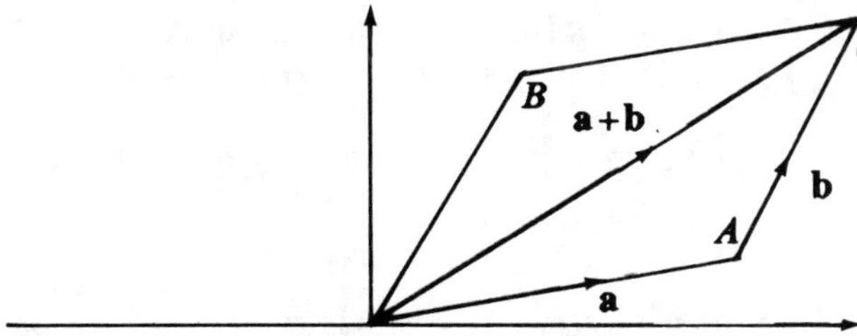

Figure 7.2.

The standard basis of $\mathbb{R}^3$ is {(1,0,0), (0,1,0), (0,0,1)}. We denote the corresponding vectors by **i**, **j**, **k**, respectively. Thus, for example, $\mathbf{i} = \overrightarrow{OE_1}$ where E_1 has coordinates (1,0,0). Then any vector **a** can be written uniquely $\mathbf{a} = a_1\mathbf{i}+a_2\mathbf{j}+a_3\mathbf{k}$; **i**, **j**, **k** are called the *standard basis vectors* and correspond to the basis $\mathbf{e}_1, \mathbf{e}_2, \mathbf{e}_3$.

When dealing with elements of $\mathbb{R}^3$, thought of as vectors, the inner-product is termed either the *scalar product* or the *dot product* and we write

$$\mathbf{a}\bullet\mathbf{b} = a_1b_1+a_2b_2+a_3b_3. \tag{7.1}$$

Thus $\mathbf{a}\bullet\mathbf{b}$ and $<\mathbf{a},\mathbf{b}>$ mean exactly the same thing. As shown in Chapter 1, $\mathbf{a}\bullet\mathbf{b} = \|\mathbf{a}\|\,\|\mathbf{b}\|\cos\theta$ where θ is the angle between **a** and **b**. The angle is well defined since OAB lies in a plane but this equality can be taken as the definition of the angle.

The important concept that distinguishes $\mathbb{R}^3$ from Euclidean spaces of other dimensions is that of the *cross product* or *vector product.*

Definition 7.1. Let **a**,**b** be two vectors. The cross product of **a** and **b**, denoted $\mathbf{a}\times\mathbf{b}$, is the vector $\mathbf{a}\times\mathbf{b} := (a_2b_3-a_3b_2, a_3b_1-a_1b_3, a_1b_2-a_2b_1)$.

This formula is more easily remembered as follows. Recall that a 2×2 determinant $\begin{vmatrix} a & b \\ c & d \end{vmatrix}$ is, by definition, $ad-bc$. Therefore, each component of $\mathbf{a}\times\mathbf{b}$ can be written as a 2×2 determinant. Recalling how a 3×3 determinant can be expanded by the first row, we see that we can write the formal expression

$$\begin{vmatrix} \mathbf{i} & \mathbf{j} & \mathbf{k} \\ a_1 & a_2 & a_3 \\ b_1 & b_2 & b_3 \end{vmatrix} \quad \text{to denote} \quad \mathbf{i}\begin{vmatrix} a_2 & a_3 \\ b_2 & b_3 \end{vmatrix} - \mathbf{j}\begin{vmatrix} a_1 & a_3 \\ b_1 & b_3 \end{vmatrix} + \mathbf{k}\begin{vmatrix} a_1 & a_2 \\ b_1 & b_2 \end{vmatrix}$$

and this latter is $\mathbf{a}\times\mathbf{b}$. We therefore use this formal 3×3 determinant to denote $\mathbf{a}\times\mathbf{b}$.

Note that $\mathbf{b}\times\mathbf{a} = -(\mathbf{a}\times\mathbf{b})$, so order is important. The unit vectors $\mathbf{i}, \mathbf{j}, \mathbf{k}$ satisfy $\mathbf{i}\times\mathbf{j} = \mathbf{k}$, $\mathbf{j}\times\mathbf{k} = \mathbf{i}$, $\mathbf{k}\times\mathbf{i} = \mathbf{j}$. For example,

$$\mathbf{j}\times\mathbf{k} = \begin{vmatrix} \mathbf{i} & \mathbf{j} & \mathbf{k} \\ 0 & 1 & 0 \\ 0 & 0 & 1 \end{vmatrix} = \mathbf{i}\cdot1 - \mathbf{j}\cdot0 + \mathbf{k}\cdot0 = \mathbf{i}.$$

If $\mathbf{a},\mathbf{b},\mathbf{c}$ are three vectors, the *scalar triple product* $[\mathbf{c},\mathbf{a},\mathbf{b}]$, in that order, is defined to be $[\mathbf{c},\mathbf{a},\mathbf{b}] = \mathbf{c}\bullet(\mathbf{a}\times\mathbf{b})$, so that

$$[\mathbf{c},\mathbf{a},\mathbf{b}] = c_1\begin{vmatrix} a_2 & a_3 \\ b_2 & b_3 \end{vmatrix} - c_2\begin{vmatrix} a_1 & a_3 \\ b_1 & b_3 \end{vmatrix} + c_3\begin{vmatrix} a_1 & a_2 \\ b_1 & b_2 \end{vmatrix} = \begin{vmatrix} c_1 & c_2 & c_3 \\ a_1 & a_2 & a_3 \\ b_1 & b_2 & b_3 \end{vmatrix}.$$

Since interchanging 2 rows of a determinant changes the sign, we have

$$[\mathbf{c},\mathbf{a},\mathbf{b}] = [\mathbf{a},\mathbf{b},\mathbf{c}] = [\mathbf{b},\mathbf{c},\mathbf{a}] = -[\mathbf{c},\mathbf{b},\mathbf{a}] = -[\mathbf{b},\mathbf{a},\mathbf{c}] = -[\mathbf{a},\mathbf{c},\mathbf{b}].$$

Thus the scalar triple product depends only on the cyclic order of the three entries. Also, since the determinant of a matrix with two equal rows is 0, we see that $\mathbf{a}\bullet(\mathbf{a}\times\mathbf{b})=0$ and $\mathbf{b}\bullet(\mathbf{a}\times\mathbf{b})=0$, that is $\mathbf{a}\times\mathbf{b}$ is orthogonal to both $\mathbf{a}$ and $\mathbf{b}$. This is of most interest when $\mathbf{a}\times\mathbf{b}\neq\mathbf{0}$.

If $\mathbf{n}$ is a vector, the set of vectors $\mathbf{x}$ such that $\mathbf{x}\bullet\mathbf{n}=\alpha$ is a *plane*. It has equation $n_1x+n_2y+n_3z = \alpha$. The vector $\mathbf{n}$ is called a *normal vector* to the plane since, for any two points A,X in the plane, $\mathbf{x}\bullet\mathbf{n}=\alpha$ and $\mathbf{a}\bullet\mathbf{n}=\alpha$ so that $(\mathbf{x}-\mathbf{a})\bullet\mathbf{n}=0$, that is, $\overrightarrow{AX}$ is orthogonal (or normal) to $\mathbf{n}$. When $\mathbf{a},\mathbf{b}$ are two vectors with $\mathbf{a}\times\mathbf{b}\neq\mathbf{0}$, the plane with equation $\mathbf{x}\bullet(\mathbf{a}\times\mathbf{b})=0$ contains both $\mathbf{a}$ and $\mathbf{b}$ (thus it contains the three points O,A,B) and is called the *plane determined by* $\mathbf{a}$ and $\mathbf{b}$. It has normal vector $\mathbf{a}\times\mathbf{b}$.

Now let $\mathbf{a},\mathbf{b}$ be two vectors. By an easy calculation it is readily shown that

$$\begin{aligned}\|\mathbf{a}\times\mathbf{b}\|^2 &= (a_2b_3 - a_3b_2)^2 + \text{two other terms} \\ &= \|\mathbf{a}\|^2\|\mathbf{b}\|^2 - (\mathbf{a}\bullet\mathbf{b})^2 \\ &= \|\mathbf{a}\|^2\|\mathbf{b}\|^2(1-\cos^2\theta).\end{aligned}$$

Thus $\|\mathbf{a}\times\mathbf{b}\| = \|\mathbf{a}\|\,\|\mathbf{b}\|\,|\sin\theta|$ where θ is the angle between $\mathbf{a}$ and $\mathbf{b}$. This leads to a geometrical interpretation of $\|\mathbf{a}\times\mathbf{b}\|$: see figure 7.3.

$\|\mathbf{a}\times\mathbf{b}\|$ = area of parallelogram with adjacent sides $\mathbf{a},\mathbf{b}$.

Using this, one can show that

$$\pm[\mathbf{a},\mathbf{b},\mathbf{c}] = \text{Volume of the parallelepiped with edges } \mathbf{a},\mathbf{b},\mathbf{c},$$
$$= (\text{area of base}) \times (\text{distance between top and base}),$$
$$= \|\mathbf{a}\times\mathbf{b}\| \, \|\mathbf{c}\| \cos\phi,$$

where ϕ is the angle between $\mathbf{a}\times\mathbf{b}$ and $\mathbf{c}$.

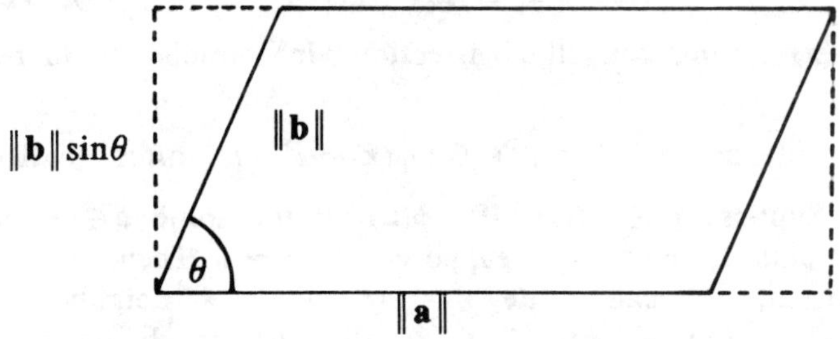

Figure 7.3.

Example 7.2. Show that $\mathbf{a}\times(\mathbf{b}\times\mathbf{c})=(\mathbf{a}\bullet\mathbf{c})\mathbf{b}-(\mathbf{a}\bullet\mathbf{b})\mathbf{c}$.

Solution. It suffices to prove that corresponding components are equal. We compute only the first components. Now

(first component of $\mathbf{a}\times\mathbf{v}$)$=a_2v_3 - a_3v_2$. Therefore,

(first component of $\mathbf{a}\times(\mathbf{b}\times\mathbf{c})$) is $a_2[b_1c_2 - b_2c_1] - a_3[b_3c_1 - b_1c_3]$

$$= b_1[a_2c_2 + a_3c_3] - c_1[a_2b_2 + a_3b_3]$$
$$= b_1[a_1c_1 + a_2c_2 + a_3c_3] - c_1[a_1b_1 + a_2b_2 + a_3b_3]$$
$$= (\text{first component of } (\mathbf{a}\bullet\mathbf{c})\mathbf{b} - (\mathbf{a}\bullet\mathbf{b})\mathbf{c}).$$

Note that $\mathbf{a}\times(\mathbf{b}\times\mathbf{c})$ is a vector orthogonal to $\mathbf{b}\times\mathbf{c}$, a normal vector to the plane containing $\mathbf{b}$ and $\mathbf{c}$ so $\mathbf{a}\times(\mathbf{b}\times\mathbf{c})$ should be a vector in this plane. This is precisely what the example shows.

7.2. GRADIENT, DIVERGENCE AND CURL

Let G be an open subset of $\mathbb{R}^3$. A function $\phi : G \longrightarrow \mathbb{R}$ is called a *scalar field*: at each point $\mathbf{x}\in G$ one associates a scalar $\phi(\mathbf{x})$. A function $\mathbf{F} : G \longrightarrow \mathbb{R}^3$ is called a *vector field*: it assigns to each $\mathbf{x}\in G$ a vector $\mathbf{F}(\mathbf{x})$.

Suppose ϕ is a scalar field and that ϕ is differentiable. Then the *gradient of* ϕ is a vector field denoted gradϕ or $\nabla\phi$ defined by $\nabla\phi(\mathbf{x})$ is the vector $\left[\frac{\partial\phi}{\partial x}, \frac{\partial\phi}{\partial y}, \frac{\partial\phi}{\partial z}\right]$ (all partial derivatives evaluated at $\mathbf{x}$). In other words, $\nabla\phi(\mathbf{x})$ is the $1\times n$ matrix $D\phi(\mathbf{x})$ considered as a vector. This concept was defined previously in Chapter 3 (§3.3) on $\mathbb{R}^n$

and we recall that $\nabla\phi$ is defined precisely by

$$\nabla\phi(\mathbf{x})\cdot\mathbf{h}=D\phi(\mathbf{x})\mathbf{h} \quad \text{for every } \mathbf{h}\in\mathbb{R}^3.$$

The symbol "∇" may be read as "del".

It was shown in Chapter 3 that if ϕ is differentiable then all directional derivatives exist and are given by $\frac{d\phi}{d\mathbf{u}}(\mathbf{x})=\nabla\phi(\mathbf{x})\cdot\mathbf{u}$. In particular $\nabla\phi(\mathbf{x})$ points in the direction in which ϕ increases most quickly.

The set of points $S:=\{\mathbf{x}\in G:\phi(\mathbf{x})=\alpha\}$ is often called a *level surface* of ϕ. Suppose $\phi\in C^1(G)$. If $\nabla\phi(\mathbf{a})\neq\mathbf{0}$ for some $\mathbf{a}\in G$, at least one of the components is non-zero; suppose $\partial\phi/\partial z\neq 0$. Then, by the Implicit Function Theorem, we can write $z=f(x,y)$ on a neighbourhood of $\mathbf{a}$. Thus, at points where $\nabla\phi\neq\mathbf{0}$, S is locally the graph of a function of two variables, and hence may indeed be called a surface.

Let $\mathbf{a}=(a,b,c)\in S$ and suppose that $\phi(\mathbf{x})=\alpha$ is equivalent to $z=f(x,y)$ on a neighbourhood of $\mathbf{a}$. The affine approximation to $f(x,y)$ near (a,b) is $f(a,b)+Df(a,b)\begin{bmatrix}x-a\\y-b\end{bmatrix}=c+\frac{\partial f}{\partial x}(x-a)+\frac{\partial f}{\partial y}(y-b)$.

This defines the tangent plane at $\mathbf{a}$ by the equation

$$z=d_1(x-a)+d_2(y-b)+c \quad \text{where} \quad Df(a,b)=(d_1,d_2).$$

Since f is defined implicitly by ϕ, the derivatives $\frac{\partial f}{\partial x},\frac{\partial f}{\partial y}$ may be calculated via the chain rule. Thus $\phi(x,y,f(x,y))=\alpha$ on a neighbourhood of (a,b) and so $\frac{\partial\phi}{\partial x}+\frac{\partial\phi}{\partial z}\frac{\partial f}{\partial x}=0$, $\frac{\partial\phi}{\partial y}+\frac{\partial\phi}{\partial z}\frac{\partial f}{\partial y}=0$. Evaluating at (a,b) gives (writing subscripts for partial derivatives)

$f_x=-\phi_x/\phi_z$, $f_y=-\phi_y/\phi_z$, where $\phi_z\neq 0$ by assumption. The equation for the tangent plane becomes

$$\phi_x(x-a)+\phi_y(y-b)+\phi_z(z-c)=0.$$

This is the standard form of the equation of a plane with normal vector (ϕ_x,ϕ_y,ϕ_z). This is a confirmation of Proposition 3.8 which says that $\nabla\phi$ is orthogonal (normal) to S.

Now let $\mathbf{F}:G\longrightarrow\mathbb{R}^3$ be a vector field with components (F_1,F_2,F_3) and suppose $\mathbf{F}\in C^1(G)$.

Definition 7.3. The *divergence* of the vector field $\mathbf{F}$ is the scalar field

$$\operatorname{div}\mathbf{F}:=\partial F_1/\partial x+\partial F_2/\partial y+\partial F_3/\partial z.$$

The *curl* of $\mathbf{F}$ is the vector field

$$\operatorname{curl}\mathbf{F}:=(\partial F_2/\partial z-\partial F_3/\partial y,\ \partial F_3/\partial x-\partial F_1/\partial z,\ \partial F_1/\partial y-\partial F_2/\partial x). \tag{7.2}$$

These definitions are recalled more easily using the so-called "del operator". Write $\nabla = \mathbf{i}\frac{\partial}{\partial x} + \mathbf{j}\frac{\partial}{\partial y} + \mathbf{k}\frac{\partial}{\partial z}$.

Then for a C^1 scalar field ϕ, $\nabla\phi=\mathbf{i}\frac{\partial\phi}{\partial x}+\mathbf{j}\frac{\partial\phi}{\partial y}+\mathbf{k}\frac{\partial\phi}{\partial z} = (\phi_x,\phi_y,\phi_z)$ coincides with the earlier definition of $\nabla\phi$. Also for a C^1 vector field $\mathbf{F}$, we may write

$$\text{div}\,\mathbf{F}=\nabla\bullet\mathbf{F}=(\mathbf{i}\frac{\partial}{\partial x}+\mathbf{j}\frac{\partial}{\partial y}+\mathbf{k}\frac{\partial}{\partial z})\bullet(\mathbf{i}\,F_1+\mathbf{j}\,F_2+\mathbf{k}\,F_3)=\partial F_1/\partial x+\partial F_2/\partial y+\partial F_3/\partial z,$$

and

$$\text{curl}\,\mathbf{F}=\nabla\times\mathbf{F}=\begin{vmatrix} \mathbf{i} & \mathbf{j} & \mathbf{k} \\ \partial/\partial x & \partial/\partial y & \partial/\partial z \\ F_1 & F_2 & F_3 \end{vmatrix}. \tag{7.3}$$

Although these manipulations are purely formal they give memorable versions of the definition.

Vector fields whose divergence is 0 may be termed *divergence free*; the terms *incompressible* and *solenoidal* are also employed, the former in applications to fluid flow, the latter in electromagnetic theory. If a fluid has velocity $\mathbf{v}(\mathbf{x})$ and there are no sources or sinks, that is, the fluid is neither being created nor destroyed, $\text{div}\,\mathbf{v}(\mathbf{x})$ measures the rate at which the amount of fluid in a fixed volume is changing. For an expanding gas $\text{div}\,\mathbf{v}$ is positive. For liquids, $\text{div}\,\mathbf{v}=0$, since liquids cannot be compressed. In electromagnetic theory, the magnetic field $\mathbf{B}$ satisfies $\text{div}\,\mathbf{B}=0$ (one of Maxwell's equations), the physical interpretation being that magnetic poles must occur in pairs.

Vector fields with $\text{curl}\,\mathbf{F}=\mathbf{0}$ are termed *irrotational* because for fluid flow, $\text{curl}\,\mathbf{v}\neq 0$ indicates the present of vortices: a paddle-wheel in the fluid would not only follow the flow but also rotate about its axis. If $\mathbf{F}$ represents a force field, $\text{curl}\,\mathbf{F}=\mathbf{0}$ is often described by saying that $\mathbf{F}$ is *conservative* because energy is conserved in such a situation.

There are various identities involving the concepts of gradient, divergence and curl. We give the simplest and most useful ones.

Proposition 7.4. *Let ϕ,ψ denote C^1 scalar fields, and let $\mathbf{F}$ denote a C^1 vector field.*

(a) If $\mathbf{F}\in C^2(G,\mathbb{R}^3)$, $\text{div}(\text{curl}\,\mathbf{F})=0$, *or,* $\nabla\bullet(\nabla\times\mathbf{F})=0$.

(b) If $\phi\in C^2(G,\mathbb{R})$, $\text{curl}(\text{grad}\,\phi)=\mathbf{0}$, *or,* $\nabla\times(\nabla\phi)=\mathbf{0}$.

(c) $\text{grad}(\phi\psi)=\phi\,\text{grad}\,\psi+\psi\,\text{grad}\,\phi$ *or,* $\nabla(\phi\psi)=\phi\,\nabla\psi+\psi\,\nabla\phi$.

(d) $\text{div}(\phi\mathbf{F})=\phi\,\text{div}\,\mathbf{F}+\text{grad}\phi\bullet\mathbf{F}$, *or,* $\nabla\bullet(\phi\mathbf{F})=\phi\,\nabla\bullet\mathbf{F}+\nabla\phi\bullet\mathbf{F}$.
(Note, both sides are scalars).

(*e*) $\text{curl}(\phi\mathbf{F})=\phi\ \text{curl}\ \mathbf{F}+\text{grad}\phi\times\mathbf{F}$, *or*, $\nabla\times(\phi\mathbf{F})=\phi\ \nabla\times\mathbf{F}+\nabla\phi\times\mathbf{F}$.
(*Note, both sides are vectors and order is important.*)

Proof. (*a*) $\text{curl}\ \mathbf{F}=(\partial F_2/\partial z-\partial F_3/\partial y,\ \partial F_3/\partial x-\partial F_1/\partial z,\ \partial F_1/\partial y-\partial F_2/\partial x)$,

so $$\begin{aligned}\text{div}(\text{curl}\ \mathbf{F})&=\partial^2F_2/\partial x\partial z-\partial^2F_3/\partial x\partial y+\partial^2F_3/\partial y\partial x-\partial^2F_1/\partial y\partial z\\&\quad+\partial^2F_1/\partial z\partial y-\partial^2F_2/\partial z\partial x,\\&=0,\quad\text{by the equality of mixed partial derivatives.}\end{aligned}$$

(*b*) is valid for the same reason as (*a*).
(*c*), (*d*) and (*e*) are nothing but the product rule for differentiation. We illustrate the arguments by proving (*c*).

$$\begin{aligned}\text{curl}(\phi\mathbf{F})&=\begin{vmatrix}\mathbf{i}&\mathbf{j}&\mathbf{k}\\\partial/\partial x&\partial/\partial y&\partial/\partial z\\\phi F_1&\phi F_2&\phi F_3\end{vmatrix}=\mathbf{i}\left[\frac{\partial}{\partial y}(\phi F_3)-\frac{\partial}{\partial z}(\phi F_2)\right]+2\text{ similar terms,}\\&=\mathbf{i}\left[\phi\frac{\partial F_3}{\partial y}+\frac{\partial\phi}{\partial y}F_3-\phi\frac{\partial F_2}{\partial z}-\frac{\partial\phi}{\partial z}F_2\right]+2\text{ terms,}\\&=\mathbf{i}\ [\phi(\partial F_3/\partial y-\partial F_2/\partial z)+(\partial\phi/\partial y\,F_3-\partial\phi/\partial z\,F_2)]\ +\ \cdot\ +\ \cdot\\&=\phi\ \text{curl}\ \mathbf{F}+\nabla\phi\times\mathbf{F}.\end{aligned}$$

Example 7.5. Newton's law of gravitation says that masses attract each other according to the inverse square law: $\mathbf{F}(\mathbf{x})=-\alpha\mathbf{x}/\|\mathbf{x}\|^3$. Calculate div **F** and curl **F**.

Solution. Write $r=\|\mathbf{x}\|$, so that $\mathbf{F}(\mathbf{x})=-(\alpha/r^3)\mathbf{x}$. By Proposition 7.4 (*d*),
$$\begin{aligned}\text{div}(-\alpha r^{-3}\mathbf{x})&=-\alpha r^{-3}\text{div}\ \mathbf{x}+\nabla(-\alpha r^{-3})\bullet\mathbf{x}\\&=-3\alpha r^{-3}+3\alpha r^{-5}\mathbf{x}\bullet\mathbf{x}\\&=0.\end{aligned}$$
Here we used the facts that $\text{div}\ \mathbf{x}=3$, $\dfrac{\partial}{\partial x}(r^{-3})=-3r^{-4}\dfrac{\partial r}{\partial x}=-3r^{-4}\dfrac{x}{r}$, by the chain rule and by differentiating $r^2=x^2+y^2+z^2$ with respect to x, with similar expressions for $\partial r/\partial y$ and $\partial r/\partial z$.
Similarly, using Proposition 7.4 (*e*),
$$\begin{aligned}\text{curl}(-\alpha r^{-3}\mathbf{x})&=-\alpha r^{-3}\text{curl}\ \mathbf{x}+\nabla(-\alpha r^{-3})\times\mathbf{x}\\&=\mathbf{0}+(3\alpha r^{-5}\mathbf{x})\times\mathbf{x},\quad\text{since curl}\ \mathbf{x}=\mathbf{0},\\&=\mathbf{0},\quad\text{since}\ \mathbf{x}\times\mathbf{x}=\mathbf{0}.\end{aligned}$$

7.3. PATHS AND CURVES

Let $\mathbf{c}:[a,b]\to\mathbb{R}^n$ be a continuous function. For each point

$t \in [a,b]$ there is a corresponding point $\mathbf{c}(t)$. We may think of t as representing time and $\mathbf{c}(t)$ as the position of a moving particle at time t. As t varies over $[a,b]$ the moving particle traces out a curve.

Definition 7.6. A continuous function $\mathbf{c} : [a,b] \longrightarrow \mathbb{R}^n$ is called a *path*; its image set $\{\mathbf{c}(t) : t \in [a,b]\}$ is called a *curve*. The path is called *simple* if $\mathbf{c}$ is one-to-one on $[a,b]$. If $\mathbf{c}$ is one-to-one on $[a,b)$ and $\mathbf{c}(b) = \mathbf{c}(a)$, then $\mathbf{c}$ is called a simple *closed* path. If $\mathbf{c} \in C^1$, $\mathbf{c}$ is called a C^1 path.

A simple path corresponds to the curve having no self intersections.

As defined here a curve may not resemble the familiar concept very closely. There are examples of continuous functions from $[0,1]$ to $\mathbb{R}^2$ whose image is all of the square $[0,1] \times [0,1]$, so-called "space filling curves". We shall not study such "curves". $\mathbf{c}(t)$ is often called the *parametric equations* of the curve; t is the *parameter*. Paths have *orientations*, the curve is traced out in the direction of increasing t.

An important concept is that of the length of paths.

Definition 7.7. Let $\mathbf{c} : [a,b] \longrightarrow \mathbb{R}^n$ be a path. For a partition P of $[a,b]$, $a = t_0 < t_1 < \cdots < t_m = b$, let $S(P) := \sum_{i=1}^{m} \|\mathbf{c}(t_i) - \mathbf{c}(t_{i-1})\|$. $\mathbf{c}$ is said to be *rectifiable* if all sums $S(P)$ are bounded above and the *arc length* of $\mathbf{c}$ is then defined to be

$$L_{\mathbf{c}} := \sup\{S(P) : P \text{ a partition of } [a,b]\}.$$

The intuition behind this definition is that the length of the curve should be well approximated by the sum of the lengths of a polygon through the points $\mathbf{c}(t_1), \mathbf{c}(t_2), \cdots, \mathbf{c}(t_m)$.

Proposition 7.8. *If* $\mathbf{c}$ *is a* C^1 *path then* $L_{\mathbf{c}} = \int_a^b \|\mathbf{c}'(t)\|\, dt.$

Proof. In the course of the proof we write $\int_a^b \mathbf{c}(t)\, dt$. This means the element of $\mathbb{R}^n$ whose components are $\int_a^b c_i(t)\, dt$. This notation was previously used in §4.4. Now let $a = t_0 < t_1 < \cdots < t_m = b$ be a partition of $[a,b]$. Then $\sum_{i=1}^{m} \|\mathbf{c}(t_i)-\mathbf{c}(t_{i-1})\| = \sum_{i=1}^{m} \|\int_{t_{i-1}}^{t_i} \mathbf{c}'(t)\,dt\|$, by the Fundamental theorem of the calculus applied to each component. Thus,

$$\sum_{i=1}^{m} \|\mathbf{c}(t_i)-\mathbf{c}(t_{i-1})\| \le \sum_{i=1}^{m} \int_{t_{i-1}}^{t_i} \|\mathbf{c}'(t)\|\,dt, \text{ by Lemma 4.27,}$$
$$= \int_a^b \|\mathbf{c}'(t)\|\,dt.$$

This shows that $L_{\mathbf{c}} \le \int_a^b \|\mathbf{c}'(t)\|\,dt$. For the converse, take an arbitrary $\varepsilon > 0$. Since $\mathbf{c}'$ is uniformly continuous on $[a,b]$ there exists $\delta(\varepsilon) > 0$ such that $\|\mathbf{c}'(t)-\mathbf{c}'(s)\| < \varepsilon$ for $|t-s| < \delta$. Let a partition of $[a,b]$ be chosen so that $|t_i-t_{i-1}| < \delta$ for each i. Then

$$\int_{t_{i-1}}^{t_i} \|\mathbf{c}'(t)\|\,dt \le \int_{t_{i-1}}^{t_i} (\|\mathbf{c}'(t_i)\|+\varepsilon)\,dt = \|\int_{t_{i-1}}^{t_i} \mathbf{c}'(t_i)\,dt\| + \varepsilon(t_i-t_{i-1}),$$

where we have used the fact that $\mathbf{c}'(t_i)$ is a constant. Thus we have

$$\int_{t_{i-1}}^{t_i} \|\mathbf{c}'(t)\|\,dt \le \|\int_{t_{i-1}}^{t_i} \mathbf{c}'(t)+\mathbf{c}'(t_i)-\mathbf{c}'(t)\,dt\| + \varepsilon(t_i-t_{i-1})$$
$$\le \|\int_{t_{i-1}}^{t_i} \mathbf{c}'(t)\| + \int_{t_{i-1}}^{t_i} \|\mathbf{c}'(t_i)-\mathbf{c}'(t)\| + \varepsilon(t_i-t_{i-1})$$
$$< \|\mathbf{c}(t_i)-\mathbf{c}(t_{i-1})\| + 2\varepsilon(t_i-t_{i-1}).$$

Adding gives $\int_a^b \|\mathbf{c}'(t)\|\,dt \le \sum_{i=1}^{m} \int_{t_{i-1}}^{t_i} \|\mathbf{c}(t_i)-\mathbf{c}(t_{i-1})\| + 2\varepsilon(b-a)$, and, as ε is arbitrary, proves that $L_{\mathbf{c}} \ge \int_a^b \|\mathbf{c}'(t)\|\,dt$.

If $\mathbf{c} : [a,b] \longrightarrow \mathbb{R}^n$ is a path and if u is an intermediate point of $[a,b]$, two paths $\mathbf{c}^{(1)} : [a,u] \longrightarrow \mathbb{R}^n$ and $\mathbf{c}^{(2)} : [u,b] \longrightarrow \mathbb{R}^n$ are obtained by cutting $\mathbf{c}$ into two pieces. It is easily shown that $\mathbf{c}$ is rectifiable if and only if each of $\mathbf{c}^{(1)}$and $\mathbf{c}^{(2)}$ is rectifiable. We sometimes wish to join curves together. Suppose one curve ends at some point $\mathbf{p}$ and another curve begins at $\mathbf{p}$. Can we define the new curve as a single path? The problem is that both paths may be defined on a common interval. To ease the task we may need to *reparametrize* the curves.

Definition 7.9. Suppose $\mathbf{c} : [a,b] \longrightarrow \mathbb{R}^n$ is a path and let $\phi : [a_1,b_1] \longrightarrow [a,b]$ be a bijective C^1 mapping. Then the composition $\mathbf{c}\circ\phi : [a_1,b_1] \longrightarrow \mathbb{R}^n$ is called a *reparametrization* of **c**.

Since ϕ is either strictly increasing or strictly decreasing on $[a_1,b_1]$ the curve is now traced out either in the same direction as before or in the opposite direction. We say that ϕ is *orientation preserving* if $\mathbf{c}\circ\phi(a_1) = a$ and $\mathbf{c}\circ\phi(b_1) = b$, and *orientation reversing* if $\mathbf{c}\circ\phi(a_1) = b$ and $\mathbf{c}\circ\phi(b_1) = a$.

If $\mathbf{c} : [a,b] \longrightarrow \mathbb{R}^n$, the path $\mathbf{p} : [0,1] \longrightarrow \mathbb{R}^n$ given by $\mathbf{p}(t) = \mathbf{c}((1-t)a+bt)$ is an orientation preserving reparameterization. Thus, when considering paths we may, if we wish, assume they are defined on $[0,1]$.

Now let $\mathbf{c}^{(1)} : [0,1] \longrightarrow \mathbb{R}^n$ and $\mathbf{c}^{(2)} : [0,1] \longrightarrow \mathbb{R}^n$ be two paths with $\mathbf{c}^{(1)}(1) = \mathbf{c}^{(2)}(0)$. The two paths can be combined into a single one. A simple way is to let $\mathbf{c}^{(3)}(t) = \begin{cases} \mathbf{c}^{(1)}(t), & \text{for } 0 \le t \le 1 \\ \mathbf{c}^{(2)}(t-1), & \text{for } 1 \le t \le 2 \end{cases}$.

This process may be continued. If each piece is a C^1 path we end up with a *piecewise-C^1 path*; the derivatives are continuous except at finitely many points. It is readily seen that the length of a piecewise C^1 path is the sum of the lengths of the constituent C^1 pieces, each of which can be evaluated by integration using Proposition 7.8.

Suppose that **c** is a C^1 path. Then $\mathbf{c}'(t)$ is often called the *tangent vector* to the curve $C = \{\mathbf{c}(t)\}$. Indeed $\dfrac{\mathbf{c}(t+h)-\mathbf{c}(t)}{h}$ is parallel to the chord from $\mathbf{c}(t)$ to $\mathbf{c}(t+h)$ so the limit should give a tangent; see Figure 3.3. When $\mathbf{c}(t)$ is interpreted as the curve traced out by a moving point, $\mathbf{c}'(t)$ is called the *velocity vector*. $\mathbf{c}'(t)$ gives the direction of travel and $s'(t) := \|\mathbf{c}'(t)\|$ is the *speed*. This interpretation follows from the definition of velocity as distance travelled divided by time taken and instantaneous velocity is the limit of this as the time interval shrinks to 0. The *acceleration* is $\mathbf{c}''(t)$. The *unit tangent vector* to a curve C at the point $\mathbf{c}(t)$ is defined to be $\mathbf{T}(t) = \mathbf{c}'(t)/\|\mathbf{c}'(t)\| = \mathbf{c}'(t)/s'(t)$. This is defined provided $\mathbf{c}'(t) \neq \mathbf{0}$. **T** points in the direction in which the curve is described as t increases. Note that $\|\mathbf{T}(t)\| = 1$.

Now we specialize to curves in $\mathbb{R}^3$; curves in $\mathbb{R}^2$ are obtained by considering $\mathbb{R}^2$ as a subset of $\mathbb{R}^3$ with third component set to 0. Let

$C = \{\mathbf{c}(t)\}$ be a curve in $\mathbb{R}^3$ and suppose that the speed is never 0. Then $\mathbf{T}(t)\cdot\mathbf{T}(t) = 1$ for all t and so $\mathbf{T}'(t)\cdot\mathbf{T}(t) = 0$. Thus $\mathbf{T}'(t)$ is *normal* to the tangent vector.

Definition 7.10. For a C^2 curve in $\mathbb{R}^3$, the unit *normal* to the curve is defined to be $\mathbf{N} = \mathbf{T}'(t)/\|\mathbf{T}'(t)\|$.

In studying the geometry of curves in $\mathbb{R}^3$ it is often convenient to reparametrize the curve using arc length as the parameter. Let $C = \{\mathbf{c}(t) : t\in[0,1]\}$ be a rectifiable C^1 curve and write $s'(t) := \|\mathbf{c}'(t)\|$ for the speed. Then $s(t) = \int_0^t s'(t)\,dt$ is the arc length function.

Lemma 7.11. *If* $\mathbf{c}$ *defines a simple* C^1 *curve then* $s(t)$ *is a strictly increasing continuous function.*

Proof. s is continuous being the integral of a continuous function. Since $s'(t) = \|\mathbf{c}'(t)\| \geq 0$ for all t, s is increasing. If s were constant on some interval I then $s'(t) = 0$ in I and therefore $\mathbf{c}'(t) = 0$ on I, hence $\mathbf{c}(t)$ would be constant on the connected set I. This is impossible when $\mathbf{c}$ is one-to-one.

This lemma shows that we can use $s^{-1} : [0,L_{\mathbf{c}}]\to[0,1]$, as an orientation preserving reparametrization of $\mathbf{c}(t)$ and we write $C = \{\mathbf{c}(s) : s\in[0,L_{\mathbf{c}}]\}$. This is nothing but a change of variable. An immediate advantage is that the speed is now $\|\mathbf{c}'(s)\| = 1$ and so $\mathbf{T} = \mathbf{c}'(s)$ is the unit tangent vector. Also, the normal vector $\mathbf{N}$ is given by $\mathbf{N} = \mathbf{T}'(s)/\|\mathbf{T}'(s)\|$ which one can write as $\mathbf{T}'(s) = \kappa\mathbf{N}(s)$, where $\kappa = \|\mathbf{T}'(s)\|$ is called the *curvature* of C.

For each s, the vectors $\mathbf{T}(s)$ and $\mathbf{N}(s)$ determine a plane known as the *osculating plane* of the curve. Its normal vector $\mathbf{B}(s) := \mathbf{T}(s)\times\mathbf{N}(s)$ is called the *binormal*. In the following we assume that second derivatives of $\mathbf{c}$ exist.

Proposition 7.12. $\mathbf{B}'(s)$ *is a scalar multiple of* $\mathbf{N}(s)$.

Proof. $\mathbf{N}$, $\mathbf{B}$ and $\mathbf{T}$ are mutually orthogonal vectors in $\mathbb{R}^3$ so span $\mathbb{R}^3$ and every vector can be written as a linear combination of these. Write $\mathbf{B}'(s) = \alpha(s)\mathbf{N}(s)+\beta(s)\mathbf{B}(s)+\gamma(s)\mathbf{T}(s)$. Then, taking dot products, we have

$$\alpha(s) = \mathbf{B}'(s)\cdot\mathbf{N}(s),\ \beta(s) = \mathbf{B}'(s)\cdot\mathbf{B}(s),\ \gamma(s) = \mathbf{B}'(s)\cdot\mathbf{T}(s).$$

Now $\beta(s) = 0$ since $\mathbf{B}(s)\cdot\mathbf{B}(s) = 1$ implies $\mathbf{B}'(s)\cdot\mathbf{B}(s) = 0$.

Differentiating $\mathbf{B}(s) = \mathbf{T}(s)\times\mathbf{N}(s)$ gives

$$\mathbf{B}'(s) = \mathbf{T}'(s)\times\mathbf{N}(s)+\mathbf{T}(s)\times\mathbf{N}'(s) = \mathbf{T}(s)\times\mathbf{N}'(s),$$

since $\mathbf{T}' = \kappa\mathbf{N}$ so $\mathbf{T}'\times\mathbf{N} = \mathbf{0}$. (We leave it to the reader to verify that dot products and cross products differentiate by the "obvious" product rule.) Therefore $\gamma(s) = \mathbf{B}'(s)\bullet\mathbf{T}(s) = (\mathbf{T}\times\mathbf{N}')\bullet\mathbf{T} = 0$, since $\mathbf{T}\times\mathbf{N}'$ is orthogonal to $\mathbf{T}$. Hence $\mathbf{B}' = \alpha\mathbf{N}$.

Definition 7.13. The function $\tau(s)$ defined by $\mathbf{B}'(s) = -\tau(s)\mathbf{N}(s)$ is called the *torsion* of C.

The torsion τ measures how much the curve twists out of the osculating plane. If the curve lies in a plane then $\tau = 0$.

Example 7.14. Find the curvature and the torsion of the helix $\mathbf{c}(t) = \{(r\cos t, r\sin t, t) : 0\le t\le 2\pi\}$.

Solution. We have $\mathbf{c}'(t) = (-r\sin t, r\cos t, 1)$,

so $s'(t) = \|\mathbf{c}'(t)\| = \sqrt{(r^2+1)}$ and $s(t) = t\sqrt{(r^2+1)} = ct$ (say).

Then $\mathbf{T}(t) = \left(\frac{-r}{c}\sin t, \frac{r}{c}\cos t, \frac{1}{c}\right)$ and $\mathbf{T}(s) = \left(\frac{-r}{c}\sin(\frac{s}{c}), \frac{r}{c}\cos(\frac{s}{c}), \frac{1}{c}\right)$.

Thus, $\mathbf{T}'(s)=(-r\cos(s/c)/c^2, -r\sin(s/c)/c^2, 0)$, and so $\mathbf{N}(s) = \mathbf{T}'(s)/\|\mathbf{T}'(s)\| = (-\cos(s/c), -\sin(s/c),0)$ and $\kappa = r/c^2 = r/(r^2+1)$.

Now $\mathbf{B} = \mathbf{T}\times\mathbf{N}$

$= (-(r/c)\sin(s/c), (r/c)\cos(s/c), 1/c)\times(-\cos(s/c), -\sin(s/c), 0)$

$=(1/c)(\sin(s/c), -\cos(s/c), r)$.

Therefore, $\mathbf{B}'(s)=(1/c^2)(\cos(s/c), \sin(s/c), 0)=-(1/c^2)\mathbf{N}(s)$ and $\tau=1/c^2$.

The other interpretation of $\mathbf{c}'(t)$ is that of the *velocity vector* of a particle with position vector $\mathbf{c}(t)$. Then $\mathbf{c}''(t)$ is its acceleration. Since $\mathbf{c}'(t) = s'(t)\mathbf{T}$, $s'(t) = \|\mathbf{c}'(t)\|$ is the speed, we have $\mathbf{c}''(t) = s''(t)\mathbf{T}+s'(t)\mathbf{T}'(t) = s''(t)\mathbf{T}+(s'(t))^2\mathbf{T}'(s)$

$$= s''(t)\mathbf{T}+(s')^2\kappa\mathbf{N}.$$

Thus, the acceleration vector has a tangential and a normal component. The normal component is zero if the curvature is 0, that is, for motion in a straight line. The reciprocal of the curvature $\rho := \kappa^{-1}$ is called the *radius of curvature*; the curvature at the point is the same as that of a circle of radius ρ.

Example 7.15. A particle moves on a circle of radius a at constant speed according to $\mathbf{c}(t) = (a\cos(\alpha t), a\sin(\alpha t), 0)$. Find its velocity and acceleration.

Solution. The velocity is $\mathbf{c}'(t) = (-a\alpha\sin\alpha t, a\alpha\cos\alpha t, 0)$, so the speed is $a\alpha$. The acceleration is

$$\mathbf{c}''(t) = (-a\alpha^2\cos\alpha t, -a\alpha^2\sin\alpha t, 0) = -\alpha^2\mathbf{c}(t).$$

Since $\mathbf{c}(t)$ is a radial vector from the centre of the circle to the point, we see that the acceleration is directed towards the centre. Since acceleration is equivalent to force there must be a force acting towards the centre, the so-called centripetal force. This force may be gravitational attraction or the particle may be attached to a string and the force is supplied by the string.

7.4. INTEGRALS ALONG PATHS

In this section the results are valid in Euclidean space of any dimension since we use only the scalar product. Since the proofs do not depend on the dimension we give them in $\mathbb{R}^n$.

Let $C=\{\mathbf{c}(t) : t\in[a,b]\}$ be a C^1 simple curve in $\mathbb{R}^n$, and let f be a continuous scalar field whose domain contains C.

Definition 7.16. The *path integral* of f along C is defined by

$$\int_C f\,ds = \int_a^b f(\mathbf{c}(t))\|\mathbf{c}'(t)\|\,dt.$$

If C is piecewise C^1 the definition may be extended by defining the integral as the sum over the pieces on which $\mathbf{c}$ is C^1.

If we recall that $s'(t) = \|\mathbf{c}'(t)\|$ defines the speed then $\int_a^b f(\mathbf{c}(t))\|\mathbf{c}'(t)\|\,dt = \int_a^b f(\mathbf{c}(t))\,s'(t)\,dt$ and one often writes ds in place of $s'(t)\,dt$ in making a change of variable in integrals ; this makes the terminology $\int_C f\,ds$ a plausible one. If $\mathbf{c}(t)$ is reparametrized using arc length as $\{\mathbf{c}(s) : 0\le s\le L\}$, then $\int_C f\,ds = \int_0^L f(\mathbf{c}(s))\,ds$.

It is implicit in Definition 7.6 that the integral is independent of the parametrization.

Theorem 7.17. *Let* $\mathbf{c} : [a,b]\to\mathbb{R}^n$ *be a path and let* $\phi : [\alpha,\beta]\to[a,b]$ *be a bijective* C^1 *mapping so that* $\mathbf{p} = \mathbf{c}\circ\phi : [\alpha,\beta]\to\mathbb{R}^n$ *is a reparametrization of* $\mathbf{c}$. *Then* $\int_a^b f(\mathbf{c}(t))\|\mathbf{c}'(t)\|dt=\int_\alpha^\beta f(\mathbf{p}(\tau))\|\mathbf{p}'(\tau)\|d\tau$.

Proof. Since $\mathbf{p}(\tau) = \mathbf{c}(\phi(\tau))$, $\tau\in[\alpha,\beta]$, $\mathbf{p}'(\tau) = \mathbf{c}'(\phi(\tau))\phi'(\tau)$, by the chain rule. Therefore $\|\mathbf{p}'(\tau))\| = \|\mathbf{c}'(\phi(\tau))\|\,|\phi'(\tau)|$ and

$$\int_{\alpha}^{\beta} f(\mathbf{p}(\tau))\|\mathbf{p}'(\tau)\|\,d\tau = \int_{\alpha}^{\beta} f(\mathbf{c}(\phi(\tau)))\|\mathbf{c}'(\phi(\tau))\|\,|\phi'(\tau)|\,d\tau.$$

Now $t=\phi(\tau)$ may be used as a change of variable in $\int_a^b f(\mathbf{c}(t))\|\mathbf{c}'(t)\|\,dt$

so this integral equals $\int_{\phi^{-1}(a)}^{\phi^{-1}(b)} f(\mathbf{c}(\phi(\tau))\|\mathbf{c}'(\phi(\tau))\|\,\phi'(\tau)\,d\tau.$

If ϕ is strictly increasing then $\phi(\alpha)=a$, $\phi(\beta)=b$ and $\phi'(\tau) = |\phi'(\tau)|$, whereas $\phi'(\tau) = -|\phi'(\tau)|$ if ϕ is decreasing.

Since $\int_{\beta}^{\alpha} = -\int_{\alpha}^{\beta}$ we see that in both cases we have

$$\int_a^b f(\mathbf{c}(t))\|\mathbf{c}'(t)\|\,dt = \int_{\alpha}^{\beta} f(\mathbf{c}(\phi(\tau)))\|c'(\phi(\tau))\|\,|\phi'(\tau)|\,d\tau\,.$$

The idea here is that changing t to $-t$ (reversing orientation) changes the sign of $\mathbf{c}'(t)$ but does not affect $\|\mathbf{c}'(t)\|$.

Example 7.18. Evaluate the path integral $\int_C (x^2+y^2)\,ds$ where C is the triangle with vertices at (0,0,0), (1,0,0) and (0,1,0), O, A, B, say.

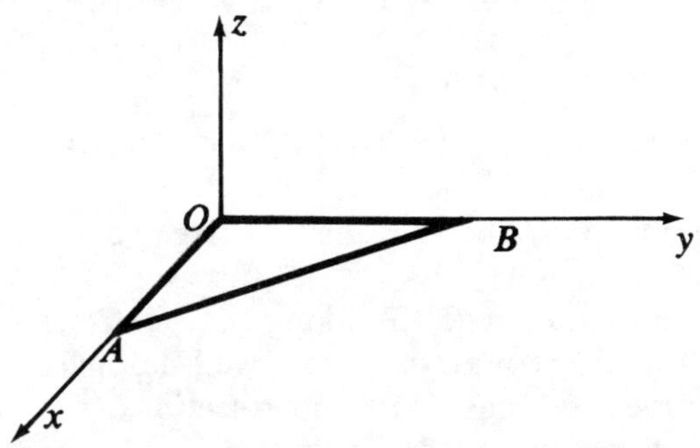

Figure 7.4.

Solution. Since C is made up of 3 pieces the integral is the sum of the integrals over the 3 pieces separately. The line segment $OA=C_1$ can be parametrized as $\mathbf{c}^{(1)}(t) := (t, 0, 0)$, $0\le t\le 1$, or we can write $x(t)=t$, $y(t)=0$, $z(t)=0$ so $\int_{C_1} (x^2+y^2)\,ds = \int_0^1 t^2\,dt = 1/3$.

The segment AB can be parameterized as $\mathbf{c}^{(2)}(t) := (1-t, t, 0)$, $0\le t\le 1$. Then $\mathbf{c}^{(2)\prime}(t) = (-1,1,0)$ so $\|\mathbf{c}^{(2)\prime}\| = \sqrt{2}$ and

$$\int_{C_3} (x^2+y^2)\,ds = \int_0^1 [(1-t)^2+t^2]\sqrt{2}\,dt = 2\sqrt{2}/3\,.$$

The segment OB can be parametrized as $\mathbf{c}^{(3)}(t) = (0, t, 0)$, $0\le t\le 1$ and

$$\int_{C_3} (x^2+y^2)\, ds = 1/3 , \quad \text{as before.}$$

Summing gives the final result $\int_C (x^2+y^2)\, ds = 2(1+\sqrt{2}) / 3$.

NOTE that we could also write $\mathbf{c}^{(1)}(x) = (x, 0, 0)$, $0 \leq x \leq 1$. Then

$$\int_{C_1} (x^2+y^2)\, ds = \int_0^1 x^2\, dx.$$

Similarly y may be used as parameter for $\mathbf{c}^{(3)}$.

We now turn to the integration of vector fields along paths.

Definition 7.19. Let $C = \{\mathbf{c}(t) : t \in [a,b]\}$ be a C^1 curve in $\mathbb{R}^n$ and let $\mathbf{F}$ be a continuous vector field (that is, a function from $\mathbb{R}^n$ into $\mathbb{R}^n$) whose domain contains C. The *line integral* of $\mathbf{F}$ along C is defined by

$$\int_{\mathbf{c}} \mathbf{F} \cdot d\mathbf{s} = \int_a^b \mathbf{F}(\mathbf{c}(t)) \cdot \mathbf{c}'(t)\, dt.$$

If C is a piecewise C^1 curve the definition may be extended by summing the integrals over the constituent pieces.

If $\mathbf{c}'(t) \neq 0$, we may write

$$\int_{\mathbf{c}} \mathbf{F} \cdot d\mathbf{s} = \int_a^b \left[\mathbf{F}(\mathbf{c}(t)) \cdot \frac{\mathbf{c}'(t)}{\|\mathbf{c}'(t)\|} \right] \|\mathbf{c}'(t)\|\, dt = \int_a^b \mathbf{F}(\mathbf{c}(t)) \cdot \mathbf{T}(t)\, \|\mathbf{c}'(t)\|\, dt.$$

This last expression may be written $\int_C (\mathbf{F} \cdot \mathbf{T})\, ds$, which may be thought of as the path integral of $\mathbf{F} \cdot \mathbf{T}$ along C. $\mathbf{F} \cdot \mathbf{T}$ is called the *tangential component* of $\mathbf{F}$. $\mathbf{F}(\mathbf{c}(t)) \cdot \mathbf{T}(t)$ is a well-defined function of $t \in [a,b]$ but, if the curve C has self intersections, it does not arise as $f(\mathbf{c}(t))$ for any function f. Nevertheless it is convenient to think of this integral as a path integral.

The line integral is not independent of parametrization but depends only on the orientation.

Theorem 7.20. *Let* $\mathbf{c} : [a,b] \to \mathbb{R}^n$ *be a* C^1 *path and let* $\mathbf{p} : [\alpha,\beta] \to \mathbb{R}^n$ *be a reparametrization of* C. *Then, for a continuous vector field* $\mathbf{F}$,

$$\int_{\mathbf{p}} (\mathbf{F} \cdot \mathbf{T})\, ds = \int_{\mathbf{c}} (\mathbf{F} \cdot \mathbf{T})\, ds \quad \textit{if } \mathbf{p} \textit{ preserves orientation, and}$$

$$\int_{\mathbf{p}} (\mathbf{F} \cdot \mathbf{T})\, ds = -\int_{\mathbf{c}} (\mathbf{F} \cdot \mathbf{T})\, ds \quad \textit{if } \mathbf{p} \textit{ reverses orientation.}$$

Proof. Let $\mathbf{p}(\tau) = \mathbf{c}(\phi(\tau))$ where $\phi : [\alpha,\beta] \to [a,b]$ is a bijective C^1

mapping. Then $\int_{\mathbf{p}} (\mathbf{F}\cdot\mathbf{T})\, ds = \int_{\alpha}^{\beta} \mathbf{F}(\mathbf{p}(\tau))\cdot\mathbf{p}'(\tau)\, d\tau$

$$= \int_{\alpha}^{\beta} \mathbf{F}(\mathbf{c}(\phi(\tau)))\cdot\mathbf{c}'(\phi(\tau))\, d\tau.$$

Let $t = \phi(\tau)$ be a change of variables. Then this integral becomes $\int_{\phi(\alpha)}^{\phi(\beta)} [\mathbf{F}(\mathbf{c}(t))\cdot\mathbf{c}'(t)]\, dt$. If ϕ is orientation preserving, $\phi(\alpha) = a$ and $\phi(\beta) = b$ and we have the first formula. If ϕ is orientation reversing, $\phi(\alpha) = b$ and $\phi(\alpha) = a$ and we get the second formula.

The absolute value of a line integral may be estimated as follows :

Proposition 7.21. *Let* **F** *be a continuous vector field whose domain contains a piecewise C^1 curve C with length L. Suppose* **F** *is bounded on C, say,* $\|\mathbf{F}(\mathbf{x})\| \le M$ *for all* $\mathbf{x} \in C$. *Then* $\left| \int_{\mathbf{c}} \mathbf{F}\cdot d\mathbf{s} \right| \le ML$.

Proof. Suppose C is a C^1 curve. Then $\int_{\mathbf{c}} \mathbf{F}\cdot d\mathbf{s} = \int_{C} (\mathbf{F}\cdot\mathbf{T})\, ds$, where **T** is the tangent vector. Since $|\mathbf{F}\cdot\mathbf{T}| \le \|\mathbf{F}\|\,\|\mathbf{T}\|$ by Schwarz's inequality and since $\|\mathbf{T}\| = 1$, this gives $\left|\int_{\mathbf{c}} \mathbf{F}\cdot d\mathbf{s}\right| \le \int_{C} |\mathbf{F}\cdot\mathbf{T}|\, ds \le \int_{C} M\, ds = ML.$

The proof for a piecewise C^1 curve is obtained by summing over the finitely many pieces.

The Fundamental theorem of the calculus asserts that $\int_{a}^{b} F'(x)\, dx = F(b) - F(a)$ when F' is integrable, which shows that the integral depends only on the end-points a,b. Interestingly there are vector fields whose line integral along a curve depends only on the endpoints of the curve. These are the vector fields that are obtained as the gradients of a scalar field. We say that **F** is a *gradient vector field* if $\mathbf{F} = \nabla\phi$ for some scalar field ϕ. ϕ is called a (scalar) *potential* of **F**; ϕ is uniquely determined up to the addition of a constant, that is, any two potentials for **F** differ by a constant. Since $\nabla\phi$ is the derivative of ϕ, the terminology *exact differential* is also used.

Theorem 7.22. *Let ϕ be a C^1 scalar field defined on an open connected set $U \in \mathbb{R}^n$ and let $C = \{\mathbf{c}(t) : t \in [a,b]\}$ be a piecewise C^1 curve lying in*

U. Then $\int_C \nabla\phi\bullet ds = \phi(\mathbf{c}(b)) - \phi(\mathbf{c}(a))$.

Proof. It suffices to give the proof assuming that **c** is C^1 by summing over the separate pieces since intermediate points cancel. Now, Let $h(t) = \phi(\mathbf{c}(t))$, a C^1 map from $[a,b] \longrightarrow \mathbb{R}$. By the chain rule, $h'(t) = \nabla\phi(\mathbf{c}(t))\bullet\mathbf{c}'(t)$ and so h' is continuous and

$$\int_C \nabla\phi\bullet ds = \int_a^b \nabla\phi(\mathbf{c}(t))\bullet\mathbf{c}'(t)\,dt = \int_a^b h'(t)\,dt$$
$$= h(b) - h(a) = \phi(\mathbf{c}(b)) - \phi(\mathbf{c}(a)).$$

Example 7.23. Let $\mathbf{c}(t) = \{(t \sin t,\ 1+\cos t,\ t^2) : 0 \le t \le \pi/2\}$ and let $\mathbf{F}(\mathbf{x}) = (yz, zx, xy)$. Calculate $\int_C \mathbf{F}\bullet ds$.

Solution. We observe that $\mathbf{F} = \nabla\phi$ where $\phi(\mathbf{x}) = xyz$.
Then $\int_C \mathbf{F}\bullet ds = \phi(\mathbf{c}(\pi/2)) - \phi(\mathbf{c}(0)) = \phi(\pi/2, 1, \pi^2/4) - \phi(0, 1, 0) = \pi^3/8$.

It would be much less pleasant to calculate $\int_0^{\pi/2} \mathbf{F}(\mathbf{c}(t))\bullet\mathbf{c}'(t)\,dt$.

A common notation is obtained by formally writing (in $\mathbb{R}^3$) $ds = (dx, dy, dz)$. Then $\int_C \mathbf{F}\bullet ds$ can be written $\int_C F_1 dx + F_2 dy + F_3 dz$. If one writes $\mathbf{c}(t) = (x(t), y(t), z(t))$, then $\mathbf{c}'(t) = (x'(t), y'(t), z'(t)$ and $\int_C \mathbf{F}\bullet ds = \int_C [F_1\,x'(t) + F_2\,y'(t) + F_3\,z'(t)]\,dt$, which partially explains the notation : $x'dt$ is often written dx, etc.

An integral written $\int_C f(\mathbf{c}(t))\,dx$ is defined to mean $\int_C \mathbf{F}\bullet ds$ where $\mathbf{F} = (f, 0, 0)$. Sometimes these may be evaluated in an obvious fashion, for example, if C is a curve from a point **a** to a point **b**, $\int_C x^3\,dx = [x^4/4]_{x=a_1}^{x=b_1} = (b_1^4 - a_1^4)/4$. This is perfectly correct here because $\mathbf{F} = (x^3, 0, 0)$ is $\nabla\phi$ where $\phi(x, y, z) = x^4/4$.

If **F** is a force field then $\int_C \mathbf{F}\bullet ds$ is the *work done* by the force in moving a particle of unit mass along C. When **F** is a gradient, the work done is the difference between the potential at the two endpoints. This is often used to define potential energy.

In fact, it is only gradient vector fields which have the property given in Theorem 7.22. Let **F** be a continuous vector field defined from an open, connected set U in $\mathbb{R}^n$ into $\mathbb{R}^n$, and let **x**,**y** be two points of U. If for every curve C in U joining **x** to **y**, the line integral $\int_C \mathbf{F}\bullet d\mathbf{s}$ is constant we say that the line integral is *independent of the path*. The question asked is for which vector fields **F** is the integral $\int_C \mathbf{F}\bullet d\mathbf{s}$ independent of the path for all pairs of points **x**,**y** in U? This is equivalent to asking which vector fields have line integral 0 around every *closed* path in U. We cannot answer the question in general but a complete answer will be given assuming that all paths are piecewise C^1. Theorem 7.22 gives a sufficient condition; the line integral is independent of the path if $\mathbf{F} = \nabla\phi$. We now show that this is also a necessary condition.

Theorem 7.24. *Let U be an open connected set in $\mathbb{R}^n$ and let $\mathbf{F} : U \to \mathbb{R}^n$ be a continuous vector field. Suppose $\int_C \mathbf{F}\bullet d\mathbf{s}$ is independent of the (piecewise C^1) path joining any two points in U. Then there exists $\phi \in C^1(U,\mathbb{R})$ such that $\mathbf{F}(\mathbf{x}) = \nabla\phi(\mathbf{x})$ for every $\mathbf{x} \in U$.*

Proof. Fix a point $\mathbf{x}^{(0)} \in U$ and define $\phi(\mathbf{x}) = \int_C \mathbf{F}\bullet d\mathbf{s}$ where C is any piecewise C^1 curve joining $\mathbf{x}^{(0)}$ to **x**. Since U is open and connected it is pathwise connected and indeed polygonally connected (Proposition 2.78) so the integral makes sense and, by hypothesis, depends only on **x** and $\mathbf{x}^{(0)}$ so does define a function of **x**. We have to show that $\partial\phi/\partial x_i$ exists and equals $F_i(\mathbf{x})$. Since U is open, there is $\delta > 0$ so that $B(\mathbf{x},\delta) \subseteq U$. Then, for $|h| < \delta$, the line segment $\{\mathbf{x}+t\mathbf{e}_i : 0 \le t \le h\}$ lies in U. (Here $\mathbf{e}_i$ is the standard basis vector.) We can obtain a piecewise C^1 path from $\mathbf{x}^{(0)}$ to $\mathbf{x}+h\mathbf{e}_i$ by first taking a polygonal path **c** from $\mathbf{x}^{(0)}$ to **x** and then the line segment from **x** to $\mathbf{x}+h\mathbf{e}_i$. Therefore,

$$\phi(\mathbf{x}+h\mathbf{e}_i) = \phi(\mathbf{x}) + \int_L \mathbf{F}\bullet d\mathbf{s},$$

where L stands for the line segment from **x** to $\mathbf{x}+h\mathbf{e}_i$. Thus we have

$$\phi(\mathbf{x}+h\mathbf{e}_i) - \phi(\mathbf{x}) = \int_0^h \mathbf{F}(\mathbf{x}+t\mathbf{e}_i)\bullet\mathbf{e}_i\,dt$$

$$= \int_0^h F_i(\mathbf{x}+t\mathbf{e}_i)\,dt = h\,F_i(\mathbf{x}+\theta\mathbf{e}_i),$$

where $0<\theta(h)<h$, by the Mean Value Theorem for integrals. Therefore, since F_i is continuous, $\lim\limits_{h\to 0}\dfrac{\phi(\mathbf{x}+h\mathbf{e}_i)-\phi(\mathbf{x})}{h}$ exists and equals $F_i(\mathbf{x})$. This is the required conclusion.

Since $\operatorname{curl}(\operatorname{grad}\phi)\equiv 0$, a necessary condition for a C^1 vector field $\mathbf{F}: U\subseteq\mathbb{R}^3\longrightarrow\mathbb{R}^3$ to be a gradient is $\operatorname{curl}\mathbf{F}\equiv\mathbf{0}$. It will be shown that this condition is also sufficient on certain domains U, but the proof of this requires us to study the connection between line integrals and integrals over surfaces.

7.5. SURFACES AND SURFACE INTEGRALS

A surface may be roughly described as some set in $\mathbb{R}^3$ which has 2 degrees of freedom. A better way to think of a surface is as a deformation of a subset of $\mathbb{R}^2$. Imagine a sheet made of rubber. This can be twisted or rolled and the resulting configuration is a surface. For example, a cylinder is obtained by rolling a rectangle and joining the edges; a torus (a doughnut) is obtained by bending a cylinder and joining the ends. If we have a region D in $\mathbb{R}^2$, a deformation of D is a continuous function $\mathbf{F}: D\longrightarrow\mathbb{R}^3$. The image set $\mathbf{F}(D)$ is a surface. However, without imposing some restrictions on $\mathbf{F}$ and D, $\mathbf{F}(D)$ may be a very complicated set, for example, it is possible to have "space-filling surfaces".

The prototype of a surface is the graph of a function $f: D\subseteq\mathbb{R}^2\longrightarrow\mathbb{R}$. Then $S = \{(x, y, f(x,y)) : (x, y)\in D\}$ is the surface sitting above D at height $z = f(x,y)$. Surfaces also arise as level sets of function $\phi:\mathbb{R}^3\longrightarrow\mathbb{R}$, $S = \{(x,y,z)\in\mathbb{R}^3 : \phi(x,y,z) = \alpha\}$ is locally the graph of a function in the neighbourhoods of points where $\nabla\phi\neq\mathbf{0}$ (§7.2). However, it is useful to consider so-called parametric surfaces defined as maps from $D\subseteq\mathbb{R}^2$ to $S\subseteq\mathbb{R}^3$.

We shall consider sets D in $\mathbb{R}^2$ whose boundary is a simple closed curve C. Such curves are called Jordan curves and the following intuitively obvious fact is important.

Theorem 7.25. (*The Jordan curve theorem.*)
Let C be a simple closed curve in $\mathbb{R}^2$. Then $\mathbb{R}^2\setminus C$ consists of two components A, B, precisely one of which is bounded. Moreover, $\operatorname{bdy}(A) = C = \operatorname{bdy}(B)$.

The bounded component is called the *interior of C* and consists of the points inside C in the "obvious" sense. Since $\mathbb{R}^2 \setminus C$ is open the component is an open set so this is consistent usage of the word interior. However, as curves may be quite complicated the proof is not easy. We do not give a proof but refer to [7] in the references.

Definition 7.26. A *domain* D in $\mathbb{R}^2$ is an open, connected set whose boundary is a simple closed curve. D is called *simply connected* if every simple closed curve in D also has its interior in D.

Thus, a simply connected domain has no holes. An annulus $A = \{(x,y) : 0 < \alpha \le x^2 + y^2 \le \beta\}$ is not simply connected since the interior of the circle $x^2 + y^2 = \alpha$ does not lie in A.

We shall restrict attention to domains whose boundary curve is also piecewise C^1.

Definition 7.27. Let D be a bounded, simply connected domain in $\mathbb{R}^2$ with a piecewise C^1 boundary curve C. Suppose there is an open set D' containing $D \cup C$ and a function $\mathbf{F} \in C^1(D', \mathbb{R}^3)$. Then $S = \mathbf{F}(D \cup C)$ is called a parametric *surface element*. If also $\mathbf{F}$ is one-to-one on $D \cup C$, S is called a *simple surface element*.

We shall impose another restriction which is that a normal to the surface exists at each point. We employ vector notation to describe the situation.

For $(u,v) \in D$ write $\mathbf{r}(u,v)$ for the position vector of the image of (u,v) in S : $\mathbf{r}(u,v) = \mathbf{F}(u,v)$. Then we write

$$\mathbf{r}(u,v) = (x, y, z) = (F_1(u,v), F_2(u,v), F_3(u,v))$$

and we may write $x(u,v)$ rather than $F_1(u,v)$ etc. For u fixed, say $u = u_0$, $v \mapsto \mathbf{r}(u_0,v)$ defines a curve in S. Varying u_0 gives a family of curves in S. Similarly $v = v_0$ defines a curve and varying v_0 gives a family of such curves. The two families form a mesh which gives a reasonable impression of the appearance of the surface S.

For the curve $\mathbf{c}(u) = \mathbf{r}(u,v_0)$, $\mathbf{c}'(u)$ is a tangent vector, which we write as $\mathbf{r}_u$ (partial derivative). Similarly $\mathbf{r}_v$ is a tangent vector to the curves with u constant. Therefore $\mathbf{r}_u \times \mathbf{r}_v$ is normal to both curves and is a *normal* vector to the surface. A *unit* normal vector is

$$\mathbf{n} = \frac{\mathbf{r}_u \times \mathbf{r}_v}{\|\mathbf{r}_u \times \mathbf{r}_v\|}.$$

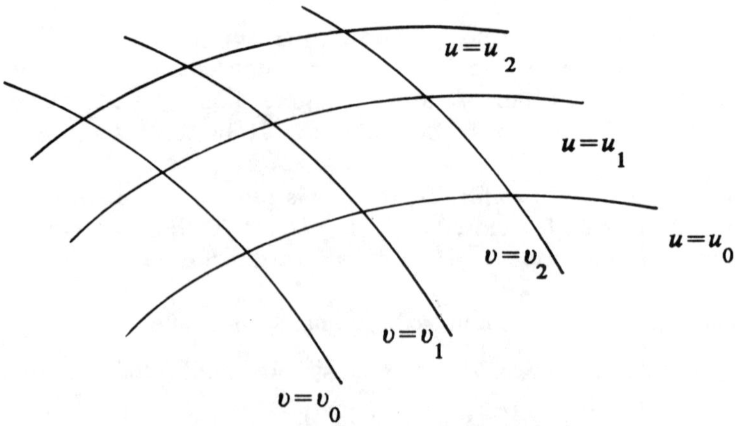

Figure 7.5.

Definition 7.28. If, in addition to the requirements of definition 7.27, $\mathbf{r}_u \times \mathbf{r}_v$ is never $\mathbf{0}$, S is called a *smooth surface element.*

Since we may write $\mathbf{r}_u = (x_u, y_u, z_u)$, $\mathbf{r}_v = (x_v, y_v, z_v)$, the normal vector $\mathbf{n} = \mathbf{r}_u \times \mathbf{r}_v$ is given by

$$\mathbf{n} = (y_u z_v - y_v z_u, z_u x_v - z_v x_u, x_u y_v - x_v y_u).$$

The components of $\mathbf{n}$ are often written J_i, $i = 1,2,3$, since J_i is a *Jacobian*. Indeed, $J_1 = \dfrac{\partial(y,z)}{\partial(u,v)}$, $J_2 = \dfrac{\partial(z,x)}{\partial(u,v)}$, $J_3 = \dfrac{\partial(x,y)}{\partial(u,v)}$ where we have written (x,y,z) for $(F_1(u,v), F_2(u,v), F_3(u,v))$.

The hypothesis in 7.28 is that at no point of S are all three Jacobians 0.

Proposition 7.29. *Let S be a smooth surface element and let $\mathbf{a} \in S$. Then there is a neighbourhood V of $\mathbf{a}$ such that $S \cap V$ can be written as the graph of a function, for example, if $J_3 \neq 0$ at $\mathbf{a}$,*

$$(x, y, z) \in S \cap V \Leftrightarrow z = \phi(x, y).$$

Proof. By hypothesis at least one of J_1, J_2, J_3 is nonzero and, by a change of notation if necessary, we may suppose $J_3(\mathbf{a}) \neq 0$. This says $\dfrac{\partial(F_1,F_2)}{\partial(u,v)} \neq 0$ at $\mathbf{a}$. Thus $\mathbf{f}(u,v) := (F_1(u,v), F_2(u,v))$ satisfies the hypotheses of the Inverse Function theorem, so $\mathbf{f}$ is locally invertible. Write $\mathbf{g}$ for the local inverse: $(u,v) = (g_1(x,y), g_2(x,y))$. Then $z = F_3(u,v) = (F_3 \circ \mathbf{g})(x,y)$ and this proves the result with $\phi = F_3 \circ \mathbf{g}$. We could equally well have invoked the Implicit Function Theorem.

This proposition justifies our calling S a *surface*: locally it is the graph of a function.

Example 7.30. Let $S = \{(x, y, f(x,y)) : (x, y) \in D\}$ be the graph of a function $f : D \subseteq \mathbb{R}^2 \longrightarrow \mathbb{R}$. Find the normal vector at a point of S.

Solution. Taking x,y as parameters we write $\mathbf{r}(x, y) = (x, y, f(x,y))$. Then $\mathbf{r}_x = (1, 0, f_x)$, $\mathbf{r}_y = (0, 1, f_y)$ and $\mathbf{n} = \mathbf{r}_x \times \mathbf{r}_y = (-f_x, -f_y, 1)$. This is the upward pointing normal. S may also be thought of as the level surface $\phi(x,y,z) := z - f(x,y) = 0$, and $\nabla\phi$ is a normal vector (§7.2) giving the same expression for $\mathbf{n}$.

Example 7.31. Let $S = \{(x,y,z) : x^2+y^2+z^2 = a^2\}$ be the sphere centred at the origin and of radius a. Then S may be parametrized by (compare spherical polar coordinates),

$$\mathbf{r}(\theta,\phi) = (a \sin\theta \cos\phi,\ a \sin\theta \sin\phi,\ a \cos\theta),\quad 0 \le \theta \le \pi,\quad 0 \le \phi < 2\pi.$$

Then $\mathbf{r}_\theta \times \mathbf{r}_\phi = a^2(\sin^2\theta \cos\phi, \sin^2\theta \sin\phi, \sin\theta \cos\theta) = a^2 \sin\theta\, \mathbf{r}(\theta,\phi)$.

This says that the normal points from 0 through $\mathbf{r}$, as is geometrically obvious .

Example 7.32. Let $S = \{(x,y,z) : x^2+y^2 = a^2,\ 0 \le z \le 1\}$ be a cylinder. A parametrization is $\mathbf{r}(\theta,z) = (a \cos\theta, a \sin\theta, z)$. Here the normal is $\mathbf{n} := \mathbf{r}_\theta \times \mathbf{r}_z = (a \cos\theta, a \sin\theta, 0)$. This says that $\mathbf{n}$ is horizontal pointing from the axis of the cylinder $x = 0$, $y = 0$ through the surface, as is evident geometrically.

Now let S be a smooth surface element. We wish to define the *area* of S. A small rectangle in the (u,v) plane is mapped into a small part of S. See figure 7.6

A point $A(u_0,v_0)$ is mapped into $A'(\mathbf{r}(u_0, v_0))$, $B(u_0+\delta u, v_0)$ is mapped into $B'(\mathbf{r}(u_0+\delta u,v_0))$. For $\delta u, \delta v$ small, it is plausible that the area of the piece of surface bounded by the curves $u = u_0$, $v = v_0$, $u = u_0+\delta u$, $v = v_0+\delta v$ is well approximated by the parallelogram with edges $A'B'$ and $A'D'$ which is close to the tangent plane. Now

$A'B' = \mathbf{r}(u_0+\delta u,v_0) - \mathbf{r}(u_0,v_0)$ and the affine approximation given by the differentiability is $\mathbf{r}_u(u_0,v_0)\delta u$. Similarly, $A'D'$ is well approximated by $\mathbf{r}_v(u_0,v_0)\delta v$. Now the area of the parallelogram $A'B'C'D'$ is $\|\overrightarrow{A'B'} \times \overrightarrow{A'D'}\|$ which is close to $\|\mathbf{r}_u \times \mathbf{r}_v\|\, \delta u\, \delta v$. This *proves* nothing but motivates the definition, imagining such small pieces to be added together. The term $\|\mathbf{r}_u \times \mathbf{r}_v\|$ can be thought of as a *local magnification factor*.

Figure 7.6.

Definition 7.33. The area of a smooth surface element $S = \mathbf{F}(D)$ is defined to be $\iint_D \|\mathbf{r}_u \times \mathbf{r}_v\|\, du\, dv = \iint_D \sqrt{J_1^2+J_2^2+J_3^2}\; du\, dv.$

Example 7.34. Let S be the paraboloid $z = x^2+y^2$, $0 \le z \le 1$. Calculate the surface area of S.

Solution. $S = \{(x, y, f(x,y)) : 0 \le x^2+y^2 \le 1\}$, where $f(x,y) = x^2+y^2$. Now D is the disk $x^2+y^2 \le 1$ in the plane $z = 0$ and by Example 7.30, $\mathbf{r}_x \times \mathbf{r}_y = (-f_x, -f_y, 1)$ so $\|\mathbf{r}_x \times \mathbf{r}_y\| = \sqrt{(f_x^2+f_y^2+1)}$.

Therefore Area $= \iint_D \sqrt{[(2x)^2+(2y)^2+1]}\, dx\, dy$. This integral is evaluated by changing the variables using polar coordinates. Thus,

$$\text{area} = \int_0^{2\pi}\int_0^1 \sqrt{(1+4r^2)}\, r\, dr\, d\theta = \int_0^{2\pi} \frac{1}{12}\,[(1+4r^2)^{3/2}]_0^1 = \frac{\pi}{6}(5^{3/2} - 1).$$

Example 7.35. Find the surface area of a sphere of radius a.

Solution We can parametrize by

$$x = a \sin\theta \cos\phi,\ y = a \sin\theta \sin\phi,\ z = a \cos\theta,\ 0 \le \phi \le 2\pi,\ 0 \le \theta \le \pi.$$

Then, $\mathbf{r}_\theta \times \mathbf{r}_\phi = a^2 \sin\theta\,(\sin\theta \cos\phi, \sin\theta \sin\phi, \cos\theta)$, as in Example 7.31, and $\|\mathbf{r}_\theta \times \mathbf{r}_\phi\| = a^2 \sin\theta$. Therefore

$$\text{area} = \int_0^{2\pi} d\phi \int_0^{\pi} a^2 \sin\theta \, d\theta = 2\pi\, a^2\, 2 = 4\pi a^2.$$

If a surface S can be decomposed into a finite number of simple smooth surface elements we extend the definition of surface area by defining the area of S to be the sum of the areas of the pieces. Additivity of the integral (Theorem 6.29) assures that this definition does not depend on the manner in which S is decomposed.

Now let S be a smooth surface element defined by $S = \{\mathbf{r}(u,v) : (u,v) \in D\}$. Let ϕ be a scalar field whose domain in $\mathbb{R}^3$ contains S. We define the *integral of ϕ over S* by

$$\iint_S \phi \, dS := \iint_D \phi(x(u,v), y(u,v), z(u,v)) \|\mathbf{r}_u \times \mathbf{r}_v\| \, du \, dv. \tag{7.4}$$

Taking $\phi = 1$ this reduces to the definition of surface area so we have Area $= \iint_S 1 \, dS$. It is convenient to write $dS = \|\mathbf{r}_u \times \mathbf{r}_v\| \, du \, dv$ which makes (7.4) easily remembered. When S is the graph of a function, $z = f(x,y)$, we find that $dS = \sqrt{(f_x^2 + f_y^2 + 1)}\, dx\, dy$. This can be interpreted as follows. Let θ be the angle between the upward normal $\mathbf{n} = (-f_x, -f_y, 1)$ and the z direction : $\mathbf{k} = (0,0,1)$. Then

$$\mathbf{n} \bullet \mathbf{k} = \|\mathbf{n}\| \|\mathbf{k}\| \cos\theta, \quad \text{that is,} \quad 1 = \sqrt{(f_x^2 + f_y^2 + 1)} \cos\theta.$$

Thus $dS = (1/\cos\theta)\, dx\, dy$. When S is a plane, θ is constant and this formula is then almost obvious from geometrical considerations.

Example 7.36. Let S be the surface defined by $z = x^2 + y$, $0 \le x \le 1$, $-1 \le y \le 1$. Evaluate $\iint_S x \, dS$.

Solution. Here $S = \{(x, y, x^2 + y) : (x,y) \in D\}$ where D is the square $0 \le x \le 1$, $-1 \le y \le 1$. Then

$$\iint_S x \, dS = \iint_D x \sqrt{[1 + (2x)^2 + 1]} \, dx \, dy = \int_{-1}^{1} dy \int_0^1 x \sqrt{[4x^2 + 2]} \, dx$$

$$= 2 \int_0^1 x \sqrt{(4x^2 + 2)} \, dx = \frac{1}{6}(6^{3/2} - 2^{3/2}) = \sqrt{6} - \sqrt{2}/3 .$$

If the graph of a function f of one variable is rotated about the

x-axis it generates the so-called solid of revolution. The surface area of the solid between the two planes $x = a$, $x = b$ is shown in courses of calculus to be $\int_a^b 2\pi|f(x)|\sqrt{1+[f'(x)]^2}\,dx$, which may also be written $\int_C 2\pi|f|\,ds$ where s denotes arc length along the curve C with equation $\mathbf{c}(t) = \{(t, f(t)) : a \le t \le b\}$. We show how this formula follows from the methods discussed above. Let S be the surface of revolution. Then S can be parametrized by

$$\mathbf{r}(x, \theta) = (x, f(x)\cos\theta, f(x)\sin\theta),\ a \le x \le b,\ 0 \le \theta \le 2\pi.$$

The surface area of S is

$$\text{area} = \iint_D \|\mathbf{r}_x \times \mathbf{r}_\theta\|\,dx\,d\theta, \text{ where } D = \{(x,\theta) : a \le x \le b,\ 0 \le \theta \le 2\pi\}.$$

Now $\mathbf{r}_x \times \mathbf{r}_\theta = (ff', -f\cos\theta, -f\sin\theta)$, so that

$$\|\mathbf{r}_x \times \mathbf{r}_\theta\| = \sqrt{(f^2(f')^2 + f^2)} = |f|\sqrt{1+f'^2}\ .\ \text{Thus,}$$

$$\text{area} = \int_0^{2\pi} d\theta \int_a^b |f(x)|\sqrt{1+f'(x)^2}\,dx = 2\pi\int_a^b |f(x)|\sqrt{1+f'(x)^2}\,dx.$$

7.6. INTEGRAL THEOREMS

We intend to show that there is a connection between line integrals and surface integrals and another between certain surface and volume integrals. We begin with a result in $\mathbb{R}^2$.

Recall that an elementary region is one that can be written $D = \{(x,y) \in \mathbb{R}^2 : a \le x \le b,\ \alpha(x) \le y \le \beta(x)\}$ or with the roles of x and y interchanged.

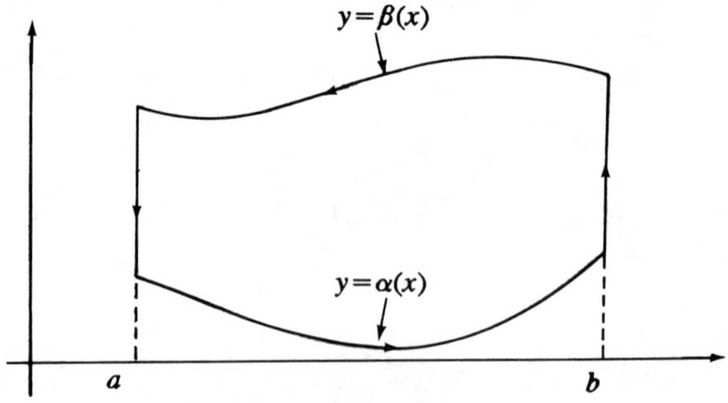

Figure 7.7.

Let C^+ denote the piecewise C^1 boundary taken in the anti-clockwise direction, as indicated by the arrows in Figure 7.7.

Lemma 7.37. *Let D be an elementary region as defined above and suppose $P,Q \in C^1(D',\mathbb{R})$ for some open set D' containing $D \cup C$. Then*

$$\int_{C^+} P\,dx = \iint_D -\frac{\partial P}{\partial y}\,dx\,dy.$$

Proof. First we have $\displaystyle\iint_D \frac{\partial P}{\partial y}\,dx\,dy = \int_a^b \left[\int_{\alpha(x)}^{\beta(x)} \partial P/\partial y\,dy\right] dx$

$$= \int_a^b [P(x,\beta(x)) - P(x,\alpha(x))]\,dx.$$

Next, the line integral consists of 4 terms corresponding to the 4 pieces of the boundary curve. Let $C_1 = \{(x,y) : y = \alpha(x),\ a \le x \le b\}$ and use x as parameter. Then $\displaystyle\int_{C_1} P(x,y)\,dx = \int_a^b P(x,\alpha(x))\,dx.$

Similarly on C_3 where $y = \beta(x)$ we may use x as parameter but the correct orientation is with x going from b to a, that is,

$$\int_{C_3} P(x,y)\,dx = \int_b^a P(x,\beta(x))\,dx = -\int_a^b P(x,\beta(x))\,dx.$$

On C_2 and C_4, x is constant so $\displaystyle\int_{C_2} P\,dx = 0 = \int_{C_4} P\,dx.$

Adding these contributions gives

$$\int_{C^+} P\,dx = \int_a^b P(x,\alpha(x))\,dx - \int_a^b P(x,\beta(x))\,dx = -\iint_D (\partial P/\partial y)\,dx\,dy.$$

Lemma 7.38. *If D is described by $D = \{(x,y) : \alpha(y) \le x \le \beta(y),\ c \le y \le d\}$ then $\displaystyle\iint_D \frac{\partial Q}{\partial x}\,dx\,dy = \int_{C^+} Q\,dy.$*

Proof Exactly similar to Lemma 7.38.

Theorem 7.39. (*Green's theorem.*)
Suppose D can be described by both of the above methods. Then

$$\int_{C^+} P\,dx + Q\,dy = \iint_D (\partial Q/\partial x - \partial P/\partial y)\,dx\,dy.$$

Proof. Simply add the results of Lemma 7.37 and 7.38.

This result is actually valid for any closed, bounded region D whose boundary consists of a finite number of simple closed rectifiable curves. We shall not prove this. A proof may be found in

Apostol's book [1]. However, we will indicate how Theorem 7.39 can be extended to cover most cases of practical interest: when D can be subdivided into a finite number of elementary regions.

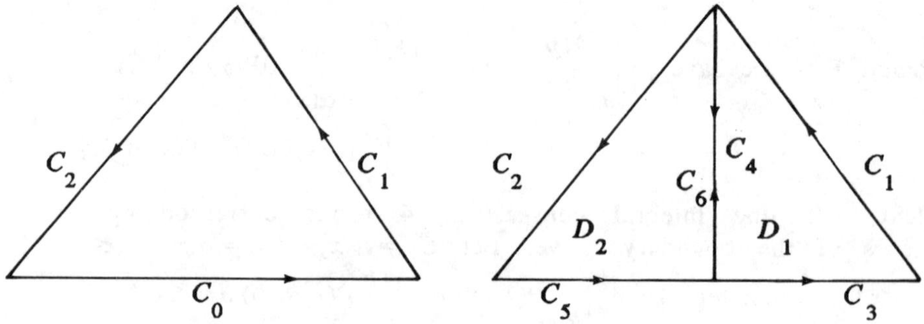

Figure 7.8.

Consider the region shown in Figure 7.8. D_1 and D_2 can each be described by both methods. Since the line segment has content zero

$$\iint_D (Q_x - P_y)\,dx\,dy = \iint_{D_1} (Q_x - P_y)\,dx\,dy + \iint_{D_2} (Q_x - P_y)\,dx\,dy.$$

Now $\displaystyle\iint_{D_1} (Q_x - P_y)\,dx\,dy = \left(\int_{C_3} + \int_{C_1} + \int_{C_4}\right)(P\,dx + Q\,dy),$

while $\displaystyle\iint_{D_2} (Q_x - P_y)\,dx\,dy = \left(\int_{C_5} + \int_{C_6} + \int_{C_2}\right)(P\,dx + Q\,dy).$

Therefore

$$\iint_D (Q_x - P_y)\,dx\,dy = \left(\int_{C_5} + \int_{C_6} + \int_{C_2} + \int_{C_3} + \int_{C_1} + \int_{C_4}\right)(P\,dx + Q\,dy).$$

However C_4 and C_6 represent the same curve but with opposite orientations. Therefore $\displaystyle\int_{C_4} = -\int_{C_6}$ and the result is shown for D.

When D has a hole we need to specify the orientation of the boundary curve more carefully. The *positive orientation* is defined as follows: if D is considered as lying in the plane $z = 0$ and a person with head pointing in the direction of $\mathbf{k}$ (upward) walks around the boundary curve C then D is on the left.

Figure 7.9 indicates the procedure for splitting such a region into elementary regions. In each subregion the boundary curve is taken anticlockwise. Then each of the common curves C_1, C_2, C_3, C_4 arise as part of the boundary curves of exactly two regions and they are taken with opposite orientations. Therefore their contributions cancel. It is readily seen that any region whose boundary consists of a finite number of line segments can be so divided.

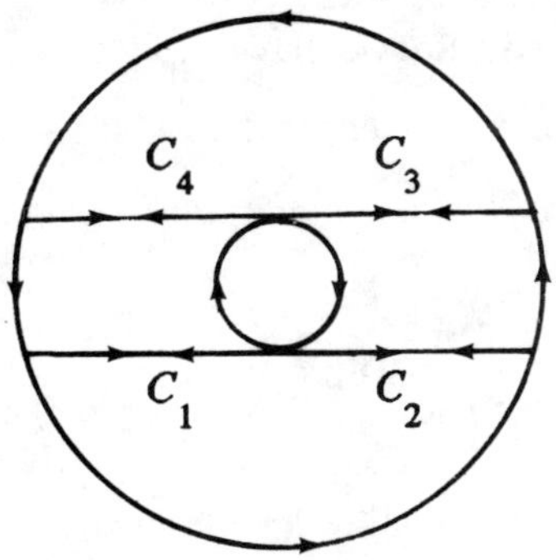

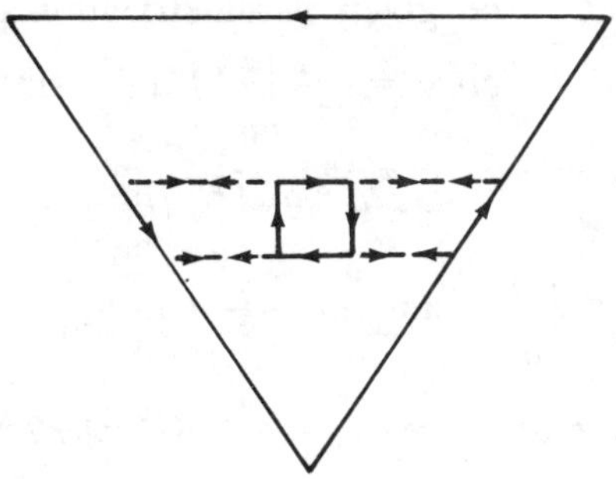

Figure 7.9.

Proposition 7.40. *Let D be a region in the plane to which Green's theorem is applicable. Then*

$$\text{area of } D = \int_{C^+} x\,dy = \int_{C^+} -y\,dx = \frac{1}{2}\int_{C^+} x\,dy - y\,dx.$$

Proof. Let $P(x,y) = -y$, $Q(x,y) = x$. Then

$$\int_{C^+} x\,dy - y\,dx = \iint_D (Q_x - P_y)\,dx\,dy = 2\iint_D dx\,dy = 2(\text{area of } D).$$

Taking firstly $P = 0$ and secondly $Q = 0$ shows the other cases.

This result shows that the area of a region is determined by its boundary curve which should not be surprising.

Example 7.41. Find the area between the x-axis and the arch of the cycloid $x = a\,\theta - a\sin\theta$, $y = a - a\cos\theta$, $0 \le \theta \le 2\pi$.

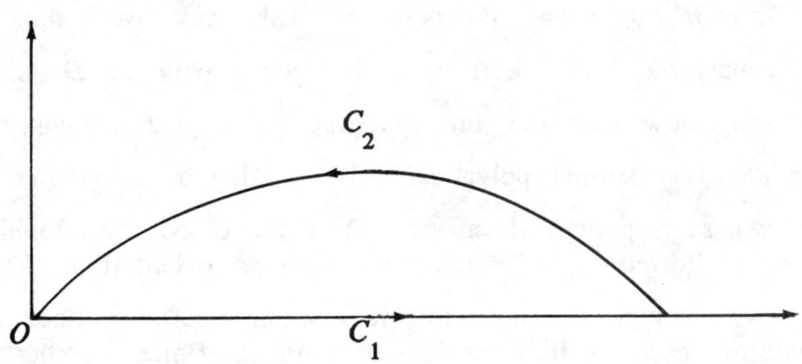

Figure 7.10.

Solution. Area $= \frac{1}{2}\int_{C_1} x\,dy - y\,dx + \frac{1}{2}\int_{C_2} x\,dy - y\,dx.$

On C_1, $y = 0$ and $dy = 0$ so $\int_{C_1} = 0.$

On C_2, the given parametrization has to be reversed. So we have

$$\text{area} = -\frac{1}{2}\int_0^{2\pi} [(a\,\theta - \sin\theta)a\sin\theta - (a - a\cos\theta)(a - a\cos\theta)]\,d\theta$$

$$= -(a^2/2)\int_0^{2\pi} (\theta\sin\theta + 2\cos\theta - 2)\,d\theta$$

Now $\int_0^{2\pi} \theta\sin\theta\,d\theta = [-\theta\cos\theta]_0^{2\pi} + \int_0^{2\pi}\cos\theta\,d\theta$ and $\int_0^{2\pi}\cos\theta = 0.$

Therefore, area $= -(a^2/2)[-2\pi] + 2\pi a^2 = 3\pi a^2.$

By means of Green's theorem we can now give a more useful answer to the question as to when a line integral in $\mathbb{R}^2$ is independent of the path (Theorem 7.24). For a vector field $\mathbf{F} : \mathbb{R}^2 \longrightarrow \mathbb{R}^2$, $\int_C \mathbf{F}\bullet d\mathbf{s}$ can be written $\int_C F_1\,dx + F_2\,dy.$

Theorem 7.42. *Let D be a simply connected domain and let $\mathbf{F} \in C^1(D,\mathbb{R}^2)$. Then $\int_{\mathbf{c}} \mathbf{F}\bullet d\mathbf{s}$ is independent of the path in D $\Leftrightarrow$ $\partial F_1/\partial y = \partial F_2/\partial x$.*

Proof. If $\int \mathbf{F}\bullet d\mathbf{s}$ is independent of the path, then by Theorem 7.24, $\mathbf{F} = \nabla\phi$ for some scalar field ϕ. Thus $F_1 = \partial\phi/\partial x$, $F_2 = \partial\phi/\partial y$ and $\partial F_1/\partial y = \partial^2\phi/\partial y\partial x$, $\partial F_2/\partial x = \partial^2\phi/\partial x\partial y$ and these are equal since $\mathbf{F} \in C^1$ gives $\phi \in C^2$. (This part does not require D to be simply connected.) Now we turn to the converse. So suppose that $\partial F_1/\partial y = \partial F_2/\partial x$ at all points of D. Take any two points $\mathbf{x},\mathbf{y} \in D$. Then, by Proposition 2.78, there is a polygonal path in D joining $\mathbf{x}$ and $\mathbf{y}$. We will show that the line integral $\int F_1 dx + F_2 dy$ has the same value over any two simple polygonal paths joining $\mathbf{x}$ to $\mathbf{y}$. Let C_1 and C_2 be any two such polygonal curves. Then let C be the closed curve formed by C_1 followed by C_2 with its reverse orientation. Various parts of the two polygons may coincide along certain line segments. The remaining parts will intersect at most a finite number of times and they will form the boundaries of a finite number of simple closed polygons $D_1, D_2, \cdots, D_m$, say. These lie in D since D is simply connected. An example is illustrated in Figure 7.11.

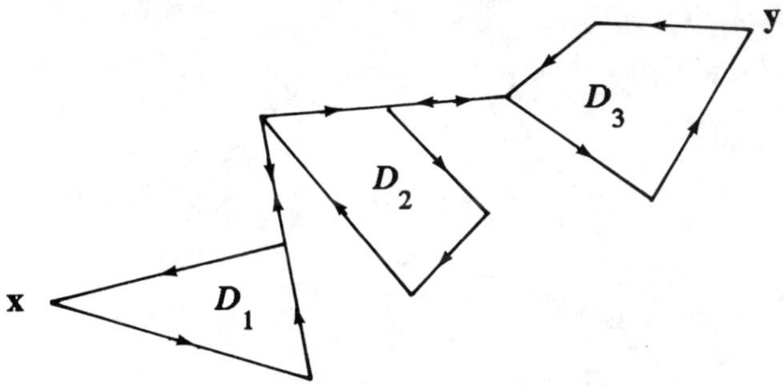

Figure 7.11.

Now the contribution to the line integral over any common line segment is 0 since the segment is traversed twice in opposite directions. For any of the regions D_i, the line integral round the boundary is, by Green's theorem,

$$\pm\iint_{D_i} (\partial F_2 / \partial x - \partial F_1 / \partial y)\, dx\, dy = 0, \text{ since } \partial F_1 / \partial y = \partial F_2 / \partial x.$$

Therefore, for $\mathbf{x}^{(0)}$ fixed, we can define ϕ by $\phi(\mathbf{x}) = \int_C \mathbf{F}\cdot d\mathbf{s}$, where C is any polygonal curve connecting $\mathbf{x}^{(0)}$ to $\mathbf{x}$. This is a function of $\mathbf{x}$ alone by what has just been shown. We may now repeat the argument of the proof of Theorem 7.24 to show that $\mathbf{F} = \nabla\phi$. This shows that the line integral is independent of the path (Theorem 7.22).

We shall next give an analogue of Green's theorem which will relate certain surface integrals to volume integrals. We shall consider a "solid" $V \subseteq \mathbb{R}^3$ with a boundary surface S which can be described in the following manner. Let D be a closed region in the x-y plane whose boundary is a piecewise C^1 curve C and let $\alpha,\beta \in C^1(D,\mathbb{R})$ satisfy $\alpha(x,y) < \beta(x,y)$ if (x,y) is an interior point of D, and $\alpha(x,y) = \beta(x,y)$ for $(x,y) \in C$. We consider the set

$$V = \{(x,y,z) : (x,y) \in D \text{ and } \alpha(x,y) \leq z \leq \beta(x,y)\}.$$

Let S_1, S_2 be the parametric surfaces described by

$$\mathbf{r}^{(1)}(x,y) = (x,y,\beta(x,y)),\ \mathbf{r}^{(2)}(x,y) = (x,y,\alpha(x,y)), \text{ for } (x,y) \in D.$$

Then S_1 is the upper surface of V, S_2 is the lower surface of V and $S = S_1 \cup S_2$ is the boundary of V.

For such a solid V we take the *positive* unit normal vector to be the one that points out of V. Thus on S_1, **n** should point up, on S_2, **n** should point down. Therefore,

$$\mathbf{n}^{(1)}=(\mathbf{r}_x^{(1)}\times\mathbf{r}_y^{(1)})/\|\mathbf{r}_x^{(1)}\times\mathbf{r}_y^{(1)}\|,\ \mathbf{n}^{(2)}=-(\mathbf{r}_x^{(2)}\times\mathbf{r}_y^{(2)})/\|\mathbf{r}_x^{(2)}\times\mathbf{r}_y^{(2)}\|.$$

In the case at hand,

$$\mathbf{n}^{(1)}=(-\beta_x,-\beta_y,1)/\sqrt{(\beta_x^2+\beta_y^2+1)},\ \mathbf{n}^{(2)}=(\alpha_x,\alpha_y,-1)/\sqrt{(\alpha_x^2+\alpha_y^2+1)}.$$

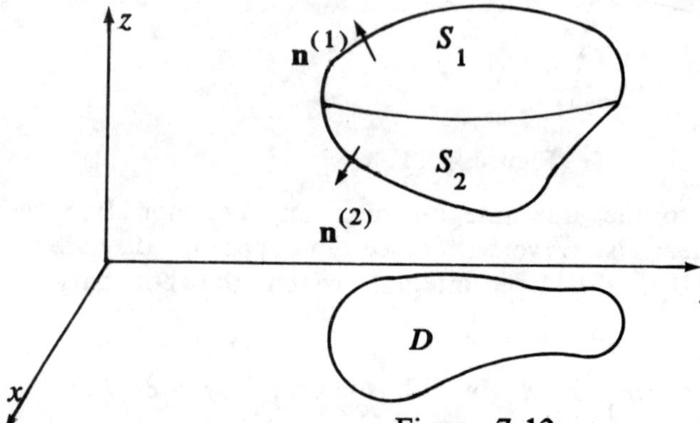

Figure 7.12.

Proposition 7.43. *Let V be as described above and let ϕ be a class C^1 scalar field on a domain containing V. Then, (writing dV in place of $dxdydz$),*

$$\iiint_V \frac{\partial\phi}{\partial z}\,dV=\iint_{S_1}\phi\, n_3^{(1)}dS_1+\iint_{S_2}\phi\, n_3^{(2)}dS_2.$$

Proof. By integrating with respect to z we have

$$\iiint_V \frac{\partial\phi}{\partial z}\,dV=\iint_D\left[\int_{\alpha(x,y)}^{\beta(x,y)}\frac{\partial\phi}{\partial z}\,dz\right]dx\,dy$$

$$=\iint_D \phi(x,y,\beta(x,y))\,dx\,dy-\iint_D \phi(x,y,\alpha(x,y))\,dx\,dy$$

$$=\iint_D \phi(\mathbf{r}^{(1)}(x,y))\,n_3^{(1)}\sqrt{(\beta_x^2+\beta_y^2+1)}\,dx\,dy\ +$$

$$+\iint_D \phi(\mathbf{r}^{(2)}(x,y))\,n_3^{(2)}\sqrt{(\alpha_x^2+\alpha_y^2+1)}\,dx\,dy$$

$$=\iint_{S_1}\phi\, n_3^{(1)}dS_1+\iint_{S_2}\phi\, n_3^{(2)}dS_2.$$

We define $\iint_S \mathbf{F}\cdot d\mathbf{S}$ to be $\iint_S (\mathbf{F}\cdot\mathbf{n})\,dS$ so that

$$\iint_S (\mathbf{F}\bullet\mathbf{n})\, dS = \iint_{S_1} (\mathbf{F}\bullet\mathbf{n}^{(1)})\, dS_1 + \iint_{S_2} (\mathbf{F}\bullet\mathbf{n}^{(2)})\, dS_2$$ and we have :

Theorem 7.44. (*Gauss's divergence theorem.*)
Suppose $V\subseteq\mathbb{R}^3$ can be described as above and in the other two similar ways. Then, for a C^1 vector field $\mathbf{F}$,

$$\iint_S \mathbf{F}\bullet d\mathbf{S} = \iint_S \mathbf{F}\bullet\mathbf{n}\, dS = \iiint_V \operatorname{div}\mathbf{F}\, dV. \tag{7.5}$$

Proof. By Proposition 7.43 and the two analogous results

$$\iint_S (\mathbf{F}\bullet\mathbf{n})\, dS = \iint_S (F_1n_1+F_2n_2+F_3n_3)dS$$

$$= \iiint_V \partial F_1/dx\, dV + \iiint_V \partial F_2/dy\, dV + \iiint_V \partial F_3/dz\, dV$$

$$= \iiint_V \operatorname{div}\mathbf{F}\, dV.$$

Note that the left side of (7.5) involves 2 integrations, while the right side involves 3 integrations and 1 differentiation.

Example 7.45. Evaluate $\iint_S \mathbf{F}\bullet\mathbf{n}\, dS$ where $\mathbf{F}(x,y,z) = (xy^2, x^2y, y)$ and S is the closed cylinder $\{x^2+y^2 = 1, -1\leq z\leq 1\}\cup\{x^2+y^2\leq 1, z = \pm 1\}$.

Solution. One may compute the surface integral directly over the three surfaces top, curved surface and bottom. However, the divergence theorem is easier here. Since $\operatorname{div}\mathbf{F} = y^2+x^2$, we have

$$\iint_S \mathbf{F}\bullet\mathbf{n}\, dS = \iiint_V (y^2+x^2)\, dx\, dy\, dz = \iint_D \left[\int_{-1}^{1} (y^2+x^2)\, dz\right] dx\, dy,$$

where D is the unit disk $D = \{x^2+y^2\leq 1, z = 0\}$. Thus,

$$\iint_S \mathbf{F}\bullet\mathbf{n}\, dS = 2\iint_D (x^2+y^2)\, dx\, dy = 2\int_0^{2\pi} r^2\, r\, dr\, d\theta = \pi,$$ where we have transformed to polar coordinates.

There is an analogue of Green's theorem which applies to surfaces and their boundary curves and known as Stokes's theorem.

Let S be a simple smooth surface element defined as points $\mathbf{r}(u,v)$ for $(u,v)\in D$. The boundary of D is a curve oriented positively with respect to D. This induces an orientation on C the boundary of S. This orientation is such that a person walking around C with his head pointing in the direction of the normal $\mathbf{n} = \mathbf{r}_u \times \mathbf{r}_v = (J_1,J_2,J_3)$ sees S on his left.

Theorem 7.46. (*Stokes's theorem.*)

Let S be a simple smooth surface element and let C denote the boundary curve of S with the induced positive orientation. Let $\mathbf{F}$ *be a vector field on S for which* curl $\mathbf{F}$ *exists. Then*

$$\iint_S (\text{curl } \mathbf{F}\cdot\mathbf{n})\, dS = \int_C \mathbf{F}\cdot d\mathbf{s}$$

Proof. We write points of S as $\{\mathbf{r}(u,v),\ (u,v)\in D\}$ and denote points on the boundary curve B^+ of D by $\{(u(t),v(t)),\ a\le t\le b\}$. Then C is described by

$$\mathbf{c}(t) := (x(u(t),v(t)),\ y(u(t),v(t)),\ z(u(t),v(t))),\ a\le t\le b.$$

Now $\displaystyle\int_C \mathbf{F}\cdot d\mathbf{s} = \int_a^b \mathbf{F}(\mathbf{c}(t))\cdot\mathbf{c}'(t)\, dt$. By the chain rule,

$$\mathbf{c}'(t) = \left(\frac{\partial x}{\partial u}u' + \frac{\partial x}{\partial v}v',\ \frac{\partial y}{\partial u}u' + \frac{\partial y}{\partial v}v',\ \frac{\partial z}{\partial u}u' + \frac{\partial z}{\partial v}v'\right).$$

We consider the term

$$\int_C F_1 dx = \int_a^b F_1\left(\frac{\partial x}{\partial u}u' + \frac{\partial x}{\partial v}v'\right) dt = \int_{B^+} F_1\frac{\partial x}{\partial u}du + F_1\frac{\partial x}{\partial v}dv,$$

where B^+ is the boundary curve of D. By Green's theorem this is

$$\iint_D \frac{\partial}{\partial u}\left(F_1\frac{\partial x}{\partial v}\right) - \frac{\partial}{\partial v}\left(F_1\frac{\partial x}{\partial u}\right) du\, dv = \iint_D \left(\frac{\partial F_1}{\partial u}\frac{\partial x}{\partial v} - \frac{\partial F_1}{\partial v}\frac{\partial x}{\partial u}\right) du\, dv,$$

the other terms cancel by the equality of mixed partial derivatives.
Now, by the chain rule once more, omitting the suffix 1 on F_1 for simplicity of notation,

$$\frac{\partial F}{\partial u} = \frac{\partial F}{\partial x}\frac{\partial x}{\partial u} + \frac{\partial F}{\partial y}\frac{\partial y}{\partial u} + \frac{\partial F}{\partial z}\frac{\partial z}{\partial u}, \tag{7.6}$$

and

$$\frac{\partial F}{\partial v} = \frac{\partial F}{\partial x}\frac{\partial x}{\partial v} + \frac{\partial F}{\partial y}\frac{\partial y}{\partial v} + \frac{\partial F}{\partial z}\frac{\partial z}{\partial v}. \tag{7.7}$$

Multiply (7.6) by $\dfrac{\partial x}{\partial v}$, (7.7) by $\dfrac{\partial x}{\partial u}$ and subtract to obtain

$$\frac{\partial F}{\partial u}\frac{\partial x}{\partial v} - \frac{\partial F}{\partial v}\frac{\partial x}{\partial u} = -\frac{\partial F}{\partial y}J_3 + \frac{\partial F}{\partial z}J_2,$$

where $J_1 = \dfrac{\partial(y,z)}{\partial(u,v)}$, $J_2 = \dfrac{\partial(z,x)}{\partial(u,v)}$, $J_3 = \dfrac{\partial(x,y)}{\partial(u,v)}$.
Therefore, the integral becomes

$$\int_C F\,dx = \iint_S \left(-\frac{\partial F}{\partial y}J_3 + \frac{\partial F}{\partial z}J_2\right) du\, dv = \iint_S \left(-\frac{\partial F}{\partial y}n_3 + \frac{\partial F}{\partial z}n_2\right) dS$$

where $\mathbf{n} = (n_1,n_2,n_3) = (J_1,J_2,J_3)/\sqrt{(J_1^2+J_2^2+J_3^2)}$ is the unit normal and $dS = \sqrt{(J_1^2+J_2^2+J_3^2)}\, du\, dv$.

Similarly we calculate the other two terms and find

$$\int_C \mathbf{F}\bullet d\mathbf{s} = \iint_S \left[-\frac{\partial F_1}{\partial y} n_3 + \frac{\partial F_1}{\partial z} n_2 - \frac{\partial F_2}{\partial z} n_1 + \frac{\partial F_2}{\partial x} n_3 - \frac{\partial F_3}{\partial x} n_2 + \frac{\partial F_3}{\partial y} n_1 \right] dS$$

$$= \iint_S \left[\left[\frac{\partial F_3}{\partial y} - \frac{\partial F_2}{\partial z} \right] n_1 + \left[\frac{\partial F_1}{\partial z} - \frac{\partial F_3}{\partial x} \right] n_2 + \left[\frac{\partial F_2}{\partial x} - \frac{\partial F_1}{\partial y} \right] n_3 \right] dS$$

$$= \iint_S (\text{curl } \mathbf{F})\bullet\mathbf{n}\, dS.$$

In order to extend Stokes's theorem to a surface that can be decomposed into simple smooth surface elements we need integrals over common parts of boundary curves to cancel, just as happened for Green's theorem. Suppose S is made up of two parts S_1 and S_2 whose positively oriented boundary curves C_1 and C_2 have one or more common parts. Each of these common portions inherits an orientation from C_1 and C_2. We say S is an *orientable surface* if these two orientations are opposite on each of the boundary curves which comprise $C_1 \cap C_2$. Otherwise, S is *nonorientable.* Similarly we can consider a finite number of simple smooth surface elements $S_1,\cdots,S_m$ with corresponding oriented boundaries $C_1,\cdots,C_m$. If each C_i has at most finitely many curves in common with all the other C_j's. and if the two orientations induced on the common parts are opposite, then S is orientable. Orientable surfaces have 2 sides: at each point of S there are two possible normal vectors $\mathbf{n}^{(1)}$ and $\mathbf{n}^{(2)} = -\mathbf{n}^{(1)}$, S is orientable if $\mathbf{n}^{(1)}$ determines a unique side of S. The Möbius band is an example of a surface with only one side so is not orientable; see Figure 7.13.

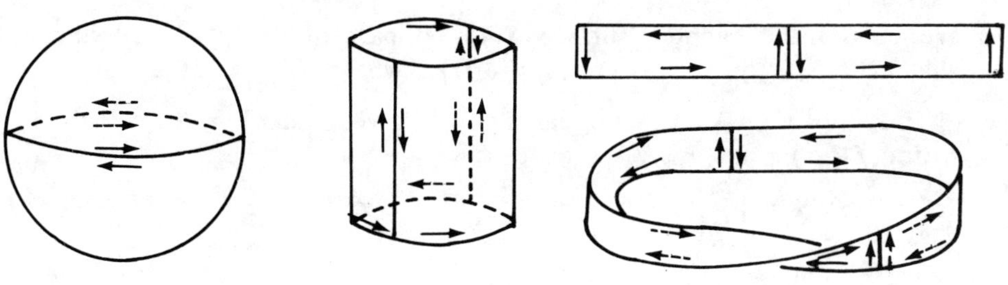

orientable nonorientable

Figure 7.13.

Stokes's theorem extends immediately to orientable surfaces because the integrals over the common parts of $C_i \cap C_j$ cancel and we are left with the curves that form the boundary of the larger surface.

If S is a *closed* surface, for example a sphere, there is no boundary curve and Stokes's theorem gives $\iint_S (\text{curl}\,\mathbf{F}\bullet\mathbf{n})\,dS = 0$.

Note that, for a given curve C there may be several possible surfaces which have C as their boundary curve. Any such surface can be called a *capping surface* for C. If S_1 and S_2 have the same boundary curve C then $\iint_{S_1} (\text{curl}\,\mathbf{F}\bullet\mathbf{n}^{(1)})\,dS_1 = \iint_{S_2} (\text{curl}\,\mathbf{F}\bullet\mathbf{n}^{(2)})\,dS_2$

since both equal $\int_C \mathbf{F}\bullet d\mathbf{s}$.

Example 7.47. Evaluate (*a*) directly, (*b*) using Stokes's theorem $\int_C -y^3\,dx + x^3\,dy - z^3\,dz$ where C is the positively oriented curve of intersection between the cylinder $x^2+y^2 = 1$ and the plane $x+y+z = 1$.

Solution. (*a*) C can be parametrized as
$x = \cos t$, $y = \sin t$, $z = 1 - \cos t - \sin t$, $0 \le t \le 2\pi$. Then

$$\int_C \mathbf{F}\bullet d\mathbf{s} = \int_0^{2\pi} (\sin^4 t + \cos^4 t)\,dt - \int_C z^3\,dz,$$

where $\int_C z^3\,dz = [z^4/4] = \left[\dfrac{(1 - \sin t - \cos t)^4}{4}\right]_0^{2\pi} = 0$. Therefore,

$$\int_C \mathbf{F}\bullet d\mathbf{s} = 4\int_0^{\pi/2} (\sin^4 t + \cos^4 t)\,dt, \text{ by symmetry,}$$
$$= 3\pi/2, \text{ by standard reduction formulae.}$$

(*b*) We choose the capping surface S to be part of the plane $x+y+z = 1$, so that $S = \{(x, y, 1 - x - y) : (x, y) \in D\}$ where D is the unit disk $\{x^2+y^2 \le 1, z = 0\}$. A simple calculation gives $\text{curl}\,\mathbf{F} = (0, 0, 3x^2+3y^2)$, so writing $f(x,y) = 1 - x - y$ we have

$$\iint_S (\text{curl}\,\mathbf{F}\bullet\mathbf{n})\,dS = \iint_D (\text{curl}\,\mathbf{F})\bullet(-f_x, -f_y, 1)\,dx\,dy$$
$$= 3\iint_D (x^2+y^2)\,dx\,dy = 3\pi/2,$$

by the same calculation as in Example 7.45.

Green's theorem allowed us to determine necessary and sufficient conditions for a 2-dimensional vector field to be a gradient. Stokes's theorem allows us to do the same for a 3-dimensional vector field. However, we shall only prove a result when the region is *convex*, rather than an analogue of the simply connected domain used earlier.

Theorem 7.48. *Let A be a convex domain in $\mathbb{R}^3$ and let $\mathbf{F}\in C^1(A,\mathbb{R}^3)$. Then $\mathbf{F}$ is a gradient vector field if and only if* curl $\mathbf{F} = \mathbf{0}$ *in A.*

Proof. If there is a scalar field ϕ such that $\mathbf{F} = \nabla\phi$ then we see that curl $\mathbf{F}$ = curl grad$\phi = \mathbf{0}$, by the identity in Proposition 7.4.
For the converse, fix a point $\mathbf{x}^{(0)}$ in A and for any point $\mathbf{x}\in A$ define $\phi(\mathbf{x}) := \int_C \mathbf{F}\bullet ds$. where C is the line segment $\{ (1-t)\mathbf{x}^{(0)} + t\mathbf{x} \}$.

Let $\mathbf{x} = (x,y,z)$ and let $\mathbf{y} = (x+h,y,z)$ where h is so small that $\mathbf{y}\in A$. Then the triangle T with vertices $\mathbf{x}^{(0)}$,$\mathbf{x}$, and $\mathbf{y}$ lies in A by convexity. Let S denote the triangular region bounded by the line segments T_0 from $\mathbf{x}^{(0)}$ to $\mathbf{x}$, T_1 from $\mathbf{x}$ to $\mathbf{y}$, T_2 from $\mathbf{y}$ to $\mathbf{x}^{(0)}$. Then

$\int_T \mathbf{F}\bullet ds = \pm\iint_S (\text{curl }\mathbf{F}\bullet\mathbf{n})\, dS = 0$, by the hypothesis that curl $\mathbf{F} = \mathbf{0}$.

Thus $\int_{T_0} \mathbf{F}\bullet ds + \int_{T_1} \mathbf{F}\bullet ds + \int_{T_2} \mathbf{F}\bullet ds = 0$, that is, $\phi(\mathbf{x}) - \phi(\mathbf{y}) + \int_{T_1} \mathbf{F}\bullet ds = 0$.

Now we may write $T_1 = \{(x+t,\ y,\ z) : 0\le t\le h\}$ and so

$$\phi(\mathbf{y}) - \phi(\mathbf{x}) = \int_0^h F_1(x+t,y,z)\, dt = h\, F_1(x',y,z),$$

where x' is some point between x and $x+h$, by the Mean Value Theorem for integrals. This shows that

$$\partial\phi/\,\partial x = \lim_{h\to 0}\, [\, \phi(x+h,\ y,\ z) - \phi(x,\ y,\ z)\,]/\,h = F_1(x,\ y,\ z).$$

Similarly $\partial\phi/\partial y = F_2$ and $\partial\phi/\partial z = F_3$.

We conclude by showing how one can prove a change of variables formula for double integrals via Green's theorem.

Let Ω and D be closed regions to which Green's theorem is applicable and let $\mathbf{g}\in C^2(\Omega,\mathbb{R}^2)$ be one-to-one from Ω onto $D = \mathbf{g}(\Omega)$ and such that $J_{\mathbf{g}}$ is never 0 on Ω. Then, by Theorem 4.33, $\mathbf{g}$ maps the boundary curve Γ of Ω onto the boundary C of D. Write coordinates in Ω as $\mathbf{u} = (u,v)$, in D as $\mathbf{x} = (x,y)$.

Proposition 7.49. *Under the hypothesis of the previous paragraph,*

$$\text{area of } D = c(D) = \iint_\Omega |J_{\mathbf{g}}(\mathbf{u})|\ du\, dv.$$

Proof. By Green's theorem,

$$\text{area of } D = \iint_D 1\ dx\, dy = \int_C x\, dy$$

Let Γ be written $\{\mathbf{u}(t)=(u(t), v(t)) : 0 \le t \le 1\}$. Then $C = \mathbf{g}(\Gamma)$ can be written $\{(g_1(\mathbf{u}(t)), g_2(\mathbf{u}(t))) : 0 \le t \le 1\}$. Therefore we have

$$\int_C x\, dy = \int_0^1 g_1(\mathbf{u}(t))\, dg_2/dt\, dt = \int_0^1 g_1\left[\frac{\partial g_2}{\partial u}\frac{du}{dt} + \frac{\partial g_2}{\partial v}\frac{dv}{dt}\right] dt,$$

$$= \pm \int_\Gamma g_1 \frac{\partial g_2}{\partial u}\, du + g_1 \frac{\partial g_2}{\partial v}\, dv$$

the sign depending on whether the orientation is preserved or reversed by $\mathbf{g}$. Now we apply Green's theorem in Ω to write this last line integral as

$$\iint_\Omega \left\{\frac{\partial}{\partial u}\left[g_1\frac{\partial g_2}{\partial v}\right] - \frac{\partial}{\partial v}\left[g_1\frac{\partial g_2}{\partial u}\right]\right\} du\, dv = \iint_\Omega \left[\frac{\partial g_1}{\partial u}\frac{\partial g_2}{\partial v} - \frac{\partial g_1}{\partial v}\frac{\partial g_2}{\partial u}\right] du\, dv,$$

the other terms cancelling by equality of the mixed partials (it is here we use the C^2 hypothesis).

This last integral is $\iint_\Omega J_{\mathbf{g}}(\mathbf{u})\, du\, dv$ so we have

$$\text{area of } D = \pm \iint_\Omega J_{\mathbf{g}}(\mathbf{u})\, du\, dv.$$

Since area is positive, the ambiguous sign must be that of $J_{\mathbf{g}}$ itself and so area of $D = \iint_\Omega |J_{\mathbf{g}}(\mathbf{u})|\ du\, dv.$

This result proves the change of variable formula for double integrals for the function $f(\mathbf{x}) = 1$. This can then be used to prove the general case using Lemma 6.49 and arguing as in Chapter 6. It is also possible to prove a similar formula in $\mathbb{R}^3$ from the divergence theorem. We shall not pursue this.

Problems

1. Prove that $\mathbf{a} \times (\mathbf{b} \times \mathbf{c}) = (\mathbf{a}\cdot\mathbf{c})\mathbf{b} - (\mathbf{a}\cdot\mathbf{b})\mathbf{c}$. Deduce that

$$\mathbf{a} \times (\mathbf{b} \times \mathbf{c}) + \mathbf{b} \times (\mathbf{c} \times \mathbf{a}) + \mathbf{c} \times (\mathbf{a} \times \mathbf{b}) = \mathbf{0}.$$

2. Let $\mathbf{F}, \mathbf{G} \in C^1(\mathbb{R}^3, \mathbb{R}^3)$. Show that

$$d/dt(\mathbf{F}(t)\bullet\mathbf{G}(t)) = \mathbf{F}(t)\bullet\mathbf{G}'(t) + \mathbf{F}'(t)\bullet\mathbf{G}(t),$$

and $$(\mathbf{F}(t) \times \mathbf{G}(t))' = \mathbf{F}(t) \times \mathbf{G}'(t) + \mathbf{F}'(t) \times \mathbf{G}(t).$$

3. Prove that, if $\mathbf{r}=(x,y,z)$, $r=\|\mathbf{r}\|$, then, for a constant vector $\mathbf{a}$, (*a*) $\nabla \times (\mathbf{a} \times \mathbf{r})=2\mathbf{a}$, (*b*) $\nabla \times \left[\frac{\mathbf{a}\times\mathbf{r}}{r^n}\right] = \frac{n(\mathbf{a}\bullet\mathbf{r})}{r^{n+2}}\mathbf{r} - \frac{(n-2)}{r^n}\mathbf{a}$.

4. Show that the curvature and the torsion of the curve $\mathbf{c}(t)=(t+\sin t, t-\sin t, \sqrt{2}\cos t)$ are given by $\kappa=\tau=\sqrt{2}/4$.

5. Show that, for any C^1 curve, $\frac{d\mathbf{N}}{ds} = -\kappa\mathbf{T} + \tau\mathbf{B}$.

6. Find the arc length of the cycloid $x=a\,\theta-a\sin\theta$, $y=a-a\cos\theta$, $0\leq\theta\leq 2\pi$.

7. Find the arc length of each of the curves :
(*i*) the hypocycloid with 4 cusps : $\mathbf{c}(t) := \{(\cos^3 t, \sin^3 t, 0) : 0\leq t\leq 2\pi\}$;
(*ii*) the curve $\mathbf{c}(t) := (2t, t^2, \ln(t))$ between $(2,1,0)$ and $(4,4,\ln(2))$.

8. Evaluate $\int_C \mathbf{F}\bullet d\mathbf{r}$ where $\mathbf{F}=(yz,2y,-x^2)$ and C is
(*i*) the straight line from $(0,0,0)$ to $(1,1,1)$;
(*ii*) the curve $x=t$, $y=t^2$, $z=t^3$, $0\leq t\leq 1$, joining $(0,0,0)$ to $(1,1,1)$.

9. Show that for any simple closed curve C and any C^1 scalar fields ϕ, ψ, $\int_C \phi\,\nabla\psi\bullet d\mathbf{s} = -\int_C \psi\,\nabla\phi\bullet d\mathbf{s}$.

10. Show that for any simple closed curve C, $\int_C \mathbf{r}\bullet d\mathbf{s} = 0$ where $\mathbf{r}$ is the usual position vector $\mathbf{r}=(x, y, z)$.

11. Show that if $\mathbf{F}$ is a *constant* vector field and C is a simple C^1 closed curve then $\int_C \mathbf{F}\bullet d\mathbf{s} = 0$.

12. Evaluate $\int_C (x^2+y^2+z^2)\,ds$ along the straight line segment from $(1,0,0)$ to $(1,3,0)$.

13. Let $\mathbf{F} = (z, x, y)$. Evaluate $\int_{C^+} \mathbf{F}\bullet d\mathbf{s}$ where C^+ is the circle $x^2+y^2 = a^2$, $z = 0$, taken in the positive sense.

14. Let $\phi(x, y) := \tan^{-1}(y/x)$ (=arctan, the inverse function). Calculate $\nabla\phi$. Let D be the annular region $D := \{0 < a^2 \leq x^2 + y^2 \leq b^2\}$ and let C be the circle $x^2 + y^2 = r^2$ where $a < r < b$. Calculate the line integral $\int_C \frac{-y\,dx}{x^2+y^2} + \frac{x\,dy}{x^2+y^2}$. Why does this not contradict Theorems 7.22?

15. Verify that $\mathbf{F} = (2x+y+3z, x+2y+2z, 3x+2y+2z)$ is irrotational, and find ϕ such that $\mathbf{F} = \text{grad}\,\phi$.

16. Let $\mathbf{a}$ be a constant and $\mathbf{r}$ be the usual position vector. Which of the following fields are conservative? Find a suitable scalar potential ϕ when appropriate.
(*i*) $(\mathbf{a}\cdot\mathbf{r})\mathbf{r}$, (*ii*) $(\mathbf{a}\cdot\mathbf{r})\mathbf{a}$, (*iii*) $(\mathbf{a}\cdot\mathbf{r})\mathbf{b} + (\mathbf{b}\cdot\mathbf{r})\mathbf{a}$.

17. Find the area of the part of the surface of the cylinder $x^2 + z^2 = 2x$ which is bounded by the planes $y = 0$ and $x + y = 2$.

18. Find the area of the surface of the cone $x^2 + y^2 = z^2 \tan^2\alpha$, $z \geq 0$ $(0 < \alpha < \pi/2)$ which lies inside the sphere $x^2 + y^2 + z^2 = 2ay$ $(a > 0)$.

19. Let $\mathbf{r} = (x,y,z)$. Evaluate $\iint_S \mathbf{r}\cdot d\mathbf{S}$ where $\mathbf{S}$ is the curved surface of the cylinder $x^2 + y^2 = a^2$, $0 \leq z \leq h$.

20. Find the surface area of the torus
$$\mathbf{r}(\theta,\phi) = ((R + \cos\phi)\cos\theta, (R + \cos\phi)\sin\theta, \sin\phi),\ 0 \leq \theta,\ \phi \leq 2\pi,$$
where $R > 1$ is a constant.

21. Find the area of the portion of the sphere $x^2 + y^2 + z^2 = 1$ that is cut off by the cone $x^2 + y^2 = z^2$, $z \geq 0$.

22. Find the area of the surface of the paraboloid $x^2 + y^2 = 2z$ which lies between the planes $z = 0$ and $z = 1$.

23. Evaluate $\int_C \frac{x\,dy - y\,dx}{x^2 + y^2 + 1}$ where C is the boundary of the part of the disc $x^2 + y^2 \leq 1$ in the first quadrant cut off by the x-axis and the line $y = x$ (*a*) directly, (*b*) by using Green's theorem.

24. Evaluate $\int_C (y - \sin x)\,dx + \cos x\,dy$, where C is the perimeter, described in the positive direction, of the triangle with vertices

(0,0), ($\pi/2$,0), ($\pi/2$,1), (*i*) directly ; (*ii*) using Green's theorem.

25. Evaluate $\int_C (x^2 y\,dx + xy^2\,dy)$, where C is the closed curve consisting of the arc of the parabola $y^2=4x$ between the points (1,2), (1, −2) and the line joining these two points, (*i*) directly, (*ii*) by using Green's theorem.

26. By using the divergence theorem, or otherwise, evaluate $\iint_S x^2y^2z\,dS$ where S is the surface of the sphere $x^2+y^2+z^2 = 1$.

27. Show that $\int_C \{z\,x+x\,y+y\,z\} = >0$, where C is the curve of intersection of the paraboloid $z = x^2+y^2$ and the cylinder $y = x^2+y^2$, (*a*) directly, (*b*) using Stokes's theorem.

28. For a solid V with boundary surface S for which the divergence theorem holds, show that the volume of V is given by

$$c(V) = \frac{1}{6}\iint_S \nabla(r^2)\bullet\mathbf{n}\,dS = \frac{1}{3}\iint_S \mathbf{r}\bullet\mathbf{n}\,dS.$$

29. Let Ω be a domain in $\mathbb{R}^3$ with boundary surface S for which the divergence theorem is applicable and let Ω' be an open set containing $\Omega \cup S$. If $u \in C^2(\Omega')$ prove that $\int_\Omega \Delta u\,dV = \int_S \frac{\partial u}{\partial \mathbf{n}}\,dS$, where Δ denotes the Laplacian $\Delta u := \frac{\partial^2 u}{\partial x^2} + \frac{\partial^2 u}{\partial y^2} + \frac{\partial^2 u}{\partial z^2}$ and $\partial u/\partial\mathbf{n}$ is the derivative of u in the direction of the outward drawn normal to the boundary.

30. Let Ω, Ω' and S be as in problem 29. For $u, v \in C^2(\Omega')$ show that

$$\iint_S u\frac{\partial v}{\partial \mathbf{n}}\,dS = \iiint_\Omega (u\,\Delta v + \nabla u\bullet\nabla v)\,dV, \text{ (Green's first identity).}$$

Deduce Green's second identity :

$$\iiint_\Omega (u\,\Delta v - v\,\Delta u)\,dV = \iint_S \left[u\frac{\partial v}{\partial \mathbf{n}} - \frac{\partial u}{\partial \mathbf{n}}\right] dS.$$

31. Show that if $u,v \in C^2(\mathbb{R}^n)$ satisfy $\Delta u=0$, $\Delta v=0$ in Ω and $u=v$ at all points of $S=\mathrm{bdy}(\Omega)$, then $u=v$ in $\mathrm{cl}(\Omega)$. Deduce that the equation $\Delta u=0$ in Ω, u given on $\mathrm{bdy}(\Omega)$, has at most one solution in $C^2(\mathbb{R}^n)$.

Hints on the problems.

5. Differentiate $\mathbf{N}(s) = \mathbf{B}(s) \times \mathbf{T}(s)$.

6. $1-\cos\theta=2\sin^2(\theta/2)$

9. Consider $\nabla(\phi\ \psi)$.

10,11. Gradient vector fields.

26. Choose $\mathbf{F}=(0, 0, F)$ for a suitable F.

29. $\Delta u = \mathrm{div}(\nabla u)$, $\partial u/\partial \mathbf{n} = \nabla u \bullet \mathbf{n}$.

30. Consider $\mathrm{div}(u\ \nabla v)$.

31. Take $w=u-v$ and apply Green's first identity to w, w.

References and Further Reading

1. T. M. Apostol, *Mathematical Analysis*, Addison-Wesley, Reading Mass., 1957.

2. R. G. Bartle, *The Elements of Real Analysis*, Wiley, New York 1964.

3. J. C. Burkill & H. Burkill, *A Second Course in Mathematical Analysis*, Cambridge Univ. Press, 1970.

4. W. H. Fleming, *Functions of Several Variables*, 2nd Ed. Springer, New York, 1977.

5. E. Hewitt & K. Stromberg, *Real and Abstract Analysis*, Springer, Berlin , 1965.

6. J. Milnor, "Analytic proofs of the "Hairy Ball Theorem" and the Brouwer Fixed Point Theorem", Amer. Math. Monthly, Vol. 85, 521-524 (1978).

7. M. H. A. Newman, *Elements of the Topology of plane sets of points*, Cambridge Univ. Press, 1951.

8. C. A. Rogers, "A less strange version of Milnor's proof of Brouwer's Fixed-Point Theorem", Amer. Math. Monthly, Vol. 87, 525-527, (1980).

9. W. Rudin, *Principles of Mathematical Analysis*, McGraw-Hill, New York, 1964.

10. J. T. Schwartz, "The formula for change of variables in a multiple integral", Amer. Math. Monthly, Vol. 61, 81-85 (1954).

11. G. F. Simmons, *Introduction to Topology and Modern Analysis*, McGraw-Hill, New York, 1963.

12. M. Spivak, *Calculus*, Benjamin, New York, 1967.

13. D. S. G. Stirling, *Mathematical Analysis*, Ellis Horwood, Chichester, 1987.

14. G. Strang, *Linear Algebra and its Applications*, 3rd. Ed., Harcourt, Brace, Jovanovich, San Diego, 1988.

15. A. E. Taylor, *Introduction to Functional Analysis*, Wiley, New York, 1958.

16. A. Torchinsky, *Real Variables*, Addison-Wesley, Reading, Mass., 1988.

17. H. Whitney, *Geometric Integration Theory*, Oxford Univ. Press, London 1959.

List of Symbols

$\Rightarrow$	implies that	$\Leftarrow$	is implied by
$\Leftrightarrow$	if and only if, or, is equivalent to		
$< x,y >$	(Euclidean) inner product of **x** and **y**		
$\Vert\cdot\Vert$	(Euclidean) norm	$\vert\cdot\vert$,	absolute value
$\in$	belongs to	$\notin$	does not belong to
$:=$	equality by definition	$\emptyset$	the empty set
$\cup$, $\cap$	unions and intersections	$\subset$, $\subseteq$	inclusions
$\sum$	summation sign	$\int$	integral sign
$\forall$	for all *or* for every	$\exists$	there exist(s)
dist(**x**,A),	distance of a point **x** to a set A		
$\rho(A,B)$,	separation of two sets		
a•**b**,	scalar (dot) product of vectors		
a$\times$**b**,	vector (cross) product of vectors		
[**a**,**b**,**c**],	scalar triple product of vectors		
curl **F**,	curl of a vector field **F**	div **F**,	divergence of **F**
$\mathbb{N}$,	Natural numbers (positive integers),	$\mathbb{Z}$,	Integers
$\mathbb{Q}$,	Rational numbers	$\mathbb{R}$,	Real numbers
A^c,	Complement of the set A	bdy(A),	Boundary of A
cl(A),	Closure of A	int(A),	Interior of A
B(**x**,ε),	ε-neighbourhood of **x** or open ball, centre **x**, radius ε	B$'$(**x**,ε),	Punctured neighbourhood of **x**
osc $f(S)$	Oscillation of f on S	ω_f(**x**),	Oscillation of f at **x**
Df	Derivative of f	$J_{\mathbf{f}}$,	Jacobian of f
∇,	del or gradient	det,	determinant
ln,	natural logarithm	exp,	exponential function
inf,	infimum=greatest lower bound	sup,	supremum=least upper bound
$U(f, P)$,	Upper sum	$L(f, P)$,	Lower sum
c(S),	Jordan content of S		

Solutions to the Problems

Chapter 1.

1. If $\mathbf{y}=\alpha\mathbf{x}$ then we have $<\mathbf{x},\mathbf{y}> = \alpha<\mathbf{x},\mathbf{x}> = \alpha\|\mathbf{x}\|^2$ and $\|\mathbf{x}\|\ \|\mathbf{y}\| = |\alpha|\ \|\mathbf{x}\|^2$ so $|<\mathbf{x},\mathbf{y}>| = \|\mathbf{x}\|\ \|\mathbf{y}\|$. Conversely, if equality holds then $b^2=4ac$ in the proof of Lemma 1.4 and we have $\|t\,\mathbf{x}+\mathbf{y}\|^2=0$ for $t=-b/2a$ in Lemma 1.4. Equality holds in the triangle inequality if and only if every inequality in the proof is an equality, that is exactly when $<\mathbf{x},\mathbf{y}> = \|\mathbf{x}\|\ \|\mathbf{y}\|$, that is, when equality holds in the Schwarz inequality and $\mathbf{y}=\alpha\mathbf{x}$ where now we must also have $\alpha>0$.

2. (*i*) $<\mathbf{x},\mathbf{x}>_{\mathbf{w}} = \sum x_i^2 w_i \geq 0$ and equals 0 only if $x_i=0$ for every *i*, since $w_i>0$. (*ii*) $<\mathbf{x}+\mathbf{y},\mathbf{z}>_{\mathbf{w}} = \sum (x_i+y_i)z_i w_i = \sum x_i z_i w_i + y_i z_i w_i$. The rest is equally simple.

3. Let $<\mathbf{x},\mathbf{y}> := (|\mathbf{x}+\mathbf{y}|^2 - |\mathbf{x}-\mathbf{y}|^2)/4$. Then

$$\begin{aligned}<\mathbf{x}+\mathbf{z},\mathbf{y}> + <\mathbf{x}-\mathbf{z},\mathbf{y}> &= (1/4)\{|\mathbf{x}+\mathbf{z}+\mathbf{y}|^2 - |\mathbf{x}+\mathbf{z}-\mathbf{y}|^2 + |\mathbf{x}-\mathbf{z}+\mathbf{y}|^2 - |\mathbf{x}-\mathbf{z}-\mathbf{y}|^2\}\\ &= (1/4)\{|\mathbf{x}+\mathbf{y}+\mathbf{z}|^2 + |\mathbf{x}+\mathbf{y}-\mathbf{z}|^2 - (|\mathbf{x}-\mathbf{y}+\mathbf{z}|^2 + |\mathbf{x}-\mathbf{y}-\mathbf{z}|^2)\}\\ &= (1/2)\{|\mathbf{x}+\mathbf{y}|^2 + |\mathbf{z}|^2 - (|\mathbf{x}-\mathbf{y}|^2 + |\mathbf{z}|^2)\},\ \text{(parallelogram law)}\\ &= 2<\mathbf{x},\mathbf{y}>.\end{aligned}$$

Now take $\mathbf{x}=\mathbf{z}$ to get $<2\mathbf{z},\mathbf{y}> = 2<\mathbf{z},\mathbf{y}>$. For arbitrary $\mathbf{u},\mathbf{v}$ let $\mathbf{x}=(\mathbf{u}+\mathbf{v})/2$, $\mathbf{z}=(\mathbf{u}-\mathbf{v})/2$ to get $<\mathbf{u},\mathbf{y}> + <\mathbf{v},\mathbf{y}> = 2<(\mathbf{u}+\mathbf{v})/2,\mathbf{y}> = <\mathbf{u}+\mathbf{v},\mathbf{y}>$. This proves property (*ii*) of inner product. For $m\in\mathbb{N}$, taking $\mathbf{x}=m\mathbf{z}$ gives

$$<(m+1)\mathbf{z},\mathbf{y}> + <(m-1)\mathbf{z},\mathbf{y}> = 2<m\mathbf{z},\mathbf{y}>.$$

Thus, if $<\alpha\mathbf{z},\mathbf{y}> = \alpha<\mathbf{z},\mathbf{y}>$ for $\alpha = m-1$ and $\alpha = m$ then it also holds for $\alpha = m+1$ and hence $<m\mathbf{z},\mathbf{y}> = m<\mathbf{z},\mathbf{y}>$ for all $m\in\mathbb{N}$ by induction on *m*. It follows that $(1/m)<\mathbf{z},\mathbf{y}> = (1/m)<m(\mathbf{z}/m),\mathbf{y}> = <\mathbf{z}/m,\mathbf{y}>$. This gives $<p\mathbf{z}/q,\mathbf{y}> = p<\mathbf{z}/q,\mathbf{y}> = (p/q)<\mathbf{z},\mathbf{y}>$ for $p,q\in\mathbb{N}$. As $<-\mathbf{z},\mathbf{y}> = -<\mathbf{z},\mathbf{y}>$, is immmediate, this proves $<\alpha\mathbf{z},\mathbf{y}> = \alpha<\mathbf{z},\mathbf{y}>$ for all rational α. Since the norm is continuous (Ex.2.24), $c<\mathbf{z},\mathbf{y}> = \lim \alpha^{(m)}<\mathbf{z},\mathbf{y}> = \lim <\alpha^{(m)}\mathbf{z},\mathbf{y}> =$ $\lim\,\{|\alpha^{(m)}\mathbf{z}+\mathbf{y}|^2 - |\alpha^{(m)}\mathbf{z}-\mathbf{y}|^2\}/4 = (|c\mathbf{z}+\mathbf{y}|^2 - |c\mathbf{z}-\mathbf{y}|^2)/4 = <c\mathbf{z},\mathbf{y}>$.

4. (*i*) (*a*) is obvious. (*b*) $\|\alpha\mathbf{x}\|_1 = |\alpha x_1| + |\alpha x_2| = |\alpha|(|x_1|+|x_2|)$. (*c*) is clear. (*d*) $\|\mathbf{x}+\mathbf{y}\|_1 = |x_1+y_1| + |x_2+y_2| \leq |x_1| + |y_1| + |x_2| + |y_2|$. The extension to $\mathbb{R}^n$ just involves writing more terms, (induction).

(*ii*) (*d*) $\|\mathbf{x}+\mathbf{y}\|_\infty = \max\{|x_1+y_1|,|x_2+y_2|\} \leq \max\{|x_1|+|y_1|,|x_2|+|y_2|\}$
$\leq \max\{|x_1|,|x_2|\} + \max\{|y_1|,|y_2|\}$.

5. $\|\mathbf{x}\|_\infty = \max\{|x_i| : 1\leq i\leq n\} \leq (\sum|x_i|^2)^{1/2} = \|\mathbf{x}\|$. In the other direction we

have $\|\mathbf{x}\| \le (\sum(\max|x_i|)^2)^{1/2} \le (n\|\mathbf{x}\|_\infty^2)^{1/2}$. For the 1-norm we have $\|\mathbf{x}\|_1 = \sum|x_i| = \sum|x_i|1 \le (\sum|x_i|^2)^{1/2}(\sum 1^2)^{1/2} = \sqrt{n}\|\mathbf{x}\|$, by Cauchy's inequality. Also $\|\mathbf{x}\|^2 = (\sum|x_i|^2) \le (\sum|x_i|)^2 = \|\mathbf{x}\|_1^2$.
Now if $\mathbf{x}$ is an interior point of a set A then $\exists\ \delta > 0$ so that $\|\mathbf{y}-\mathbf{x}\| < \delta$ implies $\mathbf{y} \in A$. Then $\|\mathbf{y}-\mathbf{x}\|_\infty < \delta/\sqrt{n}$ implies $\|\mathbf{y}-\mathbf{x}\| < \delta$ so $\mathbf{x}$ is an interior point of A in the $\|\ \|_\infty$ norm too. The others cases are exactly similar.

6. Let $\phi(t) = t^{1-p}a^p + (1-t)^{1-p}b^p$ for $0 < t < 1$. Then ϕ has a unique critical point when $t^{-p}a^p = (1-t)^{-p}b^p$, that is when $t/(1-t) = a/b$, or $t = a/(a+b)$ and this is a minimum. ($\phi \to \infty$ at the endpoints since $1-p < 0$.) The minimum value is $a(a+b)^{p-1} + b(a+b)^{p-1} = (a+b)^p$.
Now $|x_i+y_i|^p \le (|x_i|+|y_i|)^p = \min\{t^{1-p}|x_i|^p + (1-t)^{1-p}|y_i|^p : 0 < t < 1\}$.

Therefore $\sum_{i=1}^n |x_i+y_i|^p \le \min\{\sum_{i=1}^n t^{1-p}|x_i|^p + \sum_{i=1}^n (1-t)^{1-p}|y_i|^p\}$

$= \min\{t^{1-p}A^p + (1-t)^{1-p}B^p\}$, where $A^p := \sum_{i=1}^n |x_i|^p$, $B^p := \sum_{i=1}^n |y_i|^p$,

$= (A+B)^p$. Taking p-th roots gives the result.
This is the triangle inequality for $\|\cdot\|_p$ while the other properties are simple so this expression is indeed a norm.

7. $A = \{(x,y) \in \mathbb{R}^2 : 0 \le x \le 1,\ 0 < y < 1\}$, a square. A diagram makes obvious that $\mathrm{int}(A) = \{(x,y) \in \mathbb{R}^2 : 0 < x < 1,\ 0 < y < 1\}$, $\mathrm{cl}(A) = \{(x,y) \in \mathbb{R}^2 : 0 \le x \le 1,\ 0 \le y \le 1\}$, $\mathrm{bdy}(A) = \{0 \le x \le 1, y=0\} \cup \{0 \le x \le 1, y=1\} \cup \{x=0, 0 \le y \le 1\} \cup \{x=1, 0 \le y \le 1\}$. For example, to prove that for $0 < x < 1$ and $0 < y < 1$, $\mathbf{x} = (x,y)$ is an interior point of A, let $\delta = \min\{x, 1-x, y, 1-y\}$. Then $\mathbf{w} = (u,v) \in \mathrm{B}(\mathbf{x},\delta)$ if $|u-x|^2 + |v-y|^2 < \delta^2$, so certainly $|u-x| < \delta$ and $|v-y| < \delta$. This shows that $u < x+\delta \le x+(1-x) = 1$ and $u > x-\delta \ge 0$, similarly for v, that is, the δ-nbd of $\mathbf{x}$ lies in A. A is neither open nor closed.

8. If $\mathbf{x} = (x,y,0) \in A$, for any $\varepsilon > 0$, the point (x,y,ε) lies in the ε-nbd of $\mathbf{x}$ but is not in A, therefore $\mathrm{int}(A) = \emptyset$. A is not closed since $(0,0,0) \in \mathrm{cl}(A) \cap A^c$.

9. $\|\mathbf{x}\| = r \Leftrightarrow$ every nbd of $\mathbf{x}$ contains points $\mathbf{y}$ with $\|\mathbf{y}\| < r$ and points $\mathbf{z}$ with $\|\mathbf{z}\| > r$. Therefore, $\{\mathbf{x} : \|\mathbf{x}\| = r\}$ are boundary points while $\{\mathbf{x} : \|\mathbf{x}\| < r\}$ are interior points. Thus, $\mathrm{int}(\overline{\mathrm{B}}(\mathbf{0},r)) = \mathrm{B}(\mathbf{0},r)$, $\mathrm{bdy}(\mathrm{B}(\mathbf{0},r)) = \{\mathbf{x} : \|\mathbf{x}\| = r\}$ and $\mathrm{cl}(\mathrm{B}(\mathbf{0},r)) = \overline{\mathrm{B}}(\mathbf{0},r)$.

10. $\mathbf{x} \in \mathrm{int}(A \cap B) \Rightarrow \exists$ a nbd of $\mathbf{x}$ in $A \cap B$ so there is a nbd of $\mathbf{x}$ in A (and in B), that is, $\mathbf{x} \in \mathrm{int}(A)$ (and $\mathrm{int}(B)$). Conversely, if $\mathbf{x} \in \mathrm{int}(A) \cap \mathrm{int}(B)$ then $\exists$ nbds U and V of $\mathbf{x}$ which lie entirely within A and B respectively. Then $U \cap V$ is an open set in $A \cap B$ so $\mathbf{x}$ is an interior point

of $U \cap V$ hence also of $A \cap B$. Now let $\mathbf{x} \in \text{int}(A) \cup \text{int}(B)$. Then either $\mathbf{x} \in \text{int}(A)$ or $\mathbf{x} \in \text{int}(B)$ (possibly both). If $\mathbf{x} \in \text{int}(A)$ there is a nbd of $\mathbf{x}$ in A hence also in $A \cup B$. Similarly if $\mathbf{x} \in \text{int}(B)$. In all, $\mathbf{x} \in \text{int}(A \cup B)$.
Example : $A = \{\text{rational numbers in } (0,1)\}$, $B = \{\text{irrationals in } (0,1)\}$. $\text{int}(A) = \text{int}(B) = \emptyset$ since any interval (nbd in $\mathbb{R}$) always contains both rationals and irrationals. [*Or* $A = (-1,0]$, $B = [0,1)$.] Results for closures? Yes. Take complements and use the De Morgan laws :

$$[\text{int}(A \cap B)]^c = [\text{int}(A) \cap \text{int}(B)]^c \quad \text{or,}$$
$$\text{cl}(A^c \cup B^c) = [\text{int}(A)]^c \cup [\text{int}(B)]^c = \text{cl}(A^c) \cup \text{cl}(B^c).$$

As A and B are arbitrary sets this proves $\text{cl}(A \cup B) = \text{cl}(A) \cup \text{cl}(B)$ for any two sets. Similarly, we have $[\text{int}(A) \cup \text{int}(B)]^c \supseteq [\text{int}(A \cup B)]^c$, that is, $[\text{int}(A)]^c \cap [\text{int}(B)]^c \supseteq \text{cl}(A^c \cap B^c)$, or $\text{cl}(A^c \cap B^c) \subseteq \text{cl}(A^c) \cap \text{cl}(B^c)$. This proves $\text{cl}(A \cap B) \subseteq \text{cl}(A) \cap \text{cl}(B)$ for any sets A,B.

11. We show $B := \text{int}(A)$ is open. Since every point of B is an interior point of A, for any $\mathbf{b} \in B$ there is a δ-nbd of $\mathbf{b}$ lying entirely in A. Then for every point $\mathbf{x}$ in the $\delta/2$-nbd of $\mathbf{b}$, the $\delta/2$-nbd of $\mathbf{x}$ lies in A so the $\delta/2$-nbd of $\mathbf{b}$ consists of interior points of A, that is, points of B and thus B is open. $\text{cl}(A)$ is closed since its complement is $\text{int}(A^c)$, an open set. A sketch in $\mathbb{R}^2$ should clarify this.

12. Suppose that V is open and $\text{int}(A) \subseteq V \subseteq A$. Every $\mathbf{v} \in V$ is an interior point of V so there is a nbd of $\mathbf{v}$ in V and therefore in A. This proves that $\mathbf{v}$ is an interior point of A so $V = \text{int}(A)$. If $A \subseteq F \subseteq \text{cl}(A)$ where F is closed, then, taking complements, $\text{int}(A^c) \subseteq F^c \subseteq A^c$ so that $F^c = \text{int}(A^c)$ or $F = \text{cl}(A)$. The intersection of all closed sets containing A is a closed set (Thm.1.17) and is the smallest one containing A, therefore equals $\text{cl}(A)$.

13. If $\mathbf{x} \in \text{bdy}(A)$ then every nbd of $\mathbf{x}$ contains points of A^c (and of A) so $\mathbf{x} \notin \text{int}(A) = [\text{cl}(A^c)]^c$, that is, $\mathbf{x} \in \text{cl}(A^c)$ (and similarly $\mathbf{x} \in \text{cl}(A)$). Conversely, if $\mathbf{x} \in \text{cl}(A) \cap \text{cl}(A^c)$ then every nbd of $\mathbf{x}$ contains at least one point of A and at least one of A^c so is a boundary point.

$$\text{bdy}(A^c) = \text{cl}(A^c) \cap \text{cl}(A^{cc}) = \text{cl}(A^c) \cap \text{cl}(A) = \text{bdy}(A)$$

and $\text{bdy}(A)$ is the intersection of closed sets so is closed.

14. Suppose $A = \{\mathbf{a}^{(1)}, \mathbf{a}^{(2)}, \cdots, \mathbf{a}^{(m)}\}$, and for $\mathbf{x} \in A^c$ let $d_i = \|\mathbf{x} - \mathbf{a}^{(i)}\|$ and put $d = \min\{d_i : 1 \le i \le m\}$. Then $B(\mathbf{x},d)$ lies in A^c and so A^c is open.

15. If $\mathbf{x} \in \text{cl}(A)$ every nbd of $\mathbf{x}$ contains at least one point of A. If every nbd also contains a point of A^c then $\mathbf{x} \in \text{bdy}(A)$. Otherwise there is some nbd which contains no point of A^c and then $\mathbf{x} \in \text{int}(A)$. The converse is immediate since $\text{int}(A) \subseteq A$.

16. Obviously $\mathbf{0} \in V^t$. To prove that V^t is open let $\mathbf{x} \in V^t$. Then $\mathbf{x} + \mathbf{a}$ is an interior point of V so there is $\varepsilon > 0$ such that $B(\mathbf{x}+\mathbf{a},\varepsilon) \subseteq V$. Then for $\mathbf{y} \in B(\mathbf{x},\varepsilon)$, $\|(\mathbf{y}+\mathbf{a}) - (\mathbf{x}+\mathbf{a})\| < \varepsilon$ so that $\mathbf{y}+\mathbf{a} \in B(\mathbf{x}+\mathbf{a},\varepsilon) \subseteq V$, that is, $\mathbf{y} \in V^t$ and $\mathbf{x} \in \text{int}(V^t)$.

17. (*a*) False. Let $A=\{\mathbf{a}\}$, a singleton. Then A is closed and $\text{int}(A)=\emptyset$ so $\text{cl}(\text{int}(A)=\text{cl}(\emptyset)=\emptyset\neq A$. An example in $\mathbb{R}^2$ is $A=\overline{B}(\mathbf{0},1)\cup\{(x,0) : 1\leq x\leq 2\}$, the disk with a handle. Here $\text{cl}(\text{int}(A))=\overline{B}(\mathbf{0},1)$. (*b*) False. An example in $\mathbb{R}$ is $A=(-1,0)\cup(0,1)$. Here $\text{cl}(A)=[-1,1]$ and $\text{int}([-1,1])=(-1,1)$. In $\mathbb{R}^2$ an example is the open unit disk with the x-axis removed.

18. Let $t_0\in E=\{t\in\mathbb{R} : \mathbf{a}+t\mathbf{h}\in G\}$, where G is open in $\mathbb{R}^n$. We have to find $\delta>0$ such that $(t_0-\delta, t_0+\delta)\subseteq E$. As G is open, there is $\varepsilon>0$ so that the ε-nbd of $\mathbf{a}+t_0\mathbf{h}$ lies in G. Take $\delta=\varepsilon/\|\mathbf{h}\|$. Then, for $|s-t_0|<\delta$, $\|\mathbf{a}+t_0\mathbf{h}-(\mathbf{a}+s\mathbf{h})\|=|t_0-s|\,\|\mathbf{h}\|<\delta\|\mathbf{h}\|=\varepsilon$ so that $\mathbf{a}+s\mathbf{h}\in G$, that is, $s\in E$. This proves that $(t_0-\delta, t_0+\delta)\subseteq E$.

19. Take $A_m=[1/m,1-1/m]$ in $\mathbb{R}$. Then $\bigcup_{m=1}^{\infty} A_m=(0,1)$. Also the sets $F_m=[1/m, m]$ have union $(0,\infty)$. For an arbitrary family, take any open set V and write $V=\bigcup\{\{\mathbf{x}\} : \mathbf{x}\in V\}$. Taking complements would give examples for intersections *or* take $V_m=(-1/m, 1+1/m)$ with intersection $[0,1]$.

20. Let $\mathbf{x}^{(k)}\longrightarrow\mathbf{x}$. Then, by the ε-N condition with $\varepsilon=1$, $\exists\ N$ such that $\|\mathbf{x}^{(k)}-\mathbf{x}\|<1$ for all $k\geq N$. By the triangle inequality we obtain $\|\mathbf{x}^{(k)}\|\leq\|\mathbf{x}^{(k)}-\mathbf{x}\|+\|\mathbf{x}\|<1+\|\mathbf{x}\|$ for all $k\geq N$. Setting
$M:=\max(\|\mathbf{x}^{(1)}\|,\cdots,\|\mathbf{x}^{(N-1)}\|, 1+\|\mathbf{x}\|\}$ gives $\|\mathbf{x}^{(k)}\|\leq M$ for all $k\geq 1$.
Example. Take $x^{(k)}=(-1)^k$ in $\mathbb{R}$. The sequence is bounded but not convergent.

21. Suppose that $x^{k(j)}\longrightarrow x$ and for $\varepsilon>0$ let N be such that $|x^{k(j)}-x|<\varepsilon$ for all $k(j)\geq N$. Suppose that $k(p)\geq N$ is fixed. For every $m\geq k(p)$ there exists $k(q)$ (which may depend on m) such that $k(p)\leq m\leq k(q)$ and hence $x-\varepsilon<x^{k(p)}\leq x^{(m)}\leq x^{k(q)}<x+\varepsilon$, that is, $|x^{(m)}-x|<\varepsilon$.

22. Suppose property 2 holds and let $\{x^{(m)}\}$ be a monotone increasing sequence that is bounded above. Then $M:=\sup\{x^{(m)}\}$ exists. By definition of supremum, for any $\varepsilon>0$ there is N such that $M-\varepsilon<x^{(N)}\leq M$. By monotonicity this gives $M-\varepsilon<x^{(N)}\leq x^{(m)}\leq M$ for all $m\geq N$, that is $x^{(m)}\longrightarrow x$. Now suppose that property 1 holds which we use (equivalently) for a monotone decreasing sequence. Suppose that S is a set that is bounded above say $|s|\leq K$ for $s\in S$. Take a fixed $s_1\in S$ and let $I_1=[s_1,K]$ and let $x^{(1)}=K$. Bisect I_1 and let I_2 be the right-hand half if this contains points of S otherwise the left-hand half. Let $x^{(2)}$ be the right-hand endpoint of I_2. Note that $s\leq x^{(2)}$ for all $s\in S$ and that

I_2 contains at least one point of S. Continue this process and consider the decreasing sequence $\{x^{(m)}\}$. This is bounded below by s_1 so converges to x (say). Since $s \leq x^{(m)}$ for all m it follows that $s \leq x$, that is, x is an upper bound for S. For $\varepsilon > 0$, take m so large that the interval I_m has length less than ε so that $I_m \subseteq [x^{(m)} - \varepsilon, x^{(m)}]$. Then I_m contains an element s_ε (say) of S and therefore $x - \varepsilon \leq x^{(m)} - \varepsilon \leq s_\varepsilon$. This proves that x is the supremum of S.

23. Let $M = \sup A$. Then for every $m \in \mathbb{N}$ there exists $a^{(m)} \in A$ such that $M - 1/m < a^{(m)} \leq M$. Hence $a^{(m)} \longrightarrow M$ so that $M \in \text{cl}(A)$ by Proposition 1.23.

24. Suppose that $\{\mathbf{x}^{(m)}\}$ has a limit point $\mathbf{x}$ and for $\mathbf{y} \neq \mathbf{x}$ let $\varepsilon = \|\mathbf{x} - \mathbf{y}\|/4$. Since $\{\mathbf{x}^{(m)}\}$ is Cauchy, there exists N so that $\|\mathbf{x}^{(k)} - \mathbf{x}^{(p)}\| < \varepsilon$ for $k,p \geq N$. As $\mathbf{x}$ is a limit point of $\{\mathbf{x}^{(m)}\}$ there exists $p \geq N$ with $\mathbf{x}^{(p)}$ in the ε-nbd of $\mathbf{x}$ and then $\mathbf{x}^{(k)}$ belongs to the 2ε-nbd of $\mathbf{x}$ for all $k \geq N$. By the choice of ε, $B(\mathbf{y},\varepsilon)$ contains no point of $B(\mathbf{x},2\varepsilon)$ so contains at most finitely many points of $\{\mathbf{x}^{(m)}\}$ and $\mathbf{y}$ is not a limit point.

25. Since $\mathbf{x}^{(m)} \longrightarrow \mathbf{x}$, every nbd of $\mathbf{x}$ contains all but finitely many $\mathbf{x}^{(m)}$, that is, infinitely many points of $\{\mathbf{x}^{(m)}\}$.

26. For any $\varepsilon > 0$ there exists N so that $\|\mathbf{x}^{(k)} - \mathbf{x}\| < \varepsilon$ for all $k \geq N$. Then for a subsequence $\{\mathbf{x}^{k(j)}\}$, $\|\mathbf{x}^{k(j)} - \mathbf{x}\| < \varepsilon$ for all $k(j) \geq N$ so converges to $\mathbf{x}$. For the second part if $\{\mathbf{x}^{(m)}\}$ does not converge to $\mathbf{x}$, then, from (1.7), $\exists\ \varepsilon > 0\ \forall\ N\ \exists\ m \geq N$ with $\|\mathbf{x}^{(m)} - \mathbf{x}\| \geq \varepsilon$. That is, there is $\varepsilon > 0$ and a subsequence $\{\mathbf{x}^{m(j)}\}$ with $\|\mathbf{x}^{m(j)} - \mathbf{x}\| \geq \varepsilon$ for all $m(j)$. No further subsequence can converge to $\mathbf{x}$, a contradiction.

27. This is a special case of Proposition 1.23. *Alternatively*, if $\|\mathbf{x} - \mathbf{a}\| = d > \delta$ let $\varepsilon = (d - \delta)/2$. Then $\exists\ N$ such that $\|\mathbf{x}^{(m)} - \mathbf{x}\| \leq \varepsilon$ for all $m \geq N$. Hence (Coroll. 1.5), $\|\mathbf{x}^{(m)} - \mathbf{a}\| \geq \|\mathbf{x} - \mathbf{a}\| - \|\mathbf{x}^{(m)} - \mathbf{x}\| > d - \varepsilon = \delta + \varepsilon$, a contradiction.

28. Fix any $\mathbf{a} \in A$. Then for $\mathbf{x} \in A$, $\|\mathbf{x}\| \leq \|\mathbf{x} - \mathbf{a}\| + \|\mathbf{a}\| \leq \text{diam}(A) + \|\mathbf{a}\|$, so A is bounded.

29. $A_m = (0,1/m)$ has $\cap A_m = \emptyset$. $C_m = [m,\infty)$ are closed (why?) and $\cap C_m = \emptyset$.

30. 0 is the only limit point of A since for $x > 0$ the $x/2$-nbd of x contains at most finitely many points of A; $x < 0$ is easier. The limit points of B are $(1,1)$, $(1,0)$ and $(1,-1)$ since $1 + 1/m \longrightarrow 1$ while $\sin(m\pi/2)$ takes only the values 1, 0, and -1; a sketch should make this plain. The limit points of C are all points in $[0,1]$ since every real number is a limit of a sequence of rational numbers. Limit points of D are $(1,0)$ and $(-1,0)$. Again a sketch is useful.

31.(*a*) For $\mathbf{u}\in U$ and for any $\varepsilon>0$ there is $\mathbf{s}\in S$ such that $\|\mathbf{u}-\mathbf{s}\|<\varepsilon$ since S is dense. Take $\varepsilon>0$ so that the ε-nbd of $\mathbf{u}$ lies in U. Then $\mathbf{s}\in U\cap S$.
(*b*) Fix $\mathbf{x}\in\mathbb{R}^n$ and $\varepsilon>0$. As $V_1\cap B(\mathbf{x},\varepsilon)$ is open and nonempty (by (*a*)) there is a ball $B(\mathbf{y}^{(1)},r_1)$ with $r_1<\varepsilon$ such that $A_1:=\overline{B}(\mathbf{y}^{(1)},r_1)\subset V_1\cap B(\mathbf{x},\varepsilon)$. [The set contains a nbd $B(\mathbf{y},r)$; take $\overline{B}(\mathbf{y},r_1)$ with $r_1<\min\{r,\varepsilon\}$.] Similarly there is a ball $B(\mathbf{y}^{(2)},r_2)$ with radius $r_2<\varepsilon/2$ such that $A_2:=\overline{B}(\mathbf{y}^{(2)},r_2)\subseteq V_2\cap B(\mathbf{y}^{(1)},r_1)$. Continue this process and let A_m be the corresponding closed ball of radius $r_m<\varepsilon/2^m$ for each $m>2$. By Cantor's theorem (Prop.1.33), $\cap A_m$ contains exactly one point $\mathbf{p}$. Then $\mathbf{p}\in\cap V_m$ and $\|\mathbf{p}-\mathbf{x}\|<\varepsilon$.
(*c*) For any $\mathbf{x}\in\mathbb{R}^n$ and any $\varepsilon>0$, $B(\mathbf{x},\varepsilon)$ must intersect $[\mathrm{cl}(X)]^c$ since $\mathrm{cl}(X)$ has no interior points. This shows that the open set $[\mathrm{cl}(X)]^c$ is dense. If $\mathbb{R}^n=\cup\{X^{(i)}:i\geq 1\}$, then certainly $\mathbb{R}^n=\cup\{\mathrm{cl}(X^{(i)})\}$. Then $\cap[\mathrm{cl}(X^{(i)})]^c=\emptyset$ by the De Morgan laws. This contradicts (*b*).
(*d*) Let $\{r_i:i=1,2,\cdots\}$ be an enumeration of $\mathbb{Q}$. If $\mathbb{R}\setminus\mathbb{Q}=\cup X_i$ then we have $\mathbb{R}=\cup(X_i\cup\{r_i\})$ and $X_i\cup\{r_i\}$ is nowhere dense. This contradicts (*c*). If $\mathbb{Q}=\cap V_i$ where each V_i is open in $\mathbb{R}$ then each $V_i\supseteq\mathbb{Q}$ so is dense, and then $\mathbb{R}\setminus\mathbb{Q}=\cup V_i^c$ would be a union of nowhere dense sets, impossible.

32. Take I_0 to be a closed interval contained in I and trisect I_0. Let I_1 be the first sub-interval (from the left) that does not contain $x^{(1)}$. Thus, $x^{(1)}\notin I_1$. Now trisect I_1 and choose I_2 in the same way so that $x^{(2)}\notin I_2$. Repeat this process and obtain intervals I_m with $x^{(m)}\notin I_m$. By the nested intervals theorem there is a point y in every interval I_m. Since $y\in I_m$ but $x^{(m)}\notin I_m$, y is none of the $x^{(m)}$. This shows that intervals are uncountable for no enumeration can include all points.

33. If $\mathbb{R}\setminus\mathbb{Q}$ were countable then so would be the finite union $\mathbb{R}=(\mathbb{R}\setminus\mathbb{Q})\cup\mathbb{Q}$, but this is uncountable.

34. The set of all rational neighbourhoods can be identified with $\mathbb{Q}^n\times\mathbb{Q}$ (centres, radii). This is countable being a finite Cartesian product of countable sets. For an isolated point $\mathbf{a}$ there is $\varepsilon>0$ such that $B(\mathbf{a},\varepsilon)$ contains no other point of $\mathbf{a}$. There is a rational nbd inside $B(\mathbf{a},\varepsilon)$ by the argument of Lemma 1.62. Different isolated points correspond to different rational nbds so the isolated points can be put in one-to-one correspondence with a subset of the rational nbds, a

countable set. The same argument applies to a family of disjoint open sets.

35. Suppose that $\sum_{E} x = K$. For $m=1,2,\cdots$, consider $E_m := \{x \in E : x > 1/m\}$. If this set contained more than mK elements then the sum would exceed K. Thus this set is finite and $\{x \in E : x > 0\} = \cup\{E_m : m=1,2,\cdots\}$, a countable union, so is countable.

36. S is clearly neither empty nor finite so suppose S is enumerable. Then we can enumerate S as a sequence $\{f^{(1)}, f^{(2)}, \cdots\}$ where each $f^{(i)}$ is a function from $\mathbb{N}$ to $\{0,1\}$ and only distinct functions are listed. Define a function $g : \mathbb{N} \longrightarrow \{0,1\}$ by $g(n) = 1 - f^{(n)}(n)$. Then g is none of the functions $f^{(i)}$ so the enumeration is not all of S, a contradiction.

37. For $x \in [0,1]$, we can write $x = \sum_{i=1}^{\infty} a^{(i)}/3^i$ where $a^{(i)} \in \{0,1,2\}$; the series converges by comparison with the geometric series $\sum_{i=1}^{\infty} 2/3^i$ whose sum is $2\frac{1/3}{(1-2/3)} = 1$. We write $x = 0.a^{(1)}a^{(2)}a^{(3)}\cdots$. Two ternary expansions correspond to the same x if and only if $x = 0.a^{(1)}\cdots a^{(m-1)}\alpha\, 2\, 2 \cdots$ and $x = 0.a^{(1)}\cdots a^{(m-1)}(\alpha+1)\, 0\, 0 \cdots$, for some $\alpha \in \{0,1\}$, and some m. If $\alpha = 0$ we choose the former, if $\alpha = 1$ we choose the latter. Then, for $x \in P$, since $x \notin (1/3, 2/3)$, $a^{(1)} \neq 1$. Similarly (by induction) one proves that $a^{(i)} \neq 1$ for every i, so P consists of all numbers whose ternary expansion contains only 0's and 2's. Now let $\phi : S \longrightarrow P$ be defined by $\phi(f) = \sum_{i=1}^{\infty} 2f(i)/3^i$. If $\phi(f) = \phi(g)$ then the two ternary expansions would be equal, so either $f(i) = g(i)$ for all i or there is m such that $2f(m) = 2g(m) + 1$. Since $f(m)$ and $g(m)$ can be only 0 or 1 this is impossible to satisfy so ϕ is one-to-one. By the above remarks it is also onto. This gives a bijection between P and the uncountable set S so P is also uncountable.

38. (*a*) True. $C \cup K$ is closed and bounded when C and K are. *Also*, if $\{V : V \in \mathcal{F}\}$ is an open cover of $C \cup K$ then it is certainly a cover of C (and of K) and therefore there is a finite subfamily $\{V_1, V_2, \cdots, V_m\}$ that covers C and a finite subfamily $\{U_1, U_2, \cdots, U_m\}$ (say) that covers K. Then $\{V_1, V_2, \cdots, V_m, U_1, U_2, \cdots, U_m\}$ covers $C \cup K$.

(*b*) True. A simpler argument than (*a*), see (*d*).

(*c*) False. The union need not be closed (problem 19) nor bounded.

(*d*) True. The infinite intersection is a closed subset of any one of the family and a closed subset of a compact set is compact.

39. By definition of infimum of a set, for any $\varepsilon>0$ there exists $\mathbf{a}=\mathbf{a}(\varepsilon)\in A$ such that $\mathrm{dist}(\mathbf{x},A)\leq\|\mathbf{x}-\mathbf{a}\|\leq\mathrm{dist}(\mathbf{x},A)+\varepsilon$. If $\mathrm{dist}(\mathbf{x},A)=0$ then there is a sequence $\{\mathbf{a}^{(m)}\}$ in A with $\|\mathbf{a}^{(m)}-\mathbf{x}\|<1/m$ so $\mathbf{a}^{(m)}\longrightarrow\mathbf{x}$ and therefore $\mathbf{x}\in\mathrm{cl}(A)$ (Proposition 1.23). Conversely, if $\mathbf{x}\in\mathrm{cl}(A)$, every ε-nbd of $\mathbf{x}$ contains a point $\mathbf{a}$ of A and so $\mathrm{dist}(\mathbf{x},A)<\varepsilon$.

40. Consider the open intervals $\{(0,1-1/m):m\geq2\}$. This is an open cover of $(0,1)$. For any finite subset, let $p=\max\{m\}$. Then the finite subset does not include $1-1/(2p)$.

41. For $\mathbf{x}\in K$ there is some set $U(\mathbf{x})$ in the family $\mathcal{F}$ such that $\mathbf{x}\in U(\mathbf{x})$ and, since $U(\mathbf{x})$ is open, there is $\delta(\mathbf{x})>0$ with $\mathrm{B}(\mathbf{x},2\delta(\mathbf{x}))\subseteq U(\mathbf{x})$. The open cover $\{\mathrm{B}(\mathbf{x},2\delta(\mathbf{x})):\mathbf{x}\in K\}$ has a finite subcover, say $\{\mathrm{B}(\mathbf{x}^{(i)},2\delta^{(i)}):1\leq i\leq m\}$. Let $\delta=\min\{\delta^{(i)}\}$. For $S\subseteq K$ with $\mathrm{diam}(S)<\delta$ take $\mathbf{x}\in S$. Then there is $\mathbf{x}^{(i)}$ such that $\|\mathbf{x}-\mathbf{x}^{(i)}\|<\delta^{(i)}$ and then, using the triangle inequality, $S\subseteq\mathrm{B}(\mathbf{x}^{(i)},\delta+\delta^{(i)})\subseteq U(\mathbf{x}^{(i)})$.

Chapter 2.

1. For any $\varepsilon>0\ \exists\ \delta>0$ so that $\mathbf{x}\in A$ and $\|\mathbf{x}-\mathbf{a}\|<\delta\Rightarrow\|\mathbf{f}(\mathbf{x})-\mathbf{f}(\mathbf{a})\|<\varepsilon/2$ and $\|\mathbf{g}(\mathbf{x})-\mathbf{g}(\mathbf{a})\|<\varepsilon/2$. Then $\|\mathbf{f}(\mathbf{x})+\mathbf{g}(\mathbf{x})-(\mathbf{f}(\mathbf{a})+\mathbf{g}(\mathbf{a})\|<\varepsilon$ by the triangle inequality. $\|\alpha\mathbf{f}(\mathbf{x})-\alpha\mathbf{f}(\mathbf{a})\|=|\alpha|\,\|\mathbf{f}(\mathbf{x})-\mathbf{f}(\mathbf{a})\|<|\alpha|\,\varepsilon/2$ for $\|\mathbf{x}-\mathbf{a}\|<\delta$.

2. The function "x^2" is continuous and so is "y^2" and "z^2". "xyz" is continuous by the result on products so $\sin(xyz)$ is a composition of continuous functions. Finally f is a sum of continuous funtions.

3. At points other than $(0,0)$ f is continuous being a quotient of continuous funtions. At $(0,0)$ we have to consider $E:=|f(x,y)-f(0,0)|=|xy^2/(x^2+y^2)|$. Writing $x=r\cos\theta$, $y=r\sin\theta$ (polar co-ordinates) gives $E=r|\cos\theta\sin^2\theta|\leq r$. Since $r=\surd(x^2+y^2)=\|\mathbf{x}\|$, $\lim_{\mathbf{x}\to\mathbf{0}}E=0$.

4. f is continuous at points other than $(0,0)$ as in 3. At $(0,0)$, using polars gives (*or* consider $f(x,mx)$)

$$|f(x,y)-f(0,0)|=\frac{r^3|\cos\theta\ \sin^2\theta|}{r^2\cos^2\theta+r^4\sin^2\theta}\leq\frac{r^3|\cos\theta\ \sin^2\theta|}{r^2\cos^2\theta}:=C(\theta)r.$$

Since $f(x,y)=0$ when $\cos\theta=0$ we can define $C(\theta)=0$ in that case. Then *for θ fixed*, $C(\theta)$ is a finite constant so that $f(x,y)\longrightarrow0$ as $(x,y)\longrightarrow0$, that is along lines. Note that $C(\theta)\to\infty$ as $\theta\to\pm\pi/2$; C is not continuous. By approaching $(0,0)$ so that $\theta\to\pi/2$ it is not clear how $C(\theta)r$ will behave. To show that f is not continuous consider f along the curve $x=y^2$ (or $x=\alpha y^2$). Then $f(y^2,y)=1/2$ for all y and so every δ-nbd of $(0,0)$ contains points where $f=1/2$ and the limit cannot be 0. The graph of f has a ridge of height 1/2 along the curve $x=y^2$. The

the ε-δ condition cannot be satisfied for any $\varepsilon < 1/2$. The limit does not exist as the heights of these ridges vary with α.

5. Use the ε-δ condition at $\mathbf{a}$ with $\varepsilon = \|\mathbf{f(a)}\|/2$. Then $\exists\ \delta > 0$ so that $\|\mathbf{f(x)}-\mathbf{f(a)}\| < \|\mathbf{f(a)}\|/2$ for $\|\mathbf{x}-\mathbf{a}\| < \delta$. Since $\|\mathbf{f(x)}\| \geq \|\mathbf{f(a)}\| - \|\mathbf{f(x)}-\mathbf{f(a)}\|$, by the corollary to the triangle inequality, we have $\|\mathbf{f(x)}\| > \|\mathbf{f(a)}\|/2 > 0$ for $\|\mathbf{x}-\mathbf{a}\| < \delta$.

6. Exactly similar to the proof of Theorem 2.1 replacing $\mathbf{f(a)}$ by $\mathbf{b}$.

7. $-f : C \longrightarrow \mathbb{R}$ is continuous so attains its maximum on C by Theorem 2.23 , say $-f(\mathbf{c}) = \max\{-f(\mathbf{x}) : \mathbf{x} \in C\}$. Thus, $-f(\mathbf{x}) \leq -f(\mathbf{c})$ for all $\mathbf{x} \in C$ so $f(\mathbf{x}) \geq f(\mathbf{c})$ and this shows that f attains its minimum at $\mathbf{c}$. *Or* mimic the proof of Thm. 2.23.

8.(*a*) For $\mathbf{a} \in A$, $\|\mathbf{x}-\mathbf{a}\| \leq \|\mathbf{x}-\mathbf{y}\| + \|\mathbf{y}-\mathbf{a}\|$, therefore

$$\inf_{\mathbf{a}\in A}\|\mathbf{x}-\mathbf{a}\| \leq \inf_{\mathbf{a}\in A}\{\|\mathbf{x}-\mathbf{y}\| + \|\mathbf{y}-\mathbf{a}\|\} = \|\mathbf{x}-\mathbf{y}\| + \inf_{\mathbf{a}\in A}\|\mathbf{y}-\mathbf{a}\|,$$

that is, $\phi(\mathbf{x}) \leq \|\mathbf{x}-\mathbf{y}\| + \phi(\mathbf{y})$. Similarly, $\phi(\mathbf{y}) \leq \|\mathbf{y}-\mathbf{x}\| + \phi(\mathbf{x})$. Combining these gives $|\phi(\mathbf{x}) - \phi(\mathbf{y})| \leq \|\mathbf{x}-\mathbf{y}\|$.
(*b*) $\mathrm{dist}(\mathbf{x},A) + \mathrm{dist}(\mathbf{x},B)$ is never 0, for if both terms are zero we would have (by Problem 1.39) $\mathbf{x} \in \mathrm{cl}(A) = A$ and $\mathbf{x} \in B$, contradicting $A \cap B = \emptyset$. Thus, f is a quotient of continuous functions; apply Proposition 2.17.

9.(*a*) Let $M = \|\mathbf{f(a)}\|$ and suppose that $M \neq 0$. Take $B(\mathbf{a},\delta) \subseteq \mathrm{int}(K)$ so that $\mathbf{f(a)} \leq \mathbf{f(x)}$ for all $\mathbf{x} \in B(\mathbf{a},\delta)$. Since, by hypothesis, $\mathbf{f}(B(\mathbf{a},\delta))$ is an open set it contains an ε-nbd of $\mathbf{f(a)}$. In particular, $\mathbf{f(a)} - \varepsilon\mathbf{f(a)}/2M$ lies in this ε-nbd so there is $\mathbf{x}$ such that $\mathbf{f(x)} = \mathbf{f(a)} - \varepsilon\mathbf{f(a)}/2M$. However, $\|\mathbf{f(x)}\| = (1-\varepsilon/2M)\|\mathbf{f(a)}\| < \|\mathbf{f(a)}\|$, a contradiction. Therefore $M = 0$.
(*b*) Let $M = \|\mathbf{f(b)}\|$. Since the image of $\mathrm{int}(K)$ is an open set, $\mathbf{f}$ is not identically $\mathbf{0}$. Therefore $M > 0$. By the same argument as in (*a*), if $\mathbf{b} \in \mathrm{int}(K)$, $\mathbf{f(b)} + \varepsilon\mathbf{f(b)}/2M = \mathbf{f(x)}$ for some $\mathbf{x}$, and $\|\mathbf{f(x)}\| > M$, impossible.

10. If $|x-a| < \delta$ then $\left|\frac{1}{x} - \frac{1}{a}\right| = \left|\frac{a-x}{xa}\right| < \delta / |ax|$. For $\delta < a/2$, $|x-a| < \delta \Rightarrow x > a/2$ and so $\left|\frac{1}{x} - \frac{1}{a}\right| < 2\delta/a^2$. Take $\delta = \min\{a/2, \varepsilon a^2/2\}$.

11. Let $\mathbf{b}$ be a limit point of K. Then there is a sequence $\{\mathbf{x}^{(m)}\}$ of points of K converging to $\mathbf{b}$. Then $\mathbf{b} \in \mathrm{cl}(A) = A$ using Prop.1.23. As $\mathbf{f}$ is continuous on A it is continuous at $\mathbf{b}$ and therefore $\mathbf{f}(\mathbf{x}^{(m)}) \longrightarrow \mathbf{f(b)}$. Since $\mathbf{f}(\mathbf{x}^{(m)}) = 0$ for every m, the limit is 0 and so $\mathbf{f(b)} = 0$ and $\mathbf{b} \in K$. *Also* $K = \mathbf{f}^{-1}\{0\}$, the inverse image of a closed set so K is closed relative to A, so $K = A \cap F$ where F is closed and K is therefore closed.

12. For any $\varepsilon > 0\ \exists\ \delta > 0$ so that $\|\mathbf{f(x)}-\mathbf{f(a)}\| < \varepsilon$ if $\mathbf{x} \in A$ and $\|\mathbf{x}-\mathbf{a}\| < \delta$. Then $\|\mathbf{g}(x)-\mathbf{g}(a_1)\| = \|\mathbf{f}(x, a_2, \cdots, a_n) - \mathbf{f}(a_1, a_2, \cdots, a_n)\| < \varepsilon$ if $x \in A_1$ and $\|(x, a_2, \cdots, a_n) - (a_1, a_2, \cdots, a_n)\| = |x - a_1| < \delta$. Problem 4 is an example.

13. Let $A = A_1 \cup A_2$ where $A_1 := \{(0,y) : 0 \leq y \leq 1\}$, and $A_2 := \{(x,y) : 0 \leq y \leq 1\}$.

Define f on A by $\mathbf{f}(x,y) = \begin{cases} (x-y,y) & \text{if } (x,y)\in A_2. \\ (0,y) & \text{if } (0,y)\in A_1 \end{cases}$

Then f maps A onto the closed triangle D bounded by the lines $x = 0$, $y = 0$, $x + y = 1$. Moreover, $\mathbf{f}^{-1}(u,v) = \begin{cases} (u + v,v) & \text{if } u>0 \\ (0, v) & \text{if } u=0 \end{cases}.$

Now f is continuous on A because f is clearly continuous on each of A_1 and A_2 separately and f is readily seen to be continuous at the common point (0,0). But $\mathbf{f}^{-1}$ is not continuous at (0,1) for points are "pulled apart". Precisely $\mathbf{f}^{-1}(\varepsilon,1,-\varepsilon) = (1,1-\varepsilon)$ while $\mathbf{f}^{-1}(0,1) = (0,1)$.
(This example was suggested by Dr. Colin McGregor.)

14. If $f(S)$ is not bounded, for every $m\in\mathbb{N}$, there is $\mathbf{x}^{(m)}\in S$ such that $|f(\mathbf{x}^{(m)})| > m$. By the Bolzano-Weierstrass theorem, $\{\mathbf{x}^{(m)}\}$ has a convergent subsequence $\mathbf{x}^{m(p)} \to \mathbf{x}$ say (but perhaps $\mathbf{x}\notin S$). Then $\{\mathbf{x}^{m(p)}\}$ is a Cauchy sequence and, by the uniform continuity of f, so is $\{f(\mathbf{x}^{m(p)})\}$. This sequence is therefore bounded, contradicting $|f(\mathbf{x}^{m(p)})| > m(p)$ for all $m(p)$.

15.(*i*) See Proposition 6.11. (*ii*) See Lemma 6.16.
(*iii*) $\{x : f \text{ is continuous at } x\} = \bigcap_{m\in\mathbb{N}} \{x : \omega_f(x) < 1/m\}$, a countable intersection of open sets. By problem 1.31, $\mathbb{Q}$ is not of this form.

16. Take an arbitrary $\varepsilon>0$. For each $\mathbf{x}\in K$ $\exists$ $n(\mathbf{x})\in\mathbb{N}$ such that $0\le f^{n(\mathbf{x})}(\mathbf{x})<\varepsilon$. As $f^{n(\mathbf{x})}$ is continuous, $\exists$ $\delta(\mathbf{x})>0$ such that $0\le f^{n(\mathbf{x})}(\mathbf{y})<2\varepsilon$ for all $\mathbf{y}\in B(\mathbf{x},\delta(\mathbf{x}))$. Therefore, by the decreasing property, $0\le f^{(m)}(\mathbf{y})<2\varepsilon$ for $\mathbf{y}\in B(\mathbf{x},\delta(\mathbf{x}))$ and all $m\ge n(\mathbf{x})$. The open cover $\{B(\mathbf{x},\delta(\mathbf{x})) : \mathbf{x}\in K\}$ has a finite subcover, $B(\mathbf{x}^{(1)},\delta^{(1)}),\cdots,B(x^{(p)},\delta^{(p)})$, say. Set $N = \max\{n(\mathbf{x}^{(1)}),\cdots,n(\mathbf{x}^{(p)})\}$. Then, for $m\ge N$ and $\mathbf{y}\in K$, there is r so that $\mathbf{y}\in B(\mathbf{x}^{(r)},\delta^{(r)})$ and so $0\le f^{(m)}(\mathbf{y})<2\varepsilon$. This shows that $f^{(m)}\to 0$ uniformly on K. Write $f^{(m)}(\mathbf{x}) = g^{(m)}(\mathbf{x}) - g(\mathbf{x})$. Then each $f^{(m)}$ is continuous and $f^{(m)}\to 0$ uniformly by what has just been shown. This proves $g^{(m)}\to g$ uniformly. (A theorem of Dini.)

17. For $0<x<1$, $1+mx \to \infty$ as $m \to \infty$ so $f^{(m)}(x) \to 0$ pointwise on (0,1). But $f^{(m)}(1/m) = 1/2$ for every m, so $\sup f^{(m)}(x)\ge f^{(m)}(1/m) = 1/2$ and $f^{(m)}$ does not converge uniformly to 0. (0,1) is not compact, see 16.

18. As in 17, $f^{(m)}(x) \to 0$ pointwise on (0,1). For $\varepsilon>0$ and $m>1/\varepsilon$, $1+mx > 1+x/\varepsilon > x/\varepsilon$ so $0 < f^{(m)}(x) < x/(x/\varepsilon) = \varepsilon$. This is true for all

$x \in (0,1)$ and all $m > 1/\varepsilon$, so $f^{(m)} \longrightarrow 0$ uniformly on $(0,1)$.

19. $f^{(m)}(x,y) = \dfrac{(1/m) + x}{(1/m) + y} \longrightarrow \dfrac{x}{y}$ as $m \longrightarrow \infty$. Now

$$\left| f^{(m)}(x,y) - \frac{x}{y} \right| = \frac{|y-x|}{|y|\,|1+my|} \le \frac{|y-x|}{|y^2|\,m} \le \frac{4}{m}$$

and the uniform convergence follows easily from this.

20. No. If $\mathbf{f}$ is not continuous at $\mathbf{x}$ the result may fail. For we may take $\mathbf{f}^{(m)} = \mathbf{f}$ for each m and note that there is a sequence $\mathbf{x}^{(m)} \longrightarrow \mathbf{x}$ such that $\mathbf{f}(\mathbf{x}^{(m)}) \not\longrightarrow \mathbf{f}(\mathbf{x})$. If $\mathbf{f}$ is continuous the result is true. Let $\varepsilon > 0$ be given. Then $\exists\ N(\varepsilon)$ such that $\|\mathbf{f}^{(m)}(\mathbf{y}) - \mathbf{f}(\mathbf{y})\| < \varepsilon/2$ for all $\mathbf{y} \in A$ when $m \ge N(\varepsilon)$. Since $\mathbf{f}$ is continuous at $\mathbf{x}$, $\exists\ \delta(\varepsilon)$ such that $\|\mathbf{f}(\mathbf{y}) - \mathbf{f}(\mathbf{x})\| < \varepsilon/2$ for $\|\mathbf{y} - \mathbf{x}\| < \delta$. Let m_1 be such that $\|\mathbf{x}^{(m)} - \mathbf{x}\| < \delta$ for $m \ge m_1$. Then for $m \ge \max\{N, m_1\}$,

$$\|\mathbf{f}^{(m)}(\mathbf{x}^{(m)}) - \mathbf{f}(\mathbf{x})\| \le \|\mathbf{f}^{(m)}(\mathbf{x}^{(m)}) - \mathbf{f}(\mathbf{x}^{(m)})\| + \|\mathbf{f}(\mathbf{x}^{(m)}) - \mathbf{f}(\mathbf{x})\|$$
$$< \varepsilon/2 + \varepsilon/2 = \varepsilon.$$

21. For $x = 0$, $f^{(m)}(x) = 0$; for $x > 0$, $f^{(m)}(x) = 1/[(1/mx) + mx] \longrightarrow 0$.

No. For $x = 1/m$, $f^{(m)}(1/m) = 1/2$.

22. Suppose $|g^{(m)}(x)| \le M$ for all $\mathbf{x} \in A$ and all $m \in \mathbb{N}$. Since $\sum \mathbf{f}^{(m)}$ is uniformly convergent, for $\varepsilon > 0$ $\exists N(\varepsilon)$ such that

$\sup \|\mathbf{f}^{(m+1)}(\mathbf{x}) + \cdots + \mathbf{f}^{(m+p)}(\mathbf{x})\| < \varepsilon$ for all $m \ge N(\varepsilon)$, and $p \ge 1$. By Abel's partial summation formula,

$$\|f^{(m+1)}(\mathbf{x})g^{(m+1)}(\mathbf{x}) + \cdots + f^{(m+p)}(\mathbf{x})g^{(m+p)}(\mathbf{x})\| < \varepsilon\,|g^{(m+1)}(x)| < \varepsilon M$$

for all $\mathbf{x} \in A$. This proves the result.

23. $\sum a_n (R')^n$ is absolutely convergent and $|a_n x^n| \le a_n (R')^n := M_n$ for $\mathbf{x} \in [-R', R']$; apply the M-test.

24. Take $g^{(m)}(x) = x^m$, $f^{(m)}(x) = > a_m$ and apply Abel's test (prob. 22) to deduce that $\sum_{n=0}^{\infty} a_n x^n$ is uniformly convergent on $[0,1]$. Since $a_n x^n$ is continuous for each n, we obtain $\sum_{n=0}^{\infty} a_n x^n$ is continuous on $[0,1]$.

Therefore $\lim\limits_{x \to 1-} f(x) = \lim\limits_{x \to 1-} \sum_{n=0}^{\infty} a_n x^n = \sum_{n=0}^{\infty} a_n = \alpha$.

Let $F(x) = \int_0^x f(t)\,dt$. Then $F(x) = \sum_{n=0}^{\infty} \frac{a_n}{n+1} x^n$, for $|x| < 1$, by term by term integration (see chapter 6) and the series is uniformly convergent on $[0,1]$ by the argument just given. As F is continuous,

$$F(1) = \lim_{x \to 1-} F(x) = \sum_{n=0}^{\infty} \frac{a_n}{n+1}.$$

25. As in Example 2.58,

$$\cos x + \cos 2x + \cdots + \cos nx = \frac{\sin\left(n+\frac{1}{2}\right)x - \sin(x/2)}{2\sin(x/2)},\quad x\neq 2k\pi,\ k\in\mathbb{Z}.$$

So for $x\in(0,2\pi)$, $\sum_{n=1}^{\infty}\cos nx$ has uniformly bounded partial sums and the sequence $\{1/n^{\alpha}\}$ decreases to 0 for $\alpha>0$, so the series is uniformly convergent by Dirichlet's test.

As $\sin^2 nx = (1-\cos 2nx)/2$, $\sum_{n=1}^{N}\frac{\sin^2 nx}{n} = \sum_{n=1}^{N}\frac{1}{2} - \sum_{n=1}^{N}\frac{\cos 2nx}{n}$. The second partial sum is convergent, therefore the given series is divergent.

26. Let $g^{(k)}(x) = 1/(k+x)$ and $f^{(k)} = (-1)^{k+1}$. Since the partials sums $F^{(m)}$ are either 1 or 0, Dirichlet's test applies. $\sum 1/(k+x)$ is divergent for every $x\geq 0$ by (for example) the well-known integral test; see any of the references [1, 2, 9, 12, 13].

27. $\sum_{n=1}^{m} x^n(1-x) = 1-x^m$ for $x\neq 1$ so Dirichlet's test applies. By Ex.2.34,

$$M_n = \left(\frac{n}{n+1}\right)^n \frac{1}{(n+1)\ln(n+1)} \geq \frac{1/e}{(n+1)\ln(n+1)},$$ since $[n/(n+1)]^n$ decreases to $1/e$, $[e=\exp(1)]$. The series $\sum\frac{1}{(n+1)\ln(n+1)}$ diverges (integral test).

28.(*a*) $\mathrm{f}^{-1}(T\cup W) = \{\mathrm{x}\in A : \mathrm{f(x)}\in T \text{ or } \mathrm{f(x)}\in W\}$
$= \{\mathrm{x}\in A : \mathrm{f(x)}\in T\}\cup\{\mathrm{x}\in A : \mathrm{f(x)}\in W\} = \mathrm{f}^{-1}(T)\cup\mathrm{f}^{-1}(W)$.

Results for intersections? Yes :

$\mathrm{f}^{-1}(T\cap W) = \{\mathrm{x}\in A : \mathrm{f(x)}\in T \text{ and } \mathrm{f(x)}\in W\}$
$= \{\mathrm{x}\in A : \mathrm{f(x)}\in T\}\cap\{\mathrm{x}\in A : \mathrm{f(x)}\in W\}$.

(*b*) If $\mathrm{y}\in\mathrm{f}(\mathrm{f}^{-1}(T))$ then $\mathrm{y} = \mathrm{f(x)}$ for $\mathrm{x}\in\mathrm{f}^{-1}(T)$. Thus, $\mathrm{y} = \mathrm{f(x)}$ where $\mathrm{x}\in A$ and $\mathrm{f(x)}\in T$ so $\mathrm{y} = \mathrm{f(x)}\in\mathrm{f}(A)\cap T$. Conversely, if $\mathrm{y}\in\mathrm{f}(A)\cap T$ then $\mathrm{y}=\mathrm{f(x)}$ for some $\mathrm{x}\in A$ and $\mathrm{f(x)}\in T$ so $\mathrm{x}\in\mathrm{f}^{-1}(T)$ and $\mathrm{y}\in\mathrm{f}(\mathrm{f}^{-1}(T))$.

To show that $\mathrm{f}^{-1}(\mathrm{f}(S))=S$, by Lemma 2.59, we have only to show that $\mathrm{f}^{-1}(\mathrm{f}(S))\subseteq S$. If $\mathrm{x}\in\mathrm{f}^{-1}(\mathrm{f}(S))$, then $\mathrm{x}\in A$ and $\mathrm{f(x)}\in\mathrm{f}(S)$ so $\mathrm{f(x)}=\mathrm{f(s)}$ for some $\mathrm{s}\in S$. As f is one-to-one, x=s and so $\mathrm{x}\in S$.

29. If $\mathrm{y}\in\mathrm{f}(A)\setminus\mathrm{f}(S)$, then $\mathrm{y} = \mathrm{f}(\mathbf{a})$ where $\mathbf{a}\notin S$, so $\mathrm{y}\in\mathrm{f}(A\setminus S)$. An example is $\mathrm{f(x)} = \mathrm{c}$, a constant function. Then $\mathrm{f}(A)\setminus\mathrm{f}(S) = \emptyset$ while $\mathrm{f}(A\setminus S) = \{\mathrm{c}\}$, for $S\neq A$. For inverse images we have

$\mathrm{y}\in\mathrm{f}^{-1}(B\setminus T) \Leftrightarrow \mathrm{x}\in A$ and $\mathrm{f(x)}\in B\setminus T \Leftrightarrow \mathrm{x}\in A$ and $\mathrm{f(x)}\notin T \Leftrightarrow \mathrm{x}\in A\setminus\mathrm{f}^{-1}(T)$.

30. E closed relative to $A \Leftrightarrow E=A\cap F$ where F is closed.

Then $A\setminus E = A\cap E^c = A\cap[A^c\cup F^c] = A\cap F^c$ which is relatively open. The converse is similar.

If T is closed relative to $\mathrm{f}(A)$, then $\mathrm{f}(A)\setminus T$ is open relative to

$\mathbf{f}(A)$ and $\mathbf{f}^{-1}(\mathbf{f}(A)\setminus T)=A\setminus\mathbf{f}^{-1}(T)$ is open relative to A by the continuity of $\mathbf{f}$. This proves $\mathbf{f}^{-1}(T)$ is closed relative to A by the first part. The converse is exactly similar.

31. A sketch reveals that a narrow cone with vertex at $\mathbf{y}$ lies in C. A nbd of $\mathbf{x}$ can be moved along the cone shrinking in size. Take a fixed t with $0<t<1$ and write $\mathbf{x}_t := (1-t)\mathbf{x}+t\mathbf{y}$. Since $\mathbf{x}\in\text{int}(C)$, there is $\delta>0$ so that $\mathbf{x}+\mathbf{z}\in C$ for every $\mathbf{z}\in\mathbb{R}^n$ with $\|\mathbf{z}\|<\delta$. Then, by convexity, $(1-t)(\mathbf{x}+\mathbf{z})+t\mathbf{y}\in C$, that is, $\mathbf{x}_t+(1-t)\mathbf{z}\in C$. This shows that the $(1-t)\delta$-nbd of $\mathbf{x}_t$ lies in C so that $\mathbf{x}_t\in\text{int}(C)$. If $\mathbf{x},\mathbf{y}$ both belong to $\text{int}(C)$ then the above applies and we see that $(1-t)\mathbf{x}+t\mathbf{y}\in\text{int}(C)$ for $0\le t\le 1$.

32. Let $\mathbf{x}=\sum\lambda^{(i)}\mathbf{a}^{(i)}$, $\mathbf{y}=\sum\mu^{(i)}\mathbf{b}^{(i)}$ where $\mathbf{a}^{(i)},\mathbf{b}^{(i)}\in A$ and $\sum\lambda^{(i)}=1=\sum\mu^{(i)}$. By setting some of the $\lambda^{(i)}$ and $\mu^{(i)}$ to zero we can suppose that $\mathbf{x}=\sum\lambda^{(i)}\mathbf{a}^{(i)}$ and $\mathbf{y}=\sum\mu^{(i)}\mathbf{a}^{(i)}$ where the sums are taken over the same set of indices and the new set of $\mathbf{a}^{(i)}$ include the old $\mathbf{a}^{(i)}$ and $\mathbf{b}^{(i)}$. Then, for $0<t<1$, $(1-t)\mathbf{x}+t\mathbf{y}=\sum[(1-t)\lambda^{(i)}+t\mu^{(i)}]\mathbf{a}^{(i)}$, $\sum[(1-t)\lambda^{(i)}+t\mu^{(i)}]=(1-t)\sum\lambda^{(i)}+t\sum\mu^{(i)}=1$, that is, $\text{co}(A)$ is a convex set. $A\subseteq\text{co}(A)$ since $\mathbf{a}=\lambda\mathbf{a}$ for $\lambda=1$. Now let C be a convex set that contains A. Let $\mathbf{a},\mathbf{b},\mathbf{c}$ be three points in C and consider $\lambda\mathbf{a}+\mu\mathbf{b}+(1-\lambda-\mu)\mathbf{c}$, where $0\le\lambda,\mu\le 1$. Now $\mathbf{x}:=\lambda\mathbf{a}/(\lambda+\mu)+\mu\mathbf{b}/(\lambda+\mu)\in C$, by convexity, and we then have $\lambda\mathbf{a}+\mu\mathbf{b}+(1-\lambda-\mu)\mathbf{c}=(\lambda+\mu)\mathbf{x}+(1-(\lambda+\mu))\mathbf{c}\in C$. The proof for any finite sum is now a routine induction argument. Therefore $\text{co}(A)\subseteq C$ so $\text{co}(A)$ is the smallest convex set containing A.

33. If $A = U\cup V$, a disjoint union of relatively open sets, then $A\setminus U=V$ and V is closed relative to A by problem 30; similarly for U. If A is disconnected then there exist sets U, V which are both open and closed relative to A whose union is A so A must be connected when only A and $\emptyset$ have this property. Conversely if U is a set that is both open and closed relative to A then $A\setminus U$ has the same property and $A=U\cup(A\setminus U)$ is a disconnection of A.

34. f attains its maximum and minimum values on $\overline{\mathbf{B}}(\mathbf{0},1)$, say $f(\mathbf{b})$ and $f(\mathbf{a})$ respectively. Then f restricted to the connected set $S = \{(1-t)\mathbf{a}+t\mathbf{b} : 0\le t\le 1\}$ is continuous so the image set is also connected. Since it contains $f(\mathbf{a})$ and $f(\mathbf{b})$ and is an interval it must coincide with all of $[f(\mathbf{a}),f(\mathbf{b})]$. Now any point in $\overline{\mathbf{B}}(0,1)$ must of necessity lie in this interval so points outside S must have the same image as some point of S and f cannot be one-to-one.

35. Pathwise connected ; the interval $(0,3]$ lies in the set.

36. (*i*) Let $\rho(\mathbf{y}) = \inf\{\|\mathbf{x}-\mathbf{y}\| : \mathbf{x}\in A\}$. Then there exists a sequence

$\{\mathbf{a}^{(m)}\}\subseteq A$ such that $\|\mathbf{a}^{(m)}-\mathbf{y}\| \longrightarrow \rho(\mathbf{y})$. By compactness of A, $\exists$ $a^{m(k)} \longrightarrow \mathbf{a}\in A$ and $\|\mathbf{a}-\mathbf{y}\| = \rho(\mathbf{y})$ by continuity of the norm. Therefore, $\text{dist}(\mathbf{a},B) = \inf\{\|\mathbf{a}-\mathbf{y}\| : \mathbf{y}\in B\} = \inf\{\rho(\mathbf{y}) : \mathbf{y}\in B\} = \rho(A,B)$.

(*ii*) $\exists$ $\{\mathbf{a}^{(m)}\}\subseteq A$, $\{\mathbf{b}^{(m)}\}\subseteq B$ such that $\|\mathbf{a}^{(m)}-\mathbf{b}^{(m)}\| \longrightarrow \rho(A,B)$.
By compactness of A,B, we can suppose (by a change of notation) that $\mathbf{a}^{(m)} \longrightarrow \mathbf{a}\in A$, $\mathbf{b}^{(m)} \longrightarrow \mathbf{b}\in B$ and then $\|\mathbf{a}-\mathbf{b}\| = \rho(A,B)$.
(*iii*) U is open since $\phi(x) := \text{dist}(\mathbf{x}, A)$ is continuous (problem 8). Moreover, $A\subseteq U$. Similarly $B\subseteq V$, an open set. Also $U\cap V=\emptyset$ by construction so this disconnects $A\cup B$.

37. Since $C\cap\text{bdy}(A)=\emptyset$, $C=[\text{int}(A)\cap C]\cup[\text{int}(A^c)\cap C]$ a union of disjoint sets that are open relative to C. Since C is connected one of these sets must be empty and either $C\subseteq\text{int}(A)$ or else $C\subseteq\text{int}(A^c)$.

38. $L(\mathbf{x}+\mathbf{y}) = \langle\mathbf{a},\mathbf{x}+\mathbf{y}\rangle = \langle\mathbf{a},\mathbf{x}\rangle + \langle\mathbf{a},\mathbf{y}\rangle = L\mathbf{x}+L\mathbf{y}$ by properties of scalar product. Similarly for $L(\alpha\mathbf{x})$.

39.(*a*) Yes. Checks are routine. The matrix representation is
$\begin{bmatrix} 1 & 1 & 1 \\ 1 & 0 & 0 \\ 1 & -1 & 0 \end{bmatrix}$. (*b*) No. Because $\mathbf{f}(a+b,0,0)=(e^{a+b}-1, a+b, a+b)\neq$ $(e^{a}-1, a, a)+(e^{b}-1, b, b)=\mathbf{f}(a,0,0)+\mathbf{f}(b,0,0)$ in general, for example it fails for $a=b=1$. $f_1(\mathbf{x})=e^{x}-1$ is not linear, (see 40).

40. $\mathbf{Lx}=(L_1\mathbf{x},\cdots,L_m\mathbf{x})$ and $\mathbf{Ly}=(L_1\mathbf{y},\cdots,L_m\mathbf{y})$. Therefore $\mathbf{L}(\mathbf{x}+\mathbf{y})=\mathbf{Lx}+\mathbf{Ly}$ if and only if $L_i(\mathbf{x}+\mathbf{y})=L_i\mathbf{x}+L_i\mathbf{y}$ for every i. Similarly for multiplication by scalars.

41.(*a*) $\|\mathbf{L}\|$ is defined by $\|\mathbf{L}\|=\sup\{\|\mathbf{Lx}\| : \|\mathbf{x}\|\leq 1\}=\sup\{\|\mathbf{Lx}\| : \mathbf{x}\in\overline{\mathrm{B}}(0,1)\}$. Let $K=\sup\{\|\mathbf{Lx}\| : \|\mathbf{x}\|=1\}=\sup\{\|\mathbf{Lx}\| : \mathbf{x}\in\mathrm{S}(0,1)\}$. Then $K\leq\|\mathbf{L}\|$ since $\mathrm{S}\subset\overline{\mathrm{B}}$. Also for any $\mathbf{x}$ in $\overline{\mathrm{B}}$ with $\mathbf{x}\neq\mathbf{0}$, $\mathbf{y}:=\mathbf{x}/\|\mathbf{x}\|\in\mathrm{S}$ so $\|\mathbf{Ly}\|\leq K$ that is $\|\mathbf{Lx}\|/\|\mathbf{x}\|\leq K$ or $\|\mathbf{Lx}\|\leq K\|\mathbf{x}\|\leq K$. This proves $\|\mathbf{L}\|\leq K$ and hence we have equality. Now for any $\mathbf{x}\neq\mathbf{0}$, $\|\mathbf{L}(\mathbf{x}/\|\mathbf{x}\|)\|\leq\|\mathbf{L}\|$ and so $\|\mathbf{Lx}\|\leq\|\mathbf{L}\|\,\|\mathbf{x}\|$. The inequality holds trivially when $\mathbf{x}=\mathbf{0}$.
(*b*) $(\mathbf{L}+\mathbf{M})\mathbf{x}\| = \|\mathbf{Lx}+\mathbf{Mx}\| \leq \|\mathbf{Lx}\|+\|\mathbf{Mx}\| \leq \|\mathbf{L}\|\,\|\mathbf{x}\|+\|\mathbf{M}\|\,\|\mathbf{x}\|$ so $\sup\{\|(\mathbf{L}+\mathbf{M})\mathbf{x}\| : \|\mathbf{x}\|=1\}\leq\|\mathbf{L}\|+\|\mathbf{M}\|$. Similarly $\|\mathbf{LMx}\|\leq\|\mathbf{L}\|\,\|\mathbf{Mx}\|$ $\leq\|\mathbf{L}\|\,\|\mathbf{M}\|\,\|\mathbf{x}\|$ and hence $\sup\{\|\mathbf{LMx}\| : \|\mathbf{x}\|\leq 1\}\leq\|\mathbf{L}\|\,\|\mathbf{M}\|$.

42. (1) $\Leftrightarrow$ (2). If $\mathbf{L}$ is one-to-one then $\mathbf{Lx}=\mathbf{0} \Rightarrow \mathbf{x}=\mathbf{0}$ so $N(\mathbf{L})=\{\mathbf{0}\}$. If $N(\mathbf{L})=\{\mathbf{0}\}$ and $\mathbf{Lx}=\mathbf{Ly}$ then $\mathbf{L}(\mathbf{x}-\mathbf{y})=\mathbf{0}$ so $\mathbf{x}-\mathbf{y}\in N(\mathbf{L})$ and $\mathbf{x}=\mathbf{y}$, that is, $\mathbf{L}$ is one-to-one. If (3) holds then $\mathbf{Lx}=\mathbf{0} \Rightarrow \mathbf{x}=\mathbf{0}$ so $N(\mathbf{L})=\{\mathbf{0}\}$. Conversely, if $N(\mathbf{L})=\{\mathbf{0}\}$ then $\mathbf{Lx}\neq\mathbf{0}$ for all $\mathbf{x}\neq\mathbf{0}$ and so $\mathbf{Lx}\neq 0$ on $\mathrm{S}=\{\mathbf{x} : \|\mathbf{x}\|=1\}$. Let $k=\min\{\|\mathbf{Lx}\| : \mathbf{x}\in\mathrm{S}\}$, $k>0$ since S is compact and $\mathbf{L}$ is continuous. Then $\|\mathbf{L}(\mathbf{x}/\|\mathbf{x}\|)\|\geq k$ for $\mathbf{x}\neq\mathbf{0}$ and by linearity this gives $\|\mathbf{Lx}\|\geq k\|\mathbf{x}\|$.

43. If $\mathbf{LQ}=\mathbf{PL}=\mathbf{I}$ then $\mathbf{PLQ}=\mathbf{P}(\mathbf{LQ})=\mathbf{P}$ and $\mathbf{PLQ}=(\mathbf{PL})\mathbf{Q}=\mathbf{Q}$ so $\mathbf{P}=\mathbf{Q}$. (Composition of functions is associative.) $\mathbf{L}$ is one-to-one (injective) for if $\mathbf{Lx} = \mathbf{Ly}$ then also $\mathbf{PLx} = \mathbf{PLy}$, that is, $\mathbf{x} = \mathbf{y}$. $\mathbf{L}$ maps $\mathbb{R}^n$ onto $\mathbb{R}^n$ (is surjective) for, given any $\mathbf{y}\in\mathbb{R}^n$, $\mathbf{L}(\mathbf{Qy})=\mathbf{y}$. Therefore $\mathbf{L}$ is a bijection and $\mathbf{L}^{-1}$ exists. Since $\mathbf{L}^{-1}\mathbf{y}=\mathbf{x} \Leftrightarrow \mathbf{y}=\mathbf{Lx}$ we have $\mathbf{L}^{-1}\mathbf{y}=\mathbf{Qy}$ so $\mathbf{L}^{-1}=\mathbf{Q}$ which proves the result since $\mathbf{Q}=\mathbf{P}$.

44. From 43, it suffices to find $\mathbf{Q}$ such that $\mathbf{LMQ}=\mathbf{QLM}=\mathbf{I}$. We let $\mathbf{Q}=\mathbf{M}^{-1}\mathbf{L}^{-1}$. Then $\mathbf{LMQ}=\mathbf{LMM}^{-1}\mathbf{L}^{-1}=\mathbf{LIL}^{-1}=\mathbf{I}$. Similarly for $\mathbf{QLM}$.

45. Note that a set C is convex if and only if $C=\mathrm{co}(C)$ by problem 32. Then, $\mathbf{L}(C)=\mathbf{L}(\mathrm{co}(C)) = \{\mathbf{L}(\sum\lambda^{(i)}\mathbf{a}^{(i)}\} = \{\sum\lambda^{(i)}\mathbf{L}(\mathbf{a}^{(i)})\} = \mathrm{co}(\mathbf{L}(C))$.

Chapter 3.

1. If $\mathbf{L}$ and $\mathbf{M}$ both satisfy the hypotheses then, for an arbitrary $\varepsilon>0$, there exists $\delta>0$ such that $\|\mathbf{f}(\mathbf{a}+\mathbf{h})-\mathbf{f}(\mathbf{a})-\mathbf{Lh}\| < \varepsilon\|\mathbf{h}\|$ and $\|\mathbf{f}(\mathbf{a}+\mathbf{h})-\mathbf{f}(\mathbf{a})-\mathbf{Mh}\| < \varepsilon\|\mathbf{h}\|$ for $\|\mathbf{h}\| \le \delta$. Then $\|\mathbf{Lh}-\mathbf{Mh}\| < 2\varepsilon\|\mathbf{h}\|$ for $\|\mathbf{h}\| \le \delta$, by the triangle inequality. For any $\mathbf{x}\in\mathbb{R}^n$, $\delta\mathbf{x}/\|\mathbf{x}\|$ has norm δ so that $\|(\mathbf{L}-\mathbf{M})\delta\mathbf{x}/\|\mathbf{x}\|\,\| < 2\varepsilon\delta$. By linearity this gives $\|(\mathbf{L}-\mathbf{M})\mathbf{x}\| < 2\varepsilon$. As ε is arbitrary this shows that $\mathbf{Lx}=\mathbf{Mx}$, that is $\mathbf{L}=\mathbf{M}$.

2. Since $\mathbf{f}$ is differentiable at $\mathbf{a}$, $\mathbf{f}(\mathbf{a}+\mathbf{h})=\mathbf{f}(\mathbf{a})+D\mathbf{f}(\mathbf{a})\mathbf{h}+\mathbf{R}$ where $\|\mathbf{R}\|/\|\mathbf{h}\| \longrightarrow 0$ as $\|\mathbf{h}\| \longrightarrow 0$. Because $\mathbf{L}$ is linear,

$$\mathbf{L}(\mathbf{f}(\mathbf{a}+\mathbf{h}))=\mathbf{L}(\mathbf{f}(\mathbf{a}))+\mathbf{L}(D\mathbf{f}(\mathbf{a})\mathbf{h})+\mathbf{LR}$$

and we have $\|\mathbf{LR}\|/\|\mathbf{h}\| \le \|\mathbf{L}\|\,\|\mathbf{R}\|/\|\mathbf{h}\| \longrightarrow 0$ as $\|\mathbf{h}\| \longrightarrow 0$.
Alternatively. By the chain rule, $D(\mathbf{L}\circ\mathbf{f})(\mathbf{a})=D\mathbf{L}(\mathbf{f}(\mathbf{a}))\,D\mathbf{f}(\mathbf{a})=\mathbf{L}\,D\mathbf{f}(\mathbf{a})$.

3. $\mathbf{g}(t+s) - \mathbf{g}(t)=\mathbf{a}+(t+s)\mathbf{e} - (\mathbf{a}+t\mathbf{e})=s\mathbf{e}$. Therefore the definition is satisfied by $D\mathbf{g}(t)=\mathbf{e}$ with $\mathbf{R}=\mathbf{0}$.
Also, $g_i(t)=a_i+te_i$ so $\partial g_i/\partial t=e_i$, and constants are continuous.

4. $\dfrac{f(h,0)-f(0,0)}{h}=\dfrac{h^3/h^2}{h}$ so $\dfrac{\partial f}{\partial x}(0,0)=1$ and $\dfrac{f(0,k)-f(0,0)}{k}=\dfrac{-k^3/k^2}{k}$ so $\dfrac{\partial f}{\partial y}(0,0)= -1$. Is f differentiable at (0,0)? NO. The candidate for $Df(\mathbf{0})$ is $(1, -1)$. Let $R=f(x,y) - [f(0,0)+(1, -1)\begin{pmatrix}x\\y\end{pmatrix}]$. Then

$|R|/\sqrt{(x^2+y^2)}=|(x^3-y^3)/(x^2+y^2)^{3/2}-(x-y)/(x^2+y^2)^{1/2}|$. Write $x=r\cos\theta$, $y=r\sin\theta$ (polars). Then $|R/r| = |\cos^3\theta - \sin^3\theta - \cos\theta+\sin\theta| \not\longrightarrow 0$ as $r \longrightarrow 0$, for example along $\theta=\pi/3$, the line $y=\sqrt{3}x$.

5. $\dfrac{f(h,0)-f(0,0)}{h}=0$ for all $h\neq 0$, therefore $\partial f/\partial x(0,0)=0$. Similarly $\partial f/\partial y(0,0)=0$. Let $R=f(x,y) - [f(0,0)+(0,0)\begin{pmatrix}x\\y\end{pmatrix}]=|xy|^{1/2}$. Then, writing

$x = r\cos\theta$, $y = r\sin\theta$, $|R|/\sqrt{(x^2+y^2)} = (\cos\theta\sin\theta)^{1/2} \not\to 0$ as $r \to 0$.

6. $\dfrac{f(h,0,0)-f(0,0,0)}{h} = 0$ for all $h \neq 0$, therefore $\partial f/\partial x = 0$ at $(0,0,0)$. By symmetry, $\partial f/\partial y$ and $\partial f/\partial z$ are also 0. Therefore the candidate for $Df(\mathbf{0})$ is $\mathbf{0}$. Let $R = f(\mathbf{x}) - [f(\mathbf{0}) + Df(\mathbf{0})(\mathbf{x})]$, so that $R = f(\mathbf{x})$ and f is differentiable at $\mathbf{0}$ if, and only if,

$$\frac{|R|}{(x^2+y^2+z^2)^{1/2}} = \frac{|xyz|}{(x^2+y^2+z^2)^{\alpha+1/2}} \longrightarrow 0 \text{ as } x^2+y^2+z^2 \longrightarrow 0.$$

Using the arithmetic-geometric mean inequality $(abc)^{1/3} \leq \dfrac{a+b+c}{3}$, (with $a = x^2$ etc.) we see that $|R|/(x^2+y^2+z^2)^{1/2} \leq (x^2+y^2+z^2)^{1-\alpha}/3\sqrt{3}$ which has limit 0 as $\|\mathbf{x}\| \longrightarrow 0$ if $\alpha < 1$. If $\alpha \geq 1$, we may take $x = y = z$ to find $|R|/(x^2+y^2+z^2)^{1/2} = x^3/(3x^2)^{\alpha+1/2} \not\longrightarrow 0$ as $|x| \longrightarrow 0$.

7. $\dfrac{f(h,0)-f(0,0)}{h} = 0$ for all $h \neq 0$, therefore $\partial f/\partial x(0,0) = 0$. Similarly $\partial f/\partial y(0,0) = 0$. For $\phi(x) = x^2$, $f(x,\phi(x)) = (x\cdot x^2)^{1/3} = x$ so $\dfrac{d}{dx}f(x,\phi(x)) = 1$. Applying the chain rule $\dfrac{d}{dx}f(x,\phi(x)) = \dfrac{\partial f}{\partial x}\cdot 1 + \dfrac{\partial f}{\partial y}\phi'(x) = 0$ at $(0,0)$. What is wrong? *Answer*: f is not differentiable at $(0,0)$ so the chain rule cannot be applied.

8. We have $w = (2x+3y)^2 + (4x-5y)^2 = 20x^2 - 28xy + 34y^2$, thus $\partial w/\partial x = 40x - 28y$, and $\partial w/\partial y = 68y - 28x$. By the chain rule,

$$\frac{\partial w}{\partial x} = \frac{\partial w}{\partial u}\frac{\partial u}{\partial x} + \frac{\partial w}{\partial v}\frac{\partial v}{\partial x} = 2u\cdot 2 + 2v\cdot 4 = 4(2x+3y) + 8(4x-5y) = 40x - 28y,$$

and, $$\frac{\partial w}{\partial y} = \frac{\partial w}{\partial u}\frac{\partial u}{\partial y} + \frac{\partial w}{\partial v}\frac{\partial v}{\partial y} = 6u - 10v = 6(2x+3y) - 10(4x-5y) = -28x + 68y.$$

9. $\dfrac{f(h,0)-f(0,0)}{h} = 1$ for all $h \neq 0$, therefore $\dfrac{\partial f}{\partial x}(0,0) = 1$. Similarly $\partial f/\partial y(0,0) = 0$. Let $R = f(x,y) - [f(0,0) + (1,0)\begin{pmatrix} x \\ y \end{pmatrix}] = \dfrac{x^3 - 3y^2x}{x^2+y^2} - x = \dfrac{-4xy^2}{x^2+y^2}$.

Then, using polars, we have $\dfrac{|R|}{\sqrt{(x^2+y^2)}} = 4\cos\theta\sin^2\theta \not\to 0$; the limit can depend on θ. Now for $r \neq 0$, we have $f(r\cos\theta, r\sin\theta) = r(\cos^3\theta - 3\cos\theta\sin^2\theta) = r\cos 3\theta$ by the formula

$$\cos(2\theta+\theta) = \cos(2\theta)\cos\theta - \sin(2\theta)\sin\theta = (\cos^2\theta - \sin^2\theta)\cos\theta - 2\sin\theta\cos\theta\sin\theta.$$

The partial derivatives of g at $(0,0)$ are $\dfrac{\partial g}{\partial r} = \lim \dfrac{g(r,0)-g(0,0)}{r} = 1$, and $\dfrac{\partial g}{\partial \theta} = \lim \dfrac{g(0,\theta)-g(0,0)}{\theta} = 0$. Then the remainder is $R(r,\theta) = r\cos 3\theta - r$ and $|R(r,\theta)|/\sqrt{(r^2+\theta^2)} \leq |\cos 3\theta - 1| \longrightarrow 0$ as $r^2 + \theta^2 \longrightarrow 0$.

If we write $\mathbf{h}(r,\theta) = (r\cos\theta, r\sin\theta)$ then $g = f\circ\mathbf{h}$ so there is no reason to suppose that f is differentiable given that g is.

10. Does $Dh(a,b)=\left[\frac{\partial h}{\partial x}(a,b),\ \frac{\partial h}{\partial y}(a,b)\right]=\left[\frac{df}{dx}(a),\ \frac{dg}{dy}(b)\right]$ satisfy the required affine approximation property? Let $R(x,y)=h(x,y)-\left[h(a,b)+\left[\frac{df}{dx},\frac{df}{dy}\right]\begin{bmatrix}x-a\\ y-b\end{bmatrix}\right.$. Then we have

$$R(x,y)=f(x)-f(a)-\frac{df}{dx}(x-a)+g(y)-g(b)-\frac{dg}{dy}(y-b)$$

and, as $|x-a|\le\sqrt{[(x-a)^2+(y-b)^2]}$,

$$\frac{|R(x,y)|}{\sqrt{[(x-a)^2+(y-b)^2]}}\le\left|\frac{f(x)-f(a)}{x-a}-\frac{df}{dx}\right|+\left|\frac{g(y)-g(b)}{y-b}-\frac{dg}{dy}\right|$$

which has limit 0 as $(x-a)^2+(y-b)^2\longrightarrow 0$ ($\Rightarrow x\to a$ and $y\to b$).

11. The function $f(x)=x^2\sin(1/x)$, $x\neq 0$, $f(0)=0$ is differentiable but f' is not continuous at 0. Take any differentiable function g and let $h(x,y)=f(x)+g(y)$. Then h is differentiable (by problem 10), but $\partial h/\partial x$ is not continuous.

11. Let $h(x,y)=\begin{cases}0, & \text{if } x \text{ is rational}\\ x^2, & \text{if } x \text{ is irrational}\end{cases}$. Then h is continuous only at (0,0) [show this for yourself]. The candidate for $Dh(\mathbf{0})$ is (0,0) and then $R(x,y)=h(x,y)$. Thus $|R|/\sqrt{(x^2+y^2)}$ is either 0 or $x^2/\sqrt{(x^2+y^2)}\le|x|$ so $|R|/\sqrt{(x^2+y^2)}\longrightarrow 0$ as $x^2+y^2\longrightarrow 0$. Thus h is differentiable at 0. *Also* h is differentiable by the result of problem 10.

12. Let $\mathbf{R}(t)=\mathbf{f}(\mathbf{a}+t\mathbf{e})-[\mathbf{f}(\mathbf{a})+tD\mathbf{f}(\mathbf{a})\mathbf{e}]$ so that $\|\mathbf{R}(t)\|/|t|\longrightarrow 0$ as $t\longrightarrow 0$. Let $\varepsilon=\|Df(\mathbf{a})\mathbf{e}-\mathbf{Me}\|$. Then there is $\delta>0$ such that $\|\mathbf{R}(t)\|<\varepsilon|t|/2$ for $|t|<\delta$. Now we can write

$$\|\mathbf{f}(\mathbf{a}+t\mathbf{e})-f(\mathbf{a})-t\mathbf{Me}\|=\|\mathbf{R}(t)+tD\mathbf{f}(\mathbf{a})\mathbf{e}-t\mathbf{Me}\|\ge|t|\,\|Df(\mathbf{a})\mathbf{e}-\mathbf{Me}\|-\|\mathbf{R}(t)\|$$
$$>|t|\varepsilon-\varepsilon|t|/2=\varepsilon|t|/2 \text{ for } |t|<\delta,$$

which proves the result.

13. Let $f(x,y)=3-2x^2-y^2$. Then $Df(1,-1)=(-4,2)$ so the affine approximation to f near $(1,-1)$ is given by the plane $z=f(1,-1)+(-4,2)\begin{bmatrix}x-1\\ y+1\end{bmatrix}$, that is $z=-4(x-1)+2(y+1)=-4x+2y+6$.

14. Let $\phi(t)=f(t\mathbf{x})$. Then $\phi'(t)=Df(t\mathbf{x})\mathbf{x}=<\nabla f(t\mathbf{x}),\mathbf{x}>$ by the chain rule. But also $\phi'(t)=p\,t^{p-1}f(\mathbf{x})$. Setting $t=1$ shows that $<\nabla f(\mathbf{x}),\mathbf{x}>=p\,f(\mathbf{x})$. For the converse, for $\mathbf{y}\neq\mathbf{0}$ let $g(t)=f(t\mathbf{y})$. Then $g'(t)=Df(t\mathbf{y})\mathbf{y}=<\nabla f(t\mathbf{y}),\mathbf{y}>=<\nabla f(t\mathbf{y}),\mathbf{y}>/t=pf(t\mathbf{y})/t$. Therefore, $g'(t)=pg(t)/t$ or $[\ln g(t)]'=p/t$ so that $g(t)=ct^p$. Setting $t=1$ shows that $c=f(\mathbf{y})$ and thus $f(t\mathbf{y})=t^pf(\mathbf{y})$.

15. Let $\mathbf{g}:\mathbb{R}^n\longrightarrow\mathbb{R}^n$ be given by $\mathbf{g}(\mathbf{x})=\mathbf{x}+\mathbf{a}$. Then $\mathbf{g}$ is differentiable and $D\mathbf{g}(\mathbf{x})=\mathbf{I}$, the identity operator. By the chain rule, $\mathbf{h}=\mathbf{f}\circ\mathbf{g}$ is differentiable at $\mathbf{0}$ and $D\mathbf{h}(\mathbf{0})=D\mathbf{f}(\mathbf{g}(\mathbf{0}))\,D\mathbf{g}(\mathbf{0})=D\mathbf{f}(\mathbf{a})\,\mathbf{I}=D\mathbf{f}(\mathbf{a})$.

16. We have $\mathbf{h}(\mathbf{x}+\mathbf{k})-\mathbf{h}(\mathbf{x})-(Df(\mathbf{x})\mathbf{k},Dg(\mathbf{x})\mathbf{k})=(R_1,R_2)$. Since

$\|(R_1,R_2)\|^2=\|R_1\|^2+\|R_2\|^2$, it follows that $\|(R_1,R_2)\|/\|\mathbf{k}\| \longrightarrow 0$ as $\|\mathbf{k}\| \longrightarrow 0$ if, and only if, $\|R_1\|/\|\mathbf{k}\| \longrightarrow 0$ and $\|R_2\|/\|\mathbf{k}\| \longrightarrow 0$. This proves that $D\mathbf{h}(\mathbf{x})=(Df(\mathbf{x}), Dg(\mathbf{x}))$ when either side is differentiable. The proof for component functions is exactly similar since f and g are the component functions of $\mathbf{h}$.

17. $\partial H/\partial x=y$, $\partial H/\partial y=x$ and these are continuous on $\mathbb{R}^2$. Compare Ex.3.3. $(fg)(\mathbf{x})=f(\mathbf{x})g(\mathbf{x})=H(f(\mathbf{x}), g(\mathbf{x}))=H(\mathbf{h}(\mathbf{x}))$, with $H,\mathbf{h}$ as in problem 16. As H and $\mathbf{h}$ are differentiable, Thm. 3.14 proves fg is differentiable and $D(fg)(\mathbf{a})=D(H\circ\mathbf{h}(\mathbf{a}))=DH(\mathbf{h}(\mathbf{a}))Dh(\mathbf{a})=g(\mathbf{a})Df(\mathbf{a})+f(\mathbf{a})Dg(\mathbf{a})$.

18. $h(\mathbf{a}+\mathbf{x})-h(\mathbf{a})=<\mathbf{f}(\mathbf{a}+\mathbf{x}),\mathbf{g}(\mathbf{a}+\mathbf{x})> - <\mathbf{f}(\mathbf{a}),\mathbf{g}(\mathbf{x})>$
$=<\mathbf{f}(\mathbf{a}+\mathbf{x})-\mathbf{f}(\mathbf{a}),\mathbf{g}(\mathbf{a}+\mathbf{x})> + <\mathbf{f}(\mathbf{a}),\mathbf{g}(\mathbf{a}+\mathbf{x})-\mathbf{g}(\mathbf{a})>$
$=<D\mathbf{f}(\mathbf{a})\mathbf{x}+\mathbf{R},\mathbf{g}(\mathbf{a}+\mathbf{x})> + <\mathbf{f}(\mathbf{a}), D\mathbf{g}(\mathbf{a})\mathbf{x}+\mathbf{S})>$ where $\|\mathbf{R}\|/\|\mathbf{x}\| \to 0$ as $\mathbf{x}\to\mathbf{0}$. Therefore, $h(\mathbf{a}+\mathbf{x}) - h(\mathbf{a}) - <D\mathbf{f}(\mathbf{a})\mathbf{x},\mathbf{g}(\mathbf{a})> - <\mathbf{f}(\mathbf{a}),D\mathbf{g}(\mathbf{a})\mathbf{x}>$
$=<D\mathbf{f}(\mathbf{a})\mathbf{x},\mathbf{g}(\mathbf{a}+\mathbf{x}) - \mathbf{g}(\mathbf{a})> + <\mathbf{R},\mathbf{g}(\mathbf{a}+\mathbf{x})> + <\mathbf{f}(\mathbf{a}),\mathbf{S})>$.
Now $|<D\mathbf{f}(\mathbf{a})\mathbf{x},\mathbf{g}(\mathbf{a}+\mathbf{x})-\mathbf{g}(\mathbf{a})> + <\mathbf{R},\mathbf{g}(\mathbf{a}+\mathbf{x})> + <\mathbf{f}(\mathbf{a}),\mathbf{S})>|/\|\mathbf{x}\| \longrightarrow 0$ as $\|\mathbf{x}\| \longrightarrow 0$ since $|<D\mathbf{f}(\mathbf{a})\mathbf{x},\mathbf{g}(\mathbf{a}+\mathbf{x})-\mathbf{g}(\mathbf{a})>| \le \|D\mathbf{f}(\mathbf{a})\|\,\|\mathbf{x}\|\,\|\mathbf{g}(\mathbf{a}+\mathbf{x})-\mathbf{g}(\mathbf{a})\|$ and $\|\mathbf{g}(\mathbf{a}+\mathbf{x})-\mathbf{g}(\mathbf{a})\| \to 0$ by continuity of $\mathbf{g}$, while
$|<\mathbf{R},\mathbf{g}(\mathbf{a}+\mathbf{x})>|/\|\mathbf{x}\| \le \|\mathbf{g}(\mathbf{a}+\mathbf{x})\|\,\|\mathbf{R}\|/\|\mathbf{x}\| \longrightarrow 0$ by differentiability of $\mathbf{f}$.

19. Let $\phi:\mathbb{R}\longrightarrow\mathbb{R}$ be defined by $\phi(t)=\begin{cases}\sin t\,/\,t, & t\neq 0\\ 1 & t=0\end{cases}$. Now ϕ is differentiable on $\mathbb{R}$: for $t\neq 0$ this is a simple consequence of the quotient rule. When $t=0$, $\dfrac{\phi(h)-\phi(0)}{h}=(\sin h - h)/h^2$ which has limit 0 as $h\longrightarrow 0$ (L'Hôpital's rule or truncate the series for $\sin h$). For $x\neq y$, $f(x,y)= - 2\sin[(x+y)/2]\sin[(x-y)/2]\,/\,(x-y)$. Then we have $f(x,y)= - \sin[(x+y)/2]\,\phi[(x-y)/2]$ and this holds also for $x=y$. Let $g(x,y)=\phi((x-y)/2)$. Then g is differentiable by the composite function theorem, similarly $h(x,y)= - \sin((x+y)/2)$ is differentiable. By problem 17, f is differentiable.

20. Let $R(h,k)=f(a+h,b+k) - [f(a,b)+h\dfrac{\partial f}{\partial x}(a,b)+k\dfrac{\partial f}{\partial y}(a,b)]$. We write this as $R(h,k)=f(a+h,b+k)-f(a+h,b)+f(a+h,b)-f(a,b)-h\dfrac{\partial f}{\partial x}(a,b)-k\dfrac{\partial f}{\partial y}(a,b)$.

Since $\dfrac{\partial f}{\partial x}(a,b)$ exists, $\lim\limits_{h\to 0}\dfrac{f(a+h,b)-f(a,b)}{h} - \dfrac{\partial f}{\partial x}(a,b)=0$. Since $\dfrac{\partial f}{\partial y}$ exists and is continuous on G, by the mean value theorem, $f(a+h,b+k)-f(a+h,b)=k\,\partial f/\partial y(a+h,b+\theta k)$, where $0<\theta<1$. Therefore,

$|R(h,k)|/\surd(h^2+k^2) \le |f(a+h,b+k)-f(a+h,b)-k\,\partial f/\partial y(a,b)|/k +$
$|f(a+h,b)-f(a,b)-h\partial f/\partial x(a,b)|/h$.

The second turn has zero limit by the above. The first term is
$|\dfrac{\partial f}{\partial y}(a+h,b+\theta k) - \dfrac{\partial f}{\partial y}(a,b)|$ and this has limit 0 by continuity of $\partial f/\partial y$.

21. $\partial\phi/\partial x=2xy+y^3$, $\partial\phi/\partial y=x^2+3y^2x$ and these are continuous. Therefore

$D_{\mathbf{u}}\phi = <\nabla\phi,\mathbf{u}>$ and, for a general direction $\mathbf{u}=(\cos\theta,\sin\theta)$, the directional derivative at (1,1) is $3\cos\theta+4\sin\theta$. This can be maximized by the methods of elementary calculus or by writing $3\cos\theta+4\sin\theta=r\cos(\theta-\phi)$ where $r^2=3^2+4^2$ and $\cos\phi=3/5$, $\sin\phi=4/5$. This expression therefore takes the maximum value 5 when $\theta-\phi=0$, that is $\theta=\phi$. This is the direction (3/5,4/5). Since $\nabla\phi(1,1)=(3,4)$ this fits with Proposition 3.18.

22. It is easy to check that $\partial f/\partial x=0$ and $\partial f/\partial y=1$ at (0,0). Let $\mathbf{u}=(\cos\theta,\sin\theta)$ be a general direction. Then we have $f(t\cos\theta,t\sin\theta)-f(0,0)=t\sin^3\theta$ so the directional derivative in the direction $\mathbf{u}$ is $\sin^3\theta$. Since $<\nabla f(0,0),\mathbf{u}>=\sin\theta$ we see that the formula fails whenever $\sin^2\theta\neq 1$, that is, except along the positive and negative y-axis. The reason is that f is not differentiable at (0,0).

23. $\phi(t)-\phi(s)=<\mathbf{u},\mathbf{f}(\mathbf{x}+t\mathbf{h})-\mathbf{f}(\mathbf{x}+s\mathbf{h})>=<\mathbf{u}, D\mathbf{f}(\mathbf{x}+s\mathbf{h})(t-s)\mathbf{h}+\mathbf{R}>$. Dividing by $t-s$ gives the result. *Alternatively* consider $\sum u_i f_i(\mathbf{x}+t\mathbf{h})$.

24. Suppose that for $\mathbf{a}\neq\mathbf{b}$ we have $\mathbf{f}(\mathbf{a})=\mathbf{f}(\mathbf{b})$ and write $\mathbf{h}=\mathbf{b}-\mathbf{a}$. Let $g(\mathbf{x})=<\mathbf{f}(\mathbf{x})-\mathbf{f}(\mathbf{a}),\mathbf{h}>$. Then $g(\mathbf{x})=\sum_{i=1}^{n}[f_i(\mathbf{x})-f_i(\mathbf{a})]h_i$ and $\nabla g(\mathbf{x})$ has j-th component $\sum_{i=1}^{n} D_j f_i(\mathbf{x})\,h_i$. By the mean value theorem $0=g(\mathbf{b})-g(\mathbf{a})=<\nabla g(\mathbf{c}),(\mathbf{b}-\mathbf{a})>=\sum_{j=1}^{n}\sum_{i=1}^{n} D_j f_i(\mathbf{x})\,h_i h_j$. This contradiction of the positivity hypothesis proves that $\mathbf{f}$ is one-to-one.

25. For $\mathbf{f}(x)=(\sin x, 1-\cos x)$, $x\in[0,2\pi]$, $D\mathbf{f}(\theta)$ is given by the matrix $\begin{bmatrix}\cos\theta\\ \sin\theta\end{bmatrix}$, that is $D\mathbf{f}(\theta)x=(x\cos\theta, x\sin\theta)$ and so $\|D\mathbf{f}(\theta)x\|=|x|$ for every θ. Since $\mathbf{f}(2\pi)=(0,0)=\mathbf{f}(0)$ this shows the first part. Now take $t\in[0,2\pi]$ and apply the inequality of Theorem 3.22 on $[0,t]$. This says $\|\mathbf{f}(t)-\mathbf{f}(0)\|\leq\|D\mathbf{f}(\theta)t\|=|t|$. Since $\|\mathbf{f}(t)-\mathbf{f}(0)\|^2=(\sin t)^2+(1-\cos t)^2=2-2\cos t$ this shows $2(1-\cos t)\leq t^2$.

26. Let $\mathbf{g}(\mathbf{x})=\mathbf{f}(\mathbf{x})-\mathbf{L}\mathbf{x}$. Then $D\mathbf{g}(\mathbf{x})=\mathbf{0}$ for all $\mathbf{x}\in G$ and so $\mathbf{g}(\mathbf{x})=\mathbf{c}$.

27. By Theorem 3.21, $f(x,y_1)-f(x,y_2)=Df(x,z)(0,y_2-y_1)$ where $z=(1-\theta)y_1+\theta y_2$, $0<\theta<1$. As $Df(x,z)(0,y_2-y_1)=(y_2-y_1)\partial f/\partial y=0$ the result is shown. The same result does *not* hold in a connected set. A counter-example is : let G be $\mathbb{R}^2\setminus\{(x,y): x\geq 0, y=0\}$, that is $\mathbb{R}^2$ with the nonnegative real axis deleted. Then G is open and pathwise connected. Define f on G by $f(x,y)=\begin{cases}x^2 & \text{if } x>0 \text{ and } y>0\\ 0 & \text{elsewhere in } G\end{cases}$. Then $\partial f/\partial y=0$ at all points of G but $f(x,y_1)=f(x,y_2)$ for all y_1, y_2 is false.

28. For $(x,y)\neq(0,0)$, $\dfrac{\partial f}{\partial x}=\dfrac{(x^2+y^2)y^\alpha-2x^2y^\alpha}{(x^2+y^2)^2}$, $\dfrac{\partial f}{\partial y}=\dfrac{(x^2+y^2)\alpha xy^{\alpha-1}-2xy^{\alpha+1}}{(x^2+y^2)^2}$.
At $(0,0)$, $\dfrac{\partial f}{\partial x}=\lim_{h\to0}\dfrac{f(h,0)-f(0,0)}{h}=0$ and, similarly, $\dfrac{\partial f}{\partial y}=0$. If $\alpha>2$,

$$\left|\frac{\partial f}{\partial x}\right|\le\frac{|y^{2+\alpha}-x^2y^\alpha|}{(x^2+y^2)^2}\le(|y|^{2+\alpha}+|x|^2|y|^\alpha)/(x^2+y^2)^2\to0 \text{ as } x^2+y^2\to0,$$

(using polars, for example) so $\partial f/\partial x$ is continuous at $(0,0)$. Continuity at other points follows from general results. Similarly $\partial f/\partial y$ is continuous, so $f\in C^1$ for $\alpha>2$.
Also, $\dfrac{\partial^2 f}{\partial x\partial y}(0,0)=\lim_{h\to0}\dfrac{\partial f/\partial y(h,0)-\partial f/\partial y(0,0)}{h}=0$ for $\alpha>1$, and
$\dfrac{\partial^2 f}{\partial y\partial x}(0,0)=\lim_{k\to0}\dfrac{\partial f/\partial x(0,k)-\partial f/\partial x(0,0)}{k}=\lim_{k\to0}k^{\alpha-3}$. Therefore,

$$\frac{\partial^2 f}{\partial y\partial x}(0,0)=\begin{cases}\infty, & \text{if } 1<\alpha<3\\ 1, & \text{if } \alpha=3\\ 0, & \text{if } \alpha>3\end{cases}$$

, and the second order partial derivatives are equal only for $\alpha>3$. As above one can prove that $f\in C^2$ for $\alpha>3$.

29. g is continuous on $[a,b]$ so attains its minimum. This cannot be at a since $g'(a)<0$ implies that $g(x)<g(a)$ on an interval $a<x<x+\delta$. Likewise the minimum is not at b since $g'(b)>0$ gives $g(x)<g(b)$ on some interval $b-\delta<x<b$. Therefore the minimum occurs at $c\in(a,b)$ and $g'(c)=0$. Suppose $\alpha<\beta$ and let γ satisfy $\alpha<\gamma<\beta$. Write $g(x)=f(x)-\gamma x$. Then g is differentiable, $g'(a)=\alpha-\gamma<0$, $g'(b)=\beta-\gamma>0$, so there exists c where $g'(c)=0$, that is $f'(c)=\gamma$. If $\alpha>\beta$ take $g(x)=\gamma x-f(x)$.

Chapter 4.

1. $F(x,y)=x^3y+y^3x-2$ satisfies $F(1,1)=0$. Also $\partial F/\partial y=x^3+3y^2x$ is continuous and equals 4 at $(1,1)$. Therefore $y=f(x)$ on a nbd of $x=1$. Then $x^3f(x)+[f(x)]^3x-2=0$ near $x=1$ and differentiating gives $3x^2f(x)+x^3f'(x)+3[f(x)]^2f'(x)x+[f(x)]^3=0$. When $x=1$, $f=1$ and we have $3+f'(1)+3f'(1)+1=0$ so $f'(1)=-1$.
Also $f'(x)=-\partial F/\partial y\,/\,\partial F/\partial x=-(x^3+3y^2x)/(3x^2y+y^3)=-4/4=-1$, (Prop. 4.10).

2. Let $F(x,y,z)=x^2+2y^2+3z^2-1$. Then $\partial F/\partial z=6z\neq0$ at the point. The implicit function theorem applies and $z=f(x,y)$ in a nbd of $(x,y)=(1/\sqrt{6},1/\sqrt{6})$. Thus, $x^2+2y^2+3[f(x,y)]^2=1$. Differentiating with respect to x and then y gives $2x+6f\cdot\partial f/\partial x=0$ and $4y+6f\cdot\partial f/\partial y=0$. At $(1/\sqrt{6},1/\sqrt{6})$ this yields $\partial f/\partial x=-1/3$ and $\partial f/\partial y=-2/3$. Now the tangent plane has equation $z=f(1/\sqrt{6},1/\sqrt{6})+(\partial f/\partial x)(x-1/\sqrt{6})+(\partial f/\partial y)(y-1/\sqrt{6})$, that is, $z=1/\sqrt{6}-(x-1/\sqrt{6})/3-2(y-1/\sqrt{6})/3$, or $x+2y+3z=\sqrt{6}$.

3. Let $F(x,y)=(x+1)^2y-xy^2-4$. Then $\partial F/\partial x=2(x+1)y-y^2$, $\partial F/\partial y=(x+1)^2-2xy$.

Clearly F is of class C^1 in all variables at all points.
(*i*) At $(-1,2)$, $F(-1,2)=0$, $\partial F/\partial y=4$, so y is defined implicitly.
(*ii*) At $(2,1)$, $F(2,1)=3\neq 0$ so y is not defined implicitly by the relation $F(x,y)=0$ at this point.
(*iii*) At $(1,2)$, $F(1,2)=0$, $\partial F/\partial y=0$, and we cannot use the theorem to say whether y is defined implicitly. Usually we would not be able to say more but, in this case, the relation is a quadratic equation for y. We will show that there is no nbd of $x=1$ on which $y=f(x)$ by showing that y cannot exist for $x=1-h$ when $h>0$ is sufficiently small. Indeed the quadratic formula shows that y cannot exist if $(2-h)^4-16(1-h)<0$. Since $(2-h)^4-16(1-h)=-16h+24h^2-8h^3+h^4$, this is certainly negative when $h>0$ is small enough.

4. Let $\mathbf{F}(x,y,z,t)=(xz-yt-1,\ xyzt-20)$. At $(1,2,5,2)$, $\mathbf{F}=(0,0)$. Clearly $\mathbf{F}$ is of class C^1 in all variables. Also $\begin{bmatrix}\partial F_1/\partial z & \partial F_1/\partial t\\ \partial F_2/\partial z & \partial F_2/\partial t\end{bmatrix}=\begin{bmatrix}x & -y\\ xyt & xyz\end{bmatrix}$ and its determinant is $\det\begin{bmatrix}1 & -2\\ 4 & 10\end{bmatrix}=18$. Therefore $z=f(x,y)$ and $t=g(x,y)$ on a nbd. Thus $x\,f(x,y)-y\,g(x,y)=1$, and $x\,y\,f(x,y)g(x,y)=20$.
Differentiating gives
$f+x\,\partial f/\partial x-y\,\partial g/\partial x=0$ and $y\,f(x,y)g(x,y)+xyf\,\partial g/\partial x+xyg\,\partial f/\partial x=0$, which simplify to $5+\partial f/\partial x-2\partial g/\partial x=0$ and $20+10\partial g/\partial x+4\partial f/\partial x=0$. Hence $\partial f/\partial x=-5$ and $\partial g/\partial x=0$. Similarly (writing z and t rather than f,g) we have $x\,\partial z/\partial y-t-y\,\partial t/\partial y=0$ and $xzt+xyt\,\partial z/\partial y+xyz\,\partial t/\partial y=0$, or $\partial z/\partial y-2-2\partial t/\partial y=0$ and $10+4\partial z/\partial y+10\partial t/\partial y=0$ and hence $\partial z/\partial y=0$, $\partial t/\partial y=-1$.
Note that $\begin{bmatrix}\partial F_1/\partial z & \partial F_1/\partial x\\ \partial F_2/\partial z & \partial F_2/\partial x\end{bmatrix}=\begin{bmatrix}x & z\\ xyt & yzt\end{bmatrix}$ and the determinant is identically zero. This does *not* show the result : the implicit function theorem gives *sufficient* conditions not *necessary* ones. However suppose that $x=p(y,t)$ and $z=q(y,t)$ on a nbd of $(y,t)=(2,2)$. Then we would have $p\cdot q=1+yt$ and $p\cdot q=20/yt$. This would imply that $1+yt=20/yt$ for all y,t in a nbd of $(2,2)$ which is impossible since this would imply $yt=4$, a curve not a nbd.

5. Suppose that $z=h(x,y)$. Then $F(x,y,h(x,y))=0$ and differentiating with respect to x and y (independent variables) gives

$$\partial F/\partial x+0+(\partial F/\partial z)(\partial h/\partial x)=0 \quad\text{and}\quad 0+\partial F/\partial y+(\partial F/\partial z)(\partial h/\partial y)=0.$$

Hence, in the notation expected, $\partial z/\partial x=(-\partial F/\partial x)/(\partial F/\partial z)$ and $\partial z/\partial y=(-\partial F/\partial y)/(\partial F/\partial z)$, where $\partial F/\partial z\neq 0$ (by implicit assumption). Similarly calculate $\partial y/\partial x$ and $\partial x/\partial z$. On multiplying together we get

$$(\partial y/\partial x)(\partial z/\partial y)(\partial x/\partial z)=\frac{(-\partial F/\partial x)(-\partial F/\partial y)(-\partial F/\partial z)}{(\partial F/\partial y)\ (\partial F/\partial z)\ (\partial F/\partial x)}=-1.$$

6. If $x\neq 0$ then $f'(x)=1/2+2x\sin(1/x)-\cos(1/x)$ by the chain rule. At $x=0$, we have $[f(h)-f(0)]/h=1/2+h\sin(1/h)\longrightarrow 1/2$ as $h\to 0$ since

$|\sin(t)| \le 1$ for all t. Here f is not one-to-one on any nbd of 0 and f' oscillates wildly so f' is not continuous at 0.

7. We have $u = x^3/y$ and $v = -xy^2/u = -y^3/x^2$, valid for $xy \neq 0$. Let $\mathbf{F}(x,y,u,v) = (x^3 - yu,\ xy^2 + uv)$. Then $\begin{bmatrix} \partial F_1/\partial u & \partial F_1/\partial v \\ \partial F_2/\partial u & \partial F_2/\partial v \end{bmatrix} = \begin{bmatrix} -y & 0 \\ v & u \end{bmatrix}$ and det $= -yu$. Therefore u,v are defined implicitly in a nbd of any point where $yu \neq 0$. The first equation shows that $x \neq 0$ also so the conditions are the same as obtained directly.

8. Let $\mathbf{g}(r,\theta) = (r\cos\theta,\ r\sin\theta)$, $r \ge 0$. When is $\mathbf{g}$ invertible? $\mathbf{g}$ is not one-to-one on $[0,\infty) \times \mathbb{R}$ because $\mathbf{g}(r,\theta) = \mathbf{g}(r,\theta + 2k\pi))$, $k \in \mathbb{Z}$. Let $\mathbf{g}$ be restricted to $[0,\infty) \times [0,2\pi)$. Then $\mathbf{g}$ is one-to-one except at $r = 0$; all of $\{0\} \times [0,2\pi)$ is mapped onto $(0,0)$. $\mathbf{g}$ is one-to-one from $(0,\infty) \times [0,2\pi)$ onto $\mathbb{R}\backslash\{0\}$ and $\mathbf{g}$ is therefore invertible when so restricted. The inverse is then given by $\mathbf{f}(x,y) = (\sqrt{(x^2 + y^2)},\ \theta(x,y))$ where $\theta(x,y)$ is the unique angle in the range $[0,2\pi)$ between the line segment from $(0,0)$ to (x,y) and the positive x-axis. $\theta(x,y) = \tan^{-1}(y/x)$ in the first quadrant $x > 0$, $y > 0$ and may be given in other quadrants as $C + \tan^{-1}(y/x)$ for suitable constants C. [Recall that $\tan^{-1} : \mathbb{R} \longrightarrow (-\pi/2,\ \pi/2)$ is the inverse function, also denoted arctan.] Now the partial derivatives of f_1 exist (except at (0,0)) so $\mathbf{f}$ is C^1 away from (0,0). [$\partial f_1/\partial x$ and $\partial f_1/\partial y$ do not exist at (0,0).] The inverse function theorem asserts that $\mathbf{g}$ is locally invertible except when $r = 0$ since $\mathbf{g}$ is C^1 and $J_{\mathbf{g}} = r$.

9. Let $\mathbf{g}(r,\ \theta,\ \phi) = (r\cos\theta\cos\phi,\quad r\sin\theta\sin\phi,\quad r\cos\theta)$, for $0 < r <, \infty$, $\theta \in [0,\pi]$, $\phi \in [0,2\pi)$.

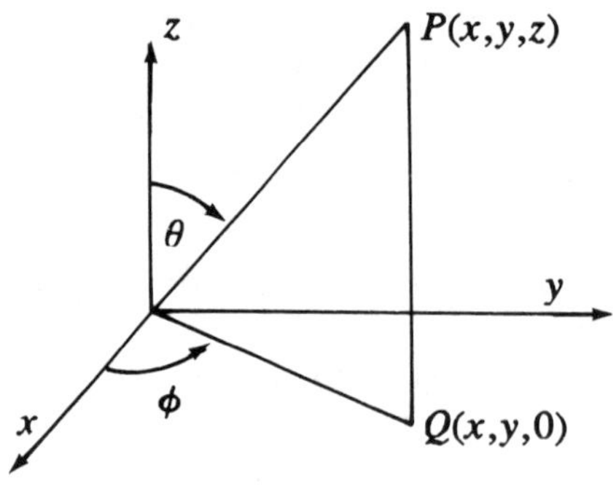

$\mathbf{g}$ is then one-to-one onto $\mathbb{R}^3\backslash\{0\}$. The inverse function is the map $\mathbf{f} : \mathbb{R}^3\backslash\{0\} \longrightarrow (0,\infty) \times [0,\pi] \times [0,2\pi)$ and is given by $f_1 = \sqrt{(x^2+y^2+z^2)}$, $f_2 = \cos^{-1}(z/\sqrt{(x^2+y^2+z^2)})$, $f_3 = \phi(x,y)$ where $\tan\phi = y/x$, for $x \neq 0$, and ϕ is defined exactly as is $\theta(x,y)$ in problem 8. Since $\mathbf{g} \in C^1$, the inverse function theorem asserts that $\mathbf{g}$ is locally invertible except when $J_{\mathbf{g}} = 0$ which is when $r^2\sin\theta = 0$, that is, $\mathbf{f}$ exists locally except along the z-axis.

10. If $\mathbf{f}^{-1}$ exists on a nbd of $\mathbf{f}(\mathbf{a})$ then $\mathbf{I}\,\mathbf{x} = \mathbf{x} = \mathbf{f}^{-1}(\mathbf{f}(\mathbf{x}))$ on a nbd of $\mathbf{a}$. Suppose $\mathbf{f}^{-1}$ is differentiable. Then, by the chain rule,

$\mathbf{I} = D\mathbf{f}^{-1}(\mathbf{f}(\mathbf{a}))\, D\mathbf{f}(\mathbf{a})$. Taking determinants, this yields $1 = 0$, impossible, so $\mathbf{f}^{-1}$ is not differentiable.

11. $D\mathbf{f}(x,y) = \begin{bmatrix} 3x^2+2y & 2x+2y \\ 2x & 1 \end{bmatrix}$ so $J_{\mathbf{f}}(1,1) = 5-8 \neq 0$. Clearly $\mathbf{f}$ is of class C^1 so by the inverse function theorem $\mathbf{f}$ has a unique local inverse. Now $D\mathbf{f}^{-1}(4,2)$ is the inverse of $D\mathbf{f}(1,1) = \begin{bmatrix} 5 & 4 \\ 2 & 1 \end{bmatrix}$ and so

$D\mathbf{f}^{-1}(4,2) = \begin{bmatrix} 5 & 4 \\ 2 & 1 \end{bmatrix}^{-1} = \dfrac{1}{-3}\begin{bmatrix} 1 & -4 \\ -2 & 5 \end{bmatrix}$. The local affine approximation to $\mathbf{f}^{-1}$ is given by $\mathbf{f}^{-1}(u,v) = \mathbf{f}^{-1}(4,2) + D\mathbf{f}^{-1}(4,2)\begin{bmatrix} u-4 \\ v-2 \end{bmatrix}$, that is,

$$\mathbf{f}^{-1}(u,v) = \begin{bmatrix} 1 \\ 1 \end{bmatrix} + \frac{-1}{3}\begin{bmatrix} (u-4) - 4(v-2) \\ -2(u-4) + 5(v-2) \end{bmatrix} = \frac{-1}{3}(u-4v+1,\ -2u+5v-5)\ .$$

12. $D\mathbf{f}(x,y,z) = \begin{bmatrix} 1 & 1 & 1 \\ y+z & x+z & x+y \\ z & xz & xy \end{bmatrix}$ so the partial derivatives are continuous at all points. Now $J_{\mathbf{f}} = \det\begin{bmatrix} 1 & 0 & 0 \\ y+z & x-y & x-z \\ yz & z(x-y) & y(x-z) \end{bmatrix} = (x-y)(x-z)(y-z)$ so $J_{\mathbf{f}}(\mathbf{a}) \neq 0$ when a,b,c are distinct and the local inverse of $\mathbf{f}$ exists. Its Jacobian is given by the reciprocal of $J_{\mathbf{f}}(\mathbf{a})$ so is $[(a-b)(a-c)(b-c)]^{-1}$. In calculating the determinant we used the fact that the determinant of a matrix is unchanged if one column (or row) is altered by subtracting a multiple of another column (or row) and we then expanded by the first row.

13. Clearly $\mathbf{f}(x,y,z) = (x-xy,\ xy-xyz,\ xyz)$ has all partial derivatives continuous on $\mathbb{R}^3$. $J_{\mathbf{f}}(\mathbf{x}) = \det\begin{bmatrix} 1-y & -x & 0 \\ y-yz & x-xz & -xy \\ yz & xz & xy \end{bmatrix} = \det\begin{bmatrix} 1-y & -x & 0 \\ y & x & 0 \\ yz & xz & xy \end{bmatrix}$, adding row 3 to row 2, so $J_{\mathbf{f}}(\mathbf{x}) = \det\begin{bmatrix} 1 & 0 & 0 \\ y & x & 0 \\ yz & z & xy \end{bmatrix} = x^2y \neq 0$ on U. If

$\mathbf{f}(x,y,z) = \mathbf{f}(u,v,w)$ then $\left.\begin{aligned} x-xy &= u-uv \\ xy-xyz &= uv-uvw \\ xyz &= uvw \end{aligned}\right\}$ and adding gives $x=u$. Then the first equation gives $y=v$ provided that $x \neq 0$ and then the third gives $z=w$, if $xy \neq 0$. This proves that $\mathbf{f}$ is one-to-one on U. To find a formula for $\mathbf{f}^{-1}$ write $\mathbf{f}(x,y,z) = (r,s,t)$. Then $x-xy=r$, $xy-xyz=s$, $xyz=t$. Adding gives $x=r+s+t$ and then $1-y=r/x=r/(r+s+t)$ so $y=(s+t)/(r+s+t)$. Hence $z=t/xy=t/(s+t)$. This shows that

$\mathbf{f}^{-1}(r,s,t) = (r+s+t,\ (s+t)/(r+s+t),\ t/(s+t))$ for $r+s+t \neq 0$ and $s+t \neq 0$ [equivalently $x \neq 0$ and $xy \neq 0$]. Then we have $J_{\mathbf{f}^{-1}}(a,b,c) = [x^2y]^{-1} = 1/[(a+b+c)(b+c)]$. The direct argument gives, writing $a+b+c=A$, $\mathbf{Df}^{-1}(\mathbf{a})=\begin{bmatrix} 1 & 1 & 1 \\ -(b+c)/A^2 & a/A^2 & a/A^2 \\ 0 & -c/(b+c)^2 & b/(b+c)^2 \end{bmatrix}$ so that

$$J_{\mathbf{f}^{-1}}(\mathbf{a})=\det\begin{bmatrix} 1 & 1 & 0 \\ -(b+c)/A^2 & a/A^2 & 0 \\ 0 & -c/(b+c)^2 & 1/(b+c) \end{bmatrix} =a/(b+c)A^2+1/A^2 =1/(a+b+c)(b+c).$$

14. $f_1(x_1,x_2) = x_1/(1+x_1+x_2)$, $f_2(x_1,x_2) = x_2/(1+x_1+x_2)$, $x_1+x_2 \neq -1$. If $\mathbf{f}(\mathbf{x}) = \mathbf{f}(\mathbf{y})$ then (1) : $x_1/(1+x_1+x_2) = y_1/(1+y_1+y_2)$ and (2) : $x_2/(1+x_1+x_2) = y_2/(1+y_1+y_2)$. If $x_1 = 0$ we see that $y_1 = 0$; if $x_2 = 0$ then $y_2 = 0$; if both x_1 and x_2 are nonzero then $\dfrac{x_1}{y_1} = \dfrac{x_2}{y_2} = \dfrac{1+x_1+x_2}{1+y_1+y_2} \neq 0$ shows that y_1 and y_2 are both nonzero. Also $x_1 = y_1x_2/y_2$ so that (1) gives $y_1(1-x_2/y_2) = 0$. Thus $x_2 = y_2$ and hence $x_1 = y_1$ also. This proves $\mathbf{f}$ is one-to-one (for $x_1+x_2 \neq -1$). Write $x_1/(1+x_1+x_2) = u_1$, $x_2/(1+x_1+x_2) = u_2$. Then adding gives $(x_1+x_2)/(1+x_1+x_2) = u_1+u_2$ and hence $x_1+x_2 = (u_1+u_2)/(1-u_1-u_2)$. Therefore $1+x_1+x_2=1/(1-u_1-u_2)$ and $x_1=u_1/(1-u_1-u_2)$, $x_2=u_2/(1-u_1-u_2)$. Since $t/(1+t)$ $(t\neq -1)$ takes every value in $\mathbb{R}\setminus\{1\}$ the inverse function $\mathbf{g}$ is defined on $\mathbb{R}^2\setminus\{u_1+u_2 = 1\}$ and

$\mathbf{g}(u_1,u_2) = (u_1/(1-u_1-u_2),\ u_2/(1-u_1-u_2))$. Finally,

$$J_{\mathbf{f}} = \det\begin{bmatrix} (1+x_2)/a^2 & x_1/a^2 \\ x_2/a^2 & (1+x_1)/a^2 \end{bmatrix}$$ where $a = 1+x_1+x_2$, so

$J_{\mathbf{f}} = [(1+x_1)(1+x_2) - x_1x_2]/a^4 = 1/a^3$. The inverse function theorem asserts that $\mathbf{f}$ is locally invertible at all points where $a\neq 0$.

15. We write $\mathbf{L}=D_{\mathbf{y}}\mathbf{F}(\mathbf{0},\mathbf{0})$, so that $\mathbf{L}^{-1}$ exists. We want to prove that for each (fixed) $\mathbf{x}$ in $\overline{\mathbf{B}}(\mathbf{0},s)$ there is a unique $\mathbf{y}$ in $\overline{\mathbf{B}}(\mathbf{0},r)$ such that $\mathbf{F}(\mathbf{x},\mathbf{y})=\mathbf{0}$; then $\mathbf{y}=\mathbf{f}(\mathbf{x})$ defines the function $\mathbf{f}$. $\mathbf{F}(\mathbf{x},\mathbf{y})=\mathbf{0}$ is equivalent to $\mathbf{y}+(\mathbf{L}^{-1}\mathbf{F}(\mathbf{x},\mathbf{y})-\mathbf{y})=\mathbf{0}$. We shall apply Theorem 4.7 to the function $\mathbf{G}(\mathbf{x},\mathbf{y}) :=\mathbf{L}^{-1}\mathbf{F}(\mathbf{x},\mathbf{y})-\mathbf{y}$, regarded as a function of $\mathbf{y}$. Now $D_{\mathbf{y}}\mathbf{G}(\mathbf{x},\mathbf{y})=\mathbf{L}^{-1}D_{\mathbf{y}}\mathbf{F}(\mathbf{x},\mathbf{y})-\mathbf{I}_{\mathbf{y}}$, by the chain rule and the linearity of $\mathbf{L}^{-1}$ and $\mathbf{I}_{\mathbf{y}}$ (the identity map of $\mathbb{R}^n \longrightarrow \mathbb{R}^n$ ($\mathbf{y}$ variables)). Therefore, $D_{\mathbf{y}}\mathbf{G}(\mathbf{0},\mathbf{0})=\mathbf{0}$ and by continuity of all the partial derivatives, there exists $r>0$ (chosen so that $(\mathbf{x},\mathbf{y})\in V$) such that $\|D_{\mathbf{y}}\mathbf{G}(\mathbf{x},\mathbf{y})\| < 1/2$ for

$\|\mathbf{x}\| \le r$, $\|\mathbf{y}\| \le r$, $i=1,2,\cdots,n$. Hence, by the remark following Thm. 3.22, $\|\mathbf{G}(\mathbf{x},\mathbf{y})-\mathbf{G}(\mathbf{x},\mathbf{z})\| < 1/2\|\mathbf{y}-\mathbf{z}\|$ for $\|\mathbf{x}\| \le r$, $\mathbf{y},\mathbf{z} \in \overline{\mathbf{B}}(\mathbf{0},r)$ [in $\mathbb{R}^n$]. Also, since $\mathbf{G}(\mathbf{0},\mathbf{0})=\mathbf{0}$ and $\mathbf{G}$ is continuous, there exists $s \le r$ such that $\|\mathbf{G}(\mathbf{x},\mathbf{0})\| < r/2$ for $\|\mathbf{x}\| \le s$. We can now apply Theorem 4.7 to $\mathbf{G}$ regarded as a function of $\mathbf{y}$ to show that for any $\mathbf{x} \in \overline{\mathbf{B}}(\mathbf{0},s)$ there is a unique $\mathbf{y} \in \overline{\mathbf{B}}(\mathbf{0},r)$ such that $\mathbf{y}+\mathbf{G}(\mathbf{x},\mathbf{y})=\mathbf{0}$. This correspondence defines the function $\mathbf{f} : \overline{\mathbf{B}}(\mathbf{0},s) \longrightarrow \overline{\mathbf{B}}(\mathbf{0},r)$ by $\mathbf{f}(\mathbf{x})=\mathbf{y}$. Since $\mathbf{0}+\mathbf{G}(\mathbf{0},\mathbf{0})=\mathbf{0}$ this definition gives $\mathbf{f}(\mathbf{0})=\mathbf{0}$ and $\mathbf{F}(\mathbf{x},\mathbf{f}(\mathbf{x}))=\mathbf{0}$ on $\overline{\mathbf{B}}(\mathbf{0},s)$. To prove continuity of $\mathbf{f}$ on $U=\mathbf{B}(\mathbf{0},s)$ let $\mathbf{u} \in U$ and $\mathbf{x} \in U$, and write $\mathbf{y}=\mathbf{f}(\mathbf{x})$, $\mathbf{w}=\mathbf{f}(\mathbf{u})$. Then $\mathbf{y}+\mathbf{G}(\mathbf{x},\mathbf{y})=\mathbf{0}$ and $\mathbf{w}+\mathbf{G}(\mathbf{u},\mathbf{w})=\mathbf{0}$ so we have

$$\begin{aligned}\|\mathbf{f}(\mathbf{x})-\mathbf{f}(\mathbf{u})\| = \|\mathbf{G}(\mathbf{u},\mathbf{w})-\mathbf{G}(\mathbf{x},\mathbf{y})\| &\le \|\mathbf{G}(\mathbf{u},\mathbf{w})-\mathbf{G}(\mathbf{x},\mathbf{w})\| + \|\mathbf{G}(\mathbf{x},\mathbf{w})-\mathbf{G}(\mathbf{x},\mathbf{y})\| \\ &\le \|\mathbf{G}(\mathbf{u},\mathbf{w})-\mathbf{G}(\mathbf{x},\mathbf{w})\| + \|\mathbf{w}-\mathbf{y}\|/2.\end{aligned}$$

Therefore, $\|\mathbf{y}-\mathbf{w}\| \le 2\|\mathbf{G}(\mathbf{u},\mathbf{w})-\mathbf{G}(\mathbf{x},\mathbf{w})\|$. As $\mathbf{G}$ is continuous on V, for any $\varepsilon > 0$ there is $\delta=\delta(\varepsilon,\mathbf{u},\mathbf{w}) > 0$ such that $\|(\mathbf{x},\mathbf{w})-(\mathbf{u},\mathbf{w})\| = \|\mathbf{x}-\mathbf{u}\| < \delta$ implies $\|\mathbf{G}(\mathbf{x},\mathbf{w})-\mathbf{G}(\mathbf{u},\mathbf{w})\| < \varepsilon$, and this proves the continuity of $\mathbf{f}$.

16. For $\mathbf{x} \in K$ there exists $\delta(\mathbf{x})$ and $k(\mathbf{x})$ such that $\|\mathbf{f}(\mathbf{z}) - \mathbf{f}(\mathbf{y})\| \le k(\mathbf{x})\|\mathbf{z} - \mathbf{y}\|$ for $\mathbf{z},\mathbf{y} \in \mathbf{B}(\mathbf{x},\delta(\mathbf{x}))$. Now suppose $\mathbf{f}$ does not satisfy a (uniform) Lipschitz condition on K. Then, for any $m \in \mathbb{N}$, there are $\mathbf{z}^{(m)}$, $\mathbf{y}^{(m)}$ such that $\|\mathbf{f}(\mathbf{z}^{(m)}) - \mathbf{f}(\mathbf{y}^{(m)})\| \ge m\|\mathbf{z}^{(m)} - \mathbf{y}^{(m)}\|$. Since $\mathbf{f}$ is continuous on K, it is bounded (by M, say) and so

$$\|\mathbf{z}^{(m)} - \mathbf{y}^{(m)}\| \le (1/m)\|\mathbf{f}(\mathbf{z}^{(m)}) - \mathbf{f}(\mathbf{y}^{(m)})\| < 2M/m.$$

By compactness, we can suppose, by passing to subsequences if necessary, that $\mathbf{z}^{(m)} \longrightarrow \mathbf{x}$. Then $\mathbf{y}^{(m)} \longrightarrow \mathbf{x}$ too and $\exists\ N \in \mathbb{N}$ so that $\mathbf{y}^{(m)}$ and $\mathbf{z}^{(m)}$ both lie in $\mathbf{B}(\mathbf{x},\delta(\mathbf{x}))$ for $m \ge N$. Then

$\|\mathbf{f}(\mathbf{z}^{(m)}) - \mathbf{f}(\mathbf{y}^{(m)})\| \le k(\mathbf{x})\|\mathbf{z}^{(m)} - \mathbf{y}^{(m)}\|$, a contradiction for m large.

17. Set $y_1 = y$, $y_2 = y'$. Then $y_1{}' = y_2$, $y_2{}' = (1 + y_2^2)^{1/2}$.

Thus $\mathbf{y}'=\mathbf{f}(\mathbf{y})$ where $\mathbf{f}(y_1, y_2) = (y_2, (1 + y_2^2)^{1/2})$; $\mathbf{f}$ is of class C^1.

Take a cylinder $C=\{|t| \le a,\ y_1^2 + y_2^2 \le b^2\}$.

$$\text{Then } M := \max\|\mathbf{f}(t,\mathbf{y})\| = \max(1 + 2y_2^2) = 1 + 2b^2.$$

By Theorem 4.30 and the comments thereafter, the solution exists on $[-\delta,\delta]$ for $\delta = \min\{a, b/M\}$. Thus we can take $\delta = b/(1 + 2b^2)$ for any b and this is greatest when $2b^2 = 1$, so that $\delta = 1/(2\sqrt{2})$.

18. $\dfrac{dy}{dt} = t^2+y^2 \ge y^2$ for all t. Therefore $(1/y^2)\dfrac{dy}{dt} \ge 1$ which may be written $d/dt[(-1/y(t))-t] \ge 0$. Thus the function inside $[\cdot]$ is increasing and $-1/y(t) - t \ge -1$, as long as y exists. This proves $y(t) \ge 1/(1 - t)$ as long as y exists. Similarly $\dfrac{dy}{dt} = t^2+y^2 \le 1+y^2$, for $0 \le t \le 1$ and hence $(1/(1+y^2))\dfrac{dy}{dt} - 1 \le 0$ or $d/dt[\tan^{-1}(y(t))-t] \le 0$. This

shows $\tan^{-1}(y(t))-t\le\tan^{-1}(y(0)) = \pi/4$ and so $y(t)\le\tan(t+\pi/4)$. This proves that the graph of y is sandwiched between that of $1/(1-t)$ and $\tan(t+\pi/4)$ and shows that y must tend to infinity somewhere between $\pi/4$ and 1.

Chapter 5.

1. $\partial f/\partial x = 4x^3 - 2(x-y)$ and $\partial f/\partial y = 4y^3+2(x-y)$. At a critical point, $4x^3 = 2(x-y)$ and $4y^3 = -2(x-y)$. Therefore $x^3 = -y^3$, that is, $x=-y$. Then $4x^3 = 2(2x)$ so $x^3 = x$ and either $x=0$ or $x^2=1$. So we have 3 critical points (0,0), (1,−1) and (−1,1). For the second derivative test, we calculate $f_{xx} = 12x^2-2$, $f_{xy} = 2$, $f_{yy} = 12y^2-2$. At (0,0), $f_{xx} = -2<0$, and the discriminant $D := f_{xx}f_{xy} - (f_{xy})^2 = (-2)(-2)-2^2 = 0$, which yields no information. However, for (x,y) small $f(x,y)-f(0,0)=x^4+y^4-(x-y)^2$ and $f(x,x) = 2x^4>0$, so (0,0) is local minimum along the line $y=x$. But $f(x,0) = x^4-x^2 = x^2(x^2-1)<0$ for $x^2<1$ so f has a local max in this direction (also $f(x,2x)$ is a local max), in all a saddle point. At (1,−1), $f_{xx} = 10$, $f_{xy} = 2$, $f_{yy} = 10$. Thus, $f_{xx}>0$, $D=100-4>0$, so this is a local MIN. (−1,1), is exactly the same as (1,−1), a local MIN.

2. $f_x=4x^3-4yz$, $f_y=4y^3-4xz$, $f_z=4z^3-4xy$ and each is 0 at (1,1,1). The Hessian matrix at (1,1,1) is $\begin{bmatrix}12 & -4 & -4\\ -4 & 12 & -4\\ -4 & -4 & 12\end{bmatrix}$. The corresponding quadratic form is $Q=12h^2+12k^2+12l^2-8hk-8kl-8lh$. Since $h^2+k^2\ge 2hk$, $k^2+l^2\ge 2kl$ and $l^2+h^2\ge 2lh$, adding gives $2(h^2+k^2+l^2)\ge 2(hk+kl+lh)$. Thus $Q\ge 4(h^2+k^2+l^2)$ and Q is positive definite so (1,1,1) is a local minimum. (This is simpler than completing the square on Q.)

3. f attains its maximum and minimum on cl(G). If these both occur on bdy(G) then f is constant on all of cl(G) and so any $\mathbf{x}\in G$ will do. Otherwise f either has an interior maximum or minimum and this must be a critical point of f.

4. At an extremum, $-(1/x^2) = \lambda$, $-(1/y^2) = \lambda$, $-(1/z^2) = \lambda$, and $x+y+z=3$. Therefore $x^2=y^2=z^2$. So we can have $x=y=z$ all equal 1, that is, (1,1,1), or we can have $x=t$, $y=-t$, $z=t$ so $t=3$, that is, (3,−3,3) and the permutations of these (by symmetry). At (1,1,1), $f=3$, at the other 3 points $f=1/3$. CAUTION : $x+y+z=3$ is a plane (unbounded) so it does *not* follow that f has a MAX or MIN. At (1,1,1) consider $E = f(1+a, 1+b, 1+c) - f(1,1,1)$ where a,b,c are *small* and the *constraint is still satisfied* so that $a+b+c=0$. Then

$$E=f(1+a,\ 1+b,\ 1+c)-f(1,1,1) = (1+a)^{-1}+(1-b)^{-1} + (1+c)^{-1} - 3$$
$$= 1-a+a^2\cdots+1-b+b^2\cdots+1-c+c^2\cdots -3$$

($\cdots$ denotes higher order terms (h.o.t.))

$$= a^2+b^2+c^2 + \text{h.o.t. (since } a+b+c=0).$$

Therefore $E>0$ so $f(1+a,\ 1+b,\ 1+c) > f(1,1,1)$, thus a local MIN.
At (3,3,–3), $E=f(3+a,\ 3+b,\ -3+c)-f(3,3,-3)$ (again $a+b+c=0$). Thus,

$$E = \frac{1}{3+a} + \frac{1}{3+b} + \frac{1}{-3+c} - \frac{1}{3} = \frac{1}{3}\left[\frac{1}{1+a/3} + \frac{1}{1+b/3} - \frac{1}{1-c/3} - 1\right]$$
$$= \frac{1}{3}\left[1-a/3+a^2/9 \cdots +1-b/3+b^2/9 - (1+c/3+c^2/9+\cdots) -1\right]$$
$$= \frac{1}{3}\left[\frac{a^2}{9}+\frac{b^2}{9}-\frac{c^2}{9}\right] + \text{h.o.t.} = \frac{1}{27}\left(a^2+b^2-(a+b)^2\right) = \frac{1}{27}(-2ab) + \text{h.o.t.}$$

since $c = -(a + b)$ to first order. Thus E can be positive or negative (a,b opposite or same sign). The other 2 points are saddles in exactly the same way. There is no *global* MAX. For example take $x=1.49$, $y=1.49$, $z=0.02$ then $f=50+\frac{2}{1.49}$. Taking z smaller f increases without bound. Also there is no *global* MIN, for example, $x = 1.51$, $y=1.51$, $z=-0.02$ gives $f=-50+\frac{2}{1.51}$. As z approaches 0, f tends to $-\infty$.

5. By the Lagrange multiplier rule, we seek solutions of

$$12x^2 = \lambda,\ 3y^2 = \lambda,\ 3z^2 = \lambda,\ x + y + z = 1.$$

We see that $4x^2=y=z$ and so $y=\pm 2x$, $z=\pm 2x$. Since we want all values positive we need $x>0$, $y=z=2x$. Then $x+y+z=1$ yields $x=1/5$, $y=2/5$, $z=2/5$. The value of f is $4(1/5)^3+(2/5)^3+(2/5)^3=4/25$. This is actually a minimum subject to the constraint since

$$f(1/5+h,\ 2/5+k,\ 2/5+l) - f(1/5,2/5,2/5)$$

is easily calculated to be $4h^3+k^3+l^3+(6/5)(2h^2+k^2+l^2)$, since $h+k+l=0$ is needed to satisfy the constraints.

We may ask about the maximum subject to the same condition. Why was there no other point to yield a maximum? The region bounded by $x\geq 0$, $y\geq 0$, $z\geq 0$, $x+y+z=1$ is a tetrahedron and f attains its maximum and minimum on this set but the global minimum is 0, the global maximum is at (1,0,0). When we consider $x>0$, $y>0$, $z>0$, $x+y+z=1$, the region is still bounded but no longer closed so maxima and minima need not exist. In the example here, there is a mimimum on the face $x+y+z=1$ with $x>0$, $y>0$, $z>0$ but the maximum is on the boundary of the region at (1,0,0). Extra constraints need to be added to find this by Lagrange's method.

6. Points on the line of intersection satisfy $y=x-1$, $z=1-4x$ so are of the form $(x, x-1, 1-4x)$. Then $(\text{distance})^2=x^2+(x-1)^2+(1-4x)^2$, which is easily minimized by elementary calculus. Alternatively, by the Lagrange multiplier rule, we seek to minimize $f(x,y,z) = x^2 + y^2 + z^2$ subject to $g(x,y,z) := x-y-1=0$ and $h(x,y,z) := 4x+z-1=0$. The equations

to be solved are $2x+\lambda+4\mu=0$, $2y-\lambda=0$, $2z+\mu=0$. Replacing λ and μ in the first equation gives $2x+2y-4z=0$ which in conjunction with $2(x-y)=2$ yields $4x-4z=2$. Since $4x+z=1$ this gives $z=-\frac{1}{2}$ and hence $x=1/2$, $y=-1/2$. The minimum distance is $\sqrt{[(1/2)^2+(1/2)^2+1^2]} = \sqrt{3}/2$.

7. We have to find the extreme points of $x^2+y^2+z^2$ subject to $x^2+y^2=90$ and $3x+y+z=40$. This is the same as finding extreme points of $90+z^2$ subject to the same constraints. By the Lagrange multiplier rule, these occur when $0=2x\lambda+3\mu$, $0=2y\lambda+\mu$, $2z=0+\mu$. Therefore,

$$x = -3\mu/(2\lambda)\,, \quad y = -\mu/(2\lambda), \quad z = \mu/2.$$

If $\mu\neq 0$ then $x=3y$ and using the constraint $3x+y+z=40$ we get $z=40-10y$. Then using $x^2+y^2=90$ gives $10y^2=90$ so $y=\pm 3$. This gives the 2 points (9,3,10) and (−9,−3,70). If $\mu=0$ then $2x\lambda=0$ and $2y\lambda=0$ and $z=0$. Since $x^2+y^2=90$, x and y are not both 0 so $\lambda=0$. Then $3x+y+z=40$ yields $3x+y=40$ so $y=40-3x$ and $x^2+(40-3x)^2=90$, that is, $10x^2-240x+1600-90=0$ or $x^2-24x+151=0$. As $b^2<4ac$ there are *no real roots*, so this case is impossible. Now $f(9,3,10)=190$, the *MIN* and $f(-9,-3,70)=580$ the *MAX*, (by geometrical considerations these exist).

NOTE 1. If one takes $x^2+y^2+z^2$ instead of $90+z^2$ one gets the same equations as above but with $(\lambda-1)$ replacing λ everywhere.

NOTE 2. The 3 equations can be considered as linear equations in $(\lambda,\mu,-1)$ in matrix notation $\begin{pmatrix} 2x & 3 & 0 \\ 2y & 1 & 0 \\ 0 & 1 & 2z \end{pmatrix} \begin{pmatrix} \lambda \\ \mu \\ -1 \end{pmatrix} = \begin{pmatrix} 0 \\ 0 \\ 0 \end{pmatrix}$.

As this has a nonzero solution, the determinant of the matrix is 0, that is $2z(2x-6y)=0$ so either $z=0$ or $x=3y$, proceed as above.

8. We want to minimize $x^2+y^2+z^2$ subject to $ax+by+cz=1$.
The min occurs when $2x=a\lambda$, $2y=b\lambda$, $2z = c\lambda$. Multiply these equations by a,b,c respectively and add : $2(ax+by+cz)=\lambda(a^2+b^2+c^2)$, so $\lambda=2/(a^2+b^2+c^2)$ and $x=a/(a^2+b^2+c^2)$, $y=b/(a^2+b^2+c^2)$, $z=c/(a^2+b^2+c^2)$.

$$\text{Therefore, } x^2+y^2+z^2 = \frac{a^2+b^2+c^2}{(a^2+b^2+c^2)^2} = \frac{1}{a^2+b^2+c^2} = \alpha \text{ (say)}$$

The minimum distance is thus $1/\sqrt{(a^2+b^2+c^2)}$. This is a MIN from the geometrical interpretation. To prove it, take $x=a\alpha+r$, $y=b\alpha+s$, $z=c\alpha+t$ (r,s,t small) and so that $a(a\alpha+r)+b(b\alpha+s)+c(c\alpha+t)=-d$, that is, $ar+bs+ct=0$. Then

$$x^2+y^2+z^2=(a\alpha+r)^2+(b\alpha+s)^2+(c\alpha+t)^2=\alpha+2a\alpha r+2b\alpha s+2c\alpha t+r^2+s^2+t^2$$
$$=\alpha+r^2+s^2+t^2 \text{ and therefore the value } \alpha \text{ is a MIN.}$$

9 (*a*) Let $f=x^2y^2z^2$, $g=x^2+y^2+z^2-r^2$. By the multiplier rule we seek solutions of $2xy^2z^2=\lambda 2x$, $2yx^2z^2=\lambda 2y$, $2zx^2y^2=\lambda 2z$. Obviously, for f to have a maximum, none of x,y,z should be 0; therefore $\lambda\neq 0$. Thus,

$y^2z^2=x^2z^2=x^2y^2=\lambda$ and so $y^2=x^2$ and $z^2=y^2$. We can clearly take all terms positive so $x=y=z$ and $3x^2=r^2$ so the maximum value of f is $(r^2/3)^3$. Since f is continuous on the closed bounded set $S:=\{g(x,y,z)=0\}$, f has a maximum and so this must be that maximum.

(*b*) Let $f=x^2+y^2+z^2$, $g=x^2y^2z^2-(r^2/3)^3$. Then we seek solutions of $2x=\lambda 2xy^2z^2$, $2y=\lambda 2yx^2z^2$, $2z=2zx^2y^2$. None of x,y,z can be zero else the constraint cannot be satisfied. Therefore $\lambda\neq 0$. But now these equations are the same as in (*a*) with λ replaced by $(1/\lambda)$ so we find $x=y=z$ and therefore $x^6=(r^2/3)^3$ and $x=r\sqrt{3}$. Then $f=r^2$.

10. Let $f=x_1^2x_2^2\cdots x_n^2$, $g=x_1^2+x_2^2\cdots+x_n^2-1$. Arguing exactly as in 9 (*a*) we see that all the x_i are equal and $f_{\max}=(1/n)^n$. Given n positive numbers $a_1,\cdots,a_n$ let $\alpha=a_1+a_2+\cdots+a_n$, and let $x_i=\sqrt{a_i/\alpha}$. Then $a_1\cdots a_n=\alpha^n x_1^2\cdots x_n^2\leq\alpha^n(1/n)^n$ (the maximum possible value, by the above). Taking n-th roots proves the inequality of the means.

11. For $x\neq 0$, $\frac{d}{dx}(x^2)^{q/2}=\frac{q}{2}(x^2)^{q/2-1}\cdot 2x=q|x|^{q-2}x$. Also, $\big|\,|x|^q/x\,\big|=|x|^{q-1}\longrightarrow 0$ as $x\longrightarrow 0$. Therefore the function $x\mapsto|x|^q$ has derivative 0 at $x=0$. Hence we have $\frac{d}{dx}|x|^q=q|x|^{q-2}x$, for all $x\in\mathbb{R}$, where we interpret $q|x|^{q-2}x$ as 0 if $x=0$. This makes the function $q|x|^{q-2}x$ continuous, and so $|x|^q$ is of class C^1. Thus Lagrange's method is applicable. Now consider $\sum_1^n a_ix_i+\lambda(\sum_1^n|x_i|^q-1)$.

At a critical point, $a_i+\lambda q|x_i|^{q-2}x_i=0$ for every i. Thus, the x_i's are all nonzero with the same sign. Take the case where $x_i>0$, for every i. Put $\lambda q=-\mu$. Then $a_i=\mu x_i^{q-1}$ and $a_ix_i=\mu x_i^q$. Therefore, $\sum a_ix_i=\mu\sum x_i^q=\mu$. Also, $a_i^p=\mu^p x_i^{pq-p}=\mu^p x_i^q$, and so $1=\sum a_i^p=\mu^p\sum x_i^q=\mu^p$, that is, $\mu=1$, and $\sum a_ix_i=1$. (note that $\mu>0$). In the other case, all x_i are negative, we get $\sum a_ix_i=-1$. The function $f(\mathbf{x})=\sum a_ix_i$ is continuous on the compact set $C:=\{\mathbf{x}:\sum|x_i|^q=1\}$. Hence f attains its max. and min. on this set, which must be ± 1. We check that $\nabla g\neq\mathbf{0}$ (the rank is 1) as needed for the application of the multiplier rule. At each point of C, at least one

of the x_i is nonzero, and therefore $(q|x_i|^{q-2}x_i, \cdots, q|x_n|^{q-2}x_n) \neq \mathbf{0}$.
To establish Minkowski's inequality write
$\sum|x_i+y_i|^p \leq \sum|x_i+y_i|^{p-1}(|x_i|+|y_i|)$ and apply Holder's inequality to get

$\sum|x_i+y_i|^{p-1}|x_i| \leq \{\sum|x_i+y_i|^{(p-1)q}\}^{1/q}\{\sum|x_i|^p\}^{1/p}$. Similarly for the second term. Since $(p-1)q = p$ this yields

$$\sum|x_i+y_i|^p \leq \{\sum|x_i+y_i|^p\}^{1/q}\left[\{\sum|x_i|^p\}^{1/p}+\{\sum|y_i|^p\}^{1/p}\right].$$

The inequality holds trivially if $\sum|x_i+y_i|^p = 0$, otherwise we may divide in the last equation, noting that $1 - 1/q = 1/p$, and get Minkowski's inequality in that case.

12. $f_x=2x-4y^3$, $f_y=-12y^2x+18y^5$ therefore (0,0) is a critical point. $f(x,0)=x^2>0$ while $f(2y^3,y)=-y^6<0$ so (0,0) is a saddle point. On any straight line $y=mx$, $f(x,mx)=(x-m^3x^3)(x-3m^3x^3)=x^2(1-m^3x^2)(1-3m^3x^2)$. Call this expression $g(x)$. Then $g'(x)=2x(1-m^3x^2)(1-3m^3x^2)+x^2(h(x))$, say, and $g''(x)=2(1-m^3x^2)(1-3m^3x^2)+x(\phi(x))$. Thus $g'(0)=0$ and $g''(0)=2$ so that 0 is a minimum of g.

13. The equations to be satisfied are (1) : $(y-z)[(z-x)-(x-y)]=\lambda 2x$, (2) : $(z-x)[(x-y)-(y-z)]=\lambda 2y$ (3) : $(x-y)[(y-z)-(z-x)]=\lambda 2z$.
Adding all 3 equations gives $2\lambda(x+y+z)=0$ so either $\lambda=0$ or $x+y+z=0$. First we consider $\lambda=0$. Then (1) gives either $y=z$ or $z-2x+y=0$. When $y=z$, (2) gives $y=x$ so all three components are equal. Otherwise we must have $z-2x+y=0$ and $x-2y+z=0$ (and $y-2z+x=0$) and these give $x-2y=y-2x$ so $x=y$. By the symmetry of the problem we see that $\lambda=0$ forces $x=y=z$ in all cases and $f=0$ at these points.
Now we consider $\lambda\neq 0$ in which case we have $x+y+z=0$. Replacing $y+z$ inside $[\cdot\cdot]$ in (1) gives $(y-z)[-3x]=2\lambda x$ so that either $x=0$ or (4) : $3(y-z)=-2\lambda$, (two similar equations by symmetry). We claim that x,y,z cannot all be nonzero. For if this is so we can add the 3 equations corresponding to (4) to get $0=\lambda$, impossible in the case under study. Since $x+y+z=0$, exactly one of the variables must be 0. By symmetry we can suppose $x=0$, $y\neq 0$, $z\neq 0$. Then (1) gives $(y-z)(z+y)=0$ so $z=\pm y$ and the points are of the form $(0,y,\pm y)$. The constraint gives $y^2=1$ so $y=\pm 1$. When $z=y$, $f=0$; when $z=-y$, $f=-y(2y)(-y)=2y^3=\pm 2$. In all, the conclusion is, maximum 2 at (0,1,−1), minimum −2 at (0,1,1), and at the other points obtained by symmetry.

14.(*i*) Let $f:\mathbb{R}\to\mathbb{R}$ have 0 as the only critical point and suppose that 0 is a local minimum. Then $f(x)\neq f(0)$ for $x\neq 0$, because otherwise, by the mean value theorem (or Rolle's theorem) $f(x)-f(0)/x = f'(\theta x)$

would imply that f' had another 0. We now claim that $f(x)>0$ for all $x\neq 0$. Indeed this holds for $x_1>0$ small and if we had $x_2>0$ with $f(x_2)<0$ then the intermediate value theorem would give $f(x_3)=0$, impossible. Similarly, the same holds for negative x.

(*ii*) For $f(x,y)=x^2+y^2(1-x)^3$, $f_x=2x-3y^2(1-x)^2$, $f_y=2y(1-x)^3$. $f_y=0$ yields either $y=0$ or $x=1$. If $y=1$ then $f_x=0$ gives $x=0$ while $x=1$ does not satisfy $f_x=0$. Thus $(0,0)$ is the only critical point. $f_{xx}=2+6y^2(1-x)$, $f_{xy}=-6y(1-x)^2$, $f_{yy}=2(1-x)^3$ so the Hessian at $(0,0)$ is $H=\begin{bmatrix}2 & 0\\ 0 & 2\end{bmatrix}$. Thus, this is a local minimum. There is no global minimum for we can take (for example) $x=2$ and y arbitrarily large.

15. Let l denote the length, w the width and h the height. Then Volume $= lwh$. Clearly the maximum subject to $l+2w+2h\leq 2$ is when $l+2w+2h=2$ (otherwise we could add a little to the length and increase the volume). By the Lagrange multiplier rule we want

$$wh=\lambda,\quad lh=2\lambda,\ \text{and}\quad lw=2\lambda.$$

For a max., $\lambda=0$ is not sensible so $l(h-w)=0$ and therefore $h=w$. Also $lh=2wh$ so $l=2w$. Thus, the dimensions are $(2w,w,w)$. The constraint yields $2w+2w+2w=2$, so $3w=1$ or $w=1/3$. Thus, the dimensions are $(2/3, 1/3, 1/3)$ and the Max volume is $(2/3)(1/3)(1/3)=2/27$ (m^3). For a cylindrical box, let r be the radius of the circular cross section and let l be the length. Then the girth is $2\pi r$. We want to minimize $V=(\pi r^2)l$ subject to $l+2\pi r\leq 2$. Obviously the maximum volume is obtained when $l+2\pi r=2$ and then $V(r)=\pi r^2(2-2\pi r)=2\pi r^2-2\pi^2r^2$. Then $V'(r)=2\pi(2r-3\pi r^2)$ and V has a maximum when $r=2/3\pi$. The maximum volume is $8/(27\pi)$, larger than the value $2/27$ for the rectangular box.

17. We want to minimize $x^2+y^2+z^2$ subject to $x^2+y^2=4$ and $xyz=2$. The simplest is to write $x=2\cos\theta$, $y=2\sin\theta$ then $z=2/4\cos\theta\sin\theta=1/\sin 2\theta$ defined provided $\sin 2\theta\neq 0$. Then $x^2+y^2+z^2=4+1/\sin^2 2\theta$ is a minimum when $\sin^2 2\theta=1$. We see that there is no maximum since θ can be taken arbitrarily small (but not equal to 0). For the Lagrange multiplier rule observe that the problem is equivalent to minimizing $f:=4+z^2$ subject to $x^2+y^2=4$ and $xyz=2$. Then we want (1): $0+\lambda 2x+\mu yz=0$, (2): $0+\lambda 2y+\mu xz=0$, (3): $2z+0+\mu xy=0$. Multiply (1) by x, (2) by y and add to get $8\lambda+4\mu=0$ so $\mu=-2\lambda$, and is nonzero by (3). Cancelling in (1) and (2) yields $x=4yz$ and $y=2xz$. Hence $x^2=xyz=2$, $y^2=xyz=2$ and $z=2/xy$. If $x=y=\pm\sqrt{2}$ then $z=1$; if $x=-y=\pm\sqrt{2}$ then $z=\pm 1$. In all cases the minimum distance is $\sqrt{5}$.

18. We seek the extrema of $f(x,y,z):=x^3+y^3+z^3$ subject to the constraints $g(x,y,z):=x+y+x-3=0$ and $h(x,y,z):=x^2+y^2+z^2-9=0$. As g,h define the intersection of a plane with a sphere, a closed bounded set, extrema exist. At an extremum $\nabla f=\lambda\nabla g+\mu\nabla h$ that is,

(1): $3x^2=\lambda+2x\mu$ (2): $3y^2=\lambda+2y\mu$ (3): $3z^2=\lambda+2z\mu$ and (4): $x+y+z=3$
(5): $x^2+y^2+z^2=9$.
The first 3 equations say that x,y,z are all roots of the quadratic equation $3t^2-2\mu t-\lambda=0$. If there is exactly one real root, r say, then $x=y=z=r$ which would imply (4) $3r=3$ and (5) $3r^2=9$, impossible. If there are 2 roots r,s (they depend on λ,μ) we have $x=r$, $y=r$, $z=s$ (or permutations of these) and then (4) $2r+s=3$ and (5) $2r^2+s^2=9$. Therefore $s=3-2r$ and $2r^2+(3-2r)^2=9$. Thus, $6r^2-12r=0$ so $r=0$ (implies $s=3$) or $r=2$ (implies $s=-1$). This gives the 2 points (0,0,3) and (2,2,−1). By symmetry we get 4 other points (3,0,0), (0,3,0), (−1,2,2), (2,−1,2).
At (0,0,3), $f=27$; at (2,2,−1), $f=15$.

Chapter 6.

1. Take $P=\{0, 1-\varepsilon, 1+\varepsilon, 2\}$. Then $L(f,P)=0(1-\varepsilon)+0(2\varepsilon)+1(1-\varepsilon)$ and $U(f,P)=0(1-\varepsilon)+1(2\varepsilon)+1(1-\varepsilon)$ so that $U-L=2\varepsilon$ and f is integrable. f cannot be a derivative because $f(0)=0$, $f(2)=1$ but there is no point c with $f(c)=1/2$ (for example) contradicting the result of Darboux (problem 3.29).

2. If f is continuous at **a** then so are $|f|$ and f^2 so these are integrable when f is. Also f,g continuous at a point **a** implies fg continuous at **a** and hence {discontinuities of fg} $\subseteq$
{discontinuities of f} $\cup$ {discontinuities of g}, a set of measure zero.

3. $F'(x)=2x\sin(1/x^2)-(2/x)\cos(1/x^2)$ for $x\neq 0$ and
$(F(h)-F(0))/h=h\sin(1/h^2)$ has limit 0 as $h\to 0$ since $|\sin(1/h^2)|\leq 1$.
F' is continuous for $x\neq 0$ by general results but not continuous at 0 since $2x\sin(1/x^2)\to 0$ but $(2/x)\cos(1/x^2)$ oscillates unboundedly. F' is not integrable because it is not bounded nor can it be redefined at 0 so as to be bounded on [0,1].

4. Write $I=\int_0^1 f$. The sum $S(f,P_n):=\frac{1}{n}\sum_{k=0}^{n} f(k/n)$ is an approximating sum for I corresponding to the partition $P_n=\{0,1/n,\cdots,n/n\}$. Since f is integrable, given $\varepsilon>0$ there is a partition P_ε such that $I\leq U(f,P_\varepsilon)\leq I+\varepsilon$. Suppose P_ε has q points in (0,1) and let P' be the refinement consisting of all points of P_n and P_ε. Then $U(f,P')\leq U(f,P_\varepsilon)$ by Proposition 6.3. Also, writing M (respectively m) for the Maximum (resp. minimum) of f on [0,1], since P' is obtained by adding at most q points to P_n, $U(f,P_n)\leq U(f,P')+q(M-m)/n$. Take n so large that $q(M-m)/n<\varepsilon$ to obtain $I\leq U(f,P_n)\leq I+2\varepsilon$. Similarly one can

show $I-2\varepsilon \leq L(f,P_n) \leq I$ for n sufficiently large. Hence $I-2\varepsilon \leq S(f,P_n) \leq I+2\varepsilon$ and this proves the result.

5. As f is integrable over E, the set of discontinuities of E has measure 0 (= content 0 for compact sets). Therefore since $c(E)>0$ there exist points in E where f is continuous. At a point of continuity $\mathbf{x}$, $f(\mathbf{x}) = \alpha > 0$ implies $f(\mathbf{y}) \geq \alpha/2$ on an ε-nbd of $\mathbf{x}$ and then $\int_E f \geq \int_{B(\mathbf{x},\varepsilon)} (\alpha/2) = (\alpha/2)c(B(\mathbf{x},\varepsilon) > 0$.

6. (*a*) Integral $= \int_0^1 dx \int_0^1 \left[\frac{2x}{(x+y)^3} - \frac{1}{(x+y)^2}\right] dy$

$$= \int_0^1 dx \left[\frac{-x}{(x+y)^2} + \frac{1}{x+y}\right]_0^1 = \int_0^1 \frac{dx}{(x+1)^2} = \frac{1}{2}.$$

(*b*) Integral $= \int_0^1 dy \int_0^1 \left[(x+y)^{-2} - 2y(x+y)^{-3}\right] dx = \int_0^1 \frac{-dy}{(x+y)^2} = -\frac{1}{2}.$

The function $f(x,y) = \dfrac{x-y}{(x+y)^3}$ is not bounded on $(0,1]\times(0,1]$, therefore not R-integrable.

7. If $\mathbf{x} \in \text{int}(A)$, there is a nbd of $\mathbf{x}$ in A and $\chi_A = 1$ on that nbd, so χ_A is continuous at $\mathbf{x}$. Similarly χ_A is 0 on a nbd of $\mathbf{x} \in \text{int}(A^c)$ so is continuous. For $\mathbf{x} \in \text{bdy}(A)$, every nbd of $\mathbf{x}$ contains points of A and points of A^c so χ_A takes both values 1 and 0 and χ_A is not continuous.

8. Let $\mathbf{x}^{(m)} \to \mathbf{x}^{(0)}$. For $m \geq 0$, let $E^{(m)}$ be the n-cube centred at $\mathbf{x}^{(m)}$ and of edge length ε. Then there exists $\delta > 0$ so that $B(\mathbf{x}^{(0)},\delta) \subseteq E^{(0)}$ and there exists N so that $E^{(0)} \supseteq B(\mathbf{x}^{(0)},\delta) \supseteq \{\mathbf{x}^{(m)}: m > N\}$. Then $\{\mathbf{x}^{(m)}\} \subseteq E^{(1)} \cup \cdots \cup E^{(N)} \cup E^{(0)}$ and so $c(\mathbf{x}^{(m)}) \leq n\varepsilon^n$, which can be made arbitrarily small.
Example : take the sequence $\{r^{(m)}\}$, an enumeration of the rationals.

9. We prove the boundary of each set is a subset of $\text{bdy}(A) \cup \text{bdy}(B)$, a set of content 0. Take $\mathbf{x} \in \text{bdy}(A \cup B)$. Then every nbd of $\mathbf{x}$ contains points of $A \cup B$ and points of $(A \cup B)^c = A^c \cap B^c$, that is either $\mathbf{x}$ belongs to A and $(A^c \cap B^c)$ or $\mathbf{x}$ belongs to B and $(A^c \cap B^c)$. In the former case $\mathbf{x} \in \text{bdy}(A)$, in the latter, $\mathbf{x} \in \text{bdy}(B)$. This proves $\text{bdy}(A \cup B) \subseteq \text{bdy}(A) \cup \text{bdy}(B)$.
Now $\text{bdy}(A \cap B) = \text{bdy}(A^c \cup B^c)$, (since $\text{bdy}(S) = \text{bdy}(S^c)$),
$\subseteq \text{bdy}(A^c) \cup \text{bdy}(B^c) = \text{bdy}(A) \cup \text{bdy}(B)$, using the first part.
Also $\text{bdy}(A \setminus B) = \text{bdy}(A \cap B^c) \subseteq \text{bdy}(A) \cup \text{bdy}(B)$, using the second part.

10. If f is integrable so is kf since they have the same set of discontinuities. By considering $f\chi_A$ in place of f it suffices to prove

the result on an n-rectangle E. For $\varepsilon > 0$ let P be a partition of E such that for all partitions P' finer than P, $| S(f,P') - \int_E f | < \varepsilon$, where $S(f,P) = \sum f(\mathbf{x}^{(m)})\, c(E^{(m)})$ is an approximating sum, and $\mathbf{x}^{(m)} \in E^{(m)}$. Let Q be a partition such that for Q' finer than Q, $| S(kf,Q') - \int_E kf | < \varepsilon$. Taking a common refinement P_1 of P and Q, noting that $S(kf,P_1) = kS(f,P_1)$, we have $| S(kf,P_1) - k\int_E f | < k\varepsilon$ and therefore $\int_E kf = k\int_E f$.

11. The boundary of A consists of pieces of graphs of functions of the form $y = f(x)$ or $x = g(y)$ and these have content 0. Thus A is Jordan measurable and $c(A) = \iint_A 1\, dx\, dy = \int_a^b \left[\int_0^{\alpha(x)} 1\, dy\right] = \int_a^b \alpha(x)\, dx.$

12. The integral is $\int_0^1 dx \int_{x^2}^{\sqrt{x}} x\, y\, dy = \int_0^1 x\,(x - x^4)/2\, dx = 1/12.$

13. $$\iint_\Omega e^{x-y}\, dx\, dy = \int_0^k e^x\, dx \int_0^{k-x} e^{-y}\, dy = \int_0^k e^x(1 - e^{x-k})\, dx$$
$$= e^k - 1 - \tfrac{1}{2}(e^k - e^{-k}) = \cosh(k) - 1.$$

14. Interchanging the order of integration, the integral is
$$\int_0^2 dy \int_0^{y^2} \frac{dx}{\sqrt{(x+y)^2}} = \int_0^2 dy \left[2\sqrt{(x+y^2)}\right]_0^{y^2} = 2\int_0^2 (\sqrt{2} - 1)\, y\, dy = 4(\sqrt{2} - 1).$$

15. The double integral can be written as the repeated integral
$$\int_{-1}^1 y^2\, dy \int_{1-\sqrt{(1-y^2)}}^{[(1+y^2)/2]^{1/2}} x\, dx = 2\int_0^1 y^2 \tfrac{1}{2}\left\{\tfrac{1}{2}(1+y^2) - [1 - \sqrt{(1-y^2)}]^2\right\} dy$$
$$= \int_0^1 y^2 \left\{\tfrac{1}{2}(1+y^2) - 2 + y^2\right\} dy - \int_0^1 2\sqrt{(1-y^2)}\, dy$$
$$= \frac{1}{2}\left(\frac{3}{5} - 1\right) + 2\int_0^{\pi/2} \cos^2\theta \sin^2\theta\, d\theta = \frac{\pi}{8} - \frac{1}{5},$$ by a standard formula.

16. Changing the order of integration, the integral is
$$\int_0^{\pi/2} dy \int_0^y \frac{\sin y}{x+y}\, dx = \int_0^{\pi/2} \sin y\, [\ln(x+y)]_0^y\, dy$$

$$= \int_0^{\pi/2} \sin y \ln(2)\, dy = \ln(2), \quad \text{since } \ln(2y)-\ln(y) = \ln(2y/y).$$

17. The integral is $\int_0^1 dy \int_0^{y^2/4a} \frac{dx}{y^2+4ax}$

$$= \frac{1}{4a}\int_0^1 [\ln(y^2+4ax)]_0^{y^2/4a}\, dy = \ln(2)/4a.$$

18. Firstly, $\int_0^1 dx \int_0^1 e^{y\ln x}\, dy = \int_0^1 (1/\ln x)[e^{y\ln x}]_0^1\, dx = \int_0^1 \frac{x-1}{\ln(x)}\, dx.$

Secondly, $\int_0^1 dy \int_0^1 x^y\, dx = \int_0^1 \left[\frac{x^{y+1}}{y+1}\right]_0^1 dy = \ln(2).$

Since $x^y = e^{y\ln x}$ is continuous on $(0,1]\times[0,1]$ and is bounded on $(0,1]\times[0,1]$ we may define $x^y = 0$ on the line $x = 0$, a set of content 0, and the double integral then exists and equals each of the iterated integrals.

19. $\iiint_V e^{x+2y+3z}\, dV = \int_0^1 dx \int_0^{1-x} dy \int_0^{1-x-y} e^{x+2y+3z}\, dz$

$$= \int_0^1 dx \int_0^{1-x} dy\, [e^{x+2y+3z}/3]_0^{1-x-y} = \frac{1}{3}\int_0^1 dx \int_0^{1-x} [e^{3-2x-y} - e^{x+2y}]\, dy$$

$$= \frac{1}{6}\int_0^1 dx\, [-3e^{2-x} + 2e^{3-2x} + e^x]\, dx = \frac{1}{6}\int_0^1 (3e - 3e^2 + e^3 - 1) = \frac{1}{6}(e-1)^3.$$

20. By Theorem 2.42 $\{f^{(m)}\}$ is uniformly bounded on $[0,1]$ say $|f^{(m)}(x)| \le M$. Let $a^{(m)} := \int_0^{1-1/m} f^{(m)}(x)\, dx$, $b^{(m)} := \int_0^1 f^{(m)}(x)\, dx$.

Then, by Theorem 6.36, $b^{(m)} \to b := \int_0^1 f(x)\, dx$.

Also $|a^{(m)} - b^{(m)}| = |\int_{1-1/m}^1 f^{(m)}(x)\, dx| \le M/m$. Hence $a^{(m)} \to b$ as $m \to \infty$.

21. By Theorem 6.41, differentiation term-by-term is correct if the differentiated series is uniformly convergent. If $\sum ma^{(m)}$ is Absolutely convergent then $\sum ma^{(m)}\sin(mx)$ is uniformly convergent on $[0,2\pi]$ by the M-test. When $ma^{(m)}$ decreases to 0, the series is uniformly convergent on $[\delta, 2\pi-\delta]$ by Dirichlet's test because the partial sums of $\sum \sin(mx)$ are uniformly bounded (compare Ex. 2.58).

22. Using polar coordinates

$$\text{area} = \iint_A r\, dr\, d\theta = \int_0^{\pi/4} d\theta \int_{2\cos\theta}^{4\cos\theta} r\, dr = \int_0^{\pi/4} 6\cos^2\theta\, d\theta$$

$$= 3\left[\theta + \frac{\sin 2\theta}{2}\right]_0^{\pi/4} = 3\frac{\pi}{4} + \frac{3}{2}.$$

23. $\iint_A xy\,dx\,dy = \iint_\Omega xy\,|J|\,du\,dv$ where Ω is bounded by the curves $u = 1$, $v = 1$, $u + v = 1$, and $J = \dfrac{\partial(x,y)}{\partial(x,y)} = 1 \Big/ \dfrac{\partial(u,v)}{\partial(x,y)}$.

Now $\dfrac{\partial(u,v)}{\partial(x,y)} = \begin{vmatrix} -y^2/x^2 & 2y/x \\ 2x/y & -x^2/y^2 \end{vmatrix} = -3$, so $|J| = \frac{1}{3}$. Also $u\,v = \dfrac{x^2y^2}{xy} = x\,y$.

Thus $\iint_A xy\,dx\,dy = \frac{1}{3}\int_0^1 dv \int_{1-v}^1 u\,v\,du = \frac{1}{3}\int_0^1 v\,(1-(1-v)^2)/2\,dv$

$$= \frac{1}{6}\left[\frac{1}{2} - \left(\frac{1}{2} - \frac{2}{3} + \frac{1}{4}\right)\right] = \frac{5}{72}.$$

Note. Although it appears as if J is never 0, in fact J is not defined at (0,0). The more powerful change of variables theorem is needed.

24. Using spherical polar coordinates,

$$\iiint z^4\,dV = \int_0^{2\pi} d\phi \int_0^{\pi} d\theta \int_0^a r^4 \cos^4\theta\, r^2 \sin\theta\,dr$$

$$= \frac{2\pi a^7}{7}\int_0^{\pi} \cos^4\theta \sin\theta\,d\theta = \frac{4\pi a^7}{35}.$$

25. Using spherical polar coordinates,

$$\iiint_V z\,dV = \int_0^{2\pi} d\phi \int_0^{\pi/3} d\theta \int_a^{2a\cos\theta} r\cos\theta\, r^2\sin\theta\,dr,$$

since the spheres intersect when $2az = a^2$ or $z = a/2$ giving $\theta = \pi/3$.

Thus, Integral $= 2\pi \displaystyle\int_0^{\pi/3} \left[\frac{(2a\cos\theta)^4}{4} - \frac{a^4}{4}\right] \cos\theta \sin\theta\,d\theta$

$$= 2\pi\left\{4a^2\left[\frac{-\cos^6\theta}{6}\right]_0^{\pi/3} + \frac{a^4}{4}\left[\frac{\cos^2\theta}{2}\right]_0^{\pi/3}\right\}$$

$$= 2\pi a^4\left[\frac{4}{6}\left(1 - \frac{1}{64}\right) - \frac{1}{8}\left(1 - \frac{1}{4}\right)\right] = 9\pi a^4/8.$$

26. Using spherical polars,

$$\iiint_V (x^2 + y^2 + z^2)^{-3/2}\,dV = \int_0^{2\pi} d\phi \int_0^{\pi} d\theta \int_b^a r^{-3} r^2 \sin\theta\,dr = 4\pi\ln(a/b).$$

27. $\displaystyle\iiint_V r\,dV = \int_0^{2\pi} d\phi \int_0^{\pi/4} d\theta \int_0^{3/\cos\theta} r\, r^2 \sin\theta\,dr =$

$$= 2\pi \int_0^{\pi/4} 3^4(\cos\theta)^{-4}\sin\theta\ /\ 4\ d\theta = \frac{27}{2}\ \pi\ [(\cos\theta)^{-3}]_0^{\pi/4} = \frac{27}{2}\ \pi\left(1 - \frac{1}{2\sqrt{2}}\right).$$

Chapter 7.

1. $\mathbf{a} \times (\mathbf{b} \times \mathbf{c}) = -(\mathbf{b} \times \mathbf{c}) \times \mathbf{a}$. Apply the result of Example 7.2. Simply add the terms obtained by applying this formula.

2. $\mathbf{F}(t)\bullet\mathbf{G}(t) = \sum_{i=1}^{3} F_i(t)\ G_i(t)$. Apply the standard sum and product rules for differentiation.
The first component of $\mathbf{F}(t) \times \mathbf{G}(t)$ is $F_2(t)G_3(t) - F_3(t)G_2(t)$ and $[F_2(t)G_3(t) - F_3(t)G_2(t)]' = F_2\,G_3{}' - F_3\,G_2{}' + F_2{}'\,G_3 - F_3{}'\,G_2$
$=$ first component of $\mathbf{F} \times \mathbf{G}' + \mathbf{F}' \times \mathbf{G}$.
The other components work in the same way.

3.(*a*) Let $\mathbf{a} = (a_1,a_2,a_3)$, $\mathbf{r} = (x,y,z)$. Then

$$\mathbf{a} \times \mathbf{r} = (a_2z - a_3y,\ a_2x - a_1z,\ a_1y - a_2x)$$

and $\text{curl}(\mathbf{a} \times \mathbf{r}) = (a_1 - (-a_1),\ a_2+a_2,\ a_3+a_3) = 2\mathbf{a}$.

(*b*)
$$\begin{aligned}\text{curl}(r^{-n}(\mathbf{a} \times \mathbf{r})) &= r^{-n}\,\text{curl}(\mathbf{a} \times \mathbf{r}) + \nabla(r^{-n}) \times (\mathbf{a} \times \mathbf{r})\\ &= r^{-n}2\mathbf{a} - nr^{-n-2}\mathbf{r} \times (\mathbf{a} \times \mathbf{r})\\ &= 2r^{-n}\mathbf{a} - nr^{-n-2}\{r^2\mathbf{a} - (\mathbf{r}\bullet\mathbf{a})\mathbf{r}\}\\ &= nr^{-n-2}(\mathbf{a}\bullet\mathbf{r})\mathbf{r} - (n-2)r^{-n}\mathbf{a}.\end{aligned}$$

4. $\mathbf{c}'(t)=(1+\cos t,\ 1-\cos t,\ -\sqrt{2}\sin t)$ so $\|\mathbf{c}'(t)\|^2=4$ and thus $ds/dt=2$ and $\mathbf{T}=(1/2)(1+\cos t,\ 1-\cos t,\ -\sqrt{2}\sin t)$. Then $d\mathbf{T}/ds=\mathbf{T}'(t)/(ds/dt)$, using the chain rule (simpler than changing the parameter to s), so that $d\mathbf{T}/ds=(1/4)(-\sin t,\ \sin t,\ -\sqrt{2}\cos t)$. Therefore, $\|d\mathbf{T}/ds\|=(1/4)\sqrt{2}$ which gives $\kappa=\sqrt{2}/4$ and $\mathbf{N}=(1/\sqrt{2})(-\sin t,\ \sin t,\ -\sqrt{2}\cos t)$. Then
$\mathbf{B}=\mathbf{T}\times\mathbf{N}=(1/2\sqrt{2})(-\sqrt{2}\cos t+\sqrt{2},\ \sqrt{2}\cos t+\sqrt{2},\ 2\sin t)$ and so $d\mathbf{B}/ds=\mathbf{B}'(t)/(ds/dt)=(1/4)(\sin t,-\sin t,\ \sqrt{2}\cos t)=(-\sqrt{2}/4)\mathbf{N}$ so $\tau=\sqrt{2}/4$.

5. We differentiate the relation $\mathbf{N}(s) = \mathbf{B}(s)\times\mathbf{T}(s)$ to get

$$\frac{d\mathbf{N}}{ds} = \mathbf{B}(s)\times\frac{d\mathbf{T}}{ds} + \frac{d\mathbf{B}}{ds}\times\mathbf{T}(s) = \mathbf{B}(s)\times\kappa\mathbf{N} - \tau\mathbf{N}(s)\times\mathbf{T}(s) = -\kappa\mathbf{T} + \tau\mathbf{B}$$

since $\mathbf{B} = \mathbf{T} \times \mathbf{N}$, and hence $\mathbf{B}\times\mathbf{N} = (\mathbf{T}\times\mathbf{N})\times\mathbf{N} = -(\mathbf{N}\bullet\mathbf{N})\mathbf{T}$.

6. We have $\mathbf{c}(\theta)=(a\,\theta-a\sin\theta,\ a-a\cos\theta)$ so $\mathbf{c}'(\theta)=(a-a\cos\theta,\ a\sin\theta)$ and $\|\mathbf{c}'(\theta)\|=\sqrt{(2a^2-2a^2\cos\theta)} = \sqrt{(4a^2\sin^2(\theta/2))} = 2a\sin(\theta/2)$. Thus we have,

$$\text{arc length} = 2a\int_0^{2\pi}\sin(\theta/2)\ d\theta = 4a[-\cos(\theta/2)]_0^{2\pi} = 8a.$$

7.(*i*) arc length $= 4\int_0^{\pi/2} \sqrt{\{[3\cos^2 t(-\sin t)]^2+[3\sin^2 t\cos t]^2\}}\,dt$

$$= 12\int_0^{\pi/2} \cos t \sin t\,dt = 12\left[\frac{\sin^2 t}{2}\right]_0^{\pi/2} = 6.$$

(*ii*) We have $\mathbf{r}'=(2, 2t, 1/t)$, and therefore $\|\mathbf{r}'\|=\sqrt{(4+4t^2+1/t^2)}=2t+1/t$. So, arc length $= \int_1^2 (2t+1/t)\,dt = [t^2+\ln(t)]_1^2 = 3+\ln 2.$

8. (*i*) $\mathbf{c}(t) = (t,t,t)$, $0\le t\le 1$, so $\mathbf{c}'(t)=(1,1,1)$ and then $\int_C \mathbf{F}\cdot d\mathbf{s} = \int_0^1 (t^2,2t,-t^2)\cdot(1,1,1)\,dt = \int_0^1 2t\,dt = 1.$

(*ii*) $\int_C \mathbf{F}\cdot d\mathbf{s} = \int_0^1 (t^5,2t^2,-t^2).(1,2t,3t^2)\,dt$

$$= \int_0^1 (t^5 + 4t^3 + 3t^4)\,dt = \frac{1}{6} + 1 - \frac{3}{5} = \frac{17}{30}.$$

9. $\phi\nabla\psi+\psi\nabla\phi=\nabla(\phi\psi)$ and $\int_C \nabla(\phi\psi)=0$ around a closed curve by Thm.7.22.

10. $\mathbf{r}=\nabla(r^2)$ and Thm. 7.22 applies. (Stokes theorem would work too.)

11. $\mathbf{F}=\nabla\phi$, where $\phi(\mathbf{x}) = F_1x+F_2y+F_3z$, apply Thm.7.22.

12. $\mathbf{c}(t)=(1, t, 0)$, $0\le t\le 3$ so $\mathbf{c}'(t)=(0,1,0)$ and $ds/dt=\|c'(t)\|=1$. Thus, $\int_C (x^2+y^2+z^2)\,ds = \int_0^3 (1+t^2)\frac{ds}{dt} = 3 + 9 = 12.$

13. Parametrize C^+ by $\mathbf{c}(\theta)=(a\cos\theta, a\sin\theta, 0)$, $0\le\theta\le 2\pi$.
Then $\mathbf{c}'(\theta) = (-a\sin\theta, a\cos\theta, 0)$. Thus,

$$\int_C (z,x,y)\cdot d\mathbf{s} = \int_0^{2\pi} (0, a\cos\theta, a\sin\theta)\cdot(-a\sin\theta, a\cos\theta, 0)\,d\theta$$

$$= \int_0^{2\pi} a^2\cos^2\theta\,d\theta = \pi a^2.$$

14. $\partial\phi/\partial x = \dfrac{(-y/x^2)}{1+(y/x)^2} = -y/(x^2+y^2)$ and $\partial\phi/\partial y = \dfrac{1/x}{1+(y/x)^2} = x/(x^2+y^2)$.
Set $x=r\cos t$, $y=r\sin t$. Then

$$\int_C = \int_0^{2\pi} -r\sin t(-r\sin t\,dt)+r\cos t(r\cos t\,dt) = \int_0^{2\pi} r^2\,dt = 2\pi r^2.$$

This is not a contradiction since ϕ is not C^1 on the interior of C. ϕ is the angle made by the vector $\mathbf{x}$ with the x-axis (vector $\mathbf{i}$). Going once around the origin on the curve C the angle increases by 2π. The derivatives of ϕ are continuous on an annulus surrounding the origin but an annulus is not simply connected.

15. curl $\mathbf{F} = \begin{vmatrix} \mathbf{i} & \mathbf{j} & \mathbf{k} \\ \partial/\partial x & \partial/\partial y & \partial/\partial z \\ 2x+y+3z & x+2y+z & 3x+2y+2z \end{vmatrix} = (2-2,\ 3-3,\ 1-1) = \mathbf{0}.$

ϕ must satisfy $\phi_x = 2x+y+3z$, $\phi_y = x+2y+2z$, $\phi_z = 3x+2y+2z$, so integrating we see that $\phi = x^2+y^2+z^2+xy+2yz+3zx$ (+ an arbitrary constant).

16. (*i*) $\text{curl}[(\mathbf{a}\bullet\mathbf{r})\mathbf{r}] = (\mathbf{a}\bullet\mathbf{r})\text{curl}\,\mathbf{r} + \nabla(\mathbf{a}\bullet\mathbf{r}) \times \mathbf{r} = \mathbf{0} + \mathbf{a} \times \mathbf{r}$.
Since $\mathbf{a} \times \mathbf{r}$ is non-zero on any open set, this vector field is not conservative.
(*ii*) $\text{curl}[(\mathbf{a}\bullet\mathbf{r})\mathbf{a}] = (\mathbf{a}\bullet\mathbf{r})\text{curl}\,\mathbf{a} + \nabla(\mathbf{a}\bullet\mathbf{r}) \times \mathbf{a} = \mathbf{0} + \mathbf{a} \times \mathbf{a} = \mathbf{0}$.
This vector field is therefore conservative. Also we observe that

$$\nabla(\mathbf{a}\bullet\mathbf{r})^2 = 2(\mathbf{a}\bullet\mathbf{r})\nabla(\mathbf{a}\bullet\mathbf{r}) = 2(\mathbf{a}\bullet\mathbf{r})\mathbf{a},$$

so a potential function is $\phi = (\mathbf{a}\bullet\mathbf{r})^2/2$.
(*iii*) $\text{curl}[(\mathbf{a}\bullet\mathbf{r})\mathbf{b} + (\mathbf{b}\bullet\mathbf{r})\mathbf{a})] = \mathbf{0} + \mathbf{a} \times \mathbf{b} + \mathbf{0} + \mathbf{b} \times \mathbf{a} = \mathbf{0}$.
Also $(\mathbf{a}\bullet\mathbf{r})\mathbf{b} + (\mathbf{b}\bullet\mathbf{r})\mathbf{a} = (\mathbf{a}\bullet\mathbf{r})\nabla(\mathbf{b}\bullet\mathbf{r}) + (\mathbf{b}\bullet\mathbf{r})\nabla(\mathbf{a}\bullet\mathbf{r}) = \nabla[(\mathbf{a}\bullet\mathbf{r})(\mathbf{b}\bullet\mathbf{r})]$
so a potential function is $\phi = (\mathbf{a}\bullet\mathbf{r})(\mathbf{b}\bullet\mathbf{r})$.

17. The cylinder may be written $(x-1)^2+z^2 = 1$. We can parametrize by $\mathbf{r}(\theta, y) = (1+\cos\theta, y, \sin\theta)$. Then the y limits are 0 and $2-x = 1-\cos\theta$. Then we have $\mathbf{r}_\theta = (-\sin\theta, 0, \cos\theta)$, $\mathbf{r}_y = (0, 1, 0)$, and

$$\text{area} = \iint \|\mathbf{r}_\theta \times \mathbf{r}_y\|\, d\theta\, dy = \int_0^{2\pi} d\theta \int_0^{1-\cos\theta} 1\, dy = 2\pi.$$

18. We may parametrize the cone by $\mathbf{r} = (zt\cos\theta, zt\sin\theta, z)$, where $t = \tan\alpha$. For these points to lie inside the sphere we must have
$(zt\cos\theta)^2+(zt\sin\theta - a)^2+z^2 \le a^2$, that is,
$\sin\theta \ge 0$ and $0 \le z \le 2t\,a\sin\theta/(t^2+1) := b\sin\theta$, say. Then
$\mathbf{r}_\theta \times \mathbf{r}_z = (zt\cos\theta, zt\sin\theta, -zt^2)$ so $\|\mathbf{r}_\theta \times \mathbf{r}_z\| = z\,t\sqrt{(1+t^2)}$.

$$\text{area} = \int_0^\pi d\theta \int_0^{b\sin\theta} t\sqrt{1+t^2}\, z\, dz = \int_0^\pi C\sin^2\theta\, d\theta = C\,\pi/2 = \pi a^2\sin^3\alpha.$$

where we wrote $C = t\sqrt{1+t^2}\, b^2/2 = 2a^2\sin^3\alpha$.

19. Since $\mathbf{r}\bullet\mathbf{n} = a$,

$$\iint_S (\mathbf{r}\bullet\mathbf{n})\, dS = a\iint_S 1\, dS = a(\text{area of } S) = a(2\pi ah) = 2\pi a^2 h.$$

20. We have $\mathbf{r}_\theta = ((R+\cos\phi)(-\sin\theta), (R+\cos\phi)\cos\theta, 0)$ and
$\mathbf{r}_\phi = (-\sin\phi\cos\theta, -\sin\phi\sin\theta, \cos\phi)$. Therefore,
$\mathbf{r}_\theta \times \mathbf{r}_\phi = ((R+\cos\phi)\cos\theta\cos\phi, (R+\cos\phi)\sin\theta\cos\phi, (R+\cos\phi)\sin\phi)$.
Thus, $dS = \|\mathbf{r}_\theta \times \mathbf{r}_\phi\|\, d\theta\, d\phi = \sqrt{(R+\cos\phi)^2(\cos^2\theta\cos^3\phi + \sin^2\theta\cos^2\phi + \sin^2\phi)}$

$$= (R+\cos\phi)\, d\theta\, d\phi.$$

Then area $= \int_0^{2\pi} (R + \cos\phi)\, d\phi \int_0^{2\pi} d\theta = 2\pi\, [R\phi + \sin\phi]_0^{2\pi} = (2\pi)^2 R.$

21. On the sphere, $dS = \sin\theta\, d\theta\, d\phi$. Here $0 \le \phi \le 2\pi$, $0 \le \theta \le \pi/4$ gives the complete region cut out. Therefore,

$$\text{area} = \int_0^{2\pi} d\phi \int_0^{\pi/4} \sin\theta\, d\theta = 2\pi\,(1 - 1/\sqrt{2}).$$

22. The paraboloid is given by $z = f(x, y) := (x^2 + y^2)/2$ for $(x, y) \in D$ where $D = \{x^2 + y^2 = 2,\ z = 0\}$. Thus, area $= \iint_S 1\, dS = \iint_D \sqrt{(1 + f_x^2 + f_y^2)}\, dx\, dy$

$$= \iint_D \sqrt{(1 + x^2 + y^2)}\, dx\, dy = \int_0^{2\pi} d\theta \int_0^{\sqrt{2}} \sqrt{(1 + r^2)}\, r\, dr \quad \text{(polars)}$$

$$= 2\pi\, [(1 + r^2)^{3/2}/\, 3]_0^{\sqrt{2}} = 2\pi(3\sqrt{3} - 1)/3.$$

23 (*a*) On C_1, $x = 0$, $dx = 0$ so $\int_{C_1} = 0$. On C_2, $x = \cos\theta$, $y = \sin\theta$,

$0 \le \theta \le \pi/4$, and $\displaystyle\int_{C_2} = \int_0^{\pi/4} \frac{\cos\theta\, \cos\theta\, d\theta\ -\ \sin\theta(-\sin\theta\, d\theta)}{1 + 1} = \int_0^{\pi/4} \tfrac{1}{2} = \pi/8.$

On C_3, $x = t$, $y = t$, $0 \le t \le 1/\sqrt{2}$, but taken with t decreasing so

$\displaystyle\int_{C_3} = \int_{1/\sqrt{2}}^{0} \frac{t\,dt - t\,dt}{2t^2 + 1} = 0.$ Thus $\displaystyle\int_C = \pi/8.$

(*b*) We have $\displaystyle\frac{\partial Q}{\partial x} - \frac{\partial P}{\partial y} = \frac{1 + y^2 - x^2}{(x^2 + y^2 + 1)^2} - \frac{(y^2 - x^2 - 1)}{(x^2 + y^2 + 1)^2}$ so by Green's theorem,

$\displaystyle\iint_D \left(\frac{\partial Q}{\partial x} - \frac{\partial P}{\partial y}\right) dx\, dy = \iint_D \frac{2}{(x^2 + y^2 + 1)^2}\, dx\, dy$. Changing to polar

coordinates gives $\displaystyle\iint_D = 2\int_0^{\pi/4} d\theta \int_0^1 \frac{r}{(r^2 + 1)^2}\, dr = \frac{\pi}{2}\left[\frac{-1}{2(1 + r^2)}\right]_0^1 = \pi/8.$

24. (*i*) Let P, Q denote the points $(\pi/2, 0)$, $(\pi/2, 1)$ respectively. Then

on OP, $y = 0$ so $\displaystyle\int_{OP} \{(y - \sin x)\, dx + \cos x\, dy\} = -\int_0^{\pi/2} \sin x\, dx = -1;$

on PQ, $x = \pi/2$ so $\displaystyle\int_{PQ} \{(y - \sin x)\, dx + \cos x\, dy\} = 0;$

on QO, $y = 2x/\pi$ so

$$\int_{QO} \{(y - \sin x)\, dx + \cos x\, dy\} = -\int_0^{\pi/2} \left(\frac{2x}{\pi} - \sin x + \frac{2}{\pi}\cos x\right) dx = 1 - \frac{\pi}{4} - \frac{2}{\pi}.$$

Adding these contributions we find that

$$\int_C \{(y - \sin x)\,dx + \cos x\,dy\} = -\left(\frac{2}{\pi} + \frac{\pi}{4}\right).$$

(*ii*) Using Green's theorem :

$$\int_C \{(y - \sin x)\,dx + \cos x\,dy\} = \int_0^{\pi/2} dx \int_0^{2x/\pi} (-\sin x - 1)\,dy$$

$$= -\frac{2}{\pi}\int_0^{\pi/2} x(1 + \sin x)\,dx = -\frac{2}{\pi}\left\{\left[\frac{\pi^2}{8}\right] + \left[-x\cos x\right]_0^{\pi/2} + \int_0^{\pi/2} \cos x\,dx\right\}$$

$$= -\frac{2}{\pi}\left(\frac{\pi^2}{8} + 1\right) = -\left(\frac{2}{\pi} + \frac{\pi}{4}\right).$$

25.(*i*) Let C_1 denote the curve with parametric equations $x = t^2$, $y = 2t$, C_2 denote the curve with parametric equations $x = 1$, $y = 2t$, $-1 \le t \le 1$. Then,

$$\int_{C_1} (x^2y\,dx + xy^2\,dy) = -\int_{-1}^{1} (t^4\cdot 2t\cdot 2t\,dt + t^2\cdot 4t^2\cdot 2\,dt) = -8\left(\frac{1}{7} + \frac{2}{5}\right),$$

$$\int_{C_2} (x^2y\,dx + xy^2\,dy) = -\int_{-1}^{1} 4t^2\cdot 2dt = \frac{16}{3}, \text{ so } \int_C (x^2y\,dx + xy^2\,dy) = \frac{16}{3} - \frac{152}{35} = \frac{104}{105}.$$

(*ii*) By Green's theorem the integral is equal to

$$2\int_0^1 dx \int_0^{2\sqrt{x}} (y^2 - x^2)\,dx = 2\int_0^1 \left(\frac{8}{3}x^{3/2} - 2x^{5/2}\right) dx = 2\left(\frac{8}{2}\cdot\frac{2}{5} - 2\cdot\frac{2}{7}\right) = \frac{104}{105}.$$

26. Let $\mathbf{F} = (0, 0, x^2y^2)$. Since $\mathbf{n} = (x,y,z)$ on the sphere $x^2+y^2+z^2=1$, $\mathbf{F}\cdot\mathbf{n} = x^2y^2z$ and, by the divergence theorem,

$$\iint_S x^2y^2z\,dS = \iint_S (\mathbf{F}\cdot\mathbf{n})\,dS = \iiint_V \operatorname{div}\mathbf{F}\,dV = 0.$$

We could also take (for example) $\mathbf{G} = (xy^2z, 0, 0)$ when $\operatorname{div}\mathbf{G} = y^2z$ and again $\iiint_V \operatorname{div}\mathbf{G}\,dV = 0$, because V is symmetrical about $z = 0$.

27. (*a*) The cylinder has equation $x^2 + (y - 1/2)^2 = 1/4$ so the curve of intersection may be parametrized by

$$x = \cos\theta/2, \quad y = 1/2 + \sin\theta/2, \quad z = x^2 + y^2 = 1/4 + \sin\theta/2.$$

Then $$\int_C = \int_0^{2\pi} \left[\left(\frac{1}{4} + \frac{\sin\theta}{2}\right)\left(-\frac{1}{2}\sin\theta\right) + \frac{\cos\theta}{2}\frac{\cos\theta}{2} + \left(\frac{1}{2} + \frac{\sin\theta}{2}\right)\cos\theta\right] d\theta$$

$$= \int_0^{2\pi} \frac{1}{4}(\cos^2\theta - \sin^2\theta)\,d\theta + \text{terms that integrate to } 0$$

$$= \int_0^{2\pi} \cos 2\theta/4\,d\theta = 0.$$

(*b*) Take $S = \{(x, y, x^2+y^2) : (x,y) \in D\}$ where $D = \{(x,y) : x^2+y^2 = y\}$

as the capping surface. Then, we have curl$(z, x, y) = (1,1,1)$ and $\mathbf{r}_x \times \mathbf{r}_y = (-2x,-2y,1)$ so that

$$\int_C (z,x,y)\cdot d\mathbf{S} = \iint_D \mathrm{curl}(z,x,y)\cdot(\mathbf{r}_x\times\mathbf{r}_y)\,dx\,dy = \iint_D (-2x-2y+1)\,dx\,dy$$

$$= \int_{-\pi/2}^{\pi/2}\int_0^{\sin\theta} (1-2r\cos\theta-2r\sin\theta)\,r\,dr\,d\theta$$

$$= \int_{-\pi/2}^{\pi/2} \frac{\sin^2\theta}{2} - \frac{2}{3}\sin^4\theta - \frac{2}{3}\sin^3\theta\cos\theta\,d\theta = 2\int_0^{\pi/2} \frac{\sin^2\theta}{2} - \frac{2}{3}\sin^4\theta\,d\theta$$

$$= \frac{2}{3}\,\frac{\pi}{2} - \frac{4}{3}\,\frac{3\cdot 1}{4\cdot 2}\,\frac{\pi}{2} = 0, \text{ by standard reduction formulae.}$$

28. $\nabla(r^2)=2\mathbf{r}$ and $\iint_S \mathbf{r}\cdot\mathbf{n}\,dS = \iiint_V \mathrm{div}\,\mathbf{r}\,dV = 3\iiint_V dV = 3c(V)$.

29. By the divergence theorem,

$$\iiint_V \Delta u\,dV = \iiint_V \mathrm{div}(\nabla u)\,dV = \iint_S (\nabla u\cdot\mathbf{n})\,dS = \iint_S \frac{\partial u}{\partial\mathbf{n}}\,dS.$$

30. Note that $\mathrm{div}(u\,\nabla v) = u\,\Delta v+\nabla u\cdot\nabla v$. Therefore, by the divergence theorem, $\iiint_V (u\,\Delta v+\nabla u\cdot\nabla v)\,dV = \iint_S u\,(\nabla v\cdot\mathbf{n})\,dS = \iint_S u\,\frac{\partial v}{\partial\mathbf{n}}\,dS.$

31. $w=u-v$ satisfies $\Delta w=0$ in V and $w=0$ on S. By Green' first identity applied to w,w, $\iiint_V (\nabla w\cdot\nabla w)\,dV = 0$. Since $|\nabla w|^2$ is continuous and nonnegative its integral is 0 only if $|\nabla w|\equiv 0$ in V. Then w is a constant on the connected set cl(V) and this constant is 0 because $w=0$ on bdy(V) so $u=v$ throughout V.

Index

Mathematics and its Applications
Series Editor: G. M. BELL,
Professor of Mathematics, King's College London, University of London

Blum, W.	Applications and Modelling in Learning and Teaching Mathematic
Brown, R.	Topology: A Geometric Account of General Topology
	Homotopy Types and the Fundamental Groupoi
Burghes, D.N. & Borrie, M.	Modelling with Differential Equation
Burghes, D.N. & Downs, A.M.	Modern Introduction to Classical Mechanics and Contro
Burghes, D.N. & Graham, A.	Introduction to Control Theory, including Optimal Contro
Burghes, D.N., Huntley, I. & McDonald, J.	Applying Mathematic
Burghes, D.N. & Wood, A.D.	Mathematical Models in the Social, Management and Life Science
Butkovskiy, A.G.	Green's Functions and Transfer Functions Handbook
Cartwright, M.	Fourier Methods: for Mathematicians, Scientists and Engineers
Cerny, I.	Complex Domain Analysis
Chorlton, F.	Textbook of Dynamics, 2nd Edition
Chorlton, F.	Vector and Tensor Methods
Cohen, D.E.	Computability and Logic
Cordier, J.-M. & Porter, T.	Shape Theory: Categorical Methods of Approximation
Crapper, G.D.	Introduction to Water Waves
Cross, M. & Moscardini, A.O.	Learning the Art of Mathematical Modelling
Cullen, M.R.	Linear Models in Biology
Dunning-Davies, J.	Mathematical Methods for Mathematicians, Physical Scientists and Engineers
Eason, G., Coles, C.W. & Gettinby, G.	Mathematics and Statistics for the Biosciences
El Jai, A. & Pritchard, A.J.	Sensors and Controls in the Analysis of Distributed Systems
Exton, H.	Multiple Hypergeometric Functions and Applications
Exton, H.	Handbook of Hypergeometric Integrals
Exton, H.	*q*-Hypergeometric Functions and Applications
Faux, I.D. & Pratt, M.J.	Computational Geometry for Design and Manufacture
Firby, P.A. & Gardiner, C.F.	Surface Topology
Gardiner, C.F.	Modern Algebra
Gardiner, C.F.	Algebraic Structures
Gasson, P.C.	Geometry of Spatial Forms
Goodbody, A.M.	Cartesian Tensors
Graham, A.	Kronecker Products and Matrix Calculus: with Applications
Graham, A.	Matrix Theory and Applications for Engineers and Mathematicians
Graham, A.	Nonnegative Matrices and Applicable Topics in Linear Algebra
Griffel, D.H.	Applied Functional Analysis
Griffel, D.H.	Linear Algebra and its Applications: Vol. 1, A First Course; Vol. 2, More Advanced
Guest, P. B.	The Laplace Transform and Applications
Hanyga, A.	Mathematical Theory of Non-linear Elasticity
Harris, D.J.	Mathematics for Business, Management and Economics
Hart, D. & Croft, A.	Modelling with Projectiles
Hoskins, R.F.	Generalised Functions
Hoskins, R.F.	Standard and Nonstandard Analysis
Hunter, S.C.	Mechanics of Continuous Media, 2nd (Revised) Edition
Huntley, I. & Johnson, R.M.	Linear and Nonlinear Differential Equations
Irons, B. M. & Shrive, N. G.	Numerical Methods in Engineering and Applied Science
Ivanov, L. L.	Algebraic Recursion Theory
Johnson, R.M.	Theory and Applications of Linear Differential and Difference Equations
Johnson, R.M.	Calculus: Theory and Applications in Technology and the Physical and Life Sciences
Jones, R.H. & Steele, N.C.	Mathematics in Communication Theory
Jordan, D.	Geometric Topology
Kelly, J.C.	Abstract Algebra
Kim, K.H. & Roush, F.W.	Applied Abstract Algebra
Kim, K.H. & Roush, F.W.	Team Theory
Kosinski, W.	Field Singularities and Wave Analysis in Continuum Mechanics
Krishnamurthy, V.	Combinatorics: Theory and Applications
Lindfield, G. & Penny, J.E.T.	Microcomputers in Numerical Analysis
Livesley, K.	Mathematical Methods for Engineers
Lord, E.A. & Wilson, C.B.	The Mathematical Description of Shape and Form
Malik, M., Riznichenko, G.Y. & Rubin, A.B.	Biological Electron Transport Processes and their Computer Simulation
Martin, D.	Manifold Theory: An Introduction for Mathematical Physicists
Massey, B.S.	Measures in Science and Engineering
Meek, B.L. & Fairthorne, S.	Using Computers
Menell, A. & Bazin, M.	Mathematics for the Biosciences
Mikolas, M.	Real Functions and Orthogonal Series
Moore, R.	Computational Functional Analysis
Murphy, J.A., Ridout, D. & McShane, B.	Numerical Analysis, Algorithms and Computation
Nonweiler, T.R.F.	Computational Mathematics: An Introduction to Numerical Approximation
Ogden, R.W.	Non-linear Elastic Deformations
Oldknow, A.	Microcomputers in Geometry
Oldknow, A. & Smith, D.	Learning Mathematics with Micros
O'Neill, M.E. & Chorlton, F.	Ideal and Incompressible Fluid Dynamics